Lens Design

Automatic and quasi-autonomous computational methods and techniques

Series Editor

R Barry Johnson a Senior Research Professor at Alabama A&M University, has been involved for over 40 years in lens design, optical systems design, electro-optical systems engineering, and photonics. He has been a faculty member at three academic institutions engaged in optics education and research, employed by a number of companies, and provided consulting services.

Dr Johnson is an SPIE Fellow and Life Member, OSA Fellow, and was the 1987 President of SPIE. He serves on the editorial board of Infrared Physics & Technology and Advances in Optical Technologies. Dr Johnson has been awarded many patents, has published numerous papers and several books and book chapters, and was awarded the 2012 OSA/SPIE Joseph W Goodman Book Writing Award for Lens Design Fundamentals, Second Edition. He is a perennial co-chair of the annual SPIE Current Developments in Lens Design and Optical Engineering Conference.

Foreword

Until the 1960s, the field of optics was primarily concentrated in the classical areas of photography, cameras, binoculars, telescopes, spectrometers, colorimeters, radiometers, etc. In the late 1960s, optics began to blossom with the advent of new types of infrared detectors, liquid crystal displays (LCD), light emitting diodes (LED), charge coupled devices (CCD), lasers, holography, fiber optics, new optical materials, advances in optical and mechanical fabrication, new optical design programs, and many more technologies. With the development of the LED, LCD, CCD and other electo-optical devices, the term 'photonics' came into vogue in the 1980s to describe the science of using light in development of new technologies and the performance of a myriad of applications. Today, optics and photonics are truly pervasive throughout society and new technologies are continuing to emerge. The objective of this series is to provide students, researchers, and those who enjoy self-teaching with a wide-ranging collection of books that each focus on a relevant topic in technologies and application of optics and photonics. These books will provide knowledge to prepare the reader to be better able to participate in these exciting areas now and in the future. The title of this series is Emerging Technologies in Optics and Photonics where 'emerging' is taken to mean 'coming into existence,' 'coming into maturity,' and 'coming into prominence.' IOP Publishing and I hope that you find this Series of significant value to you and your career.

Lens Design

Automatic and quasi-autonomous computational methods and techniques

Donald Dilworth
Optical Systems Design, Inc.

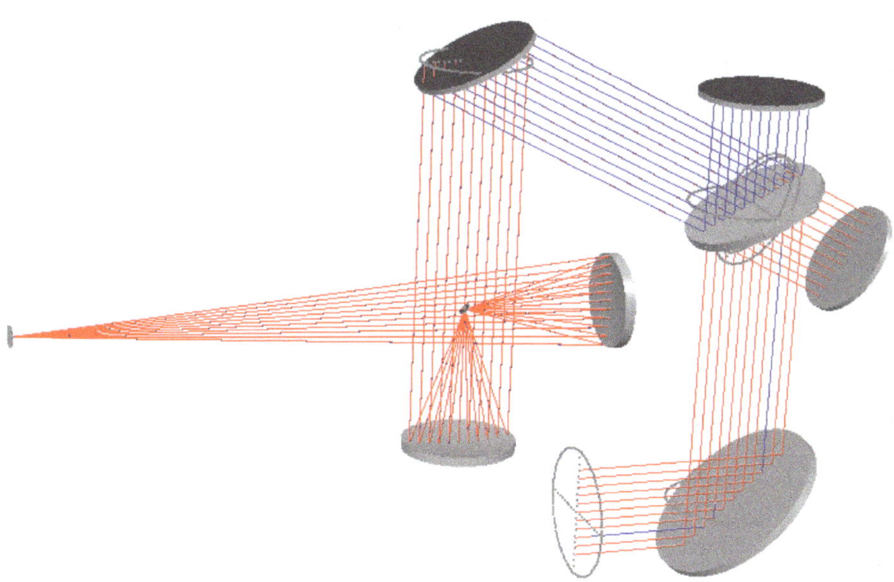

IOP Publishing, Bristol, UK

ISBN 978-0-7503-1611-8 (ebook)
ISBN 978-0-7503-1609-5 (print)
ISBN 978-0-7503-1610-1 (mobi)

DOI 10.1088/978-0-7503-1611-8

Version: 20180701

IOP Expanding Physics
ISSN 2053-2563 (online)
ISSN 2054-7315 (print)

British Library Cataloguing-in-Publication Data: A catalogue record for this book is available from the British Library.

Published by IOP Publishing, wholly owned by The Institute of Physics, London

IOP Publishing, Temple Circus, Temple Way, Bristol, BS1 6HG, UK

US Office: IOP Publishing, Inc., 190 North Independence Mall West, Suite 601, Philadelphia, PA 19106, USA

I cannot help fearing that men may reach a point where they look on every new theory as a danger, every innovation as a toilsome trouble, … and that they may absolutely refuse to move at all for fear of being carried off their feet.

—Alexis de Tocqueville, 1840

Contents

Foreword

The use of computers in lens design dates back over 60 years. In that time, computers have increased in speed by many orders of magnitude and in memory capacity by even more. This has led to enormous improvements in the capabilities of lens design programs. Analysis of lens performance has not changed significantly for several decades, but optimization has—and this book is, at its root, about optimization.

Optimization algorithms may be divided into three categories: local, regional and global. The first of these categories refers to algorithms that proceed from the given starting point to the nearest local minimum. Regional algorithms attempt to escape from this local minimum and find a nearby one that is better. As the name implies, global algorithms attempt to search the entire design space and provide a solution that is better than any other alternative.

Don Dilworth has made significant contributions to all three categories of optimization algorithms which are included in his program, SYNOPSYS. His advance over damped-least-squares (DLS) is called the pseudo second derivative (PSD) approach. This algorithm uses consecutive derivative matrices to approximate the second derivative matrix and uses this to calculate an improved damping factor for each variable. The result is a tremendous increase in convergence speed from initial designs that are far from optimal. Dilworth's program also has an algorithm that is often able to start with a lens that does not pass all the required rays and tweak it until it does before starting optimization.

In the regional optimization category, SYNOPSYS starts with a standard simulated annealing algorithm, but combines it with PSD to make it far more effective than simulated annealing is in other programs. Masaki Isshiki's global optimization with escape function algorithm is also implemented, but this reviewer does not have sufficient experience with other programs' implementations to offer a comparison. Regional optimization features unique to SYNOPSYS are 'automatic element insertion' and 'automatic element deletion', which either insert or delete a lens element at the optimal location. The former algorithm can be run in a way that is very similar to Florian Bociort's saddle point algorithm.

New global optimization algorithms, DSEARCH and ZSEARCH, which Dilworth has recently added to SYNOPSYS, are most impressive. DSEARCH starts from a very rough description of a lens (the object, wavelengths, F/#, and number of elements) along with any other required constraints, and produces several candidate designs that are often close to final designs. ZSEARCH does the same thing for zoom lenses. Both algorithms can provide lens designs even if the designer has no idea of what an initial configuration might look like. Dilworth has published a paper with Dave Shafer, a renowned lens designer, comparing the results of DSEARCH with Dave Shafer's design of a well-designed eleven-element lens. The paper is an entertaining discussion of man versus machine. DSEARCH was able to quickly find solutions with eleven, ten, nine, and, remarkably, only eight elements. Once he knew that there was a potential solution space having fewer elements,

Shafer was also able to find designs, but took much more time to do so. He found one design that DSEARCH missed (with default options), but the algorithm came up with several other designs which surpassed those of this famous lens designer.

One could say a lot more about optimization in general, and Dilworth's contributions in particular, but this should be enough to give you a flavor. Reading this book will teach you more design tricks and insights. Many examples are provided that readers can run on their own computer and experiment with by changing parameters and other commands. I am certain that you will enjoy and profit from this book.

Dr Steve Eckhardt
Eckhardt Optics LLC
White Bear Lake, MN
March 2018

Series Editor's foreword

Mr Donald C Dilworth began developing in the late 1960s a lens design and analysis software package called SYNthesis of OPtical SYStems (SYNOPSYS™). Over the past 50 years, by himself, he created, developed, and has continued to improve and expand the capabilities of this program to meet the demanding needs of lens design professionals throughout the world. During this period, other lens design programs were developed for both internal corporate use and general use by the optics community. Many of these programs left the marketplace while a modest few have stood the test of time. All current commercial lens design programs have excellent capabilities that meet the general needs of the optical designer; however, over the past couple of decades, optics has become broadly pervasive, and the demand for people capable of accomplishing optical design rapidly outstripped the supply of trained and talented lens designers. To overcome this lack of supply, lens design software providers worked hard to make it easier for essentially untrained people to use their programs and obtain meaningful designs. One important capability that has been explored by the software providers is to incorporate a means to allow the user to input basic parameters of the optical system and then allow the software program to attempt to create design(s) that meet the designer's requirements. In this book, Mr Dilworth presents how to design and analyze lens systems, and his innovative and unique developments in automated lens design that have had, and will continue to have, significant utility in mitigating the growing lack of professional lens designers as well as improving the efficiency and creativity of the professional lens designers. Achieving the goal of autonomous optical design continues to be an emerging technology for lens design software providers; however, Mr Dilworth has made great strides to accomplish quasi-autonomous lens design.

Arguably, during the 1960s and 1970s Berlyn Brixner (Los Alamos National Laboratory) was the first to suggest and demonstrate starting a design using just flat plates and allow the program to 'explore' alternative paths to generate a design solution. As computer power became available at affordable costs and research in optimization techniques evolved, lens design program developers began to incorporate what is commonly called *global optimization.* The objective is to allow the program to search for a solution that is hopefully superior to that following normal optimization methods. Indeed, the various approaches often did find better designs and often new design forms the human designer had not envisioned. Such a search process is quite time- and resource-consuming, with the result potentially being a design that is not practical. Nevertheless, the global optimization tool has been of great benefit to the beginning lens designer as well as the professional. Further refinement of potential designs can be obtained by using the simulated annealing feature available in all current programs.

While lens design software developers focused on methods to improve global optimization along similar lines, Mr Dilworth additionally explored and exploited several 'out-of-the-box' approaches to achieving, what might be called, *quasi-autonomous lens design.* There are four such innovations that provide lens designers additional tools that have the potential to create solutions to their design requirements in less time, with more ease, and by exploring new configurations the SYNOPSYS program produces.

The first two innovations to aid the designer were developed initially about 1990, and provided artificial intelligence (AI) capabilities utilizing natural language processing and expert systems. The natural language AI feature provides a very flexible way to interact with SYNOPSYS, and to perform certain tasks not readily accomplished with the normal command syntax or spreadsheet input. Among these tasks are displaying and changing certain lens parameters, defining new commands or character strings, and obtaining plotted parametric curves involving up to three different quantities. The input consists of English sentences comprising subjects, verbs, and conditions. The vocabulary contains many hundreds of words and the flexibility for the user is vast. This is a very powerful and useful tool for designers.

In general, an 'expert systems' program is one that employs a tree-structured logic wherein decisions are derived from the responses of a number of experts in a particular field to a lengthy debriefing. Remarkable performance has been achieved in some areas, rivaling that of a human expert. In contrast, the SYNOPSYS program takes a somewhat different approach in that the expert system feature (XSYS) is presented with a number of finished lens designs that are the products of expert designers and represent the state-of-the-art. Using these lenses as models, the program analyzes the optical properties in great detail, determining for each lens the first and higher-order properties, and the aberrations that are present in the beam before and after each element. In so doing, the program 'learns' how a particular optical problem was solved by its designer. The more examples provided to XSYS, the more it learns. When presented with a new problem or a lens that is not well corrected, XSYS can determine if the current problem resembles one for which it knows a solution and then attempts to use bits and pieces of lenses in its expert-designed-lenses library to create various potential configurations for the user to review, select one to optimize, and then analyze. This feature is very creative and often finds unexpected configurations that work better than having followed a conventional design path. Of course, not having a library of relevant lenses will mitigate the utility of XSYS. However, if it is available, XSYS can provide the lens designer with a powerful tool to identify potential lens configurations for further optimization and analysis.

As mentioned previously, a bit over a quarter of a century ago, as computer resources became more capable, the search for the optimum solution of a lens design moved from optimizing a 'point' design to allowing the software program to explore an enormous variety of possible configurations in solution space. Such a search process takes, in general, a huge amount of time in contrast to point design optimization. This is known as global optimization and often includes simulated annealing. A variety of approaches have been developed by researchers to locate the 'true' global optimum solution (minimum merit function) while at the same time minimizing the convergence time. Arguably, progress has been made, but the time needed for finding candidate solutions often remains excessive. Early in the current decade, in an effort to dramatically reduce the search time for candidate design solutions, Mr Dilworth developed a quasi-autonomous method to search for candidate fix-focused lens configurations which he named DSEARCH™ (Design SEARCH). This method is not intended to produce a finished lens design; its purpose is to identify attractive starting points, and each of the designs can then be subjected to further optimization, first with transverse intercepts, then with OPD targets, and lastly with MTF targets in the merit function. Although the solutions lie in a multidimensional hyperspace that cannot be visualized, it is common to visualize the solution space as a rough mountain range where the highest peak corresponds to a flat-plate starting design. The objective is to find the lowest valley, which is certainly a challenge. Since any direction from the mountain peak is downward, the question is which direction to go. Mr Dilworth reckoned that an efficient selection of directions to try would be a 'binary' approach where the flat-plate lenses are assigned either a positive or negative power. For an N element lens, this implies 2^N starting optical configurations or directions to consider. When the valley in a particular direction is found, the question still remains: is the this the best solution in this direction? To overcome this, simulated annealing is used to explore the surrounding hyperspace to determine if a lower valley exists. Of course, a RANDOM mode can be used to attempt to find the best solution, but it can be far more demanding of time. If the initial flat-plate design is specified to have N elements, and in the event that the best solutions found are not good enough, an automatic element insertion option is available and the prior solution lenses can be reprocessed. Should the solutions be 'too good', an automatic element deletion option can be used to reprocess the lenses to look for adequate solutions having $N-1$ elements.

The final innovation to mention is the evolution of DSEARCH to accomplish finding potential solutions for zoom lenses. The design of zoom lenses is significantly more challenging than the design of fixed-focused lenses. Mr Dilworth named this new feature ZSEARCH™ (Zoom SEARCH) and it was presented at the 2016 SPIE Meeting in San Diego, CA. ZSEARCH works essentially like DSEARCH except that the searching process is much more complex, and the results are rather amazing.

Mr Dilworth continues to develop improvements in his quest for *quasi-autonomous lens design* to address the issue of the growing dearth of experienced lens designers. The inclusion of additional features in the search and final design selection process that consider manufacturability and cost of the candidate lenses should be most beneficial to lens designers and their employers.

Working alone on his SYNOPSYS program for the past half century, Mr Dilworth has made exceptional contributions in computer-based optical design software, which is used worldwide, and in his development of the aforementioned innovations to assist both novice and expert designers in their work to design high-performance optical systems in minimal time.

Those learning lens design will find the material in this book helpful in mastering how to design lens systems regardless of which lens design software is used. All readers should benefit from the knowledge and wisdom Mr Dilworth has gained over the past half-century. Unlike the traditional books on lens design, the interactive nature of this book provides readers with an unprecedented opportunity to work examples themselves and further explore what happens if they modify the examples while using a provided version of SYNOPSYS. This is an excellent method to learn the subject and sharpen one's skills. Optics has become pervasive throughout most technological areas, and the design and manufacture of optical systems is the fundamental foundation.

R Barry Johnson, DSc, FInstP, FOSA, FSPIE
Series Editor, *Emerging Technologies in Optics and Photonics*
Huntsville, AL

Preface

Why this book? A new paradigm. The lens design landscape. The lens design tree.

When a friend suggested I write a book about lens design, my first reaction was: Why? There are a number of excellent books out there already. Who needs another one?

My second thought was: Why me? I have never taken a course in optics or lens design, and (I hate to admit it) I have never even read most of the aforesaid books. What can I possibly add to the corpus?

My friend pointed out that I have been designing lenses for more than 50 years, and it is likely I have picked up a few pointers along the way. True enough.

That led to my third thought: this might be a very short book. I picked up one of the best of the current texts, *Lens Design Fundamentals* by Rudolph Kingslake and R Barry Johnson (2010, Bellingham, WA: SPIE), and read descriptions of how one does certain tasks *that you do not have to do anymore*. The classic authors and designers had to make do with very primitive tools and were faced with a staggering amount of manual labor. So they devised shortcuts, found insights where they could, and invented ways to obtain an approximate answer with relatively little work. What I mean by 'little' can be understood in the context of Kingslake's comment:

> When someone applied for a position in our department at Kodak, I would ask him if he could contemplate pressing the buttons of a desk calculator for the next forty years, and if he said 'yes', I would hire him.

In this computer age, most of those tools and shortcuts are no longer needed. In fact, some of them actually get in the way and are better avoided. The classic texts will not tell you that. This book will.

When I first entered this field, there were no books on computer-aided lens design. I first traced rays on a computer filled with vacuum tubes. I had to figure everything out for myself. This turns out to be an advantage, it puts me in a position to ask: what do you *really* need to know in order to design lenses? Not all that much, it turns out. Certainly, you do not need all the primitive tools used by the old masters, and if I had to use those tools and do all that manual labor, I would have left this rewarding field years ago. Some will be offended that I give little ink to those techniques in this book—but I want to teach what is useful today, not what people had to learn a generation ago.

Today we live in a new world. I estimate that perhaps 90% of what I read in the classic texts is now irrelevant. I could write on two pages most of what one needs to know when designing lenses, assuming one has a modern PC and the most powerful of today's lens design programs (and a basic understanding of math and physics). A good friend, who teaches lens design at a major university, once boasted that her student could design a five-element lens in only five days. Modern software can design a seven-element lens in less than one second. My friend did not know that. Things are changing.

Many brilliant minds have contributed to this field, and prudence says we should not just discard the knowledge and skills handed down to us by the old masters. So how can I justify saying that very little of that knowledge is relevant today? Here is how: I challenged my friend to design a 90-degree eyepiece, essentially perfect, with the classic tools she had long mastered. She gave up after 100 h of work. Using software that I have developed over more than 50 years, I was able to solve that problem in only 15 minutes—using almost none of those tools. I think experience speaks for itself. We are thinking out of the box here, and the results are wonderful. This is a game-changer.

Before I go too far, however, I have to caution the reader that learning to use lens design software is not for everyone. For one thing, it requires some familiarity with optics and physics. The software is not going to anticipate and solve every problem for you. The aptitude necessary to become a good engineer is not to be found in every human head, which is why engineering schools have entrance exams. You will be obliged to read the instruction manuals and make an effort to understand them. I sometimes receive questions from very naïve beginners showing they are starting without this familiarity and are confused by many of the concepts presented in the manuals and in this book. There is a reason people attend college and obtain a degree in the sciences. With that background, those concepts are simple and clear enough. I urge those without a technical education to take steps to acquire it before diving into this fascinating field.

Some will object that the methods I present in this book rely too much on chance. A throw of the dice. Trial and error. They want the process to be deterministic, guided at every turn by expert knowledge. There is much to be said for that point of view—but consider a conversation I once had with Kingslake (an expert if ever there was one):

> I said: 'In your book, you show how one could alter a certain radius of curvature in steps and plot an aberration curve, then change a certain thickness and plot a second curve. Where those curves cross is a better design—but how did you know you should change *this* radius and *that* thickness?'
>
> He replied: 'That's the only thing that worked. *I tried everything else*.'

It seems that even the experts had to resort to trial and error much of the time, if one makes an objective appraisal of how they worked. What I have done is to make it work much better and much faster. How much better? A colleague once volunteered, 'These search methods work so well it's *scary*!' He knew what he was talking about; he had tried them himself.

But, wait a minute. Two pages? A friend objected: 'We cannot just teach the students to *push a button*!' True enough. There is a whole lot of optics written between the lines on those pages, and the student really should know what happens when you push it. Kingslake observed:

'We are losing the ability to design a lens through an effort of the intellect.'

He is right[1]. Grab a qualified person off the street, sit him in front of a modern lens design program, and ask: what does he need to know in order to use it effectively? I hope to put the answers on the pages of this book.

[1] But we have also lost the ability to make flint arrowheads and ride horses, and I do not care.

Acknowledgements

I want to thank the many loyal and very smart customers using my lens design program who have suggested new features and improvements. Together, we are more effective than I could be by myself.

I especially want to thank Dr R Barry Johnson, whose encouragement and ability to see outside the box has been instrumental in the progress of my career and led me to write this book.

Lastly, I wish to thank my wife Sarah, who, despite knowing nothing about optics, continues to believe that what I am doing has merit.

Author biography

Donald C Dilworth

Mr Dilworth has been designing lenses since 1961, when he participated in the development of the optical navigation system for the Apollo project, the American program to land men on the Moon. He is a graduate of MIT, where he majored in physics, but is self-taught in the fields of optical engineering and lens design. His later work includes designing one of the spy cameras that flew over the Soviet Union during the Cold War, and he has worked in industry for many years, designing infrared FLIR systems (which can see in total darkness), and numerous kinds of lens systems including a submarine periscope. He formed his own company, Optical Systems Design, Inc. in 1976 and has been developing and marketing the lens design program SYNOPSYS™ (**SYN**thesis of **OP**tical **SYS**tems) ever since.

IOP Publishing

Lens Design

Automatic and quasi-autonomous computational methods and techniques

Donald Dilworth

Chapter 1

Preliminaries

I am going to use a lens design program called SYNOPSYS[1] (**SYN**thesis of **OP**tical **SYS**tems) in this book, partly because it will perform all the lessons quickly and easily, and also because it happens that I wrote it—but the principles are valid for any code with comparable features. The best way to learn these new techniques and become familiar with the software is to work through many different examples. With this in mind, I have written a set of chapters that exercise many features of the code. To save the labor of typing all of the lens files and optimization MACros in these examples, a copy will be found in the folder DBOOK, which you can download as instructed below (a 'MACro' is a file containing input commands and data that are recognized by the program).

To install and run the program you will need two files as well as the example folder, all located at www.osdoptics.com:

SYNOPSYS200_v15.zip,
InstallSYNOPSYSdll.msi, and
DBOOK.

Scroll down on that web page and click the 'Download' button, which links to the first one; then unzip the zip file and run the msi file that it contains. Then download and double-click the msi file with the dlls, so you will have those files too.

The program creates a folder C:\SYNOPSYS, and under that a folder USER. Copy the DBOOK folder under C:\SYNOPSYS as well. Now you should have two directories under C:/SYNOPSYS: USER and DBOOK. The program expects to find this directory structure, so do not move things anywhere else.

The program is frequently updated, and you always want the current revision, so check the website frequently for updates. Sign up for the mailing list and you will stay informed.

[1] SYNOPSYS™ is a trademark of Optical Systems Design, Inc., a Maine, USA, corporation.

doi:10.1088/978-0-7503-1611-8ch1

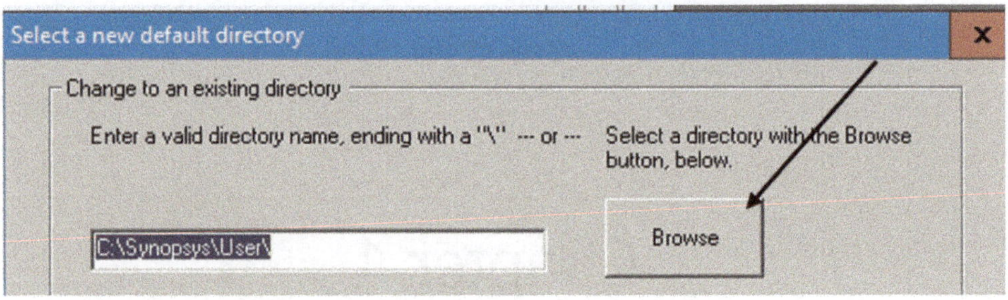

Figure 1.1. The change directory dialog.

When you start the program, it will tell you if you do not have a hardware key. In that case, just select the option to run in demo mode, with a limit of 12 surfaces.

Then type the characters **CHD** (**CH**ange **D**irectory) into the Command Window (**CW**). A dialog opens. Click the 'Browse' button, shown in figure 1.1.

Select the directory 'DBOOK' (which you should have copied per the above instructions) and click 'OK'. This becomes your new default directory.

Then type **HELP TM** in the CW, and the Tutorial Manual will open. It would be a very good idea to print each chapter and go through the text carefully. I do not want you to become stuck on a simple problem simply because you do not yet know how to use the program. The chapters in this book contain many examples that assume you know the basics—and I will coach you here and there when something is important but perhaps not obvious—but you should be acquainted with the basic features of the program and know how to use them before you go much further.

When you have finished for the day, exit from SYNOPSYS by typing **EXIT** in the CW or by clicking the 'X' in the upper-right corner of the frame. The program cleans up many temporary files when it exits, and if you terminate or crash the program, that last step will not take place.

Many of the chapters below refer to lens files and MACros that you will want to open. The former are text files with the extension '.RLE' that contain a description of a lens, and the latter are lists of commands with the extension '.MAC'. When they are appropriate to the lesson, you will be informed by entries such as (**C40L1**), which means the lens is found in file **C40L1.RLE** and can be opened with the command **FETCH C40L1**, and for MACros with entries like (**C40M1**), which refers to a MACro called **C40M1.MAC** and can be opened in an editor with the command **LM C40M1** (**LM** means Load MACro).

Some of the instructions have you change the MACros in certain ways after you open them, as you make decisions and work on the lens. It is a good idea, however, to first rename the MACro, so the original is not overwritten when you run it. Click the button |-N, and a working copy will be saved with the default name. That way, if you want to review that chapter again, you can open the original file and start out with the contents as it should be rather than as you changed it. (When you click the '**Run MACro**' button, , the file is first saved, and then executed.)

1.1 Why is lens design hard?

Although the art of lens design has matured and now has the benefit of many powerful tools that were not available when I started—and certainly not known to the old masters—it is not, and probably never will be, a simple endeavor. The difficulty arises primarily because of the geography of the lens design landscape. One must not only design a great lens, but also must take into account the dimensions of the housing, the properties, cost, transmission, and availability of the glass to be used—while at the same time avoiding lenses whose tolerances are so tight that nobody can build them. Before you even reach those considerations, you have to find a lens construction that works. Well. That is hard to do.

Why is it hard? Because we are dealing with a design space of many dimensions, where many variables and the image quality are related to each other in a nonlinear manner, and where clumsy boundary conditions apply to most variables. Few engineering applications have to reckon with such difficulties.

Traditional methods have long relied on having a good starting point, a design not too far from the desired goals, and then working to improve it. If the starting point was indeed a good one and your skills were sharp enough, you could in that way arrive at a great design. However, one rarely has such a starting point, and only a few of us have the required skills. Thus the job is hard for most, and difficult even for the experts, most of the time. The core problem here is the fact that, except in a few simple cases, there is no *closed-form* solution to the lens design problem. That means there is no formula you can simply plug numbers into and obtain a great design. You have to think, try things, learn from experience, and iterate. My goal in writing the program has been to put as much of the burden on the computer as possible, freeing you from the most tedious of traditional tasks. I hope you will thank me once you see how much easier the job is now, with these new tools.

1.1.1 The lens design landscape

I often describe the lens design landscape as a mountain range with peaks and valleys all over the place. In that scenario, your job is to find the lowest valley, which corresponds to the lowest *merit function* (**MF**). The MF is usually defined as the sum of the squares of a set of quantities that represent how far the design is from its ultimate goals; the MF would be zero if all the targets were met exactly—which almost never happens. The lowest valley overall is the best, or 'optimum' design, because it has the lowest MF. How can one find it?

One way is to start at the top of the highest peak, from where you can see all the valleys, select a direction, and head downhill. That is the principle behind **DSEARCH**[2], a tool you will use in many of these chapters. The highest peak, in this view, is a lens with all plane-parallel surfaces; that design can go anywhere. DSEARCH selects a variety of directions according to its own logic, and then heads downhill, evaluating the quality in the lowest valley it comes to in each direction. The algorithm is discussed more fully in appendix B.

[2] DSEARCH™ is a trademark of Optical Systems Design, Inc., a Maine, USA corporation.

Figure 1.2. Graphical illustration of the lens design tree.

Another way to visualize the task is to imagine climbing a tree, as shown in figure 1.2. Here, one can start at the bottom and climb up—but which branch to take? See the problem? There are usually many solutions to a given task, of roughly equal quality, and when you run a lens optimization program[3], you are climbing up whatever branch you happen to be on. A different starting point will go up a different branch. When you reach the end of the branch you are on, you are at a *local minimum*. Just running the optimization program yet again will not move you from that branch to a better one. You need other tools. How can you get to a different branch? How many branches are there? Those are perplexing questions, and although the techniques described in this book enable one to explore the design tree quickly and easily, they have, even now, not been completely answered.

Here is a helpful way to look at it. Figure 1.3 shows the statistics of the merit function resulting from optimizing a total of 5000 random starting points for a typical lens design job. The peaks give the number of cases that arrived at a particular value of the MF, and each one corresponds to a different local minimum. Since there is no obvious way to escape from one to another with classical tools, it is evident that the designer has a tough job. The solution you want is way over at the left end of the curve. That is the main reason why lens design is hard: *too many local minima*!

[3] An *optimization program* is computer code that uses an algorithm to deduce how to change the design variables in a way that drives the value of the merit function down while observing boundary conditions.

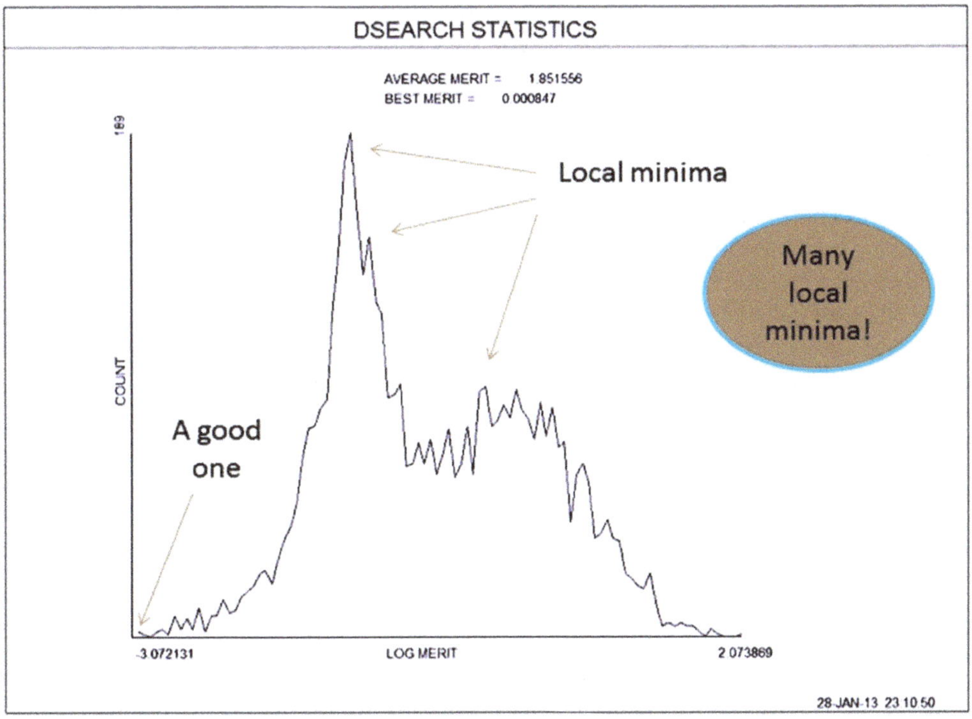

Figure 1.3. Statistics of the results of searching 5000 random branches of the lens design tree.

1.1.2 Simulated annealing

Most lens design programs today offer what is termed *simulated annealing*, a process that involves making small random changes to each of the design variables and then optimizing, over and over. That technique can jump sideways from one branch to another, although usually not very far. Nonetheless, it is surprisingly effective and is one of the most important tools of the trade today. You will use it in many of the examples below.

Other tools are more powerful yet. SYNOPSYS offers several search routines for exploring the tree. DSEARCH starts at the bottom and can go anywhere, and the tools called **AEI** and **AED**[4] can deterministically select the best place to insert or delete an element. Those tools can jump to an entirely different branch and usually can find a better solution than can a human expert.

1.1.3 Global optimization

Most lens design programs today also offer a form of 'global optimization', which can find a variety of solutions—but most of those programs are not practical because of the very long time required to return their results, often measured in hours or

[4] AEI™ and AED™ are trademarks of Optical Systems Design, Inc., a Maine USA corporation.

days. DSEARCH, on the other hand, executes very quickly, using the algorithm described in appendix B, and that is why we use it in these examples.

We now enjoy a new paradigm. It used to be the case than an expert would spend days, weeks, sometimes years making small improvements to a classic design form, always guided by experience, insight, theory, and a large dose of dogged labor. If he succeeded, he was rightly proud of the achievement. Today we do things differently. Instead of inching up a single promising branch of the tree, day after day, hoping the result will justify the effort, we use software that examines hundreds or thousands of branches in a matter of minutes or seconds and returns a set of candidate design forms the user can then evaluate and try to adapt to his current requirements. Those candidate lenses are often so well corrected that they need little or no improvement, and a few are sometimes already at what we call the 'diffraction limit', a condition in which the only significant limit to the resolution is due to the finite wavelength of light. In that case, further improvement to the lens may be unnecessary, and one is concerned mainly with mechanical properties, tolerances, the cost of the glass, and so on. This new paradigm is a true revolution, and its importance is only beginning to be recognized. That is perhaps the main reason I agreed to write this book: to spread the word.

A corollary of this deciduous metaphor is the problem of just *how* to write this book. There are many examples in the chapters below that utilize the search tools, and in many cases the reader is instructed to run the simulated annealing program after optimizing the lens. However, those methods involve random changes at some stage, and by their very nature those changes are different every time through. Thus, unless special precautions are taken, the user would likely obtain a result different from that shown here when he works that example. The end point of a run on DSEARCH is very sensitive to the initial conditions and the particular random changes in each annealing step and would be different every time. This would be disconcerting to the reader, who expects to see on the monitor what is on the paper in front of him.

To resolve this difficulty as far as possible, I have programmed one of the mode-control switches in SYNOPSYS (number 98) to recycle a single random number series whenever randomness is called for (this is like opening a book of random numbers to the same page every time). That switch was turned on when these lessons were prepared, and if the user also has the switch on, he will most likely obtain the same results as those shown here. I say 'most likely', because other effects can influence the path up the design tree as well.

1.1.4 Chaos in lens design

You can obtain different results if you change some of the other mode-control switches, assign different weights to the requirements, or even use a different release of the software, which is updated regularly. The glass tables change as vendors add or delete glasses from their catalogs. Even a tiny change at any stage of the process can alter the path through the tree and lead to a different end point; the branch of mathematics called *chaos* theory deals with situations like this, and lens design involves more than just a tree: it is a *chaotic* tree. However, the saving grace is the fact that, for most problems, many of the branches are equally good. All you have to do is to find a good one. That is good to know. Otherwise most of us would be out of a job. Chapter 27 gives an interesting discussion of the chaos inherent in lens design.

We caution the reader that, except when following the examples below, one does *not* want switch 98 turned on. The beauty and power of these new tools comes precisely because the results are *different* every time. If you lose your keys, it makes no sense to look for them in the same place over and over. We often run DSEARCH several times when working a design job, selecting the best results from each run, thereby exploring a great many branches of the design tree. Usually the program finds several rather different lens constructions with almost identical performance. Then we can base our selection on considerations of packaging, cost, and so on.

We also observe that, in a previous age, when a designer had to steer a lens to the solution with sometimes heroic effort, when he finally arrived at a good one it is likely he would declare success and look no further. Why go through all that work again? However, with modern search tools, the program can suggest many potential solutions very quickly, which adds a whole new dimension to the process. It sometimes finds great designs that an expert would never have thought of, simply because he already found one that seemed good enough and quit looking.

1.2 How to use this book

The chapters that follow present a variety of lens design problems and show how to find a solution using the search tools and other features of the software. In most cases, they advise you to `FETCH` a given lens, load a given MACro into an editor and run it, and then modify the input to deal with problems that show up. Lens design software has a reputation for solving the problem you asked for, not the one you *should have* asked for. We make our living figuring out how to ask the program to solve the problems we did not anticipate.

Doing so requires modifying the input file, and you are reminded to first *rename* the MACro before you make any changes, by clicking the |-N| button. It would be unpleasant if you wanted to review an earlier chapter, but the input files were all different and the solution was not the same as before.

Also, you will see instructions like 'optimize and then anneal (**50, 2, 50**)' in many chapters, and it is important that you carry out all of the instructions *in the exact order* given. Not because that order or those data are especially important—indeed, you are encouraged to experiment with them—but *not the first time*. Other data combinations will almost always return a different lens, and that is one of the virtues of the search tools described herein. You might find solutions that are better than we give here—but they will not be the *same* solutions, and to properly follow the instructions and learn how to deal with problems that show up, you really want them to be the *same* problems. That will happen only if you follow the steps outlined below exactly.

Chapter 2

Fundamentals

Paraxial optics; Snell's law; Lagrange invariant; thin-lens equation; pupils

There is widespread misunderstanding about lenses and images. A customer of mine once objected when I designed an eight-element lens that satisfied his requirements and made a good image; he wanted it done with a singlet (one element)! More enlightened people will understand that, to obtain a good image, one usually needs more than one lens element. Why is that?

That is the whole domain of lens design: the cause and cure of image defects, which come in a wide variety of types, can be difficult to correct, have to be carefully balanced—but are actually quite easy to understand.

2.1 Paraxial optics

To set the stage properly, let us first discuss *paraxial* optics, also called *first-order* or *Gaussian* optics[1]. That is the domain where the well-known Snell's law is simplified. Consider the lens in figure 2.1.

A ray of light comes in, is refracted at both surfaces, and heads to the image plane. It does not hit the desired image point however, and the image is blurred. The lens surfaces here are portions of spheres, which is by far the most common and usually the most economical situation with lenses.

Now, consider what would happen if we reduce the aperture to a very small value, as in figure 2.2.

The ray now hits much closer to the center of the image. (Galileo, not understanding lens aberrations, found that by 'stopping down' his lens (reducing the

[1] Strictly speaking, *paraxial* optics refers to analysis with infinitesimal aperture and field. When you scale up the latter, to analyze real systems, it becomes Gaussian optics. The latter can also be analyzed by replacing nI with $n\tan(i)$, as mentioned in reference 1 in the bibliography.

doi:10.1088/978-0-7503-1611-8ch2

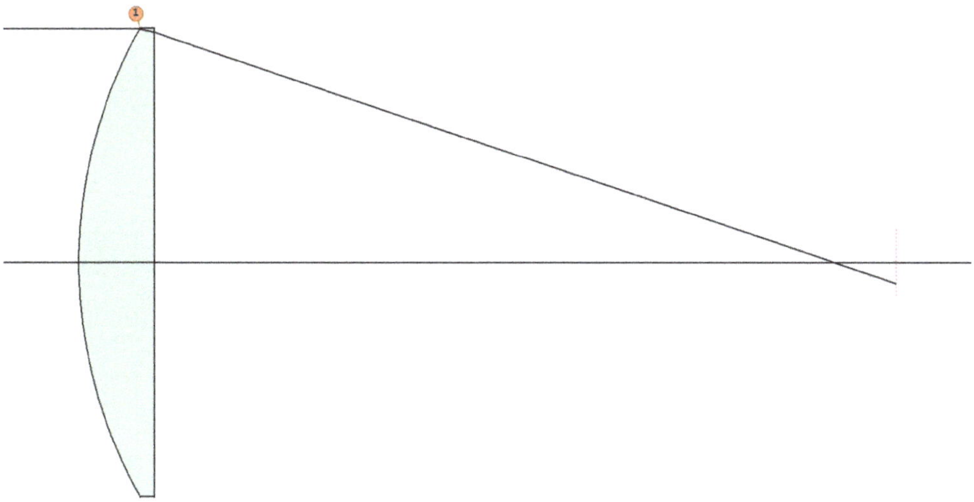

Figure 2.1. A simple lens.

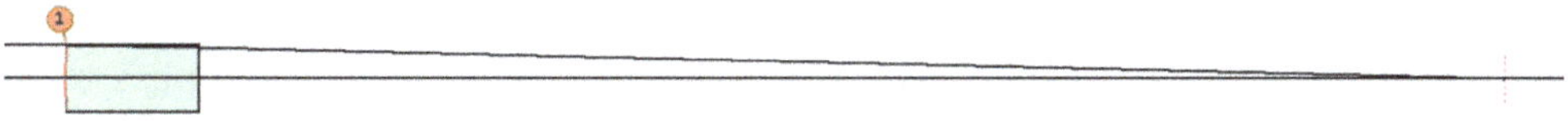

Figure 2.2. The lens with reduced aperture.

aperture diameter) he could sharpen the image). Of course, less light enters when you do that, but there is a lesson here: if the aperture goes to zero, the angles also go to zero, and the sine of a very small angle equals the angle itself. So Snell's law changes from

$$n' \sin i' = n \sin i$$

to

$$n'i' = ni,$$

where n is the *index of refraction* before the surface, n' is the index following it, and i and i' are the angles of the ray relative to the surface normal (a line perpendicular to the surface at the ray point) before and after refraction. All transparent materials slow down the passage of light relative to the speed in air, and the ratio of that speed to the speed inside the material is what we mean by the index of refraction. This geometry is shown in figure 2.3.

So what? Well, it turns out that if the aperture and field both approach zero, the imaging properties of the lens can be described very simply. All the *paraxial* rays, as they are then termed, go to the desired image point, with no aberrations. The value of paraxial optics is that it points the way to the design goals: if real rays wind up just where the paraxial rays do, the lens is close to perfect. When we speak of *aberrations*, we mean the departure of real rays from the image point given by paraxial rays. (The term is often generalized to mean anything that is wrong with a lens, but there is a

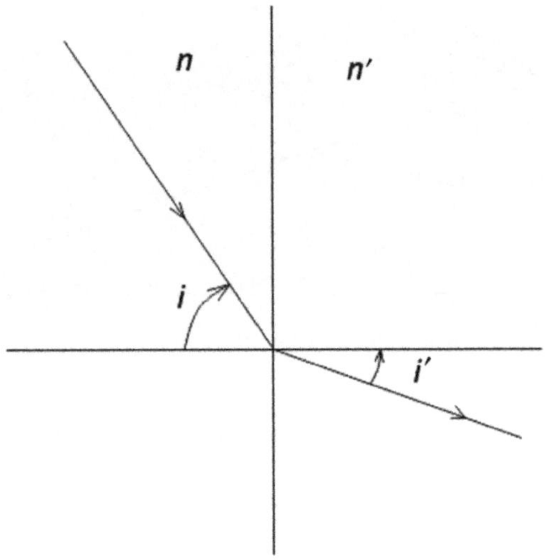

Figure 2.3. Geometry of Snell's law.

family of aberrations with characteristic properties and standard names, which we will refer to frequently in this book.) The concept of paraxial optics applies to the common situation in which the lens has no tilted or decentered elements and has a flat image surface; otherwise, the concept may not strictly apply—but it is nonetheless one of the fundamentals you should know about.

It is very simple: divide both the aperture and field by a large number, perform a paraxial raytrace—yielding the paths of two rays, using the simpler form of Snell's law. Then multiply the *results* by the same large number. That gives you the paraxial ray paths through the lens, which are sometimes displayed as shown in figure 2.4.

Classic texts usually show the formulas for tracing a paraxial ray through a lens. However, all the lens design programs already have those formulas in their code and there is no reason for you to learn them—and certainly no reason ever to trace a paraxial ray by hand. It is enough to know that paraxial rays do not really exist but are a useful concept to help you determine just how good a lens is.

The ray shown in red in figure 2.4 is the *marginal ray*, which comes in from the on-axis object at the edge of the beam, and the ray in blue is the *chief ray*, which starts at the full-field object and comes in at the center of the beam. (We will discuss the fine points of this definition later.)

According to paraxial optics, light from an object at the edge of the field of view forms an image whose distance from the optical axis is called the *Gaussian image height* (**GIHT**). That distance is just the tangent of the entering chief-ray angle times a constant, called the *focal length* of the lens, or **FOCL**. Another fundamental property is the *F/number*, or **FNUM**, defined as the focal length divided by the aperture diameter. (Those properties are valid for a lens where the object and image are both in air. Your software will correct for cases where this is not the case.)

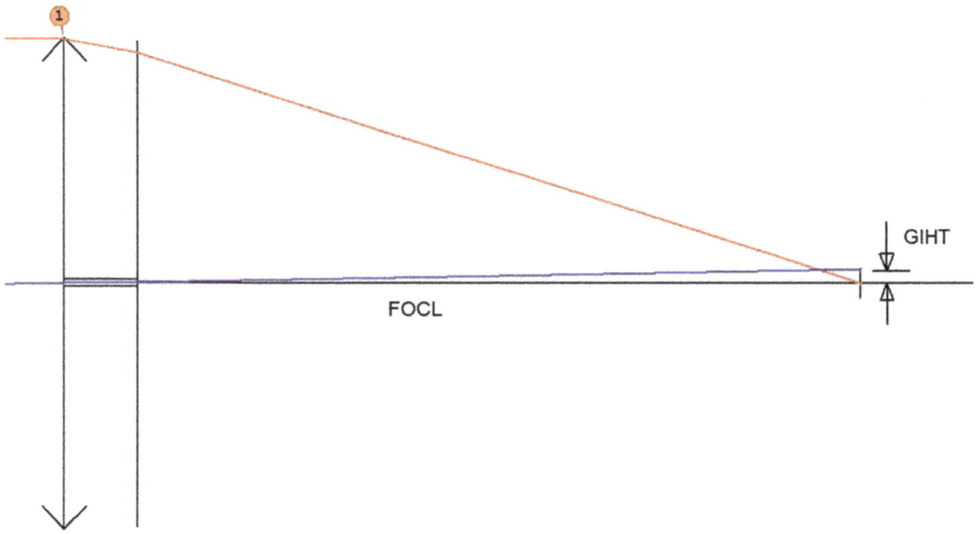

Figure 2.4. Paraxial rays, shown schematically.

The above lens is described by a specification listing like this, produced by the command **SPEC**:

```
ID SINGLET
LENS SPECIFICATIONS:

SYSTEM SPECIFICATIONS
OBJECT DISTANCE      (THO)          INFINITE  FOCAL LENGTH       (FOCL)          8.0886
OBJECT HEIGHT       (YPP0)          INFINITE  PARAXIAL FOCAL POINT              12.2688
MARG RAY HEIGHT     (YMP1)            2.0000  IMAGE DISTANCE     (BACK)          7.6733
MARG RAY ANGLE      (UMP0)            0.0000  CELL LENGTH        (TOTL)          0.6299
CHIEF RAY HEIGHT    (YPP1)            0.0000  F/NUMBER           (FNUM)          2.0222
CHIEF RAY ANGLE     (UPP0)            1.0000  GAUSSIAN IMAGE HT(GIHT)            0.1412
ENTR PUPIL SEMI-APERTURE              2.0000  EXIT PUPIL SEMI-APERTURE           2.0000
ENTR PUPIL LOCATION                   0.0000  EXIT PUPIL LOCATION               -0.4153

WAVL (uM) .6562700 .5875600 .4861300
WEIGHTS   1.000000 .0010000 1.000000

COLOR ORDER     2    1    3
UNITS                                INCH
APERTURE STOP SURFACE (APS)           1     SEMI-APERTURE       2.00898
FOCAL MODE                           ON
MAGNIFICATION               -8.08861E-12
GLASS INDEX FROM SCHOTT OR OHARA ADJUSTED FOR SYSTEM TEMPERATURE
SYSTEM TEMPERATURE =    20.00 DEGREES C
POLARIZATION AND COATINGS ARE IGNORED.
SURFACE DATA

SURF          RADIUS       THICKNESS    MEDIUM              INDEX       V-NUMBER

  0         INFINITE       INFINITE    AIR
  1          4.18015        0.62992    N-BK7              1.51679      64.17 SCHOTT
  2         INFINITE       7.67331S    AIR
IMG         INFINITE
```

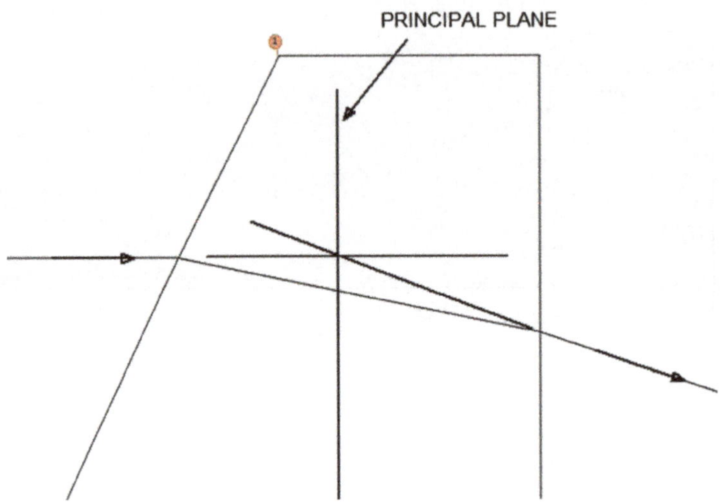

Figure 2.5. Definition of the second principal plane.

The focal length in this case is given as 8.0886 inches. Now it becomes a little more complicated. If you start at the image plane and measure back that distance, you wind up—where? The answer depends on the shape of the lens, and often it is somewhere inside it. If the lens is very weak and has zero thickness (a 'thin lens'), you wind up at the lens itself. Otherwise you go to one of the *principle planes* of the lens. Draw a line along the marginal ray coming into the lens, and another backwards along the refracted ray coming out, as shown in figure 2.5. Where those lines meet is the *second principal plane,* also defined as the plane of unit linear magnification. A related construction yields the first principal plane, and there are in addition two points called *nodal points* that give the points of unit *angular* magnification. We mention these only so the students will know the terms in case they talk to an older generation of lens designers. Personally, I have never had any need for them or reason to look at the values. That is one of the 90% of topics I think one can ignore.

One does sometimes need to look at the *paths* of the paraxial rays, however, and the listing looks like this, produced by the command **PXT P**:

ID SINGLET

PARAXIAL RAYTRACE DATA

SURF	Ymarg	U'marg	Imarg	Ychief	U'chief	Ichief
OBJ	0.00000	2.000E-12		-1.74551E+10	0.01746	
1	2.00000	-0.16302	0.47845	0.00000	0.01151	0.01746
2	1.89731	-0.24726	-0.16302	0.00725	0.01746	0.01151
IMG	0.00000			0.14119		

GIHT	FOCL	FNUM	BACK	TOTL	DELF
0.14119	8.08861	2.02215	7.67331	0.62992	0.00000

This listing shows the path and angles of the two paraxial rays as well as a first-order analysis of the singlet. (The angles listed are actually the *tangents* of the angles, in keeping with paraxial practice.) Note that the FNUM can easily be derived from the final value of the marginal-ray angle $U'_{\text{marg.}}$:

$$\text{FNUM} = -0.5/U'_{\text{marg.}},$$

thus, −0.5/−0.247 26 = 2.0222.

This is useful when a customer specifies the FNUM and you want to know the angle so you can specify a curvature solve, a topic we will introduce in a later chapter.

2.2 Lagrange invariant, thin-lens equation

Another topic we should mention here is called the *Lagrange invariant*. This is a simple concept:

$$\lambda = y_B\, n\, u_A - y_A\, n\, u_B,$$

where y_A is the y-coordinate of the paraxial ray A, and y_B that of the chief ray, B. Angles u_A and u_B also refer to the paraxial rays, and all quantities are taken at the same surface, somewhere in the lens, where the index of refraction is n. The importance of this equation is profound. Once a beam of light enters a lens, the value of λ is fixed. There is no way to change its value, except by blocking some of the light, and this makes it a great tool for deciding whether a given imaging task is possible. (The quantity λ is sometimes called the 'etendue' of the lens.) If another element changes the value of, say u_A, you can be sure that the other variables in this equation will automatically adjust themselves in such a way that λ remains constant. We have received more than one request from a hopeful customer to design a lens that violates this rule. (We had to politely decline.) From this simple rule one can deduce, for example, that if a telescope has an eyepiece that magnifies the field by, say 100×, then the exit pupil, where the observer puts his eye, will be 1/100 of the diameter of the objective lens or mirror. This is a powerful concept. One can derive from this the fact that the paraxial magnification of a system is given by $m = n_0u_0/n_ku_k$, where the subscript 0 refers to object space and k is the final surface.

Consider a telescope pointed at a clear blue sky. Ignoring reflection and transmission losses for the moment, what is the brightness of the sky as seen through the telescope? The surprising answer is, as long as the pupil of the eye is filled, the image is *exactly as bright as the sky itself*, regardless of the magnification. Why, you ask? If the telescope magnifies by, say 10×, then the exit pupil is 1/10 of the diameter of the objective lens. So the amount of light passing through the exit pupil from a given patch of sky is 100 times what would pass through the eye of an observer looking directly at that patch. However, that patch also looks 10 times larger through the eyepiece, which spreads the light out over an area on the retina that is 100 times larger. So the flux/unit area/unit solid angle is unchanged. A telescope pointed at a star, on the other hand, magnifies the apparent *brightness* of the image by the area ratio, since the size of the image on the retina is not increased by the telescope. (Stars are too far away to be resolved, so the image is always a disk whose size is given by the wavelength and F/number, if the telescope is free of aberrations, and is

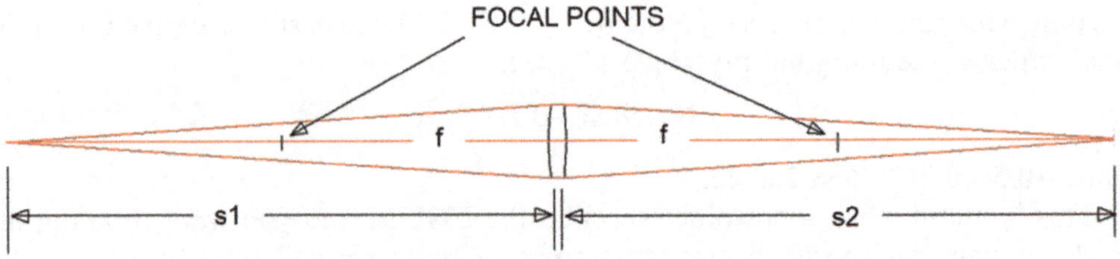

Figure 2.6. Geometry illustrating the thin-lens equation.

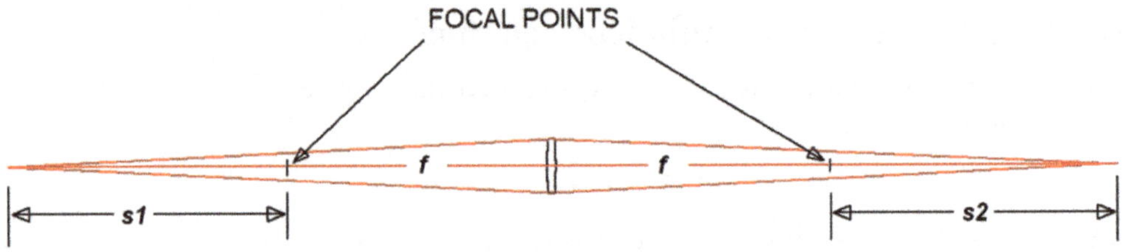

Figure 2.7. Thin-lens equation in Newtonian form.

independent of magnification.) Of course, atmospheric turbulence usually affects the star image as well, so in practice it is somewhat larger.

Among the basic principles you should know about are the thin-lens formulas, which come in two forms. Given the thin lens in figure 2.6, where the focal points are at a distance f from the lens and an object at distance s_1 is imaged at distance s_2, the following relations hold:

$$-1/s_1 + 1/s_2 = 1/f,$$

where s_1 and s_2 are measured from the principal planes (which coincide in a thin lens).

If one measures s_1 and s_2 from the *focal points* instead of from the principle planes, one obtains the Newtonian form, shown in figure 2.7:

$$-s_1 s_2 = f^2.$$

2.3 Pupils

Now we have to define what we mean by an 'exit pupil'. This idea is also quite simple. There is, in many lens systems, a surface whose purpose is to restrict the diameter of the beam of light that goes through, called the 'aperture stop'. Consider the lens in figure 2.8.

Surface 9 (shown by the arrow) in this case is a dummy surface declared the *stop* of the lens. The elements to the left make an image of that surface that is visible to someone looking in from the left (in 'object space'). That image is called the 'entrance pupil' and is where light has to be *aimed* so that it will actually go through the stop. Similarly, the image of the stop, formed by the elements to the right, as seen from the right (in 'image space') is the *exit pupil*, and light appears to come from that

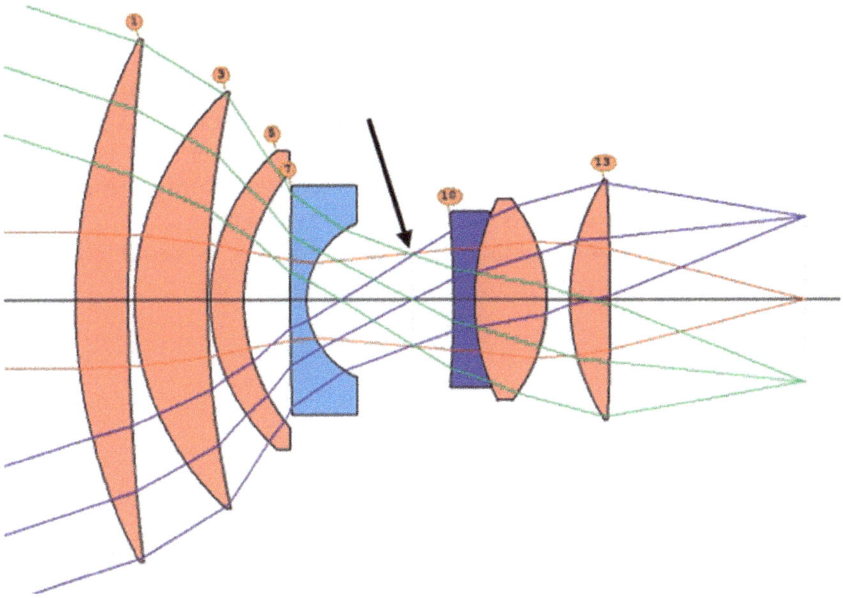

Figure 2.8. Lens with an aperture stop on surface 9.

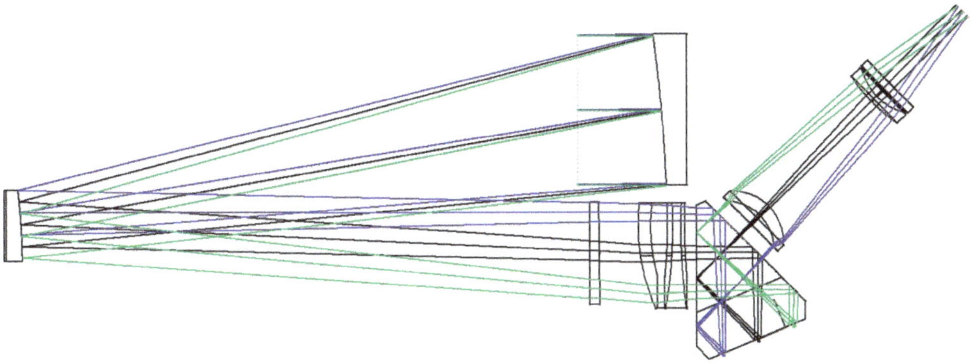

Figure 2.9. A telescope with the stop on surface 1 and an exit pupil formed at the location of the observer's eye.

location. In case it is not obvious, a 'dummy surface' is one with no change in index on either side, so light goes straight through. (Those are useful for locating the stop and sometimes for convenience when employing tilted or decentered coordinates.)

In a telescope with an eyepiece, the exit pupil is just the image of the objective, located a short distance beyond the eyepiece, where the observer will put his eye. All the light that is collected by the objective lens therefore passes through that exit pupil and into the eye (ignoring absorption and reflection losses). The stop in this case is at the objective, since that is the most expensive element and one wants to utilize the entire aperture. An example is shown in figure 2.9.

In this system, the primary mirror is an off-axis paraboloid that collects the light. But the eye point, the exit pupil at the upper right, is just an image of the mirror. So the observer's eye will receive all the light that gets through (assuming the *size* of the pupil of the eye is the same or larger than that of the exit pupil, so all the light can enter). The distance from the last lens element to the observer's eye is called the 'eye relief', and one must make sure it is large enough to allow comfortable viewing, and in some cases even larger, to allow for the use of eyeglasses.

The concept of the *pupil*, while simple in principle, is more complicated in practice, and will be dealt with in more detail in chapter 22.

IOP Publishing

Lens Design
Automatic and quasi-autonomous computational methods and techniques
Donald Dilworth

Chapter 3

Aberrations

Seidel aberrations; spherical aberration; coma; astigmatism; rayfan curves; correction methods; Abbe sine condition; higher-order aberrations, spot diagrams; wavefront aberrations; OPD; Fermat's principle; MTF; knife-edge trace; encircled energy; chromatic aberration; Strehl ratio

Now we get into the meat of what lens design is all about: minimizing aberrations that degrade image quality and are almost never zero.

They come in families and are analyzed in one of two ways. The classic approach was to calculate what are called 'third-order' aberrations, an important topic before the middle of the twentieth century, before computers were available, when one wanted to know how good a lens was. (By 'lens' we mean a collection of one or more lens elements, each with its own aberration contributions.) In 1856, Ludwig von Seidel published the equations for calculating a very approximate estimate of the quality of an optical image. In those days, even a rough estimate was better than trying to trace a large enough set of rays to obtain a more accurate answer (using log tables!). However, it was not a very good estimate, and the art of lens design split into two schools. In England, designers such as H Dennis Taylor and Charles Hastings would often work up a third-order solution (which they knew would not be very accurate), and then *grind and polish* that design so they could measure the image quality directly. It was actually faster to make the lens in those days than to trace a representative set of rays. (The physical lens was a great *analog* computer!) In Germany, designers such as Joseph Max Petzval and Ernst Abbe insisted on tracing rays, getting the design right before grinding any glass. Sometimes it took months of ray tracing.

So what, exactly, is a third-order aberration? Well, if paraxial optics uses the first term in the expansion of the sine function (the sine is then equal to the angle), the next term in that expansion is a function of the third power of the angle. Deriving the raytrace equations with this added term yields a different answer than the paraxial result, and that difference is called the third-order aberration. The value can be

doi:10.1088/978-0-7503-1611-8ch3

expressed as a power series involving the position of a ray in the aperture, (ρ, θ, in polar coordinates) and the object point in the field (HBAR), and one should be acquainted with the concept. We can write the image errors in x and y (E_x, E_y) as follows:

$$E_x = \mathrm{SA3}\rho^3 \sin(\theta) + \mathrm{CO3}\rho^2\mathrm{HBAR}\sin(2\theta) + \mathrm{SI3}\rho\mathrm{HBAR}^2 \sin(\theta)$$

$$E_y = \mathrm{SA3}\rho^3 \cos(\theta) + \mathrm{CO3}\rho^2\mathrm{HBAR}(2 + \cos(2\theta)) + \mathrm{TI3}\rho\mathrm{HBAR}^2 \cos(\theta) + \mathrm{DI3HBAR}^3.$$

SA3 is the third-order spherical aberration, CO3 is the coma, SI3 and TI3 are the sagittal and tangential astigmatism, and DI3 is the distortion.

Here we see that the error due to spherical aberration (SA3) is constant over the field (no HBAR dependence) and varies as the third power of aperture; coma (CO3) varies as the second power of aperture and linearly with field; and astigmatism (TI3 and SI3) varies as the square of the field height and linearly with aperture. Note that by 'distortion' (DI3), which varies as the third power of field, we mean the error in the *location* of the image of a point object; the term does not refer to the *sharpness* of the image, as it sometimes does when used by a layman. All of the third-order aberrations apply only to rotationally symmetric systems.

Although these third-order concepts can be useful when analyzing image errors, they play almost no role when one designs a lens today. (Some older designers still use the methods of Kingslake: changing parameters one at a time and looking at how the third-order aberrations change. Then iterating. Over and over. I do not recommend this procedure.) Since tracing rays with modern computers is so fast, we now deal almost exclusively with the errors of real rays and utilize algorithms that drive the errors as close to zero as possible, largely ignoring third-order aberrations. (There are rare exceptions, which will be noted in the chapters that follow.)

There are also higher orders, fifth, seventh, and so on, which have their own dependences on pupil and field positions. We mention them only in passing since they are no longer of much interest to the lens designer. (Why are there only *odd* orders you ask? Because the expansion of the sine function has only odd orders.) When one alters a lens, it is the lowest-order aberrations that change the fastest, a fact that makes higher orders more difficult to correct. Fortunately, that task is handled rather well when one defines the merit function with a suitable set of real rays.

3.1 Ray-fan curves

To show the aberrations of a lens, we create so-called *ray-fan* curves, which show the image errors of a fan of rays. Before you can trace them, however, you have to be sure you are tracing the *correct* rays. This brings up the subject of *stops* and *pupils*, which deserves a chapter of its own, chapter 22, and which we will only touch on for now. Assuming you have properly defined the lens and the pupil, the ray fans tell you a great deal about the image-forming properties of the lens.

If you trace a fan of rays through a lens, you obtain a characteristic pattern in the image. A *sagittal fan*, or **SFAN**, is a set of rays across the aperture in the x-direction, shown on the left in figure 3.1, while a *tangential fan*, or **TFAN**, goes from bottom to top, shown on the right.

In practice, since a lens is usually symmetrical about the y–z plane (the *meridional* plane), one usually traces only one-half of the sagittal fan (because the other half is identical but inverted). If one now plots the x-position of a given ray at the image against the x-position of that ray in the aperture, one obtains, for a lens with undercorrected third-order spherical aberration, an SFAN curve such as shown in figure 3.2.

In this case, the curve is very close to a third-power function of the pupil position, as one might expect. This shape is characteristic of third-order spherical aberration (so named because it is the most prominent defect in the image formed by a *spherical* telescope mirror; a *paraboloidal* mirror would make a perfect image at the center of the field because it is free of spherical aberration). Figure 3.3 shows the display on the monitor of a singlet lens along with ray-fan curves at three field points. (This display, provided by the **SketchPAD**[1] feature of the program, is opened with the

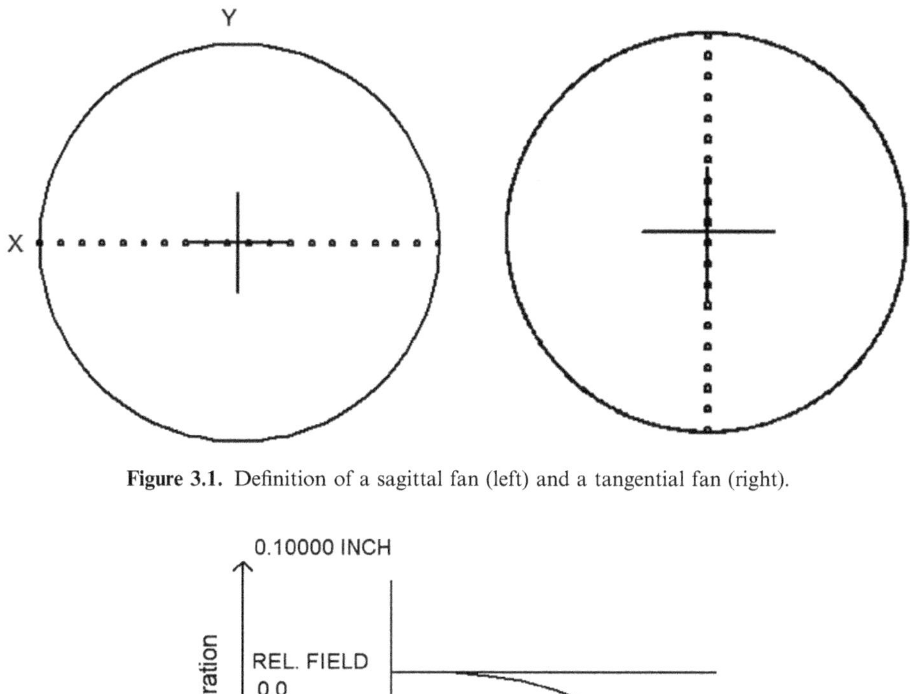

Figure 3.1. Definition of a sagittal fan (left) and a tangential fan (right).

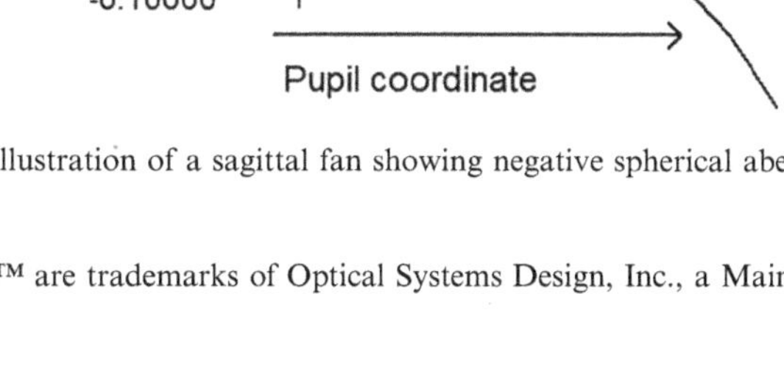

Figure 3.2. Illustration of a sagittal fan showing negative spherical aberration, SA3.

[1] PAD™ and SketchPad™ are trademarks of Optical Systems Design, Inc., a Maine, USA corporation.

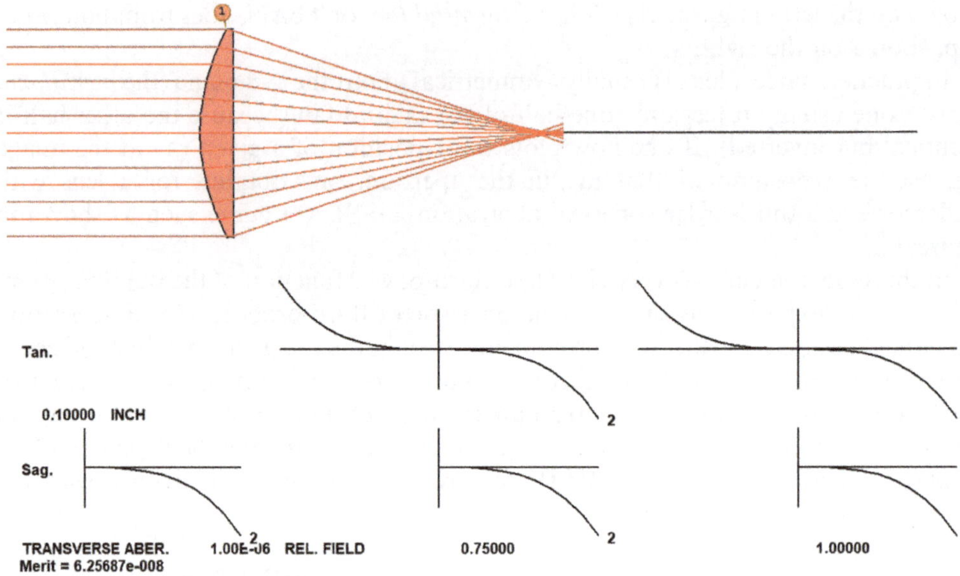

Figure 3.3. SketchPAD display of a lens in which spherical aberration predominates.

command **PAD**.) The curves are all very similar, since SA3 is the predominant aberration in this case and it does not vary with field. The lower three curves are SFANs, while the upper two are TFANS, plotted at the relative field points noted, where full field is at 1.0.

One can ask the software for the values of these aberrations with the command **THIRD**:

```
SYNOPSYS AI>THIRD

ID SINGLET

THIRD-ORDER ABERRATION ANALYSIS

FOCAL LENGTH  ENT PUP SEMI-APER  GAUSS IMAGE HT
       8.089              2.000             0.141

THIRD-ORDER ABERRATION SUMS
             SPH ABERR        COMA  TAN ASTIG  SAG ASTIG     PETZVAL DISTORTION
                 (SA3)       (CO3)      (TI3)      (SI3)      (PETZ)  (DI3(FR))
              -0.13229    -0.00395   -0.00109   -0.00050    -0.00020 -4.422E-06

PARAXIAL CHROMATIC ABERRATION SUMS
              AX COLOR   LAT COLOR  SECDRY AX SECDRY LAT
                 (PAC)       (PLC)      (SAC)      (SLC)
               0.03062   3.849E-05    0.00942  1.184E-05
```

The value of SA3 is here −0.13229, which is just what is shown on the SFAN curves.

A paraboloidal mirror has aberrations of its own, as shown in figure 3.4, except at the center of the field, where it is perfect. This is the common Newtonian configuration.

In this case, there is no spherical aberration, but the tangential fan off-axis shows a U-shaped curve, which is characteristic of *coma*, **CO3**. (The fold mirror blocks some of the light, as indicated by the dots at the center of the fan curves.) Figure 3.5 shows the dependences of SA3 and CO3: SA3 varies as the third power of aperture, CO3 varies as the second power of aperture and linearly with field, as given by the

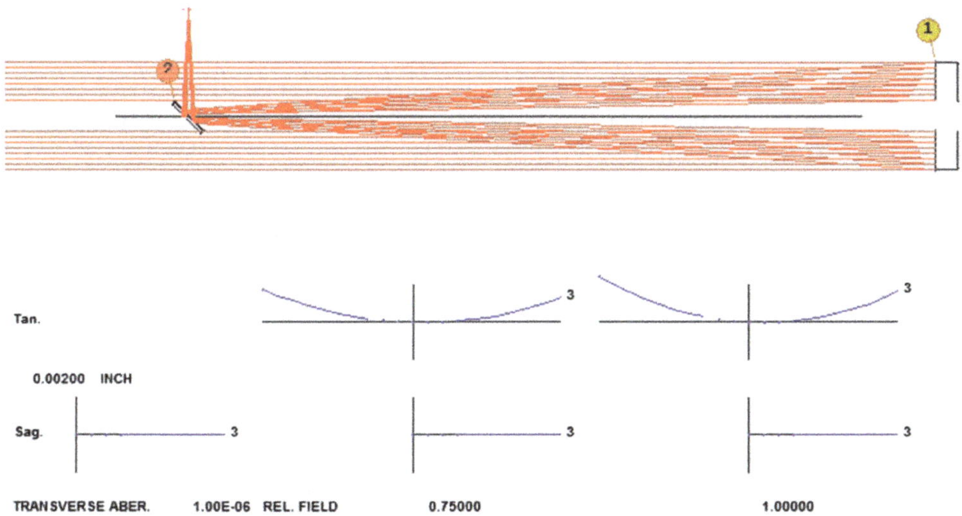

Figure 3.4. A typical Newtonian configuration, showing no spherical aberration at the on-axis point and coma off-axis.

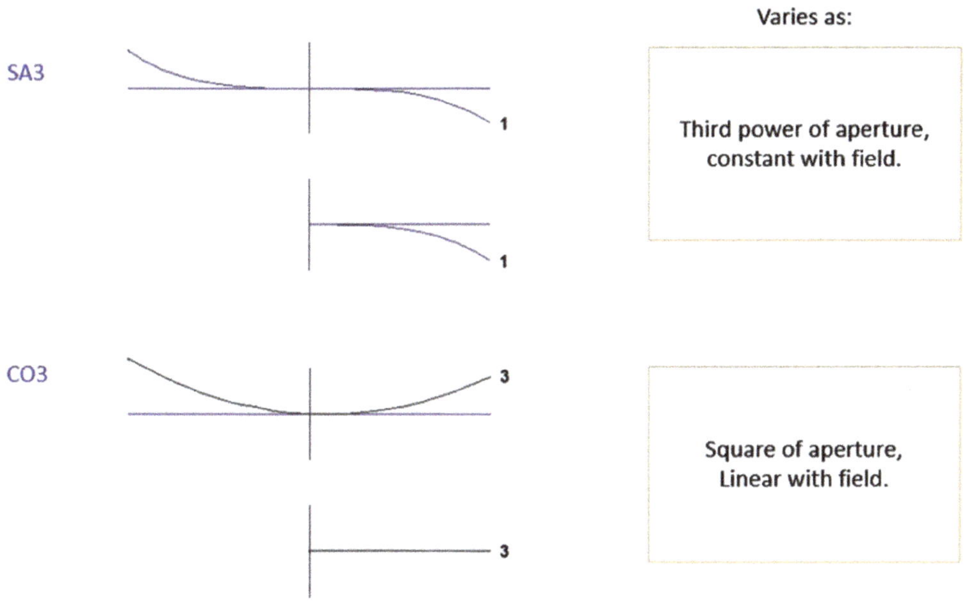

Figure 3.5. Illustration of the shape of ray-fan curves indicating spherical aberration and coma.

above equations and illustrated in figure 3.5. Coma can be viewed as a variation of magnification from one zone to another in the lens.

Other third-order aberrations include sagittal astigmatism (**SI3**), tangential astigmatism (**TI3**) and Petzval curvature (**PETZ**). TI3 is shown in figure 3.6; SI3 is similar but affects the SFAN instead of the TFAN. Petzval curvature affects the flatness of the image surface and is related to both of them. (In case you missed the point, a perfect image would show up as a straight horizontal line on these plots.)

The Petzval curvature is an interesting concept. If all the *other* aberrations of a lens were zero, the sharpest image would be found on a surface with that curvature. If the astigmatism is not zero, the best image surface is found in the space between the S and T surfaces, and is not very close to the Petzval surface, as one might guess. So it does not tell you much in most cases. The formula is very simple:

$$\text{PETZ} = -\sum \frac{1}{n_j f_j},$$

where n_j is the index of each element and f_j is the focal length. From this one can learn that two positive elements cannot correct PETZ to zero. Adding a negative lens (which has negative f) makes it possible, and this insight led to the original Cooke triplet, which has a negative element between two positive ones. An example is shown in figure 3.7 on the upper left. A refractive system tends to produce a Petzval surface curved inwards towards the lens, while a reflective mirror of positive power does the opposite, a fact that is sometimes useful. By the way, the *curvature* of a surface is just the inverse of the radius of curvature.

In the past, designers took pains to correct these aberrations, and some still do—but I have learned an important lesson that I want to pass on: when you design a lens, *correct only those things that matter*, not what you learned in an old textbook. **In most cases, you only care about two things: is the image sharp, and is it in the right place?** Third-order aberrations usually get in the way and are not worth calculating, in my opinion. In almost all cases, you need third-order aberrations to balance higher orders, so trying to make them go to zero is a mistake. They do not tell you much about a lens, *with some important exceptions* (I do not want to overstate my case). It turns out that **any lens that is** ***exactly symmetric*** **has no odd-order field aberrations**, so the coma is zero. A lens construction derived from a symmetrical

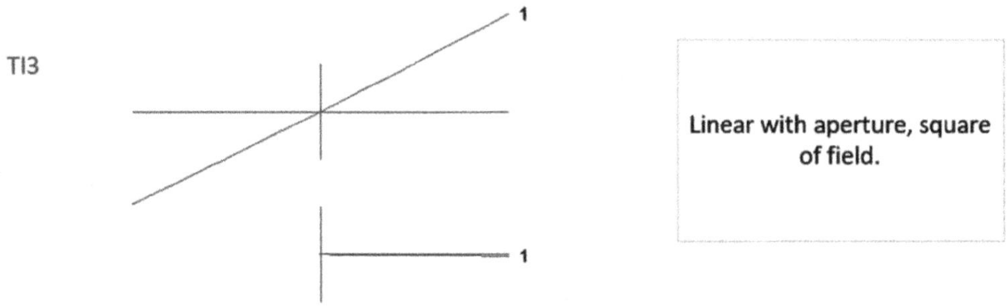

Figure 3.6. Ray-fan curves showing tangential astigmatism.

shape gains some of the same benefit, even if it is no longer strictly symmetric. Examples are shown in figure 3.7.

Another application of this principle is the case of two paraboloids with a common focal point, as shown in figure 3.8. This combination has no spherical aberration (because the paraboloids have none) and no coma (because the coma from the first is exactly cancelled out by the second). Does this seem unlikely? Well, Nature is not usually so kind to us lens designers, but it turns out that **the coma from**

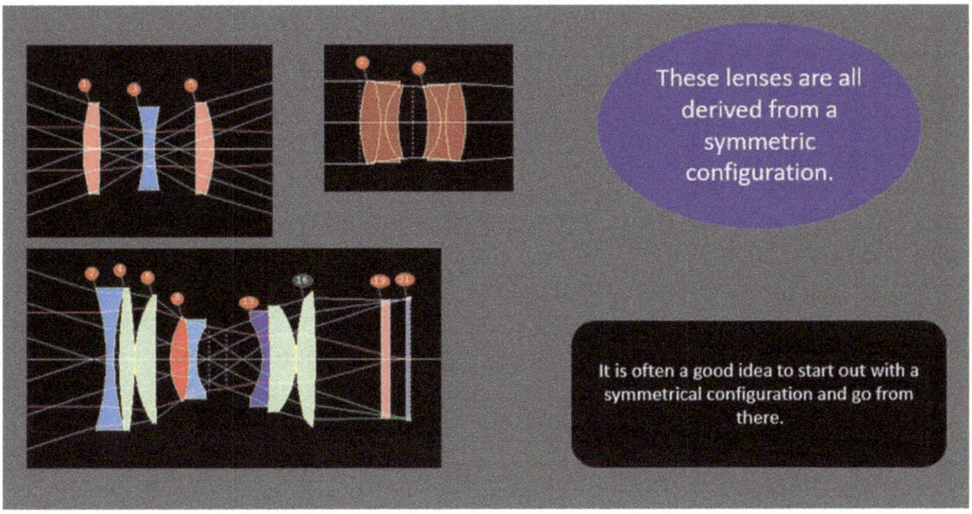

Figure 3.7. Examples of lens configurations derived from symmetrical forms.

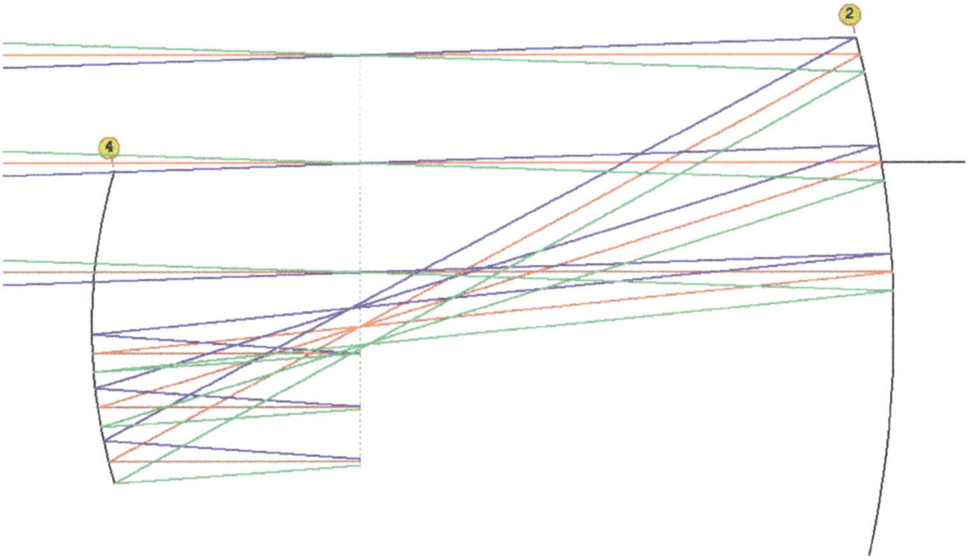

Figure 3.8. Arrangement of confocal paraboloids; this form has no spherical aberration or coma. The output is collimated. One cuts an off-axis portion of the two mirrors in practice.

a paraboloid at a given image height is a function only of the F/number, not the focal length. Scale a mirror up by a factor of two, and the coma doubles—at the new full-field image height. However, at the old image height, which is one-half of the new one, the coma is one-half of that at the new image, or just the previous value. Perfect balance! So, yes, there is some value in learning about third-order aberrations, but not much beyond these simple cases. If a customer insists the lens must have, for example, 'no astigmatism', you can safely ignore him. That would rule out many excellent designs in which third-order astigmatism balances out higher orders—which is a concept he does not know about.

The methods used by the old masters for reducing aberrations are still valid, if cumbersome—but today the computer does most of the work for you, so those methods are of little use anymore. That said, here are some basic relationships you should know about.

Be aware that it is impossible to correct spherical aberration in a single element with spherical surfaces if the light comes in collimated (from an object at infinity) unless the power is zero. One can 'bend' the lens, and the aberration changes, but there is no bending that makes it go to zero. Three different bendings of a singlet are shown in figure 3.9, the second at the best position, none of them perfect. (It is, however, sometimes possible to correct spherical aberration by bending a lens if the light that comes in is *not* collimated.)

3.2 Abbe sine condition

Let us also take a deeper look at coma. We want the focal length of all rays in the aperture to be the same, for a given object point. Figure 3.10 shows why this is not the case for the Newtonian telescope. The marginal ray has to go further than the chief ray, so the focal length along that ray is longer and that is why you get coma. Simple.

Bending a lens also bends a nodal surface, and when it makes a sphere about the image point, the coma goes away. This is the basis for the **Abbe sine condition**. In figure 3.11 we corrected SA3 by adding an aspheric term to surface 1 and corrected CO3 by bending the lens.

When we satisfy the requirement that

$$\sin(\theta) = A/F,$$

the condition is satisfied and there is no coma. (This equation is valid when the object is at infinity.) A lens that satisfies this condition is termed an *aplanat*. In practice, one corrects the aberrations of real rays, not third-order aberrations, and the sine condition is satisfied automatically if one succeeds.

A corollary of this rule is the fact that the minimum possible F/number of a lens that is corrected for coma is 0.5. By the rule we gave earlier, FNUM = $-0.5/UA$, paraxially, and FNUM = $-0.5/n\sin(\theta)$ for real rays. (The quantity $n\sin(\theta)$ is known as the *numerical aperture* of the lens.) Since θ can never exceed 90 degrees, QED. In practice, one rarely approaches an angle as steep as 90 degrees. If someone asks you for a lens at F/0.25, you can be sure it will not have a very wide field (because the

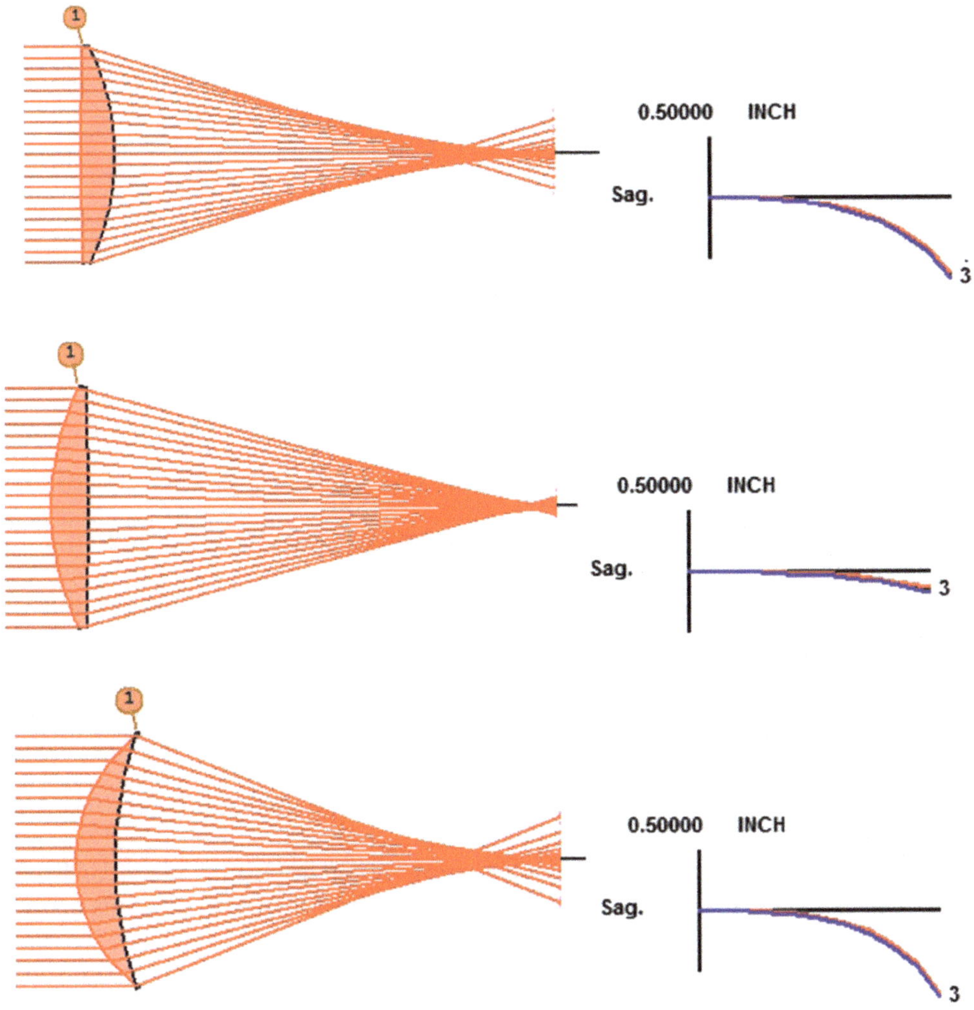

Figure 3.9. Illustration of the variation in spherical aberration with lens bending.

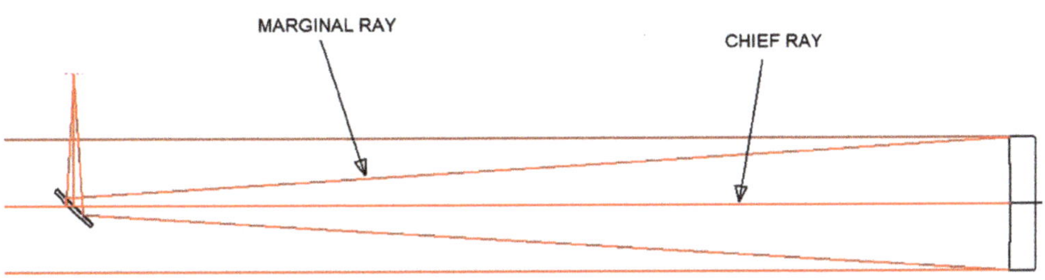

Figure 3.10. Illustration of the origin of coma in a Newtonian telescope.

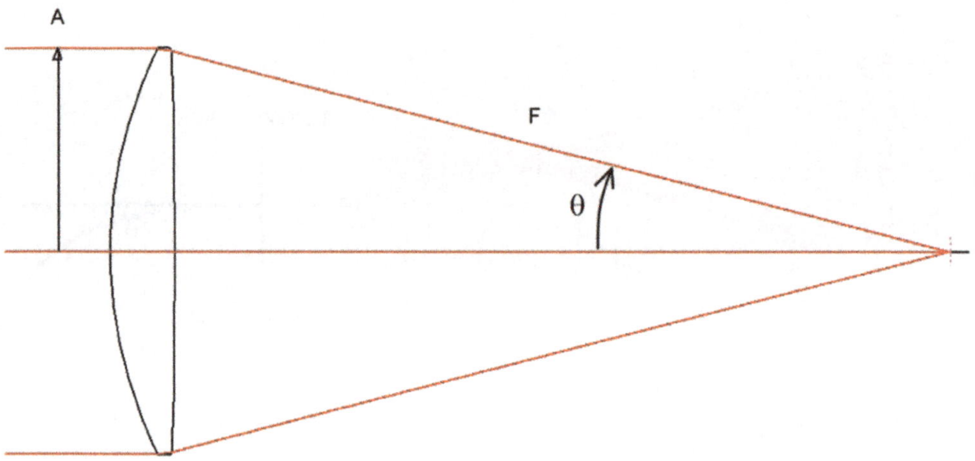

Figure 3.11. An aspheric singlet bent to eliminate coma by satisfying the Abbe sine condition.

coma would be horrible; the focal length would change by a factor of two from the central to the marginal zone!).

3.3 Higher-order aberrations

We have discussed third-order aberrations, but of course there are also higher orders which, in principle, one can calculate by taking successive terms in the expansion of the sine function. In practice, this becomes algebraically intractable very quickly, and the effort to derive and calculate the values is one of the immense tasks our forebears were stuck with but are now seldom warranted.

However, the *effect* of higher-order aberrations is still important and is easily seen in the ray-fan diagrams. In figure 3.12 one can see that the paraxial defocus is zero (because the curve starts out horizontal at the axis), and the lens has negative third-order spherical aberration (it turns down) and positive fifth-order (it turns up again). This is a common situation when the optimization program balances one aberration against the others.

If you subtract the third-order aberrations from the actual ray-fan data, what is left is a good indication of the influence of the higher orders. Generations ago, designers realized that when you change a lens parameter, it is the low-order aberrations that change the fastest. The higher the order, the more slowly it changes—and this is the main reason why high orders are more difficult to correct.

Figure 3.13 shows the curve for a lens with multiple orders. Here you see defocus, third, fifth, seventh, and ninth-order spherical aberration, all balanced so the lens exhibits just over 0.03 waves peak-to-peak wavefront error, as shown in the OPD plot in figure 3.14 (OPD means optical path difference). This lens operates in the deep UV at 0.226 μm, at a speed of F/0.625, and this is typical of the balance achieved in microlithography lenses of 20 or more elements, which must be functionally perfect. Clearly, balancing higher-order aberrations is a fundamental goal in lens design and, fortunately, modern software takes care of that task rather

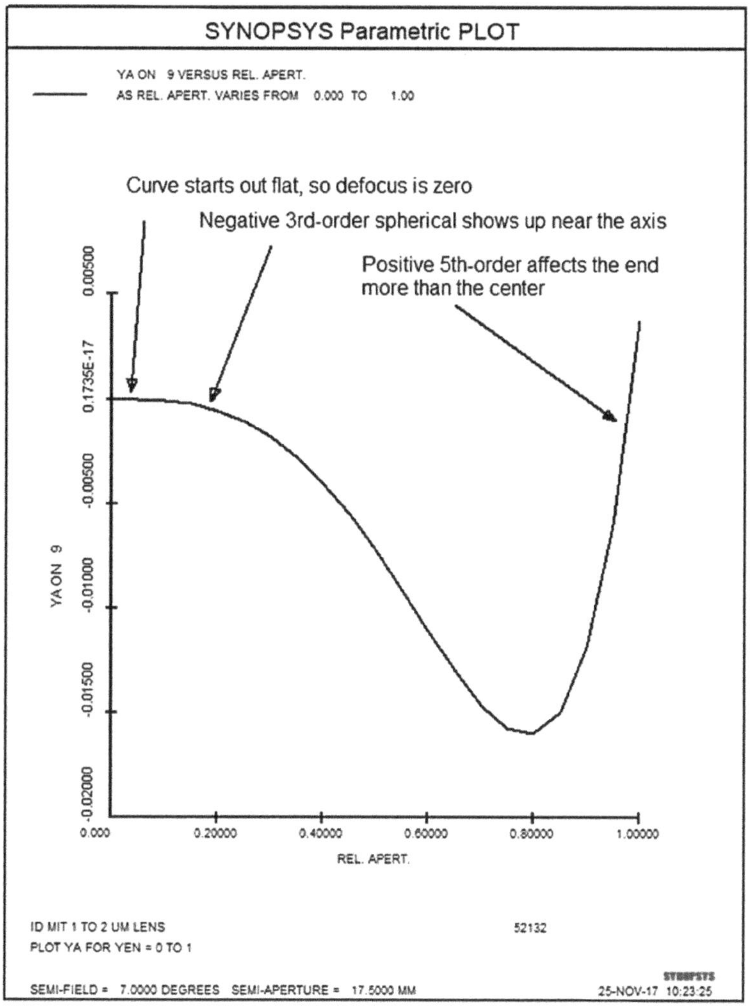

Figure 3.12. Reading aberration contributions from the ray-fan curves.

well. Lenses such as this represent the pinnacle of the lens designer's art—and constructing them is no less of a challenge, with tolerances so tight that each element must first be centered in its cell and then the cell centered on a rotary table monitored with a laser beam. When completed, the price for such a lens is measured in seven figures or more. The PC on which you run your optimization program would never exist except for these very exceptional lenses, themselves designed on just such a machine.

Whenever you see curves like these, it is a sign that you probably should ask for a grid number larger than the default 5 on-axis and 3 elsewhere (which you obtain with ready-made MF number 6, which you will use in some of the chapters below). Otherwise you might happen to obtain great correction *at* the requested rays but wild swings in between. This can happen with multielement lenses, and especially if aspherics or DOEs are present.

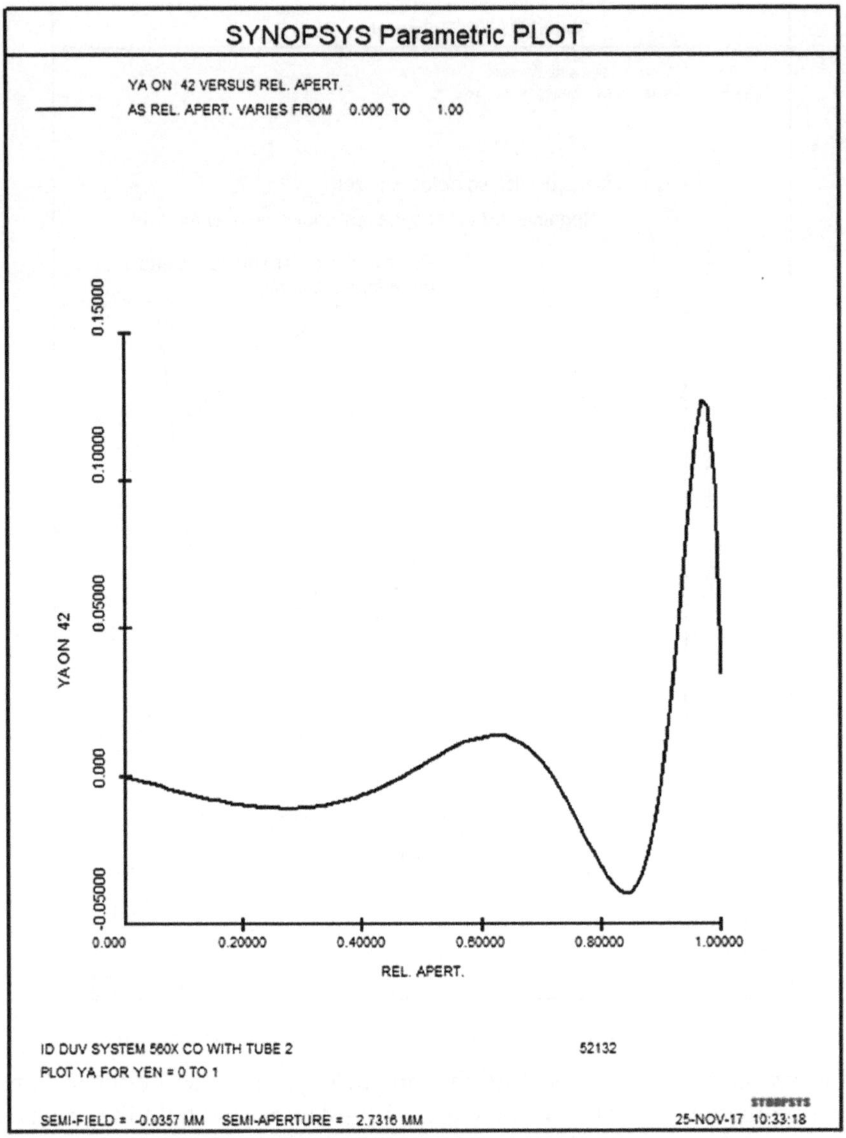

Figure 3.13. Aberration curves for a lens with many orders of spherical aberration.

We should repeat here that the meaning of the aberration balance of a lens depends on exactly how one defines the beam that is traced at each field point. This is the function of the *pupil definition*, and there are subtleties involved. A discussion can be found in chapter 22.

3.4 Spot diagrams

Today, with fast modern PCs, one can often obtain a good picture of the quality of an image by tracing a suitable number of real rays and looking at the geometrical

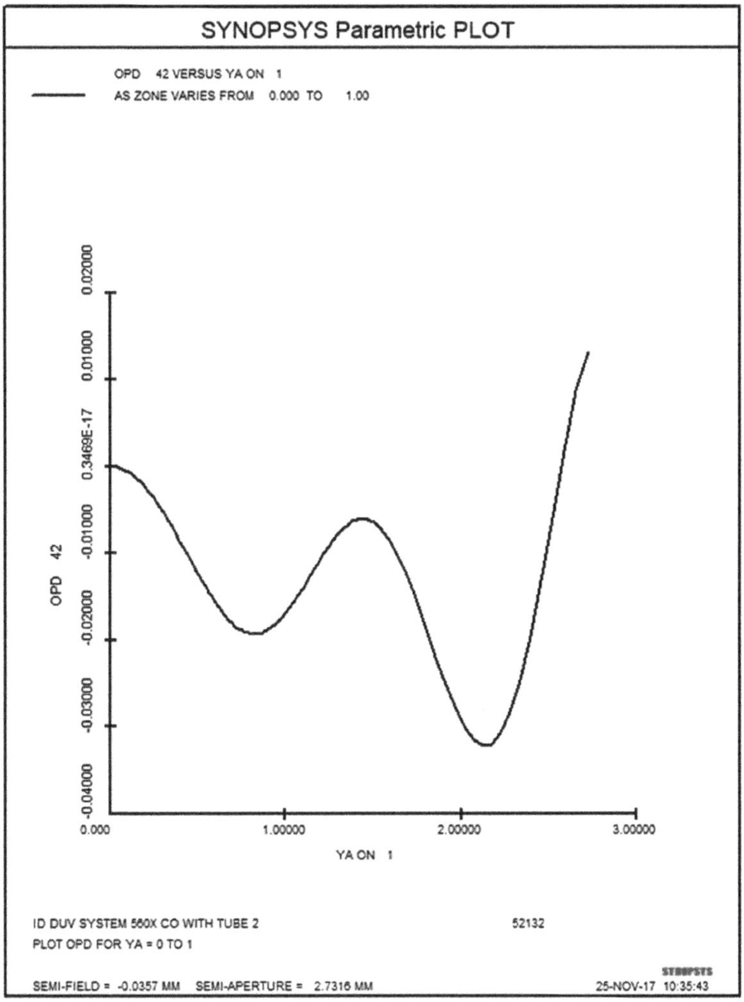

Figure 3.14. OPD aberration for the lens with multiple orders of aberration.

image errors. One common method involves creating a *spot diagram*. Consider the lens in figure 3.15. Look at the rays coming out, headed toward the image plane from the full-field object, shown in blue.

Let us magnify the pattern made by a grid of rays when they reach the image plane, shown on the right in figure 3.16.

This is certainly not a very sharp image. It was calculated by tracing 1731 rays, in three colors, an effort that would have been impractical only a generation ago (imagine doing that with a Marchand calculator!). This kind of picture gives you a good indication of the image quality, especially if it is not a particularly good image. (The structure evident in the spot diagram is a consequence of the ray pattern in the aperture; here it is a uniform square grid. Other ray patterns alter the appearance, but the overall effect is much the same.)

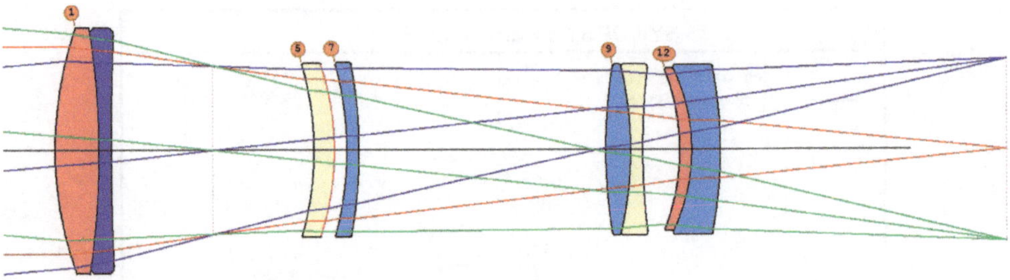

Figure 3.15. Example lens for calculating a spot diagram.

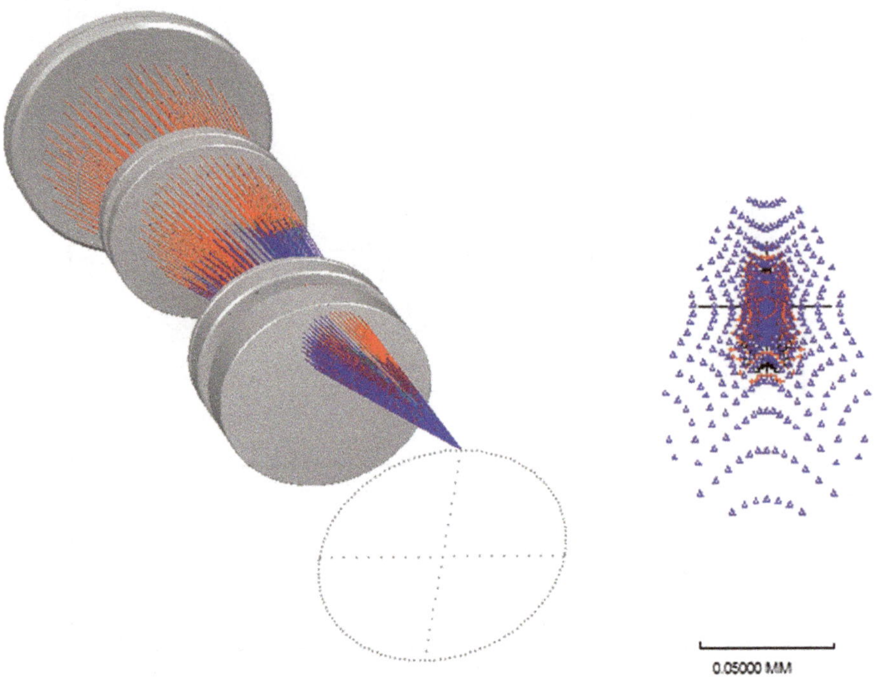

Figure 3.16. Geometry of a spot diagram.

If it is a *very* good image, on the other hand, one needs to look at the wavefront.

3.5 Wavefronts and aberrations: the OPD

Even if an image were geometrically perfect, it would not focus to a single point due to the finite wavelength of light. Instead of a single point, one obtains a *diffraction pattern*. Considering again the Newtonian telescope—and ignoring the obscuration for the moment—the pictures in figure 3.17 show, from left to right, the geometric image on-axis, the diffraction pattern on-axis, the geometric image at the edge of the field, and the diffraction pattern there. (This telescope has an aperture diameter of 10 inches and a focal length of 80 inches, and thus an F/number of 8. The semi-field angle is 0.5 degrees.)

The red circle on the first picture gives the size of the first dark ring in the diffraction pattern, and the geometric blur is much smaller than that. (It is not zero in the picture due to the size of the symbol representing each ray.) If a lens is corrected so well that the geometric blur is comparable to or smaller than the diffraction pattern, one must usually try to control the optical path difference (**OPD**) of a ray instead of (or in addition to) the geometric blur. You will find examples of this kind of correction in subsequent chapters. In this case, the image must be evaluated in a way that takes diffraction into account as well. The OPD is just the difference in total path length of a given ray compared to the path of the central ray (the *chief ray*). In a perfect lens, those paths would be identical, a condition known as *Fermat's principle*.

Look at the OPD fans for this telescope, shown in figure 3.18. Here we plot the OPD of each ray instead of the transverse aberration shown in the above examples.

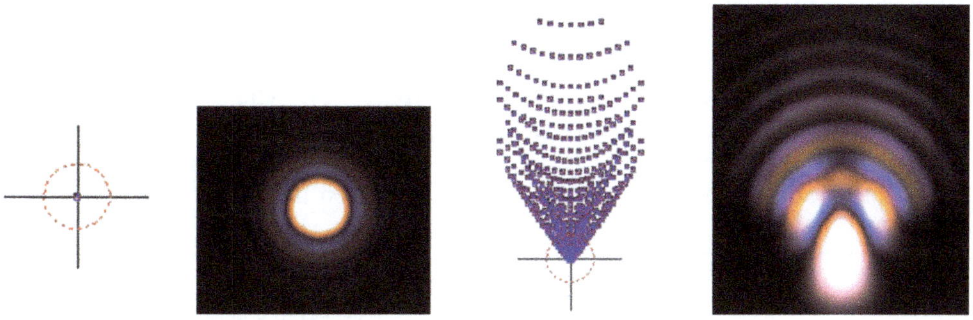

Figure 3.17. Image characteristics of a Newtonian telescope. Diffraction effects enlarge a geometrically perfect image.

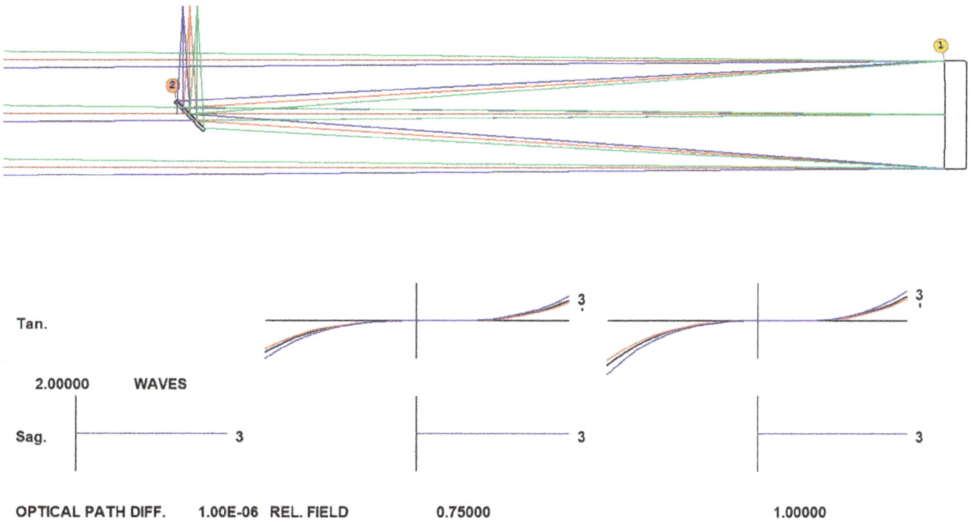

Figure 3.18. Newtonian telescope with OPD fans shown on the bottom of the display.

Again, the on-axis aberrations are zero, but now the TFAN at the edge of the field has a different shape. This is what coma looks like when OPDs are plotted. The relationship is simple: **the OPD plot is the *integral* of the transverse plot** (which we call the **TAP** plot). It can be viewed as giving the *shape* of the wavefront forming that image, relative to a perfect spherical wavefront converging to that image point. You can see here that the wavefront has an S-shaped curve, relative to perfect, which is characteristic of coma.

3.5.1 Image analysis

Any program that can calculate the spot diagram or the diffraction pattern can usually calculate any of a number of other quantities related to image quality. The choice of which kind of analysis to perform depends on the purpose of the lens, what kind of technology will utilize the image (CCD, film, IR sensor) and the preferences of the people involved. The principal techniques are the following:

1. ***The MTF** (modulation transfer function)*. This is a curve showing the frequency content in the image of a point source, often interpreted as the contrast in a sine-wave image of a given frequency. (The two are equivalent if there are no pupil aberrations.) Looking at a sine-wave target whose spatial frequency at the image is 100 cycles/mm, the on-axis MTF shown in figure 3.19 would have a contrast of about 0.37. When the actual sine-wave image

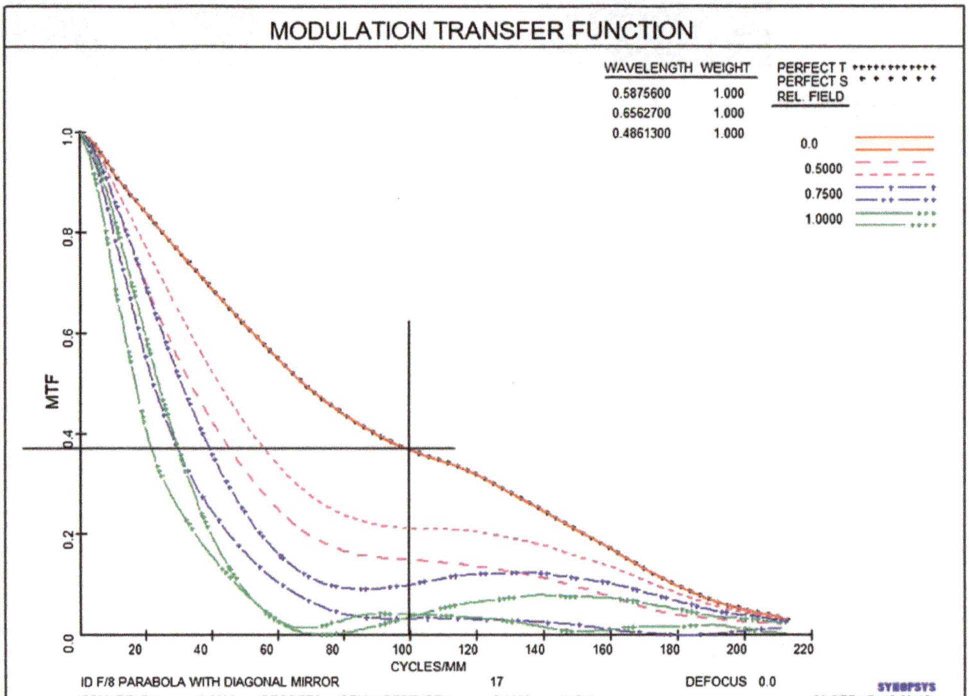

Figure 3.19. MTF curves for several field points with a Newtonian telescope. The on-axis MTF is about 0.37.

is plotted, shown in red in figure 3.20, one can measure the contrast, as (MAX – MIN)/(MAX + MIN). Here, the result is 0.037, in good agreement with the prediction. This MTF was calculated for the Newtonian telescope above, with the obscuration taken into account. At low spatial frequencies, on the left end of the curve, the MTF is high, and toward the right it goes to zero. The zero point is called the cutoff frequency, and an image requiring a higher frequency cannot be resolved with this system. The MTF is sometimes called the optical transfer function, or OTF, if other components in the system also affect the resolution. In that case, the OTF is composed of both the MTF and the physical transfer function (PTF) the latter due to all other sources of degradation.

An obscuration materially affects the image structure and the MTF. Here are the MTF and diffraction point-spread function plots of this F/8 telescope for various sizes of the obscuration, shown in figures 3.21–3.23. These results show why one usually tries to minimize the obscuration as much as possible: image quality declines as the obscuration grows. In each case here, the *geometric* image is perfect, but the diffraction pattern is not if the obscuration is nonzero. The top left MTF is as good as possible for this F/number and wavelength.

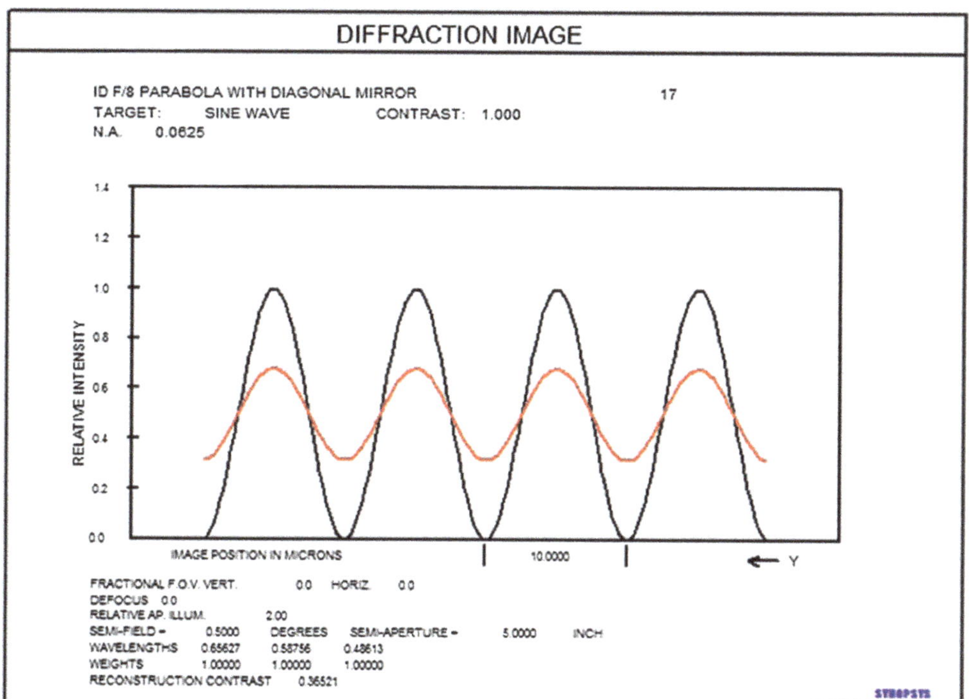

Figure 3.20. Characteristics of the on-axis image of the Newtonian telescope, showing a contrast of about 0.37 for the image profiled in red.

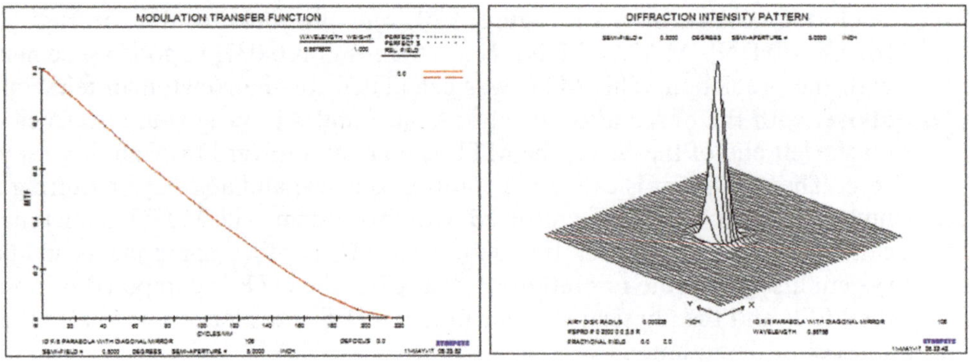

Figure 3.21. MTF and image structure of the telescope ignoring the obscuration.

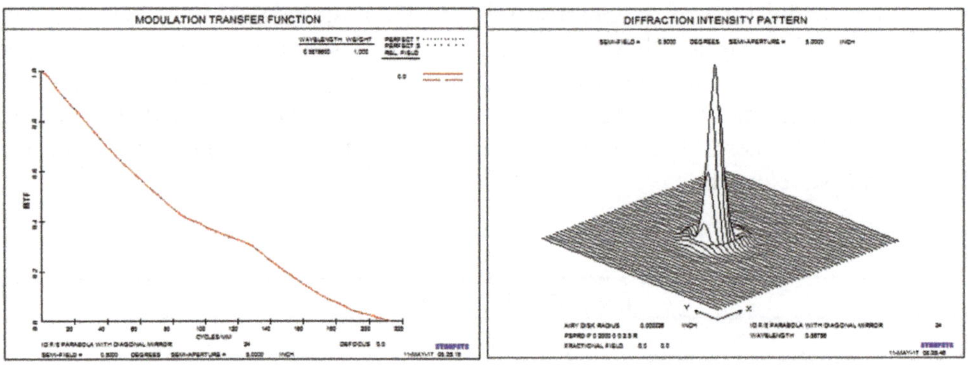

Figure 3.22. MTF and image when the obscuration is 25% of the aperture diameter.

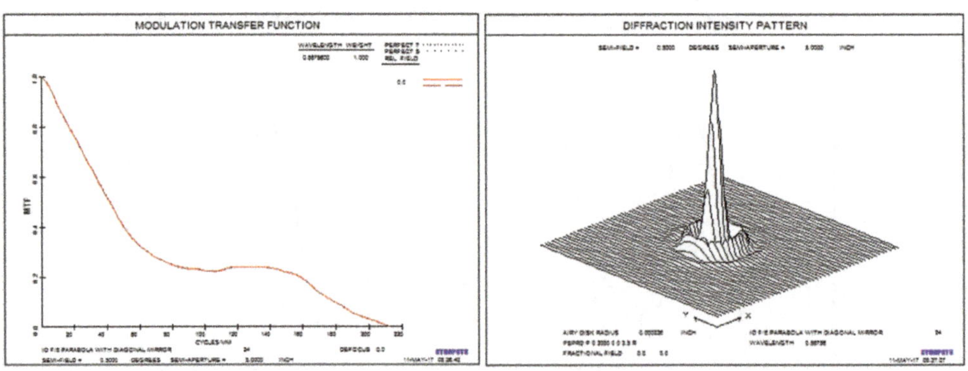

Figure 3.23. MTF and image when the obscuration is 50% of the aperture.

2. **The knife-edge trace**. Figure 3.24 shows the image at the edge of the field of the Newtonian telescope (ignoring the obscuration).

   One can pass a knife edge over this image, from bottom to top, and plot the energy that is uncovered, making the curve in figure 3.25. Some lens

Figure 3.24. Diffraction pattern of the Newtonian telescope off-axis.

designers find this kind of analysis useful, and it is popular with lens makers since it is conveniently measured.

3. **Encircled energy**. Sometimes one plots the energy through a circle as the radius of the circle increases. The result of course depends on where one centers the circle. Here is an example, in figure 3.26, at the center of the field of the telescope. The diffraction rings are clearly apparent.
4. **Wavefront variance**. If one traces a large grid of rays equally spaced over the entrance pupil, a simple calculation yields the variance, denoted as σ^2:

   $\sigma^2 = \langle OPD^2 \rangle - \langle OPD \rangle^2$, where $\langle \ldots \rangle$ denotes an average over the set of rays.

   This quantity is frequently specified as a design goal. Zero would be perfect.
5. **Strehl ratio**. Another common measure of lens quality is the Strehl ratio (**SR**), often used when lens performance is close to perfect. It is defined as the ratio between the peak intensity in the diffraction pattern divided by the intensity it would have if all aberrations were zero. If the image is perfect, the ratio is 1.0.

   The SR can be calculated in two ways: approximately with an exponential function of the wavefront variance, or by calculating the diffraction point-spread function (**PSF**) with and without aberrations and taking the ratio of the peaks.

   If the OPD errors are small, the exponential approximation is reasonably accurate. This works because, for small errors, the SR does not depend strongly on which aberrations are present. Finding the ratio via the PSF is somewhat more rigorous, but if the OPD errors are not small, the shape of the pattern can become complex and it may not be obvious just where the peak is. Since it is useful only when errors are small in either case, the exponential approximation is widely used.

   The SR can be calculated more quickly than can the diffraction MTF, and since it correlates strongly with the latter, it is a useful measure of image

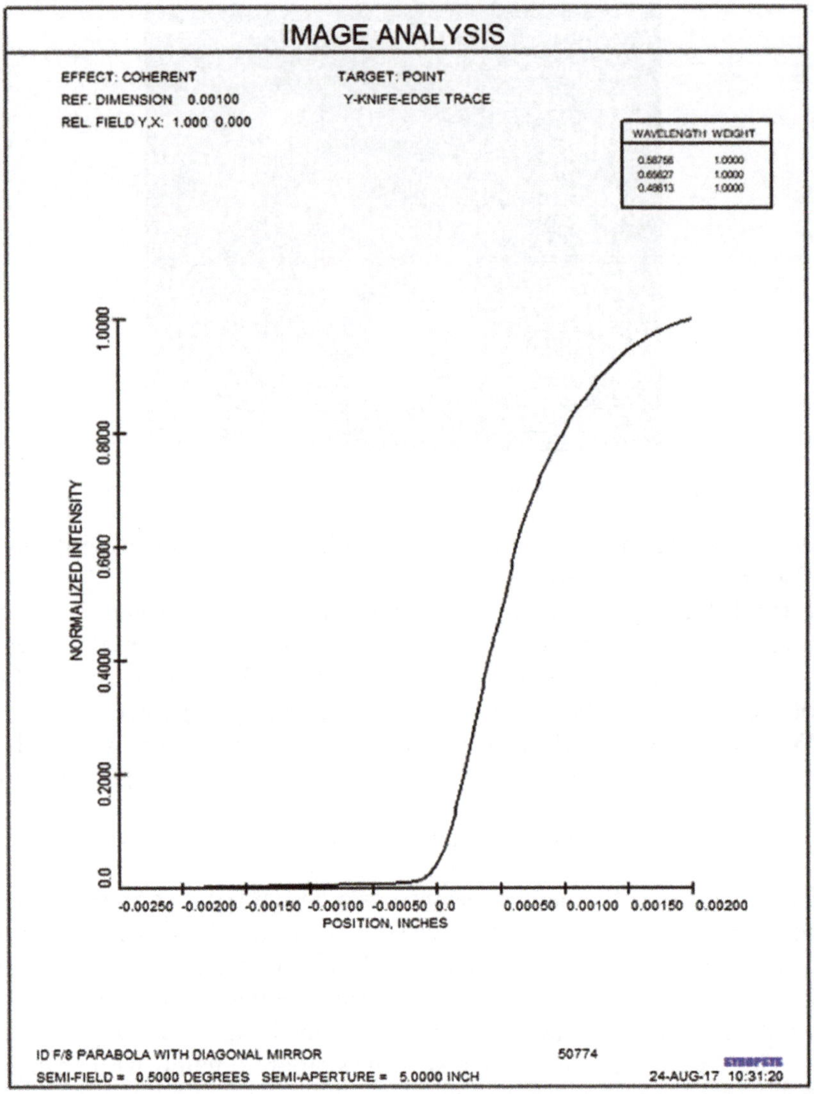

Figure 3.25. Knife-edge trace of the diffraction pattern.

quality. Figure 3.27 shows two curves: the SR and the diffraction MTF at 100 c mm^{-1} as a function of defocus for a paraboloid mirror with a central obscuration. For small errors, the two curves track rather well.

Here is how to make the above picture:

1. **FETCH NEWTONIAN**
2. Open a MACro editor and enter the lines

```
VARIANCE P 0 600 0 0
ORD = FILE 3
```

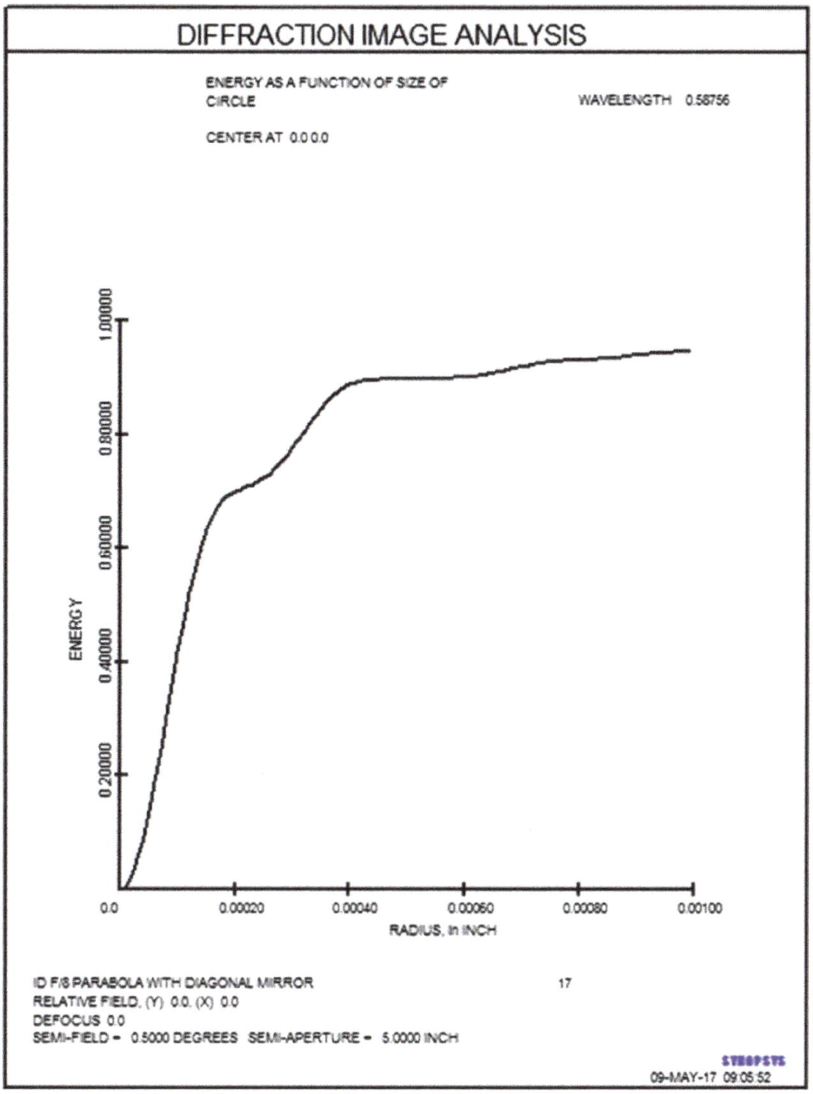

Figure 3.26. Encircled-energy plot for the Newtonian telescope on-axis.

3. Open a second editor and enter the lines

```
MTF P 0 100
ORD = FILE 1
```

4. Run the first MACro, and then type, in the CW:

```
MULTI DO MACRO FOR 3 TH = 10 TO 10.01
```

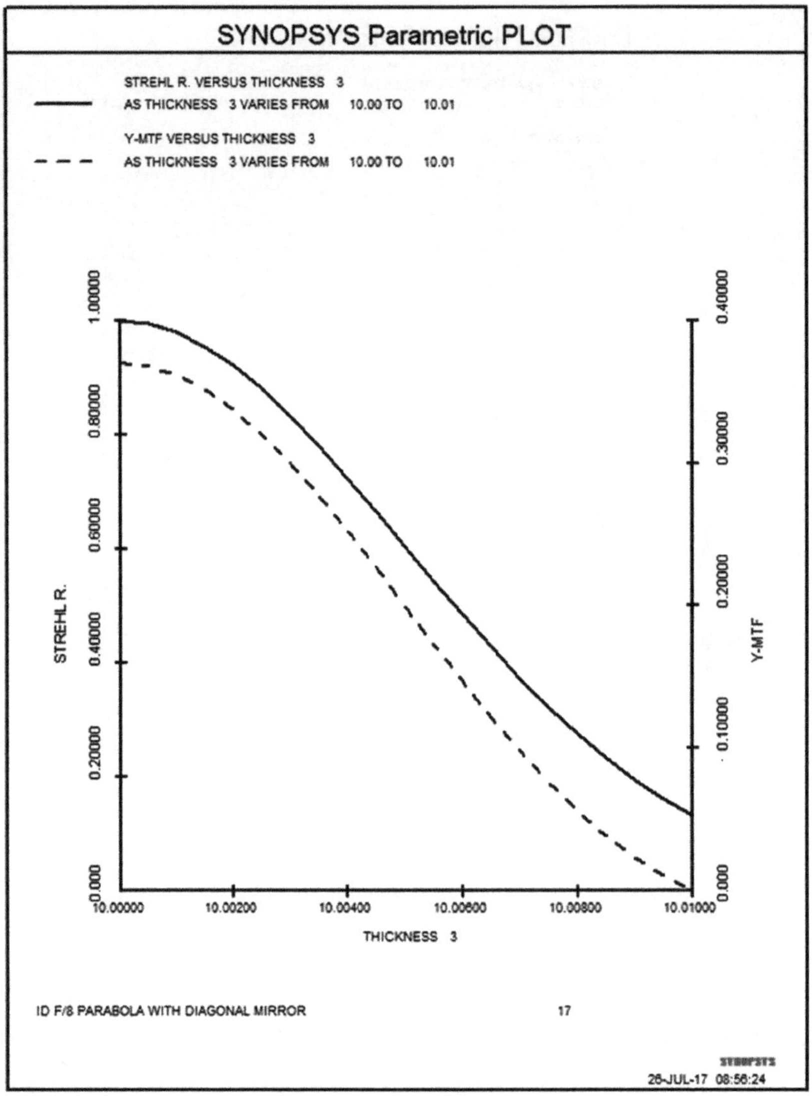

Figure 3.27. The Strehl ratio and MTF at 100 c mm^{-1} as a function of defocus. The SR degrades at about the same rate as the MTF, which makes it a useful measure of quality in the design phase.

5. Run the second MACro, and then type

```
ADD DO MACRO FOR 3 TH = 10 TO 10.01
```

6. Type **END**.

3.6 Chromatic aberration

Since the index of refraction of all glasses changes with wavelength, it follows that the size, shape, and location of the image also vary, and an important goal of the lens designer is usually to minimize this effect along with the other aberrations. When choosing the exact wavelengths to correct and the goals, one must take into account the spectral sensitivity of the sensor and the properties of the light source.

When designing optics for the visual spectrum, one must weight the aberrations according to the sensitivity of the eye. Figure 3.28 shows this sensitivity, which is a function of the light level; in bright light one sees in color according to the curve on the left, while in dim light one sees a monochromatic image following the curve on the right, the peak of which is shifted toward the blue end of the spectrum. The three triangles at the bottom are located at the at the C, d, and F Fraunhofer lines, with wavelengths of 0.6563, 0.5876, and 0.4861 μm, which are often chosen when designing optics for visual use. The reasons are historical, having to do with the availability of monochromatic gas lamps that emit at those wavelengths for use in laboratory testing.

The ray fans in figure 3.29 are calculated for these wavelengths, as is the spot diagram to the right. In this image, you see that the error at the image in color 2 is fairly small, while color 3 shows a larger error.

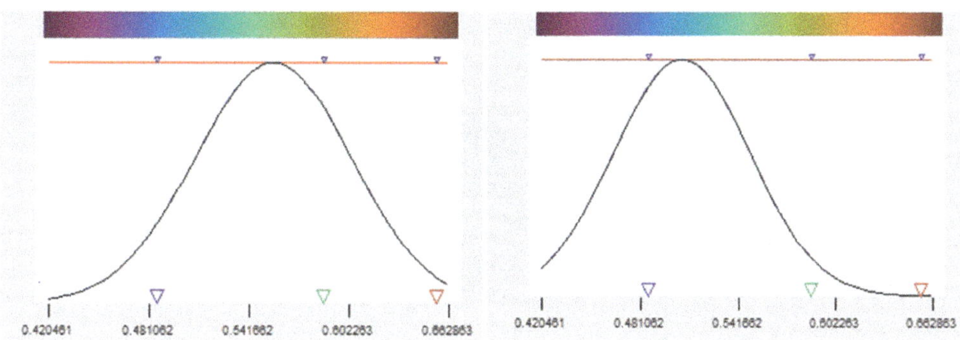

Figure 3.28. Eye sensitivity, bright light on the left and dim light on the right.

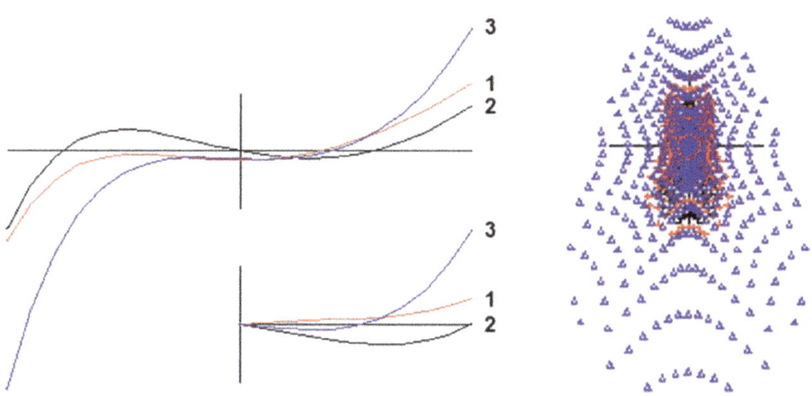

Figure 3.29. Effects of chromatic aberration on image structure. This image is fairly well corrected for axial color, but spherical aberration in blue light (color 3) is greater than in red. This effect is called *sphero-chromatism*.

The actual size of the image can be deduced fairly well from the shape and size of the curves on the left. This kind of analysis can be performed with many fewer rays than can a spot diagram and is useful for evaluating the progress of a lens design in the initial stages. Indeed, it is fun to watch the curves change during optimization, as the software makes tradeoffs and balances one aberration against another. In these curves, we see both *primary axial color*, or **PAC**, and *secondary axial color*, or **SAC**. The software calculates PAC by taking the difference between the transverse image intercepts at the marginal ray in the long and short wavelengths, while SAC is calculated by taking the differences at the long and central wavelengths, all in the third-order approximation.

Lenses also can suffer from *primary lateral color*, or **PLC**, which is the change in image height between the long and short wavelengths, along with *secondary lateral color*, or **SLC**, which is the change between the long and central. An example of an image with both coma, PCL, and SLC, is shown in figure 3.30.

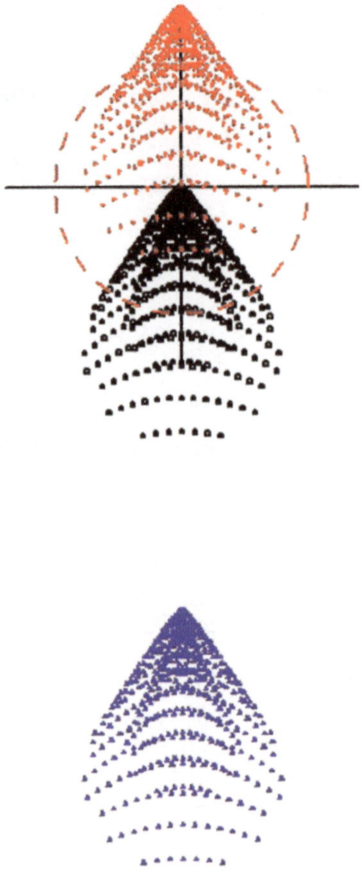

Figure 3.30. Example of off-axis image showing both coma and lateral color. The dotted red circle shows the size of first dark ring in the diffraction pattern for this system, for comparison.

Chromatic aberration is usually corrected by combining two or more elements made of glasses whose *dispersion*, the change of index with wavelength, differs substantially. Examples will be given in chapter 12 of lenses for which chromatic aberration correction is of particular importance and even *secondary* color is a major issue. That is the case when longer and shorter wavelengths focus in the same place, but intermediate wavelengths do not.

3.6.1 Cemented doublets

A classical approach to color correction was to *cement* the elements of a doublet together. A fringe benefit of that solution was the reduction of reflection losses, since two air–glass interfaces are replaced with one glass–glass interface. In the days before antireflection (AR) coatings were available, this was an important advantage. Today, we have very good AR coatings and that benefit is not as important—but there is another that should not be overlooked: one can avoid total internal reflection with a cemented interface in those cases where rays arrive with a steep angle. Figure 3.31 shows a lens with a cemented interface at surface 11. If we change that to a thin air gap between the elements, rays near the edge of the beam at full field will not trace, due to total internal reflection (**TIR**) at surface 11, as shown in figure 3.32. The search programs DSEARCH and ZSEARCH, which are used in many of the chapters below, do not investigate systems with cemented surfaces, but you can request the Automatic Element Insertion (AEI) program to include them in its search algorithm. This can sometimes open new solution regions one would likely not discover otherwise. If TIR problems are not a problem, on the other hand, you

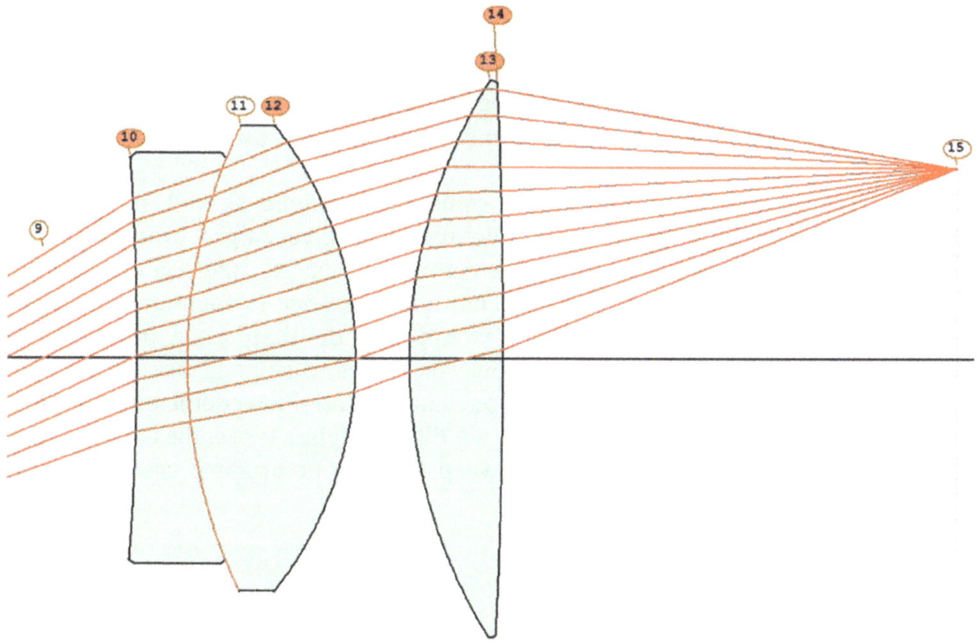

Figure 3.31. Example cemented interface at surface 11. All rays in the tangential fan trace properly.

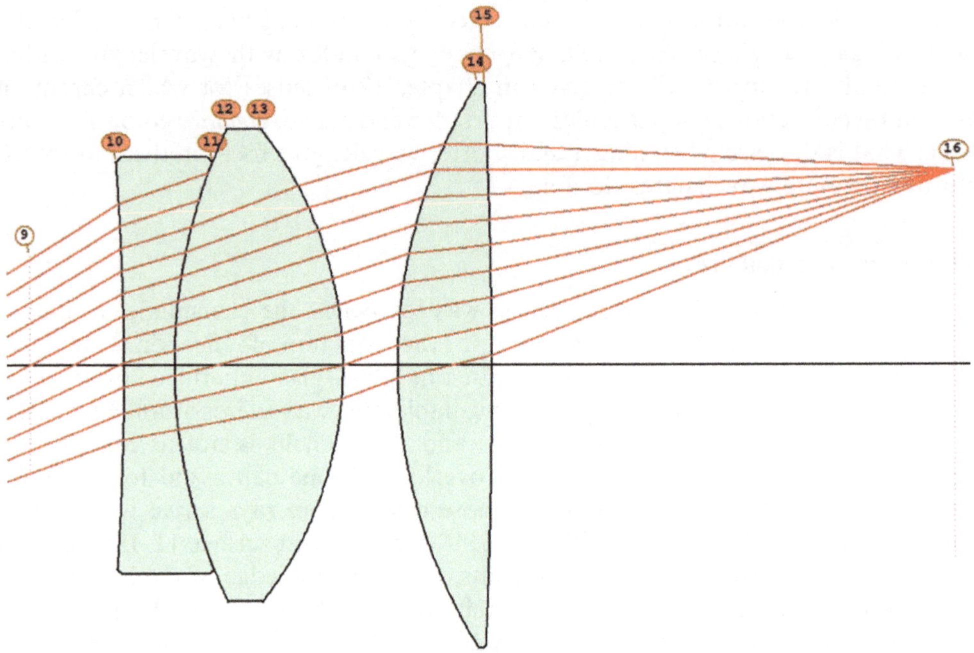

Figure 3.32. Changing the cemented interface to a glass–air interface causes total internal reflection (TIR) for some rays, which therefore cannot get through.

always want airspaced rather than cemented elements. Each case then provides two additional degrees of freedom for the optimization program, and the MF is usually better as a result.

Where does TIR come from? From Snell's law. If a ray is refracted such that it comes out of the lens at exactly 90 degrees to the surface normal, just grazing the surface itself, we say it is at the *critical angle*. Any steeper, and it is reflected back into the lens.

There is sometimes another reason for cementing elements as well. If an element has strong optical power, the centering tolerance can be very tight. However, if two strong elements of opposite power can be cemented together, the combination will have less power and a looser tolerance. The tight tolerance of each element is held during the cementing process, where it is easier to do than when mounting the elements separately into a lens cell. However, be careful when asking for large elements to be cemented; if the temperature changes, and if the coefficient of thermal expansion (TCE) is different between the two different glass types, the cemented pair undergoes mechanical stresses that can warp the two or in some cases break the cement bond or the glass itself.

3.6.2 Secondary color

Secondary color has a history of being especially difficult to correct, and some older designers may insist that one needs at least one material with abnormal dispersion

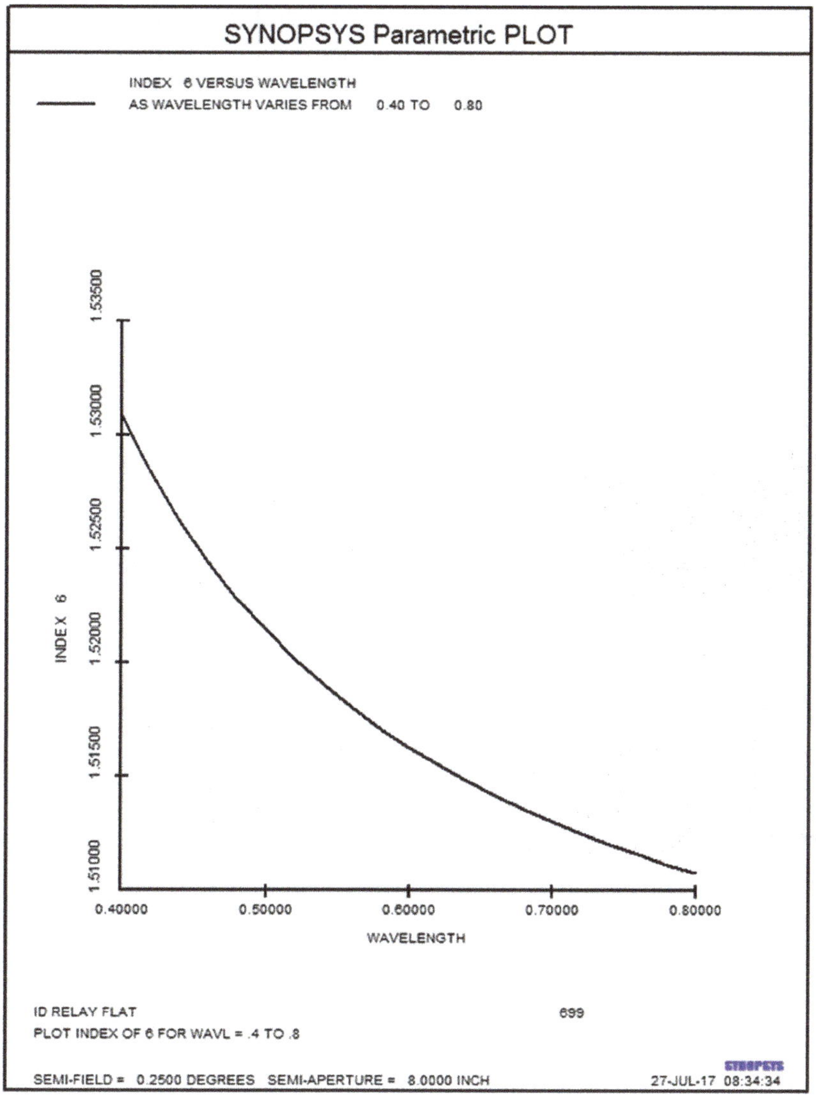

Figure 3.33. Index of glass BK7 as a function of wavelength.

characteristics—as was the case decades ago. Calcium fluoride was a popular choice, especially for microscope objectives, where the lenses are physically very small and natural crystals were available.

However, modern glass companies now have a wider assortment of glass types, and today one can correct secondary color with ordinary glasses if they are properly selected. Figure 3.33 shows how the index of glass type BK7 varies over the wavelength range 0.4–0.8 μm. The curve is nonlinear, and that is the cause of much of the problem.

By combining positive and negative elements with different dispersions (different *slopes* of the index curves), one can make the change in focal point stationary with wavelength at a chosen wavelength. However, because the glass curves also tend to exhibit different *curvatures*, the correction is imperfect at other wavelengths. Chapters 12 and 34 explain how one can minimize this effect by clever choice of three different glass types, and chapter 47 shows a design in which the program has found an excellent combination all by itself.

IOP Publishing

Lens Design
Automatic and quasi-autonomous computational methods and techniques
Donald Dilworth

Chapter 4

Using a modern lens design code

Sign conventions; stop and pupil definitions; wide-angle pupil; ray aiming; paraxial solves; the WorkSheet; the process of lens design

4.1 Using the software

One cannot teach lens design today without also teaching the use of the software, since the latter actually does most of the work, but before you can use it, you have to learn how. If you have previously used a different program, you must first *unlearn* the procedures that apply there, before learning the capabilities of SYNOPSYS. The effort will be amply rewarded, however, as the power of the new tools becomes apparent.

Different programs have different sign conventions, and for SYNOPSYS they are as shown in figure 4.1.

This is a left-handed coordinate system, which is the default.

We spoke of paraxial optics in chapter 2, and a correct analysis requires a rigorous definition of the first-order properties of the entering beam. These are defined by four quantities, which we call YPP0, TH0, YMP1, and YPP1, defined in figure 4.2. All these dimensions are in the y–z plane. (A similar set of definitions can be assigned in the x–z plane, for cases when the lens or field is not rotationally symmetric. We will ignore those in this book.) All of these first-order pupil parameters can be entered explicitly by the user or implied by something else.

You can tell the program to calculate a value for YMP1 so that the marginal ray hits the stop surface exactly at the edge of the clear aperture, or calculate YPP1 so the real chief ray hits the center of the stop, and even adjust the object height YPP0 to yield a desired GIHT; these options are explained below. There is also an option for describing a wide-angle object, for which the input angle can exceed 90 degrees; this is object type **OBD**, discussed in chapters 41 and 45.

doi:10.1088/978-0-7503-1611-8ch4

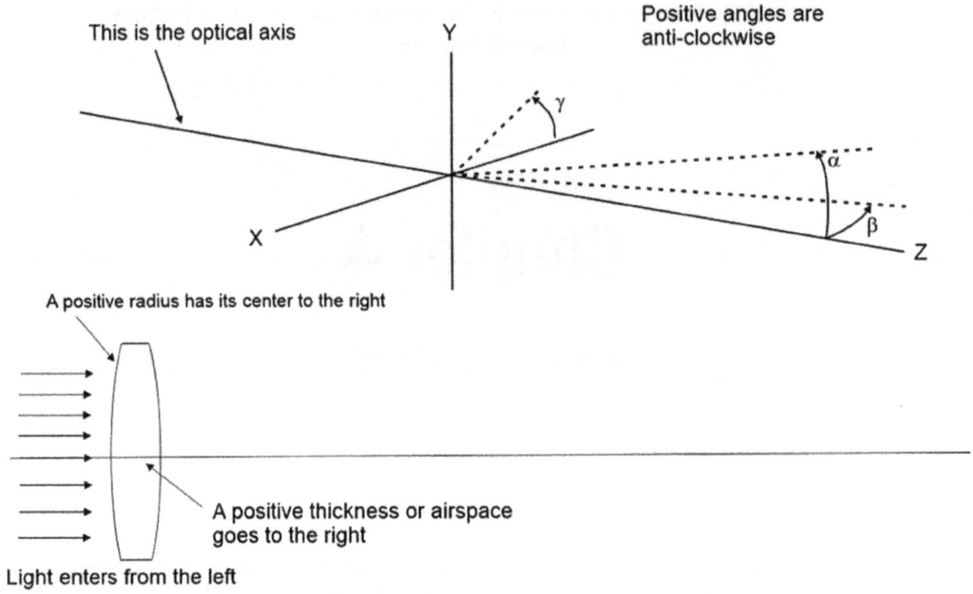

Figure 4.1. Sign conventions of lens geometry.

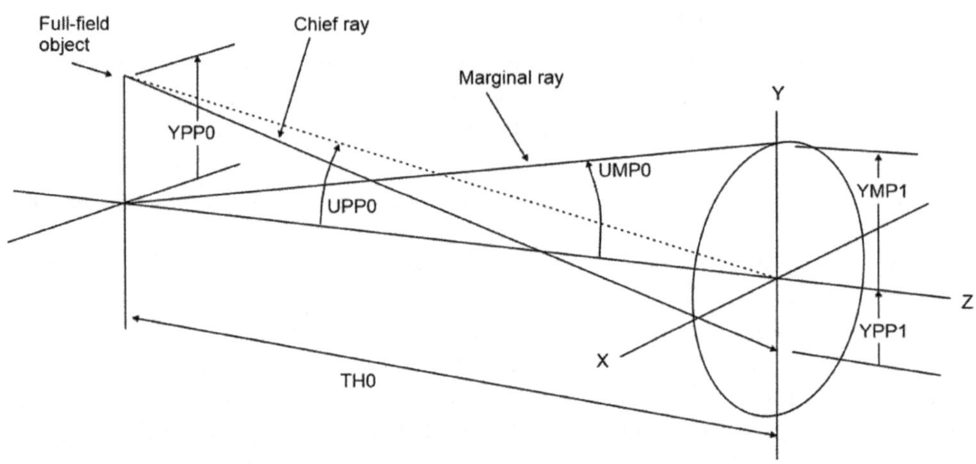

Figure 4.2. Definition of object specifications.

In figure 4.3, the lens has a stop where indicated, and the paraxial chief ray, which follows the path appropriate to the *paraxial* value of YPP1, goes to the center of the stop surface as it should.

However, in this lens, to get a *real* ray, drawn with a dotted line in figure 4.3 (and which obeys Snell's law exactly) to go there, it has to be decentered coming in or else it will miss the center of the stop. So a real YPP1 will often differ from the paraxial. If you declare a surface to be a real stop, the program then finds the real chief ray by iteration whenever it is required. (To declare it, enter a *negative* surface number on the **APS** input line in the RLE file.)

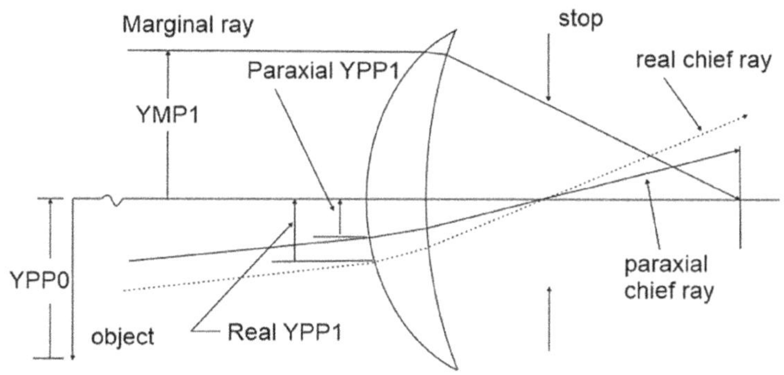

Figure 4.3. The case where a real chief ray does not coincide with the paraxial.

4.1.1 Wide-angle pupil options

We briefly mention here the three wide-angle pupil (WAP) options, called WAP 1, 2, and 3; a more detailed discussion is given in chapter 22. The default is WAP 0, which makes no adjustments to the entering paraxial beam.

WAP 1 keeps the diameter of the entering beam constant for all field angles, measured normal to the chief ray. This is described in chapter 41.

If WAP 2 is turned on, the program will calculate a value of YMP1 so the marginal ray hits the stop surface at the edge; an example is given in chapter 22.

However, sometimes the stop is not the only surface blocking the beam, and one must consider *vignetting* when rays that would otherwise get through the stop are blocked elsewhere. One then has the choice of simply letting the image evaluation programs delete those vignetted rays, or else defining the entrance pupil so it is reduced in size to encompass only the rays that actually do get through. That last option uses WAP 3, the geometry of which is summarized in figure 4.4.

If WAP 3 is turned on, the pupil size is reduced as shown on the lower right. That option should be saved for analyzing a finished design, since it is usually not practical when a design is still being optimized, and even then, only when vignetting is an issue at surfaces other than the stop.

4.1.2 Ray aiming

The above discussion relates to what is broadly termed *ray aiming*. When analyzing an image, one must ensure that an appropriate bundle of rays is traced. Some programs implement this by creating a grid at the stop and then finding, by iteration, a ray that hits every point in that grid. While this works, it is very slow. SYNOPSYS instead can define the pupil by iterating only five rays to find the outline and then filling the pupil so defined with a uniform grid. This greatly speeds things up since there is no iterating once the pupil has been found.

So which pupil option should you use? The simplest one that gets the job done. The paraxial pupil is the fastest, a real pupil is slower, and the WAP options are slower still. It is unwise to always specify a sophisticated option, such as WAP 3, when it is wholly unnecessary. So start with paraxial and look at the path of the chief ray (the central ray in the bundle) to see if it passes near the center of the stop. If not, activate the real pupil option. Then check whether the stop is properly filled. If not, consider the WAP 2 option. We will revisit this topic in chapter 22.

4.1.3 Paraxial solves

One other item we should mention here is that of paraxial *solves*, an important concept that will be used frequently. When a solve is defined, the program will calculate the actual curvature or thickness so as to satisfy a paraxial requirement, and you do not then give it a value yourself. Fetch the lens **SINGLET.RLE**.

Suppose we want the singlet, shown in figure 4.5, to operate at an F/number of 4.0; at the moment, it is about F/2. Enter a change file as follows:

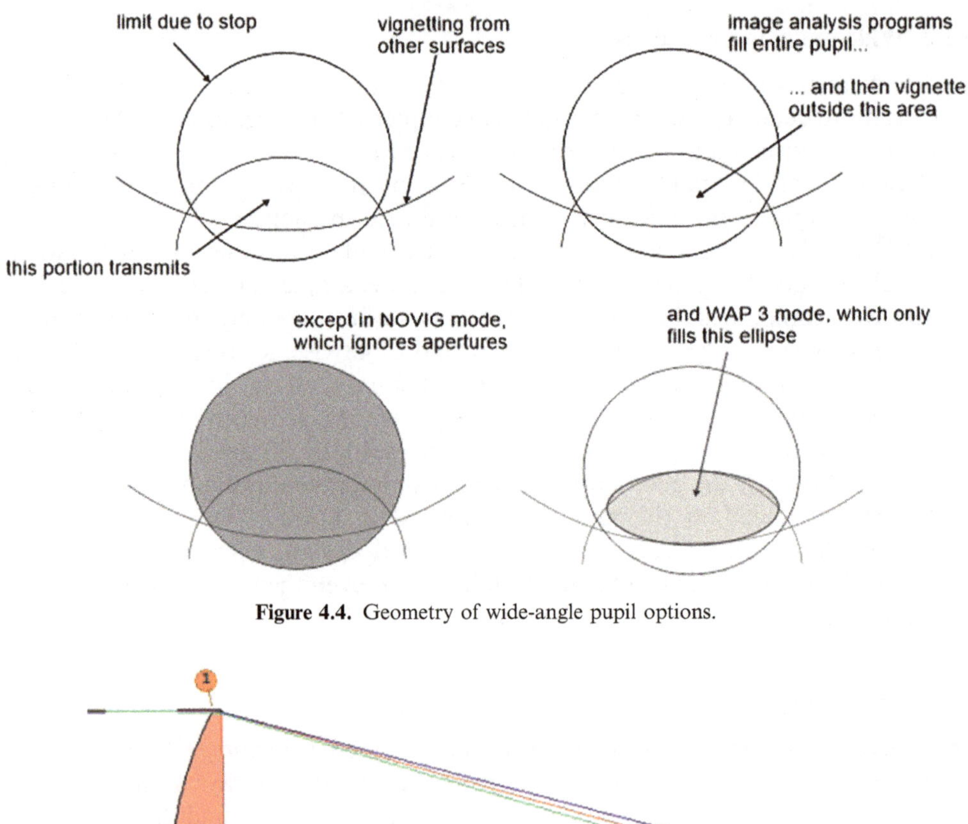

Figure 4.4. Geometry of wide-angle pupil options.

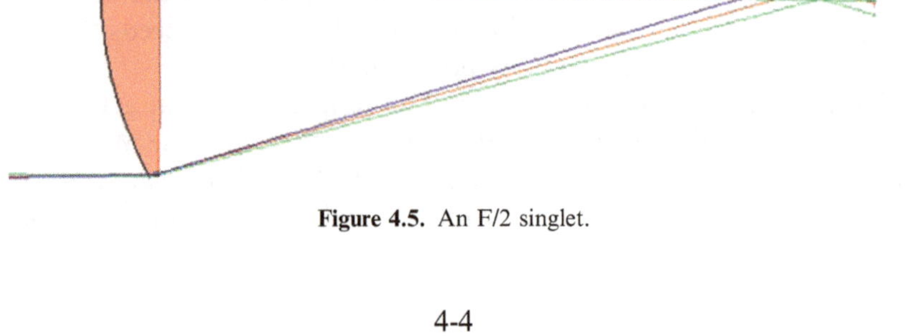

Figure 4.5. An F/2 singlet.

```
CHG
2 UMC -.125
END
```

The lens is changed, as shown in figure 4.6.

The paraxial angle is just 0.5/FNUM, with a negative sign since the marginal ray is heading downwards.

From the table below, you see that a UMC solve will control the paraxial angle of the marginal ray. This lens also has a YMT solve on surface 2, so the image plane also moves when the F/number is changed. The several kinds of solves are listed in table 4.1; a complete discussion is given in the help file of the program.

4.1.4 The WorkSheet

You made the change to the singlet above by typing a CHG file in the Command Window (**CW**), which is the window where you can type commands and where your printed output shows up. However, you will frequently make use of the WorkSheet[1] for editing a lens.

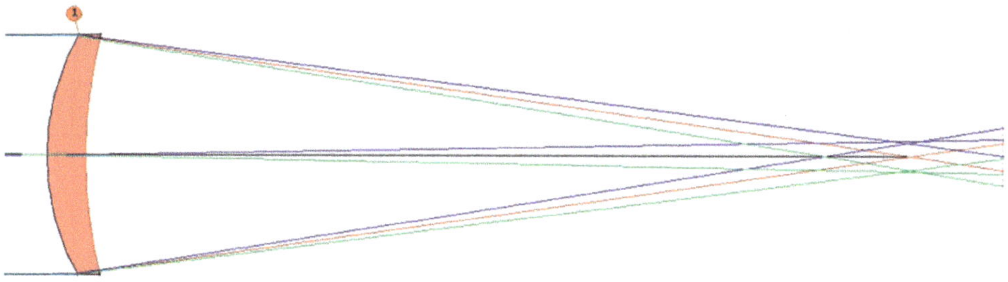

Figure 4.6. Singlet with F/number controlled by a UMC solve, yielding F/4.

Table 4.1. Curvature and thickness solves.

	Curvature solves:
UMC <u>**NB**</u>	U is a paraxial angle
UPC <u>**NB**</u>	Y is a paraxial height
YMC <u>**NB**</u>	M is the marginal ran
YPC <u>**NB**</u>	P is the principal ray (the chief ray)
APC	C designates a curvature solve
CCC	T is a thickness solve
YMT <u>**NB**</u>	Thickness solves
YPT <u>**NB**</u>	

[1] WorkSheet™ is a trademark of Optical Systems Design, Inc., a Maine, USA corporation.

In the CW, type **WS** to open the WorkSheet, and then click on surface 1 in the lens drawing. The edit pane shows the lens data for surface 1, as in figure 4.7.

Move the 'Bending' slider to the left and right, watching the ray fans, and stop when the curves are as flat as possible. You will do most of your lens editing in the WorkSheet, because it has useful features like this one that are not available in a spreadsheet. The image is improved by this bending, as shown in figure 4.8.

WS has sliders for the curvature, thickness, and bending; you can slide an element between its neighbors, and even assign a slider to any numeric quantity in the edit pane. It also links to the **Lens Layout Tool** (LLT), an example of which is shown in figure 4.9. There you can click and drag the black squares (the 'handles') and the lens construction changes accordingly, even changing the bending and powers of the elements. To see how this works, type **HELP LLT** in the CW.

4.2 The process of lens design

Much has changed since the classic texts on lens design were written. Even the more recent authors, when they mention the power and importance of computer optimization, dwell at length on the details of how to do a first-order layout, how to calculate the magnifying effect of the second group in a telephoto lens, stress the importance of coming up with a good third-order design before going to the computer, and so on. However, except for the first-order requirements, almost all of that is now irrelevant.

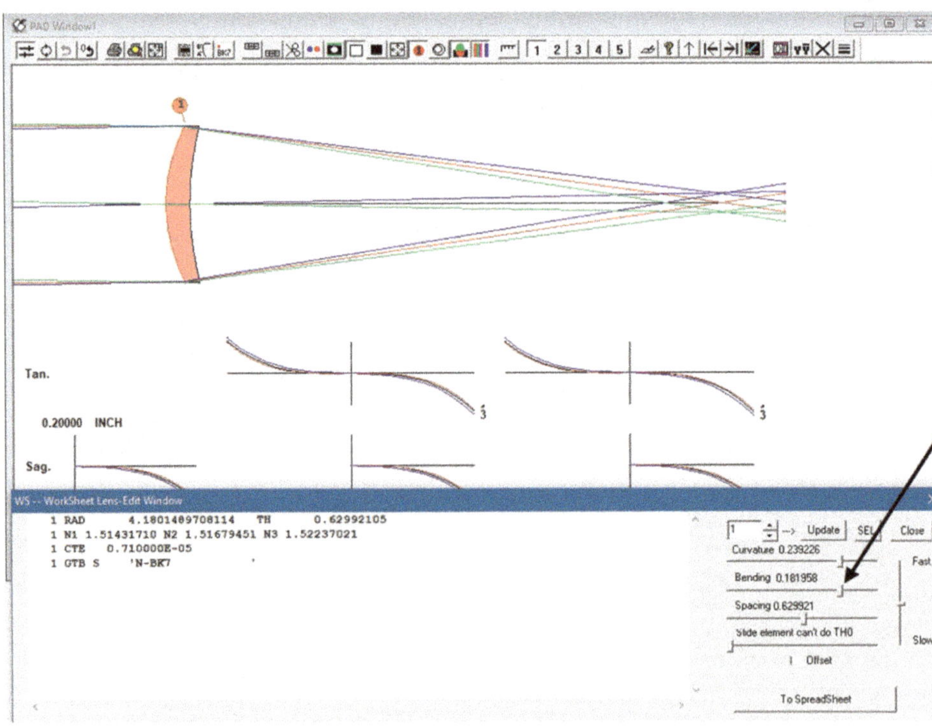

Figure 4.7. Using the WorkSheet slider to change lens bending.

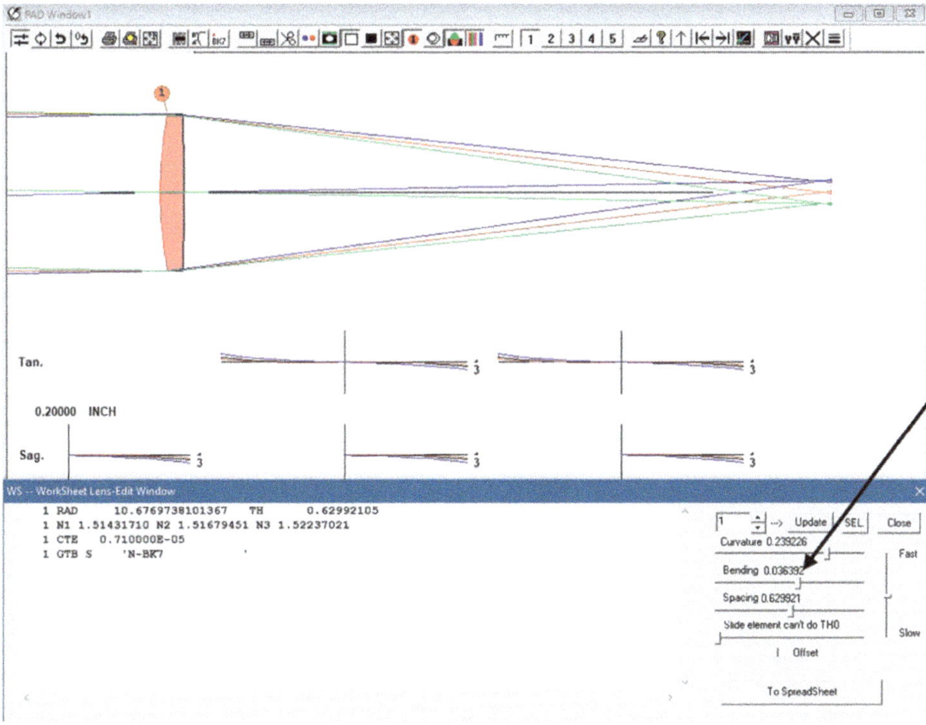

Figure 4.8. Lens with bending adjusted to minimize spherical aberration.

Not to overstate my case, it is still necessary for you to understand the purpose of the design, the mechanical envelope, the project budget, transmission requirements, the effect of the environment on the polished lens surfaces, and so on—and to have an understanding of the problems faced by the fabricators because of tradeoffs you make early on. No computer program can make those decisions for you. You will make many design choices based on your understanding of the whole project, not just the image quality requirements—but the latter are arguably the most important single consideration, since if not satisfactory, none of the other matters matter.

Lens designers in the past always tried to fit a new project into the context of their experience, often using a previous design, either from their files or from a publication somewhere, as a starting point for a new one. There are many classic forms, such as the Dagor, double-Gauss, Petzval lens, and so on, and designers who could simply modify one of those well-known forms, with an eye to some new requirements, often succeeded in a reasonable amount of time. That was the old-school method. If there was no classic design that would do, the problem was vastly more difficult.

Today, you can still do things that way if you want to—but today a new approach is called for, outlined below:

1. Understand the problem. If the customer wants to violate a law of optics, you might persuade him to adopt a more reasonable set of goals. If he cannot, you must decline the job. If the goals are possible but his approach is not feasible, lengthy consultation may yield a consensus on the right way

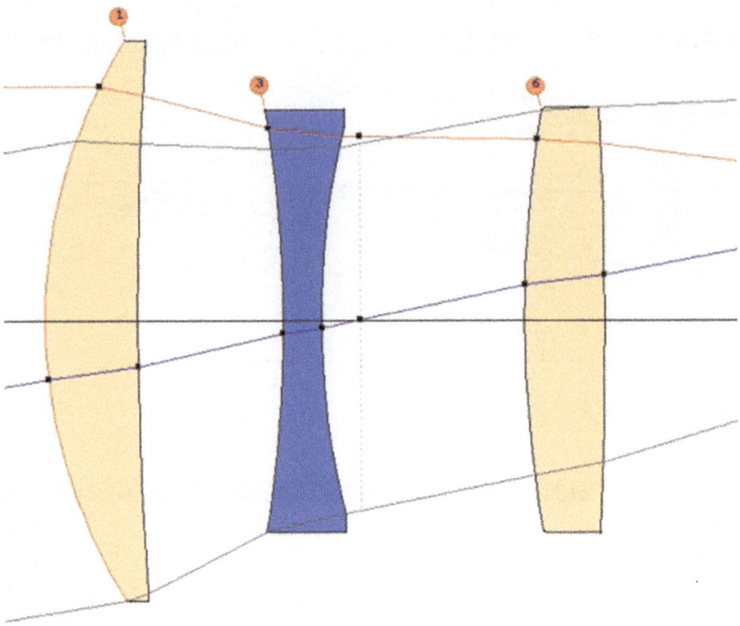

Figure 4.9. The **LLT** showing edit handles on the surfaces of a triplet lens.

to proceed. It is unethical to agree to work on a problem and collect a fee when you know it is very unlikely you will succeed. Your reputation depends on total honesty. If you cannot do the job, say so. Often the customer will express his goals with incorrect terminology, such as asking for a certain focal length when he really means back focus distance. Be sure you and he agree on just what the goals really are.

2. Work out a first-order design. Yes, it is still necessary for you to determine the basic requirements, such as focal length, F/number, image location, and whatever else is important to the customer. No computer can do that part of the job for you, and those results will be your input to the computer.
3. Once the goals are laid out and the boundary conditions expressed, unless the job is very simple try to enter those goals into one of the search programs DSEARCH, ZSEARCH, or FFBUILD[2]. Create and run that input, and then evaluate all of the returns that look practical. Optimize each of the better ones, adjusting the merit function definitions as you discover issues that are not well addressed in the lenses returned by the program.
4. Most of your effort will consist of modifying the merit function. It has been said we do not design lenses anymore; we design merit functions. Go cautiously, increasing weights here and there if a particular image error is not well balanced with the others. If a design does not converge as well as you would like, try another of the designs returned by the search programs.

[2] FFBUILD is a trademark of Optical Systems Design, Inc., a Maine, USA corporation.

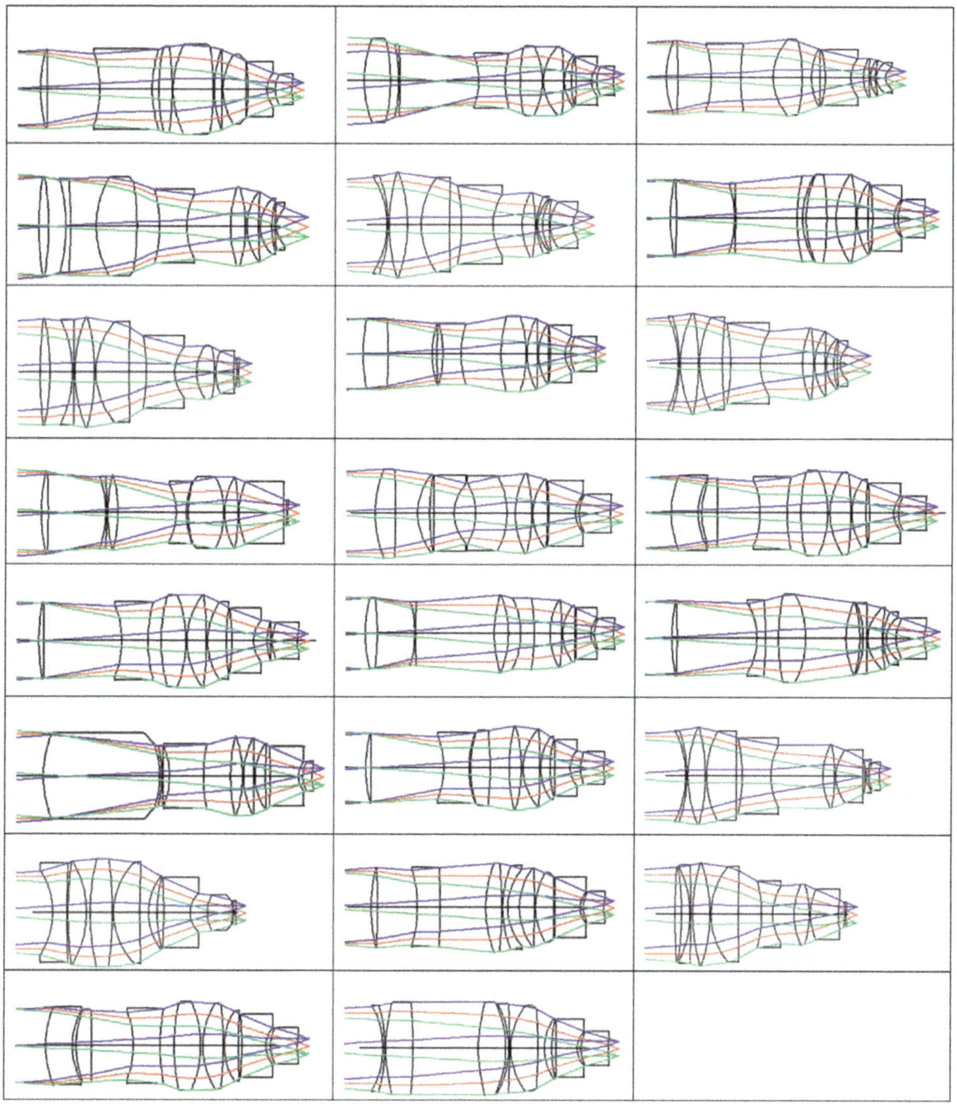

Figure 4.10. A variety of lenses found by DSEARCH for an imaging problem. Each of these lenses is as good as or better than one found by an expert using classic tools.

They only consider what they have been given as input, and sometimes there are other goals not easily described there. The SPECIAL AANT section of the search input can often include those extra goals, if you know of them in advance.

5. Unless your design is absolutely wonderful, run the search program again, perhaps several times (with switch 98 turned off!). Then the program will investigate a different set of branches of the design tree every time when it reaches the annealing stage; or you can change the input data slightly,

which will also search in different directions. Then you can compare the results from each run to see if there is a better configuration in the new set. In this way, I once came up with 23 designs that were as good as or better than one found by an expert using older tools[3]. (He was astonished.) Chaos can be an asset as well as a problem. Figure 4.10 shows an example of the variety of lenses that were found by running DSEARCH many times, varying some of the input parameters slightly. It is apparent that there is more than just one excellent solution.

6. When the design is nearing completion and if resolution near the diffraction limit is desired, add some OPD targets to the MF. (The search programs DSEARCH and ZSEARCH can both consider OPDs right from the start, with user input OPD or TOPD, if you know in advance they will be required.) When you add OPD targets, take into account the very different weighting factors that are then appropriate. A one-wave error is usually much better than a 1 mm (or 1 inch!) transverse image error, so the weights should be accordingly lower.
7. If a design looks satisfactory, assign real glass types with the automatic real glass program **ARGLASS**, which can be run from the menu **MRG**. Make a checkpoint, and carefully select the glass properties, cost, chemical stability, and so on. If your choices are too narrow, the lens might not be as good as before. In that case, restore your checkpoint and try easing up on the requirements. That is how tradeoff studies are done, and now you can tell the customer he can obtain a certain resolution with inexpensive glasses but can do better if he is willing to pay more. Just as the designer must consider the project goals and constraints early on, the reverse is also true: the project has to be flexible enough to accommodate new information that becomes apparent during the design phase. If ARGLASS does not return a good enough glass selection, try the more powerful GSEARCH[4]. Examples of its use are given in chapters 38 and 47.
8. Always try to match the design to the tooling of the selected vendor. This saves the cost of new tools, and, crucially, has an effect on the tolerance budget. Surfaces that are matched to existing testplates are easier to hold to tight tolerances because they can be measured more quickly. If a surface cannot be matched—because no tooling is close enough to the design value—*do not* simply give a radius tolerance and assume the vendor will make a new tool that is within that tolerance. Although this is commonly done, it has a serious consequence: the statistics of a batch of lenses will come out wrong. One wants the *mean value* of the lens parameter to equal the design value, and the standard deviation should be a function of the tolerance—as the budget assumes. However, if the vendor makes new tooling, the mean value itself will be different from the design value, so the lenses will in turn get that mean value instead of the correct one, and the statistics of the entire budget

---

[3] Dilworth D C and Shafer D 2013 Man versus machine; a lens design challenge *Proc. SPIE* **8841** 88410G-1.
[4] GSEARCH is a trademark of Optical Systems Design, Inc., a Maine, USA corporation.

are thrown off. The design can be matched to a testplate list automatically with **TPMATCH**, which can be launched from the MMT dialog (**MMT**). Section 50.1 gives an example.

9. When the above steps are completed, it is time to perform the tolerance analysis. The main tolerance program in SYNOPSYS is called BTOL, and by following the instructions in the User's Manual you can derive a budget for the entire lens that will ensure that a desired performance level is reached. The resulting tolerances can then be added automatically to the element drawing program ELD and on assembly drawings made with DWG. The statistics can be verified with the Monte-Carlo program (MC). Both BTOL and MC can account for the effects of adjustments that you plan to perform on the finished lenses. In the event that tolerances come out impossibly tight after all this work, it is time to look into tolerance desensitization. Read section 10.13 of the User's Manual, add the appropriate requirements to the MF, and iterate. An example of this extra step is given in chapter 10.
10. If the lens is to be used in the thermal infrared, be sure to examine the *narcissus* effect. Chapter 30 explains how to control it. It is also a good idea to evaluate the ghost-image properties of your lens. That too can be controlled in the merit function, but first you have to know about it. Chapter 36 explains how.
11. In the event that your project has important requirements that cannot be input to the search programs, you have to work somewhat harder. Remember, those programs work by evaluating a number of potential configurations, usually starting with elements of very weak positive or negative power. If necessary, you can emulate that approach yourself. Just as there are many good designs returned by the search programs, it is likely there will be several that you can find in this manner. Thus the process may take less time than you expect.

Chapter 5

The singlet lens

First steps with the software; aberrations of a singlet

The best way to learn lens design is by actually doing it, and you can start by reading the chapters below, which present a variety of problems and their solutions, working the problems as instructed. Most of these lessons involve lens input files (with the extension 'RLE'), and MACro files (with the extension 'MAC'). To save typing, you can open those files as instructed in section 1.1.

Whenever you encounter a mnemonic whose meaning you do not know, be sure to look it up in the help file. You will become a power user in no time.

5.1 Entering data for the singlet

Here is how to enter a lens with the editor.

When the program opens, type, in the Command Window (**CW**),

```
EE
```

to open the Excellent Editor, and type the text as shown in figure 5.1.

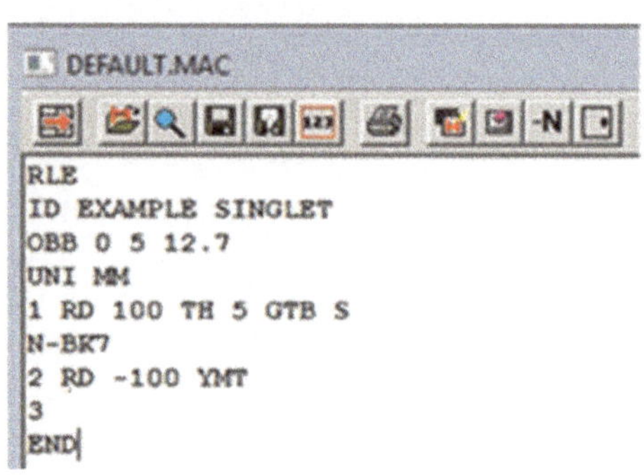

Figure 5.1. Entering a lens file in the EE editor.

doi:10.1088/978-0-7503-1611-8ch5

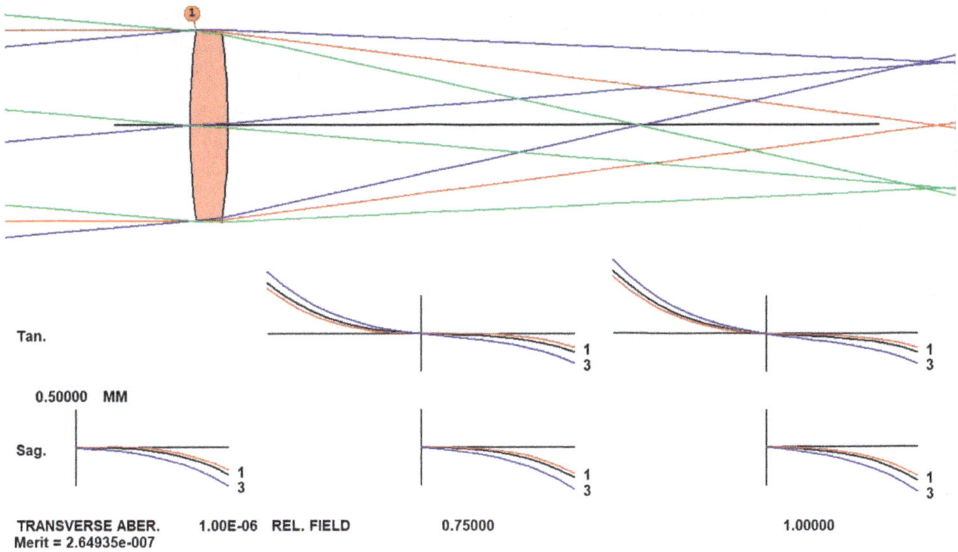

Figure 5.2. The SketchPAD display of the singlet lens.

Click the run button, , and then click the PAD button in the top toolbar, , or type **PAD**. The lens picture shows up along with TAP ray-fan curves, as shown in figure 5.2.

The meaning of the entries in the RLE file should be self-evident, except perhaps for the OBB line, which declares the object coordinates: the three arguments give the entering marginal ray angle UMP0 (which is zero for an infinite object), 5 degrees semi-field angle UPP0, and 12.7 mm for the semi-aperture YMP1. This is all explained in the menus **MPW** (Menu, Pupil Wizard) and **MOW** (Menu, Object Wizard), which you should examine before going much further. Like all input files, the RLE file must end with the **END** line.

This lens is in the DBOOK directory with the name **C5L1.RLE**. Type **FETCH C5L1** to open it.

Some people prefer to enter lens data with the SpreadSheet, which you can open with the command **SPS** or with the button in the top toolbar—but we find that entering the data with the EE editor as shown above is many times faster and easier. To enter this simple singlet with the SpreadSheet, one must execute no less than 17 different mouse clicks and type into several boxes. Users of other programs sometimes think the spreadsheet should be open all the time—as it is with those codes—but once you learn how to use the WorkSheet, you will not want to go back. Use SPS when it makes sense and close it afterwards.

The screenshot in figure 5.2 shows the **SketchPad** display, which interacts with WS, and if your display does not show those formats, click the button on the PAD toolbar to restore the default display settings. PAD is the main graphical

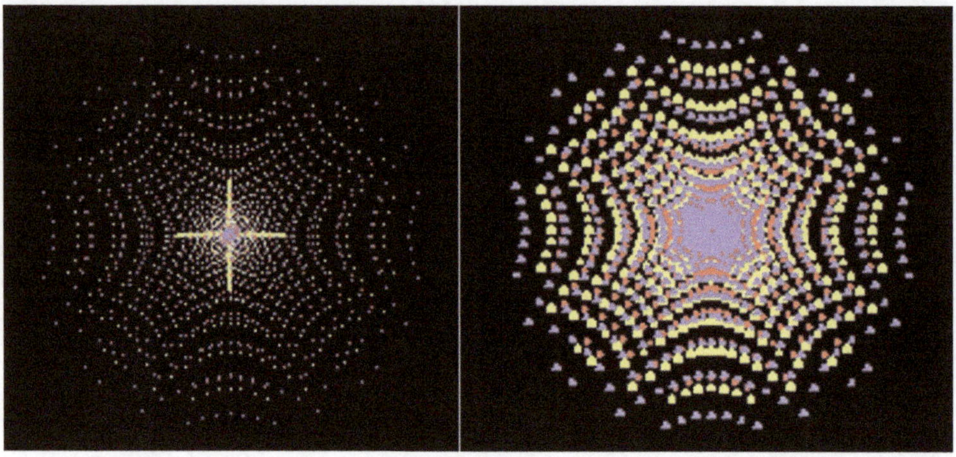

Figure 5.3. Spot diagrams. Switch 27 is on for the left panel and off for the right.

interface of the program, and it can display any combination of lens profile drawings, perspective drawings, paraxial drawings, ray fans, OPD fans, astigmatism curves, or spot diagrams. Explore the buttons on the PAD toolbar to learn how to manage the display. PAD can update itself during optimization too, making it easy to monitor the progress of your design.

Let us examine the performance of this lens, which of course is not very good. What does the image look like? Type **`MGI`** to open the Geometric Image menu, select the 'Multicolor' option on the **SPT** section, and then click that button. Your spot diagram shows up, shown on the left in figure 5.3. Now press the <Enter> key, and the **MGI** dialog opens again. Click the switches button, turn off switch 27, and click 'Apply'. Now click the **'SPT'** button again. The picture on the right of figure 5.3 appears. There are many mode switches that affect how the program works, and you can experiment with them to learn how to customize the program for your needs.

Now type, in the CW,

```
PLOT BACK FOR WAVL = .4 TO .8.
```

Figure 5.4 shows how the back focus changes with wavelength, evaluated with the artificial-intelligence program. (By the way, you always have to press the **<Enter>** key after typing a command or AI sentence.) You have just typed an English-language sentence that is understood by **AI**. We will use AI again in other lessons.

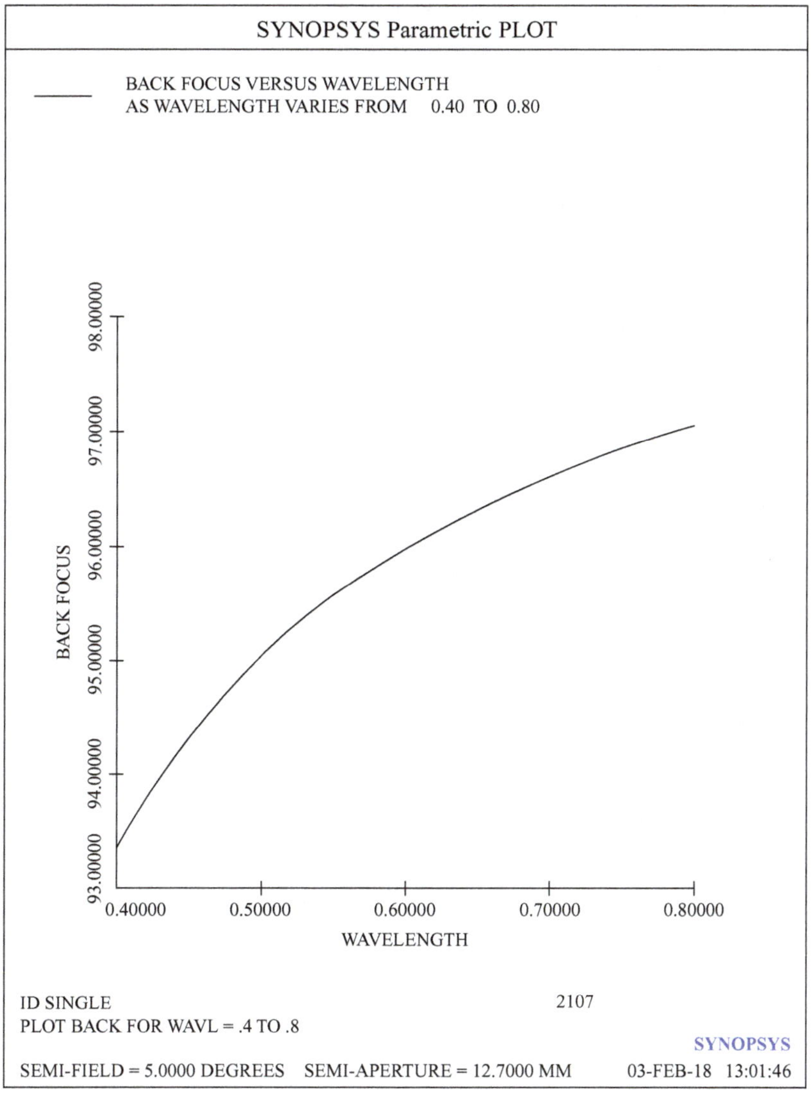

Figure 5.4. Example of an AI plot, showing how the back focus varies with wavelength.

IOP Publishing

Lens Design

Automatic and quasi-autonomous computational methods and techniques

Donald Dilworth

Chapter 6

Achromatizing the lens

Color correction with differing dispersions

Now we will add a second element and try to control the chromatic aberration of the singlet from chapter 5 (**C5L1**). First, click the 'Checkpoint' button on the PAD toolbar . It is often a good idea to save a checkpoint before changing a lens. That way, if things do not work out, you can instantly go back, with the button. (You can also press the <**F3**> key to cycle back through previous versions, but we prefer the button since then we have total control.) Now click the WorkSheet button , click in the drawing on surface 1, and then 2, and see how the RLE data for the selected surface show up in the edit pane. This is where you will do most of your lens editing.

Now add a second element. Click the '**Insert Element**' button on the toolbar near the top of the window. Then click in the lens drawing on the PAD display just behind the singlet. A second element shows up, shown in figure 6.1.

The program has removed the YMT solve from surface 2 and added an index pickup on surface 3 (**`3 PIN 1`**). Click on surface 3, and type **`3 GLM 1.6 44`** in the edit pane, as shown in figure 6.2. Then click the 'Update' button.

We are going to vary the glass types in order to correct color, and we do not want surface 3 to always pick up the index values from surface 1. So we assigned it a glass model of its own, somewhere in the flint part of the glass map, to override the pickup. Now type, in the edit pane,

```
4 UMC -.125 YMT
```

and click 'Update' again. The lens gets a curvature solve on surface 4, yielding an F/number of 4.0, and surface 5 will be at the paraxial focus because of the YMT solve—simple enough. Make a new checkpoint and close WS.

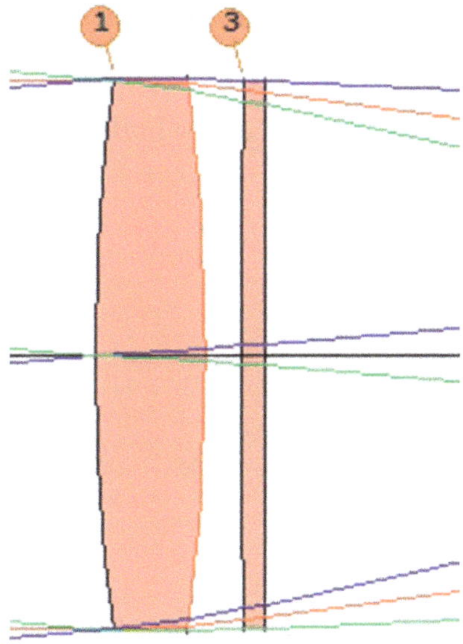

Figure 6.1. A second element has been added with the WorkSheet.

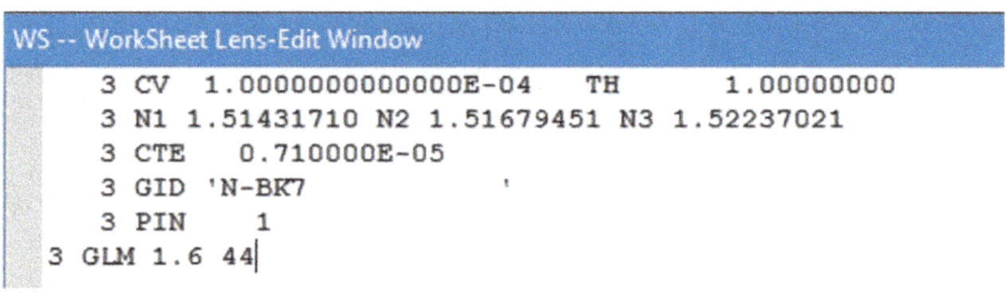

```
WS -- WorkSheet Lens-Edit Window
   3 CV  1.0000000000000E-04   TH       1.00000000
   3 N1 1.51431710 N2 1.51679451 N3 1.52237021
   3 CTE   0.710000E-05
   3 GID 'N-BK7            '
   3 PIN    1
3 GLM 1.6 44
```

Figure 6.2. The WorkSheet edit pane for surface 3, with GLM data added.

Now type **EE** to go back to the editor and click the 'Clear MACro' button to erase what you typed earlier. You are going to make an optimization MACro. Type in the editor the following lines:

```
PANT
VLIST RAD ALL
VLIST TH ALL
VLIST GLM 1 3
END

AANT

END
SNAP
SYNO 20
```

Optimization in SYNOPSYS requires a parameter file (**PANT**), a merit function file (**AANT**), and then some optimization commands. Put the cursor in the blank line after the AANT command as shown by the arrow above and click the '**Ready-Made Raysets**' button on the editor toolbar. This opens a dialog where you can select one of nine prepared merit functions. Number 6 is the default and is the one we want, so just click the '**Back to MACro' editor** button. Now your MACro looks like this:

```
PANT
VLIST RAD ALL
VLIST TH ALL
VLIST GLM 1 3
END

AANT
AEC
ACC
GSR .5 10 5 M 0
GNR .5 2 3 M .7
GNR .5 1 3 M 1

END
SNAP
SYNO 20
```

The editor has created a simple set of optimization requests. You can read the meaning of these lines in the help file. For example, if you type in the CW

```
HELP AEC
```

you will learn about the Auto Edge Control monitor (**AEC**). Now just *select* the characters '**GSR**' in the editor window and look down at the tray, near the bottom of the screen, shown in figure 6.3. If you select a word in the editor that is a command or a common entry in the merit function, the program displays the format of that entry in the tray.

The TrayPrompt gives you an instant summary of the syntax of the GSR ray-grid request. Press the <**F2**> key when you see something in the tray if you want more information, and you go right to the help file on that topic. That would probably be a good thing to do right now.

Figure 6.3. The TrayPrompt[1] when the characters 'GSR' are selected in the editor.

[1] TrayPrompt™ is a trademark of Optical Systems Design, Inc., a Maine, USA corporation.

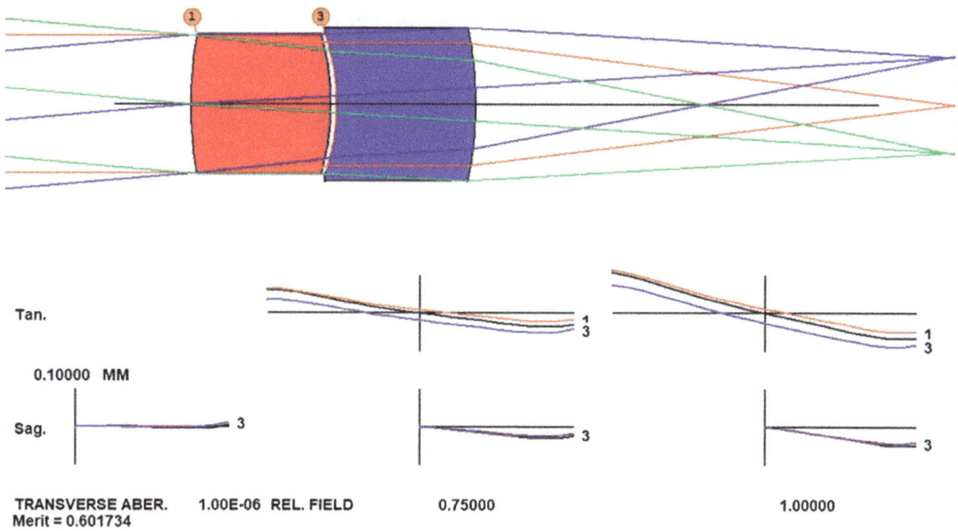

Figure 6.4. Results of the first optimization of the doublet.

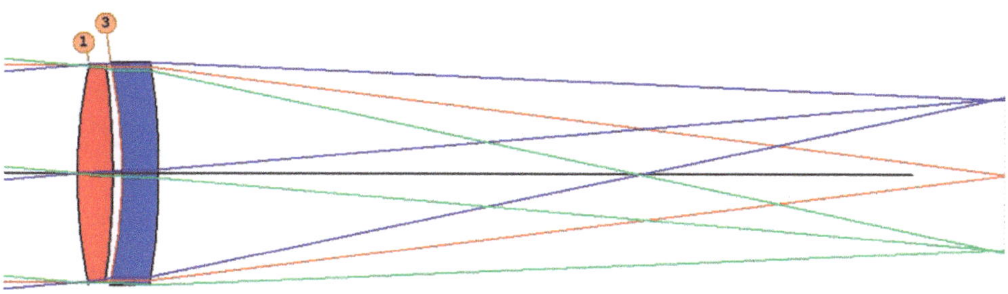

Figure 6.5. Result when a smaller target is assigned to the ACC monitor.

So much for the basics. Now click the 'Run' button on the editor and watch the display change, as in figure 6.4.

Oops! What happened? The image looks much better, but the lenses are far too thick. There is a lesson here: the program will do *absolutely anything* it can to bring down the value of the merit function, which is mostly just the sum of the squares of the items in the AANT file. Sometimes it does things that you do not want (and did not tell it about)—but it is just doing its job. The AANT file includes an **ACC** monitor, which is intended to prevent lenses from becoming too thick, which they often want to do, but the default limit is one inch, or 25.4 mm. That is what you got. Now edit the AANT file, changing that monitor to **`ACC 4 1 1`**, and run it again, giving a target of 4 mm. The result is much better, as shown in figure 6.5.

Type **`HELP MONITORS`** to read about the 12 different kinds of monitors you can use to keep your systems reasonable.

Now ask AI to show the color correction again, but this time define that sentence as a *symbol.* Type the line below in the CW:

```
QQ: PLOT BACK FOR WAVL = .4 TO .8
```

Here, we have defined the characters '**QQ**' to equal the rest of that input line. (One to three characters followed by a colon and a space define a symbol.) Now just type **QQ,** and the program processes that sentence. The color correction is much better, as shown in figure 6.6.

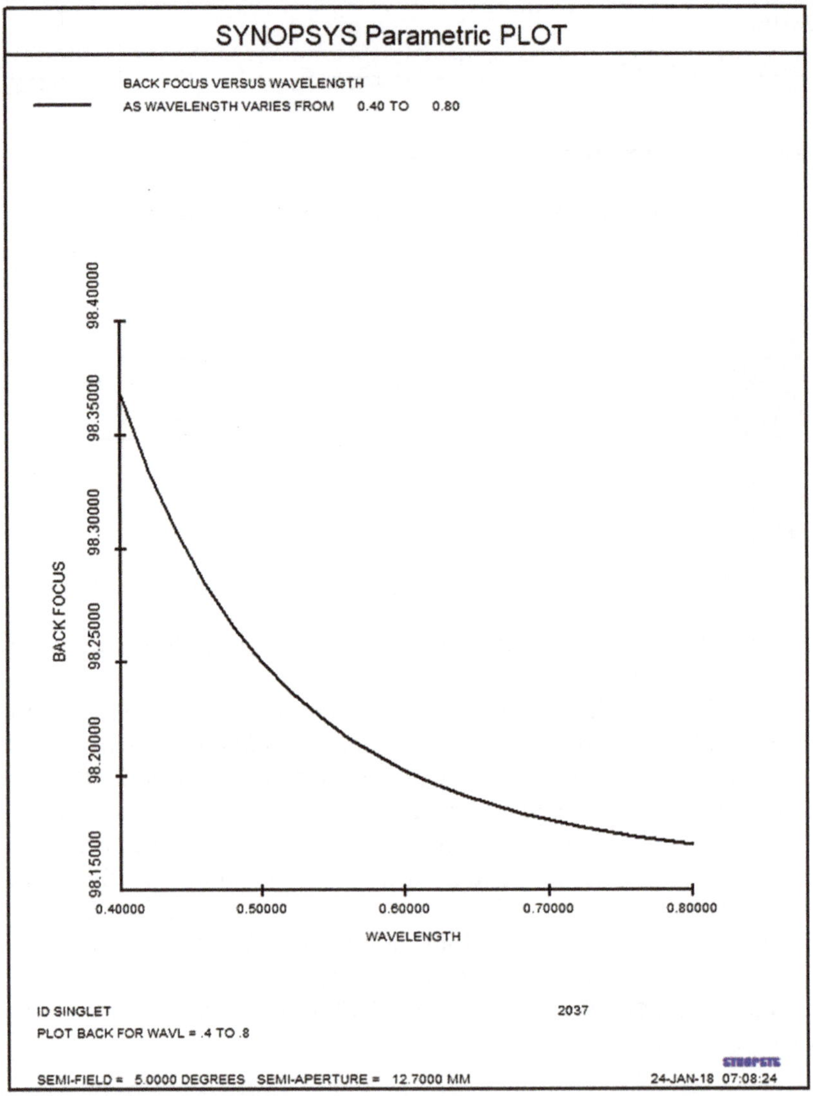

Figure 6.6. Color correction curve after optimization. It is not perfect, but other aberrations are much larger and the program came to this solution.

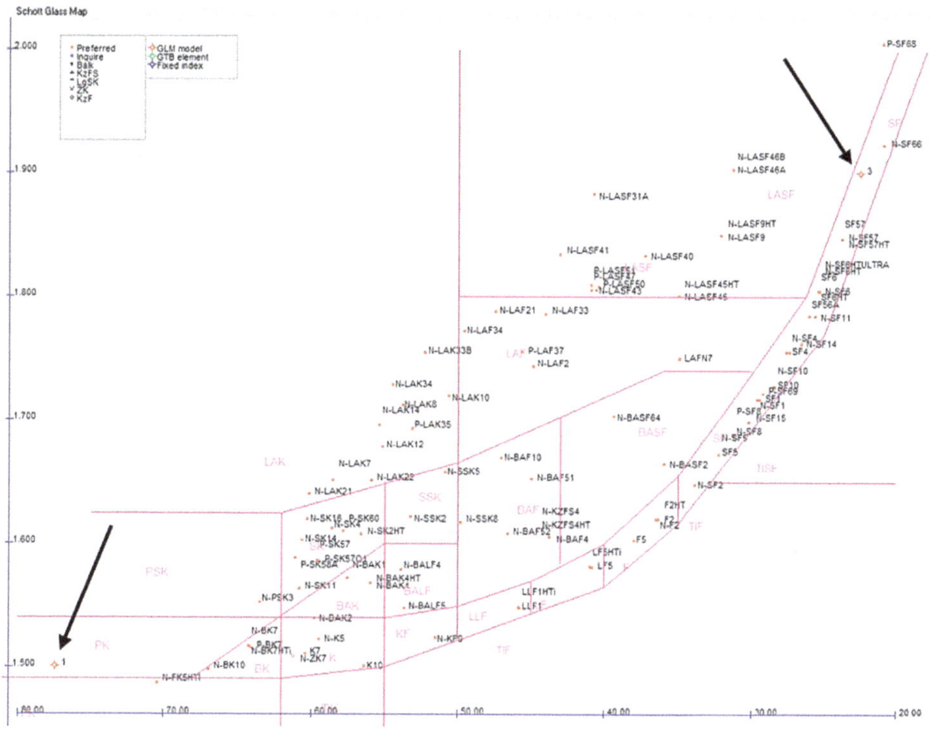

Figure 6.7. Model glass selections.

The program has balanced the color correction against the change in spherical aberration with wavelengths, so color itself is not corrected completely. Lens design is all about compromise. When you define a symbol this way, you only have to type it once and it remains defined until you exit from the program.

What kind of glass did the program come up with? Type **MGT** (Menu, Glass Table) in the CW, select the Ohara catalog, and click 'OK'. Surface 1 has been assigned a glass model in the BAL section, and surface 3 is in the PBH section, as you can see in the glass-map picture in figure 6.7. Chapter 12 will develop a three-element apochromat design with even better color correction—but first, let us save this lens.

Type the command:

```
SAVE MYDOUBLET.
```

That will save an RLE file of the lens data on disk with the name MYDOUBLET.RLE, so you can open it again later if you wish.

You can also store your lens in the lens library. Put this lens in location 4 with the command **STORE 4.** The command **PLB** will list the contents of the library, and

MLB will open a dialog that gives you another way to access it. The library is a good place to store up to ten lenses that you are currently working on, and you can save any number of lenses with the **SAVE** command.

(When you run this exercise yourself, you may obtain somewhat different results, since the exact place where you added the second element will likely differ. That underscores the fact that any change in initial conditions, however small, sends the program in a different direction.)

IOP Publishing

Lens Design
Automatic and quasi-autonomous computational methods and techniques
Donald Dilworth

Chapter 7

PSD optimization

Optimization starting with plane-parallel plates; using the PSD III algorithm

Modern software can optimize a lens many times faster than can a human expert using classical tools. This lesson gives a notable example.

There has long been contention in the lens design industry between theorists and 'number crunchers'. The former endeavor to understand their lens and steer the design in a cogent way following their deep knowledge of aberration theory. The latter employ enough optics knowledge to establish the goals in a cogent manner, but then turn the job of meeting those goals over to the computer. Today, for many problems, the number crunchers can far outpace the theorists. And it is not even close. That is why much of what theorists struggled to understand is no longer as important as it once was.

Figure 7.1 shows a design problem that starts with a very bad lens, where all surfaces are flat, all thicknesses and airspaces are equal, and all glasses are in the middle of the glass chart. Let us see how a good optimization algorithm can quickly turn that bad design into a rather good one.

Here is the optimization MACro (**C7M1**):

```
AWT: 0
OFF 67
RLE
ID START FROM FLAT
UNI MM
OBB 0 20 12.7
1 TH 5 GLM 1.6 50
2 TH 5 AIR
3 TH 5 GLM 1.6 50
4 TH 5 AIR
5 TH 5 GLM 1.6 50
```

doi:10.1088/978-0-7503-1611-8ch7

```
6 TH 5 AIR
7 TH 5 GLM 1.6 50
8 TH 5 AIR
9 TH 5 GLM 1.6 50
10 TH 5 AIR
11 TH 5 GLM 1.6 50
12 TH 5 AIR
13 TH 5 GLM 1.6 50
14 TH 50 AIR
15
APS 1
END

STO 9
TIME
QUIET
PANT
VY 1 YP1
VLIST RAD 1 2 3 4 5 6 7 8 9 10 11 12 13 14
VLIST TH ALL EXCEPT 14
VLIST GLM ALL
END

AANT
AEC
ADT 7 .01 1
ACC
M 33 2 A GIHT
GSR AWT 10 5 M 0
GNR AWT 2 3 M .7
GNR AWT 2 3 M 1

END

DAMP 1000
SYNO 100
LOUD
TIME

RMS M 0 600
Z1 = FILE 1
RMS M .5 600
Z2 = FILE 1
RMS M 1 600
Z3 = FILE 1
= (Z1 + Z2 + Z3)/3.0
```

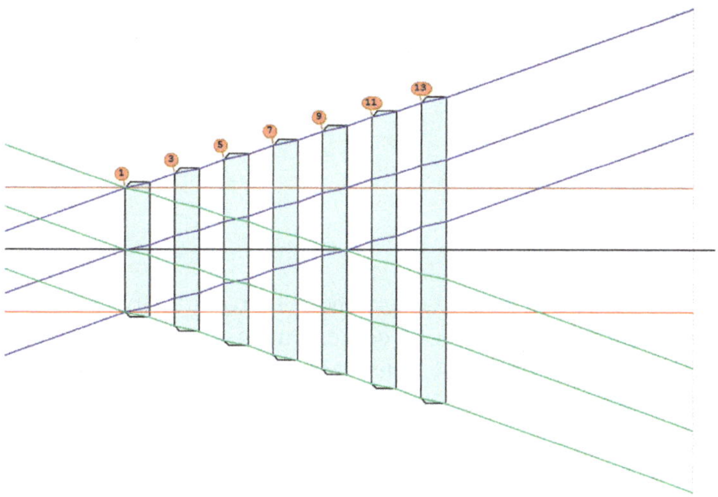

Figure 7.1. Starting design, with all flat surfaces.

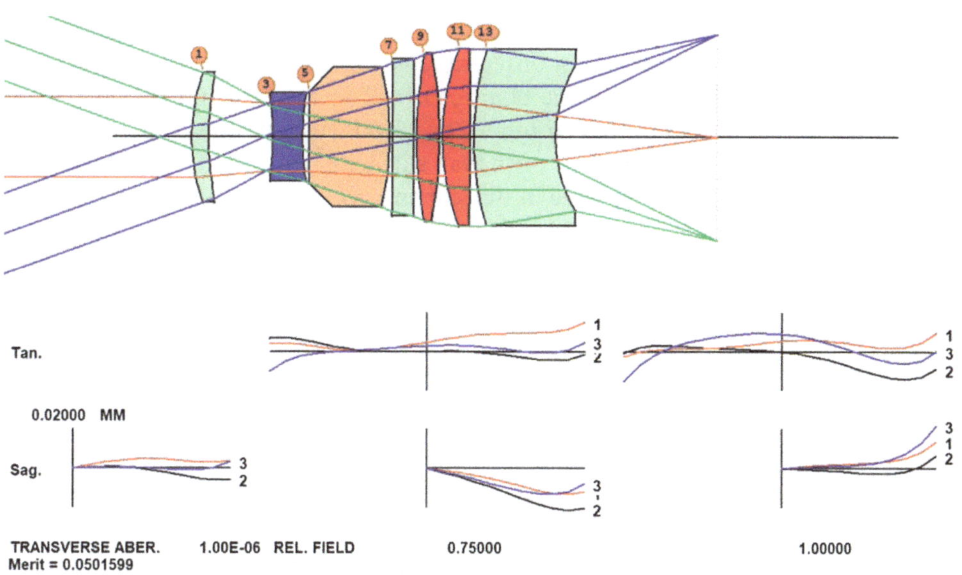

Figure 7.2. Result of PSD III optimization of the very bad starting lens.

Run this MACro, and you obtain the lens (**C7L1**) in figure 7.2 after less than one second.

Whenever you see an entry in these MACros you do not understand, be sure to look it up in the User's Manual immediately. For example, the above file contained the line **`ADT .01 1.`** This is a useful item to consider including; sometimes the results are better with it and sometimes better without. It says that the ratio of the diameter to thickness of lens elements should be at least 7, in this case with a low weight. Including this specification, or varying the parameters, will often cause the search programs to explore different branches. You will want to experiment with it often.

In the CW you see the lines

```
...
= (Z1 + Z2 + Z3)/3.0
 The composite value is      0.00636407
```

from the end of the MACro. Those use the AI program to calculate the average RMS spot size at three field points. It comes out to just over 6 µm.

Now let us change things a bit. The second parameter on the ray-generation directives currently is 0. This applies an aperture-dependent weighting on each ray, and if we increase the value to, say 0.2, then rays near the center of the pupil will be weighted somewhat more heavily than rays at the edge. (Remember to click the button **-N** so you will not overwrite the original file.) Then edit the MACro, changing the value of the symbol AWT:

```
AWT: 0.2
```

and then run it again. You obtain a different lens. (The symbol **AWT** appears in the AANT file and is replaced with the characters 0.2 in this case.) The score is only slightly higher, at 0.0069 mm. It is another very good lens, and this exercise shows an important insight: when you start with flat surfaces, the PSD algorithm can go *anywhere*—and a very slight change in the starting point or requirements can send it down a different path. This is an example of the chaos we mentioned in chapter 1. At this point we would normally run the simulated annealing program, by clicking on the button in the top toolbar, temperature 55, cooling 2, passes 50 (**55, 2, 50**). This brings the score down to 6.0 µm. A good lens indeed, shown in figure 7.3 (**C7L2**)!

This lesson illustrates the chaos inherent in lens design, which is the principle behind the design search feature, **DSEARCH**. That program creates a set of candidate designs by assigning initial powers according to the bit in a binary number, as explained in appendix B. Type **HELP DSEARCH** in the Command Window to learn about it. Each case ascends the design tree in a different direction, and one expects that at least a few of the branches it explores will turn out to be just what you are after.

It is sometimes helpful to run the optimization program more than once when starting out with a lens that is far from a good configuration. The reason has to do with the way glass model variables are implemented. Since the lenses in this chapter start out with flat surfaces, it is not obvious right away what the final powers will turn out to be, and along the way the program will likely move the Abbe numbers over to the crown or flint boundary of the glass map, where they become pinned. However, after more passes, the powers may well have changed, and a glass might then work better if it left that boundary. Starting the optimization again frees all

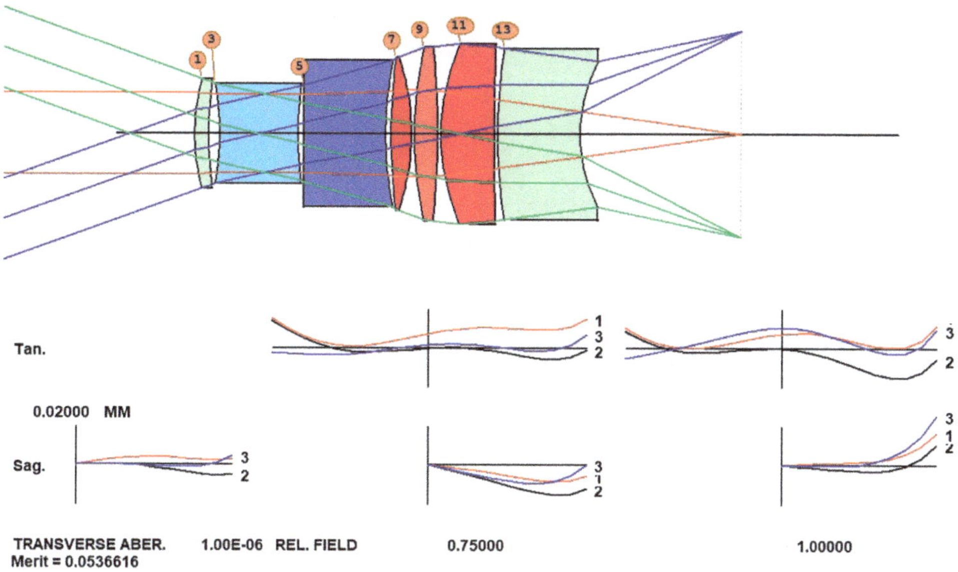

Figure 7.3. Lens optimized and annealed with higher aperture weighting.

GLM variables that have become pinned to a boundary so they can leave it if necessary. Running the simulated annealing program does the same thing, freeing GLM variables before it alters the lens. This topic is discussed more fully in chapter 26.

IOP Publishing

Chapter 8

The amateur telescope

A variety of small telescope designs

This chapter is intended for those interested in designing or analyzing the kind of small telescopes popular with amateur astronomers.

8.1 The Newtonian telescope

Probably the most popular kind of amateur telescope is the Newtonian, about the simplest system there is except for a shaving mirror. Here is the input RLE file for a typical telescope (**C8L1**):

```
RLE
ID F/8 PARABOLA WITH DIAGONAL MIRROR                    '
WAVL .6562700 .5875600 .4861300
 APS        1
 GLOBAL
 UNITS INCH
 OBB 0.000000 0.50000  5.00000  0.00000  0.00000  0.00000  5.00000
 MARGIN    0.050000
 BEVEL    0.010000
   0 AIR
   1 RAD  -160.0000000000000  TH  -70.00000000 AIR
   1 CC   -1.00000000
   1 AIR
   1 EFILE EX1   5.050680   5.050680   5.060680   0.000000
   1 EFILE EX2   4.900000   4.900000   0.000000
   1 EFILE MIRROR  2.000000
   1 REFLECTOR
   2 EAO   1.34300000   1.90000000   0.00000000  -0.10000000
   2 CV   0.0000000000000  TH   0.00000000 AIR
   2 AIR
   2 DECEN   0.00000000   0.00000000   0.00000000  100
   2 AT   45.00000004   0.00000000  100
   2 EFILE EX1   1.950000   1.950000   1.960000   0.000000
```



```
    2 EFILE EX2    1.950000    1.950000    0.000000
    2 EFILE MIRROR  -0.300000
    2 REFLECTOR
    3 CV    0.0000000000000   TH    10.00000001 AIR
    3 AIR
    3 DECEN    0.00000000    0.00000000    0.00000000   100
    3 AT    45.00000004    0.00000000   100
    3 TH    10.00000001
    3 YMT    0.00000000
    4 CV    0.0000000000000   TH    0.00000000 AIR
    4 AIR
  END
```

This is the system, as shown by the PAD display, in figure 8.1. At the moment, the telescope has a fold mirror but has not been assigned an obscuration.

The field of view was set at 1/2 degree from the axis by the OBB line:

```
OBB 0.000000 0.50000 5.00000 0.00000 0.00000 0.00000 5.00000
```

where the arguments are as shown in figure 8.2. The second entry is the semi-field angle.

To show this information in the TrayPrompt, simply open the WorkSheet and *select* the characters ‘OBB’ in the editor. The program then looks up the format for you. In this input,

- **ump0** is the incident marginal ray angle, zero for an object at infinity. (The OBB format is used mostly for that situation.)

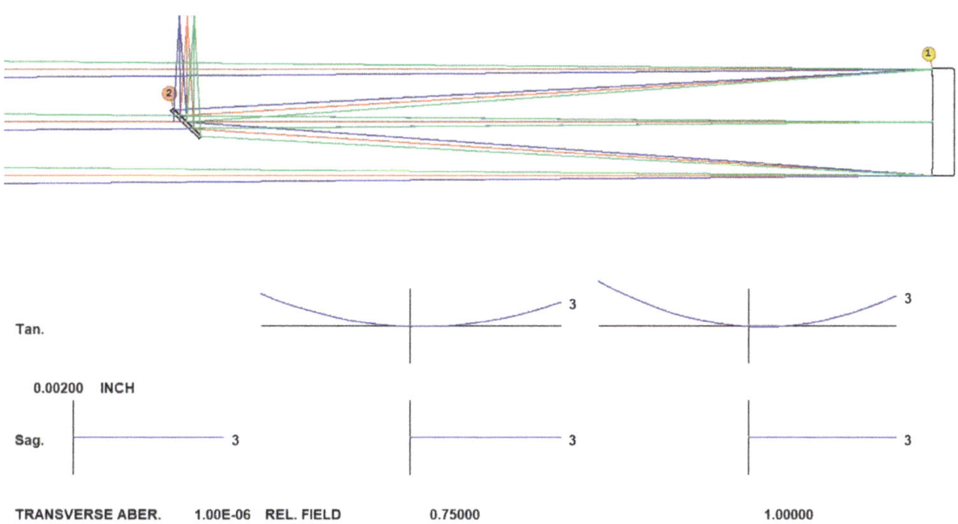

Figure 8.1. The Newtonian telescope.

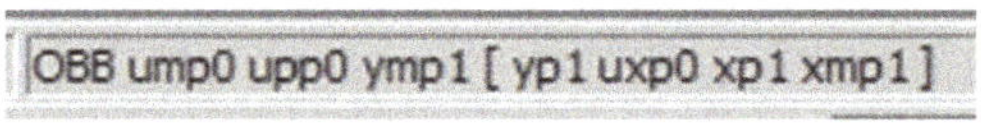

Figure 8.2. The TrayPrompt display showing the arguments of the OBB object format.

- **upp0** is the incident chief-ray angle, here 0.5 degrees.
- **ympl** is the incident marginal ray height, here 5 inches, making a 10 inch diameter entering beam.

The other arguments are all irrelevant here; **ypl** is the chief-ray height on surface 1, zero since that surface is the stop, and the rest refer to the *x–z* plane, which we ignore here since the input is axially symmetric. If you are confused already, just open the Object Wizard[1] (**MOW**) to see everything spread out and explained for you there.

The RLE input is designed to be easy to read without any explanation. Surfaces 1 and 2 are declared reflectors, and the conic constant on the primary mirror is –1.0, making it a paraboloid. The **EFILE** data are somewhat cryptic, however; those data are used to define the edge geometry of lenses, and for mirrors the substrate thickness. They make no difference in the paths of rays, but a proper edge makes a nice picture and is important if you want to make drawings of the mirror for the shop. This subject is discussed in more detail in chapter 40.

The above file is produced in response to the commands **LEO** (**LE**ns **O**ut), or **LE** (**L**ens **E**dit). It contains a complete description of the system, including all of the defaults—and you could leave those out if you want to enter the data yourself. In that case, the file would be shorter. **LEO** prints the file in the CW, while **LE** loads it into the editor.

The image is, of course, perfect on-axis, but there is lots of coma, a well-known defect of this simple system. How serious is that coma? In PAD, select view 2, (click that number in the PAD toolbar 1 2 3 4 5) and then click the 'PAD Bottom' button. In the dialog that opens, select the 'OPD Fan Plots' option and then 'OK', as shown in figure 8.3. The OPD aberrations are shown in figure 8.4.

Yes, there appears to be about two waves of coma at the edge of the field.

Here is how to obtain a listing:

```
SYNOPSYS AI>OPD                 ! The next command will be in OPD mode

SYNOPSYS AI>TFA 5 P 1 ! tangential fan, five rays, primary color, full field

ID F/8 PARABOLA WITH DIAGONAL MIRROR
TANGENTIAL RAY FAN ANALYSIS

FRACT. OBJECT HEIGHT                HBAR       1.000000    GBAR        0.000000
COLOR NUMBER                           2
 REL ENT PUPIL    WAVEFRONT ABERR
      YEN              OPD (WAVES)
______________________________________________
     -1.000             -2.355059
     -0.800             -1.271960
     -0.600             -0.583027
     -0.400             -0.200234
     -0.200             -0.035356
      0.200             -0.005883
      0.400              0.035526
      0.600              0.212506
      0.800              0.613233
      1.000              1.325667
```

[1] Object Wizard™ is a trademark of Optical Systems Design, Inc., a Maine, USA corporation.

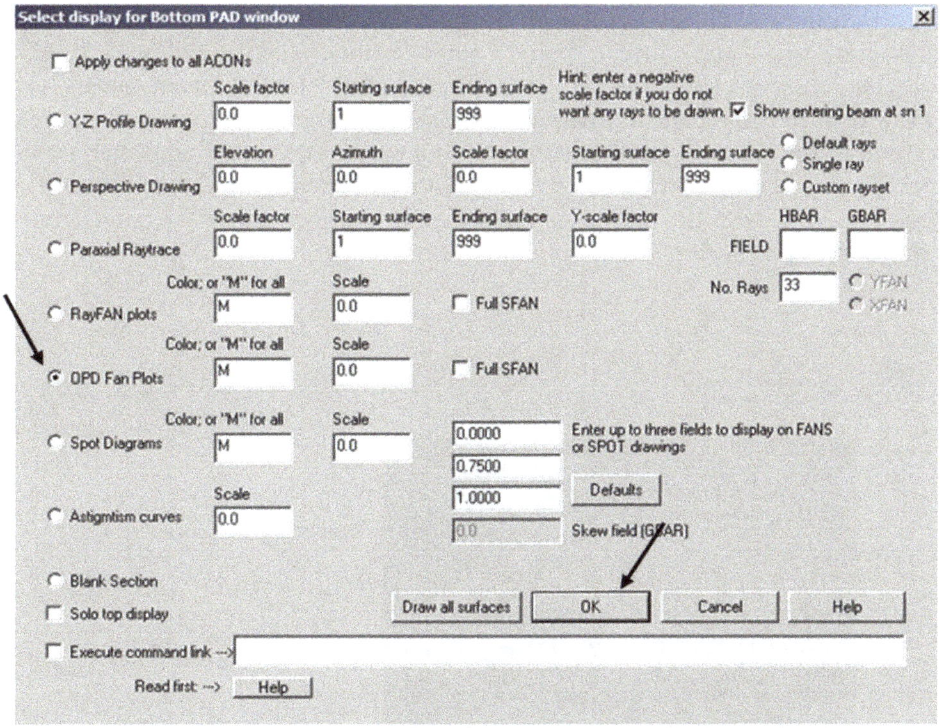

Figure 8.3. Option to select OPD plots in the PAD display.

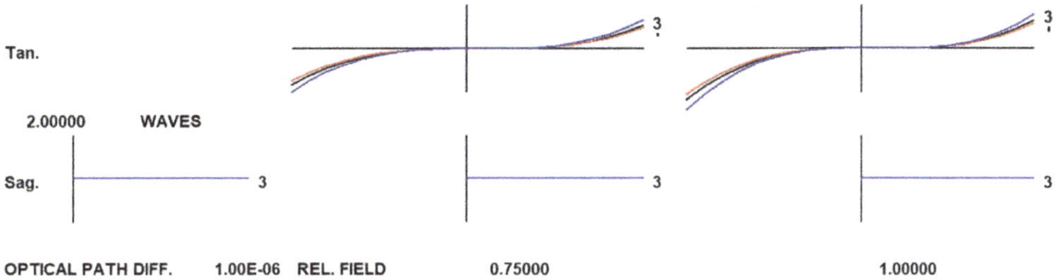

Figure 8.4. Wavefront aberrations of the telescope.

As usual, you can obtain the same thing without typing the command yourself if you want to. Go to the dialog **MRR** (Menu, Real Rays) or navigate the menu tree, and make your selection there. In this case, typing the command is faster, so I usually do it that way.

Let us look at this image in some detail. There are several tools for the purpose, but I like the Image Tools (MIT) dialog. Type **MIT,** and then make the selections shown in figure 8.5.

This is a rather beautiful example of third-order coma.

Just for fun, try this with the 'Geometric', and then 'Diffraction' selection in the 'Effects' section. The 'Coherent' analysis is somewhat smoother. It uses a two-dimensional FFT algorithm, while the 'Diffraction' method evaluates the diffraction integral, clipping at about six times the radius of the Airy disk. This image is slightly larger than that—and since a point source is always coherent with itself, the 'Coherent' choice is generally best for that case and definitely better here.

How does the image quality vary with the value of the conic constant? Close MIT and look at the PAD display. Click the 'Checkpoint' button, , and then open the WorkSheet. Click on surface 1 (or enter that number in the number box and click 'Update'). Now, with the mouse, *select* the *entire* number giving the conic constant, as shown in figure 8.6.

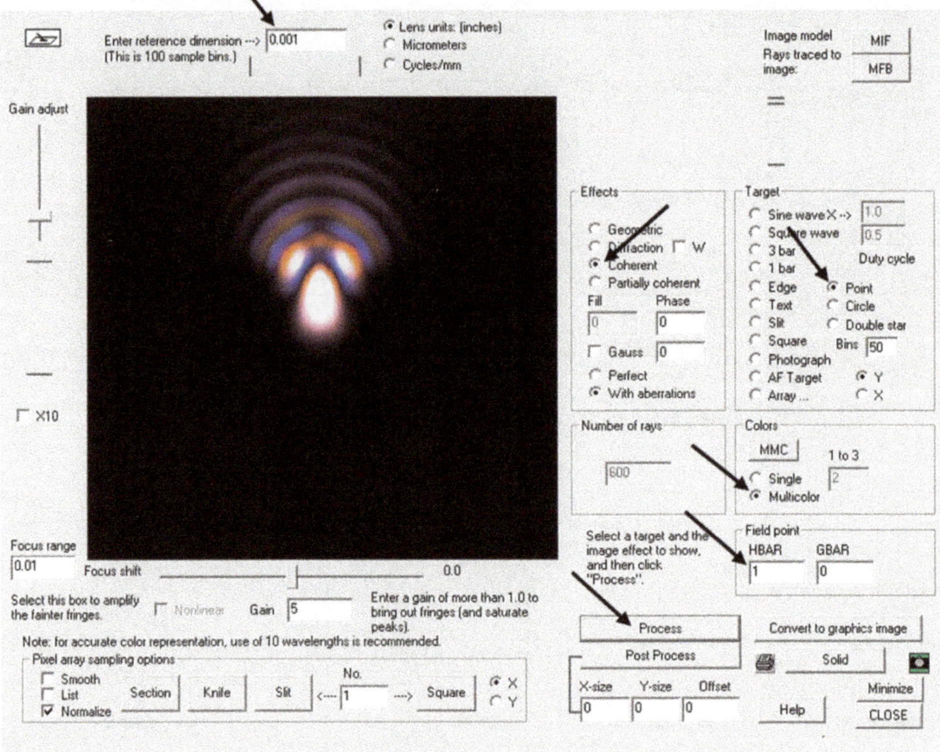

Figure 8.5. MIT display showing off-axis coma of the telescope.

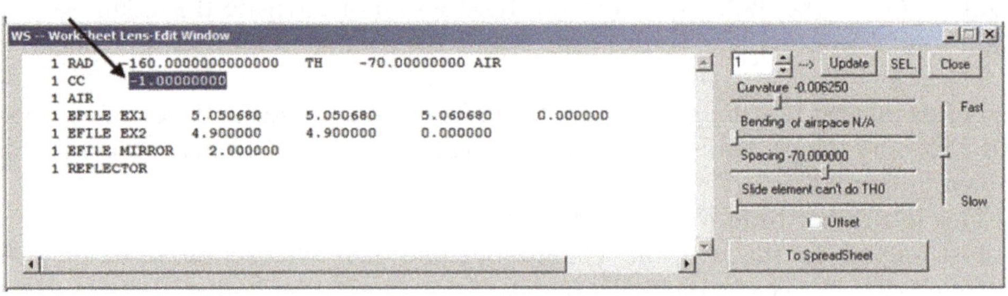

Figure 8.6. Selecting text in the WS edit pane. This assigns the top slider to that quantity with the SEL button.

Then click the '**SEL**' button. The top slider now controls that string of numbers. Slowly drag it to the left and right, watching the PAD display. Those sliders give you a convenient way to evaluate the effects of changing nearly anything in the lens. Restore your checkpoint.

Now evaluate the image quality on-axis, using a different tool—but first you need to tell the program about the obscuration, which was not given in the input file. With **WS** still open, type into the edit pane

```
1 CAI 1.4
```

and then click the 'Update' button. (**CAI** means Clear Aperture, Inside.) Now, a hole shows up in the primary mirror. Click the 'Checkpoint' button again. Type **CAP** in the CW, and you see the CAI data listed:

```
SYNOPSYS AI>CAP

ID F/8 PARABOLA WITH DIAGONAL MIRROR

CLEAR APERTURE DATA
(Y-coordinate only)

SURF    X OR R-APER.      Y-APER.     REMARK        X-OFFSET       Y-OFFSET   EFILE?
-------------------------------------------------------------------------------------
  1          5.0007                  Soft CAO                                    *
  1          1.4000                 *User CAI                                    *
  2          1.3430        1.9000   *User EAO        0.0000         -0.1000      *
  3          1.2378                  Soft CAO
  4          0.7006                  Soft CAO
```

This system has mostly default apertures, although there is now a user-entered inside aperture (a CAI) on surface 1 along with a decentered outside elliptical aperture (EAO) on surface 2, the diagonal mirror. Let us create a footprint plot on the primary mirror. Navigate with the menu tree to **MFP** (or type MFP in the CW). Make the selections in figure 8.7 and click 'Execute'.

Now you see the inside aperture, which is free of rays, shown in figure 8.8. Here is a handy trick: suppose you do not know where the rays are being vignetted (that happens sometimes with complex lenses). Here is how to find out: first press the <Enter> key to go back to **MFP**. (When you have used a dialog to run anything, you go right back there when you press that key. That is a time-saver). Now, click the 'Switches' button, and click the radio button to turn on switch 21. SYNOPSYS has about 100 mode-control switches, and that one causes several features to display the surface number where a ray stops. Click 'Apply', and then run the footprint command again. It makes a new picture, and in the center you see numerals where each ray would have gone, showing that it stopped at surface 1, as shown in figure 8.9.

Now it is time for some image analysis. Go to the MOP dialog, either with the menu tree or with the command **MOP** (MTF **OP**tions). Select the 'Multicolor' option for the MTF and click the 'MTF' button. The result is shown in figure 8.10. (The grid lines are shown because we turned on switch 87, and the data points are indicated because we turned off switch 27.)

The obscuration has indeed produced a dip in the MTF at mid frequencies, as everyone knows should happen with a Newtonian telescope.

Let us discuss that elliptical aperture on surface 2. In the WS, select surface 2 and then click the button to open the **Aperture dialog**. The option '**User-entered elliptical aperture**' is selected; click that button to display another dialog where you can change the numbers if you wish. Diagonal mirrors are usually edged with an elliptical shape, and that is where you can enter the data. Or you can just edit the numbers in the WS edit pane, once you identify them.

8.2 The Schmidt–Cassegrain telescope

The classical Schmidt telescope is a great illustration of the power you obtain when you understand basic optics. That form consists of a spherical mirror and a spherical focal surface, with the stop at the common center of curvature. Since the system is symmetric about that center, there is no unique optical axis anywhere and every field point has the same aberrations. Correct the spherical aberration, and the system has

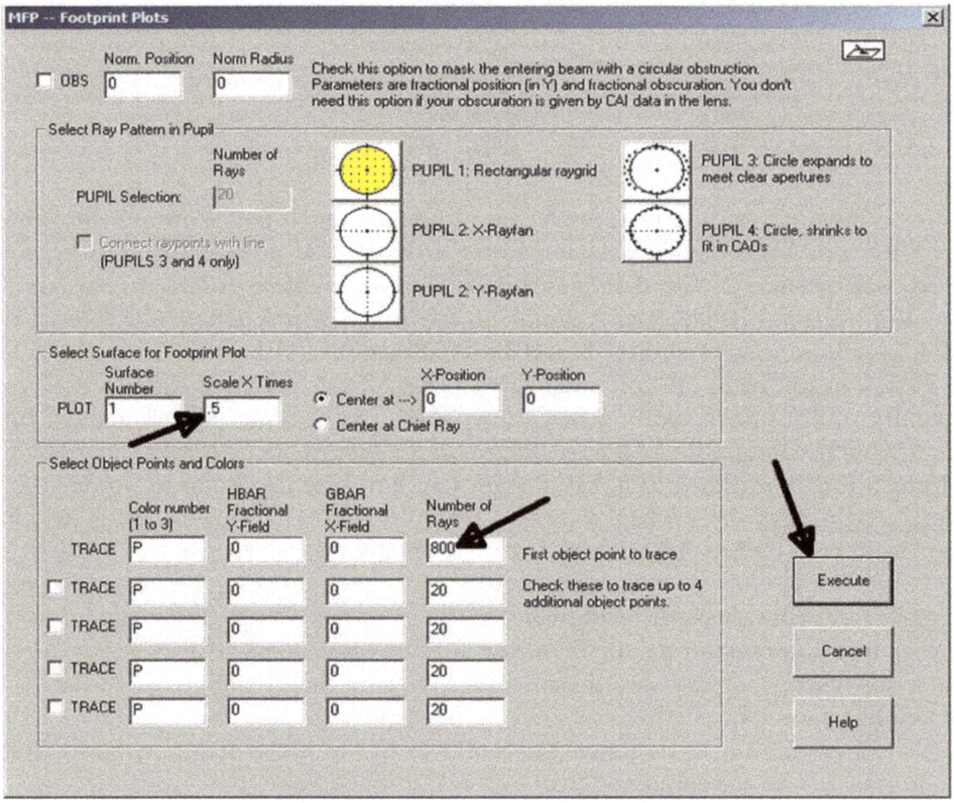

Figure 8.7. Selecting options for a footprint plot.

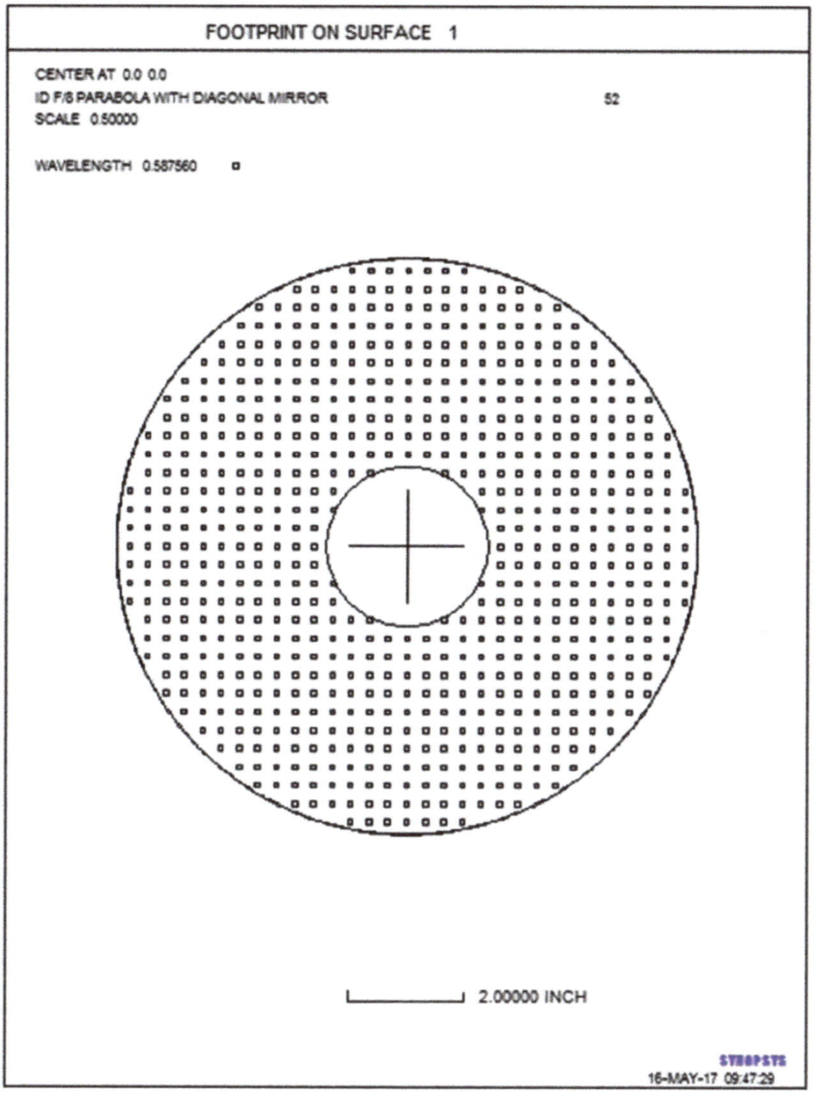

Figure 8.8. Footprint plot showing the obscuration on the primary mirror.

no off-axis aberrations either. This is a brilliant insight—but you have to correct the spherical aberration first, which is done with a thin aspheric plate located at the common center. You do not obtain perfect correction when you do that, since off-axis beams see the corrector foreshortened, but it is pretty good nonetheless. A serious drawback is the fact that you wind up with a curved image surface, which is tricky to manage with glass photographic plates. Still, this is a classic design form that is widely used in astronomy. It has the drawback that the primary mirror is much larger than the entrance pupil, which is at the corrector, if the field of view is wide, as it usually is.

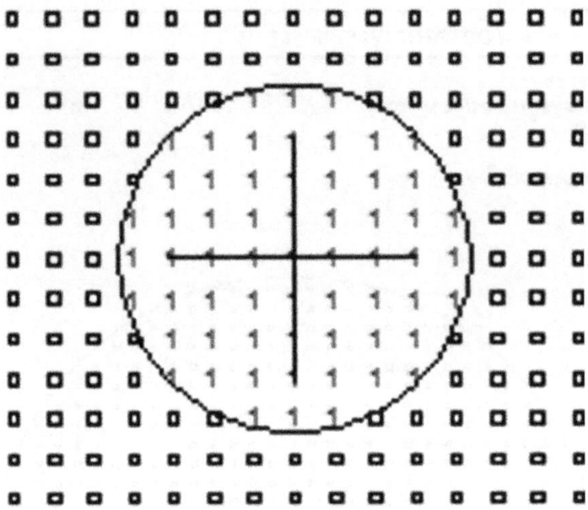

Figure 8.9. Portion of the footprint plot when switch 21 is turned on, showing where vignetted rays were stopped.

Adding a secondary mirror opens up more possibilities, and the system is then called a Schmidt–Cassegrain telescope. This is a highly corrected form used with a small field. The lens file below gives the input for this example (**C8L2**):

```
RLE
ID CC SCHMIDT CASS ZERNIKE
 FNAME 'SCT.RLE                                              '
WAVL .6562700 .5875600 .4861300
 APS               1
 GLOBAL
 UNITS INCH
 OBB  0.000000  0.40800  5.00000  0.00000  0.00000  0.00000  5.00000
 MARGIN        0.050000
 BEVEL        0.010000
   0 AIR
   1 CV       0.0000000000000    TH       0.25000000
   1 N1 1.51981155 N2 1.52248493 N3 1.52859442
   1 GTB S     'K5              '
   1 EFILE EX1      5.050000      5.050000      5.060000      0.000000
   1 EFILE EX2      5.050000      5.050000      0.000000
   2 CV       0.0000000000000    TH      20.17115161 AIR
   2 AIR
   2 ZERNIKE      5.00000000       0.00000000       0.00000000
    ZERNIKE    3     -0.00022795
    ZERNIKE    8      0.00022117
    ZERNIKE   15 -2.00317788E-07
    ZERNIKE   24 -3.81789104E-08
    ZERNIKE   35 -3.47468956E-07
    ZERNIKE   36  3.76974435E-07
   2 EFILE EX1      5.050000      5.050000      5.060000
   3 CAI      1.68000000       0.00000000        0.00000000
   3 RAD    -56.8531404724216    TH    -19.92114987 AIR
   3 AIR
```

```
  3 EFILE EX1      5.204230     5.204230     5.214230     0.000000
  3 EFILE EX2      5.204230     5.204230     0.000000
  3 EFILE MIRROR     1.250000
  3 REFLECTOR
  4 RAD     -23.7669696838233   TH     29.18770982 AIR
  4 CC      -1.54408563
  4 AIR
  4 EFILE EX1      1.555450     1.555450     1.555450     0.000000
  4 EFILE EX2      1.545450     1.545450     0.000000
  4 EFILE MIRROR    -0.243545
  4 REFLECTOR
  4 TH        29.18770982
  4 YMT        0.00000000
   BTH        0.01000000
  5 CV        0.0000000000000   TH      0.00000000 AIR
  5 AIR
END
```

Note how the vignetted rays are identified on the fan plots in PAD in figure 8.11. Switch 21 is honored there as well.

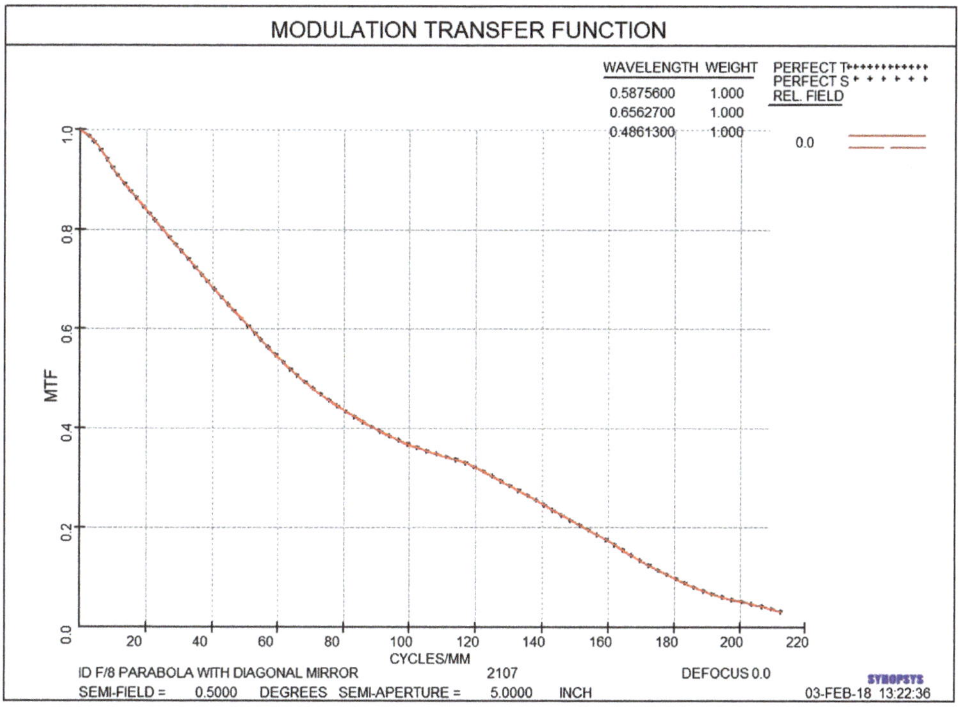

Figure 8.10. MTF curve when the obscuration is accounted for.

On the SPEC listing you see that surfaces 2 and 4 are aspheric, denoted by the 'O' after the radius column:

```
SYNOPSYS AI>SPEC

ID CC SCHMIDT CASS ZERNIKE                  272            07-JAN-13   12:40:34
LENS SPECIFICATIONS:

SYSTEM SPECIFICATIONS
-------------------------------------------------------------------------------
OBJECT DISTANCE     (TH0)          INFINITE  FOCAL LENGTH          (FOCL)   98.1614
OBJECT HEIGHT      (YPP0)          INFINITE  PARAXIAL FOCAL POINT           29.1777
MARG RAY HEIGHT    (YMP1)            5.0000  IMAGE DISTANCE        (BACK)   29.1877
MARG RAY ANGLE     (UMP0)            0.0000  CELL LENGTH           (TOTL)    0.5000
CHIEF RAY HEIGHT   (YPP1)            0.0000  F/NUMBER              (FNUM)    9.8161
CHIEF RAY ANGLE    (UPP0)            0.4080  GAUSSIAN IMAGE HT(GIHT)         0.6992
ENTR PUPIL SEMI-APERTURE             5.0000  EXIT PUPIL SEMI-APERTURE        2.0218
ENTR PUPIL LOCATION                  0.0000  EXIT PUPIL LOCATION           -10.5157

WAVL (uM) .6562700 .5875600 .4861300
WEIGHTS    1.000000 1.000000 1.000000
COLOR ORDER     2    1    3
UNITS                                INCH
APERTURE STOP SURFACE (APS)             1    SEMI-APERTURE      5.00000
FOCAL MODE                             ON
MAGNIFICATION                -9.81862E-11
GLOBAL OPTION                          ON
BTH OPTION ON, VALUE =            0.01000
GLASS INDEX FROM SCHOTT OR OHARA ADJUSTED FOR SYSTEM TEMPERATURE
SYSTEM TEMPERATURE =    20.00 DEGREES C
POLARIZATION AND COATINGS ARE IGNORED.
SURFACE DATA
--------------------------------------------------------------------------------
SURF          RADIUS       THICKNESS    MEDIUM              INDEX       V-NUMBER
--------------------------------------------------------------------------------
  0         INFINITE        INFINITE    AIR
  1         INFINITE         0.25000    K5                 1.52248      59.49 SCHOTT
  2         INFINITE O      20.17115    AIR
  3        -56.85314       -19.92115    AIR               <-
  4        -23.76697 O      29.18771S   AIR
IMG         INFINITE

KEY TO SYMBOLS
______________
A  SURFACE HAS TILTS AND DECENTERS      B  TAG ON SURFACE
G  SURFACE IS IN GLOBAL COORDINATES     L  SURFACE IS IN LOCAL COORDINATES
O  SPECIAL SURFACE TYPE                 P  ITEM IS SUBJECT TO PICKUP
S  ITEM IS SUBJECT TO SOLVE             M  SURFACE HAS MELT INDEX DATA
T  ITEM IS TARGET OF A PICKUP

SPECIAL SURFACE DATA
-------------------------------------------------------------------------------
 SURFACE NO.    2 -- ZERNIKE POLYNOMIAL
 APER. SIZE OVER WHICH ZERNIKE COEFF. ARE ORTHOGONAL (AP)        5.000000
 TERM        COEFFICIENT       ZERNIKE POLYNOMIAL
    3          -0.000228        2*R**2-1
    8           0.000221        6*R**4-6*R**2+1
   15       -2.003178E-07       20*R**6-30*R**4+12*R**2-1
   24       -3.817891E-08       70*R**8-140*R**6+90*R**4-20*R**2+1
   35       -3.474690E-07       252*R10-630*R8+560*R6-210*R4+30*R2-1
   36        3.769744E-07       924*R12-2772*R10+3150*R8-1680*R6+420*R4-42*R2+1

_____________
 SURFACE NO.    4 -- CONIC SURFACE
 CONIC CONSTANT (CC)       -1.544086
 SEMI-MAJOR AXIS (b)       43.682407   SEMI-MINOR AXIS (a)      -32.221087

THIS LENS HAS NO TILTS OR DECENTERS
SYNOPSYS AI>
```

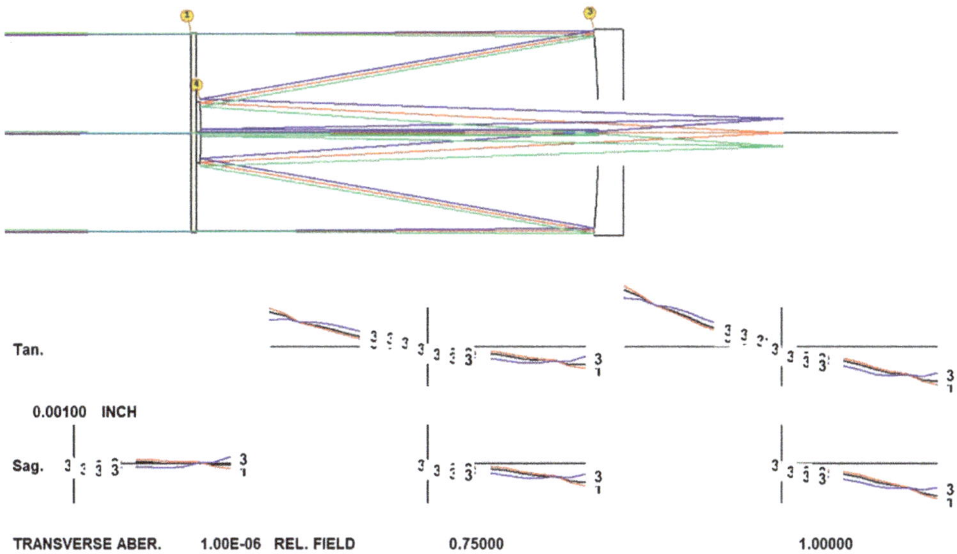

Figure 8.11. PAD display showing where vignetted rays were stopped because switch 21 is turned on.

Surface 2 is defined as a Zernike polynomial aspheric. Let us see what that surface looks like. Type

```
ADEF 2 PLOT
```

and you obtain the plot in figure 8.12.

The black curve shows how the surface departs from the closest-fitting sphere (**CFS**), which in this case is very close to flat. That tells you how to figure the corrector, once you have generated a surface with the radius of the CFS.

The ray-fan curves in PAD show the system free from coma and spherical aberration, although there is a tiny bit of spherochromatism. The striking thing is the strong field curvature, indicated by the nearly parallel S- and T-fan curves shown in figure 8.11. A young viewer with a good eyepiece would see a sharp image all over the field, since young eyes can accommodate for the focus shift. Without refocusing, on the other hand, an older viewer would be faced with about 1.5 waves of defocus. Let us see what that would look like over the field.

This time, start at the menu tree (**`EZ Menus`** in the top toolbar) and go down to MDI (Diffraction Image Analysis), where you have many choices. Select MPF, near the bottom (or just type **`MPF`** in the CW). Select '**Show visual appearance**' and click '**Execute**'. You obtain the results shown in figure 8.13.

The image at the lower left is the on-axis image, and it is essentially perfect, while the upper right shows the image at the edge of the field. That is not too sharp. Let us examine it in a different format. Back in MPF, select the option to '**Show as surface**', and change the 'Height' from the default 1 to 0.

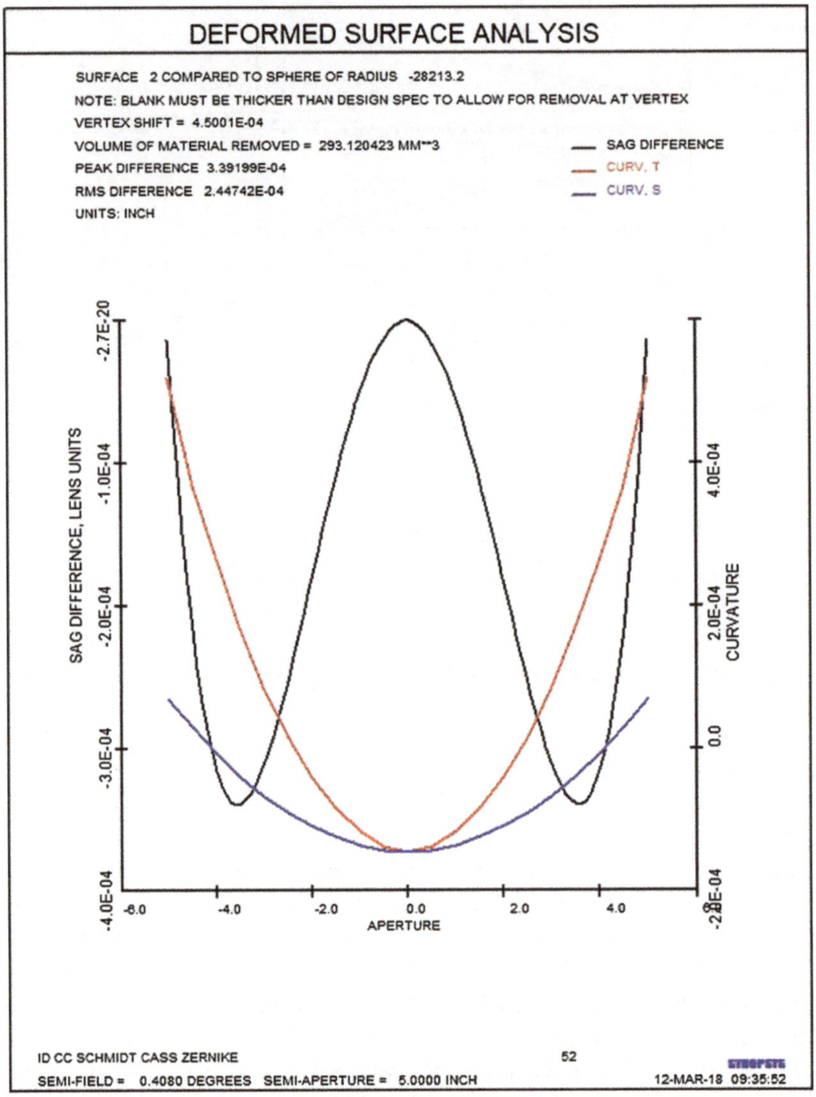

Figure 8.12. Analysis of Zernike surface showing difference in sag with respect to a reference sphere and curvatures in *x* and *y*.

Indeed, the image at the edge of the field is pretty smeared out, as shown in figure 8.14. Of course, a young observer would see a much sharper image than shown here, since he can adjust the focus of his eyes.

You can edit the Zernike terms most easily just by changing the value in the WS, but there is also a dialog that lists them by polynomial, which you can reach from the WS by clicking the '**Curvature Dialog**' button and following the trail as you did the Aperture dialog above. You arrive at the dialog shown in figure 8.15, where you can change things if you want.

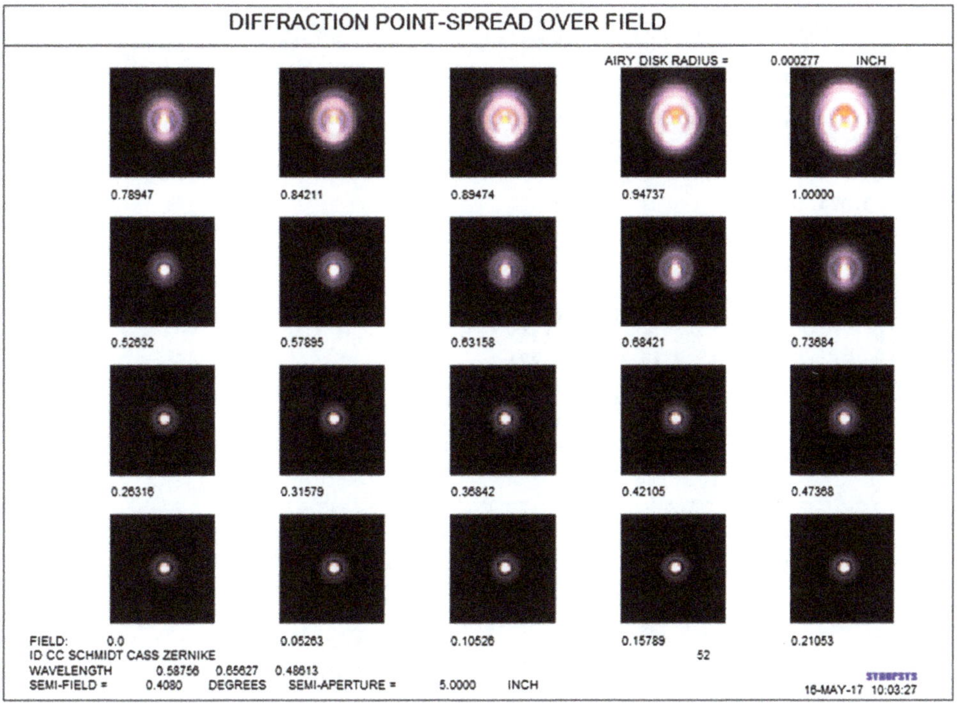

Figure 8.13. Analysis of diffraction point-spread over the field of a Schmidt–Cassegrain telescope.

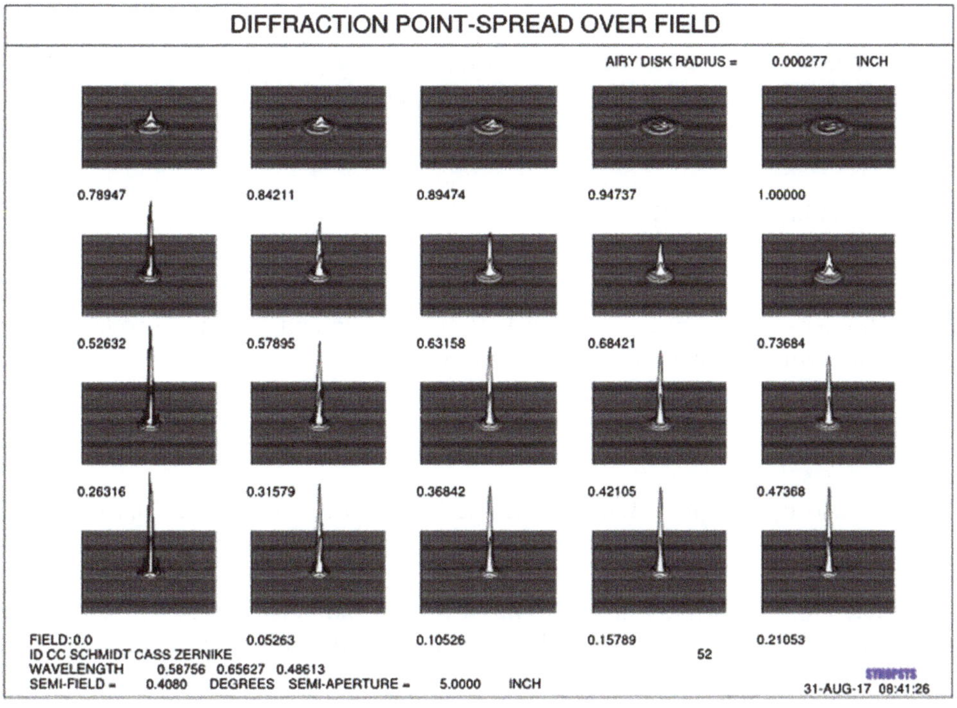

Figure 8.14. Oblique-perspective plots of the PSF over the field.

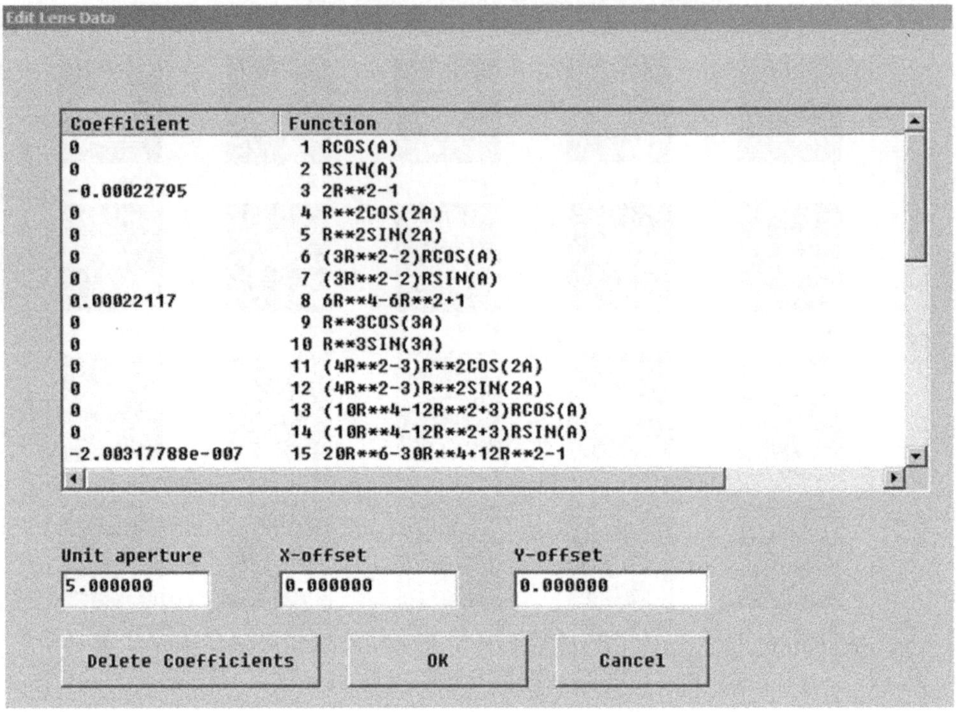

Figure 8.15. Zernike coefficients shown with the Curvature dialog, where the values can be edited.

To design this kind of system, vary the Zernike terms with the general-purpose 'G' variables. For example, the PANT entry **VY 2 G 8** would vary term number 8 on surface 2. The definition of the G terms depends on the current shape definition on the surface.

8.3 The relay telescope

This example is a relay telescope that the author built in his basement many years ago. An early version was described in *Sky & Telescope* in 1977, but this one has an extra relay lens and is better corrected. It comes bundled with the program as file 4.RLE, and you can open it with the command

```
FETCH 4
```

You can also open **MWL** (Menu, Window, Lens) to see all of the lens files in the current user directory, with a preview pane for any file you click on, and select the file there.

The telescope, shown in figure 8.16 has a 16 inch diameter mirror, and all surfaces are spherical, making it easy to build compared to an aspheric design.

By now you should be able to enter, modify, and evaluate designs like this without much coaching. The interesting point of this design is the use of a Mangin

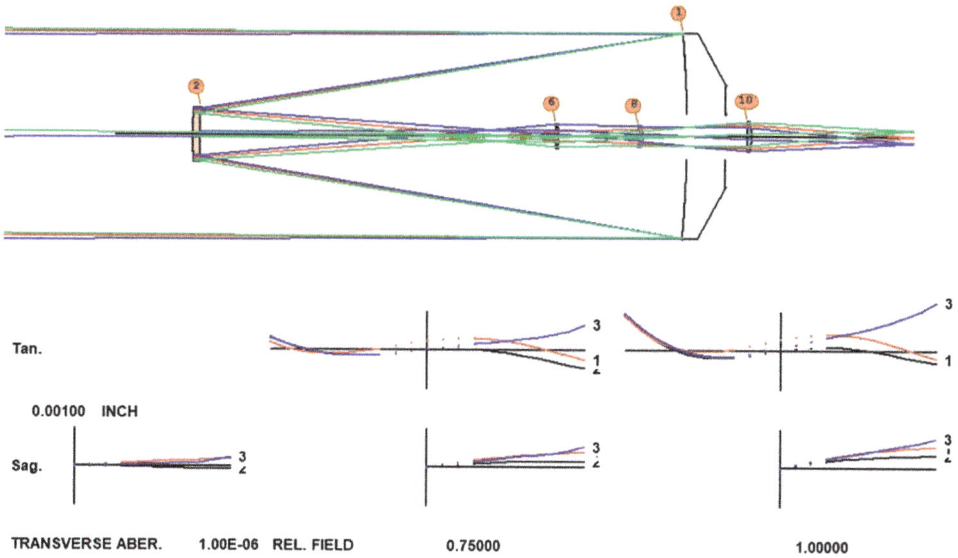

Figure 8.16. The relay telescope.

mirror, from surfaces 2 to 4. Surface 3 is a reflector, and surface 4 is coincident with surface 2, so the light goes through the element twice, reflecting from the back side at surface 3. With this kind of element, both spherical aberration and secondary color can be well corrected. This telescope has given myself and my friends many years of excellent viewing of the heavens.

When the file is opened, type **LEO** in the CW to examine the input file. Note how the Mangin is set up:

```
2 N1 1.58014165 N2 1.58312558 N3 1.58994952
2 CTE    0.640000E-05
2 GTB S     'SK12                 '
2 EFILE EX1       2.110000      2.110000      2.110000      0.000000
2 EFILE EX2       2.110000      2.110000      0.000000
3 RAD     536.5921599999994    TH        0.50000000
3 N1 1.58014165 N2 1.58312558 N3 1.58994952
3 CTE    0.640000E-05
3 GID 'SK12                 '
3 EFILE EX1       2.110000      2.110000      2.110000      0.000000
3 EFILE EX2       2.110000      2.110000      0.000000
3 REFLECTOR
3 PTH    -2       1.00000000       0.00000000
3 PIN     2
4 TH      23.05965000
4 AIR
4 EFILE EX1       2.110000      2.110000      2.110000
4 PCV     2       1.00000000       0.00000000
  ...
```

In case it is not obvious, surface 4 is the same as surface 2, and both must be present since the light by default goes through all surfaces in sequence. (There is also a nonsequential mode, which is more complicated and is never used for designs as simple as this.) Surface 3 has a pickup of thickness 2, with a sign change, and surface 4 picks up the curvature of 2. When you design systems like this, be sure to assign the appropriate pickups, so the geometry will stay correct as the design variables change.

The shape of the primary mirror, which was ground to a cone shape on the back, was entered with **EFILE** data, which are used to describe edges of elements. In PAD, click the button, to open the **Edge Wizard** (or type **MEW**, Menu, Edge Wizard), and select surface 1, if it was not selected in WS, as shown in figure 8.17.

This dialog is where you will define up to five points on lenses and mirrors, identified by the small diagram. In the case of a reflective surface, two of the edit boxes are assigned to the mirror thickness (here 3 inches) and cone angle on the back (here 28 degrees). Point E in this case marks the start of the cone, 4 inches from the axis. Click the '**Next el.**' button, and the program jumps to the first side of the next lens. Click in the PAD display to select other surfaces and see how the parameters A through E define the shape of the lens edges. Then click the button to read about all the things you can do with the edge definition, or EFILE data. We will discuss the Edge Wizard again in chapter 40.

8.4 How good is good enough?

When designing telescopes such as these, one naturally wants to know when they are good enough, and there are some simple rules of thumb that you should know

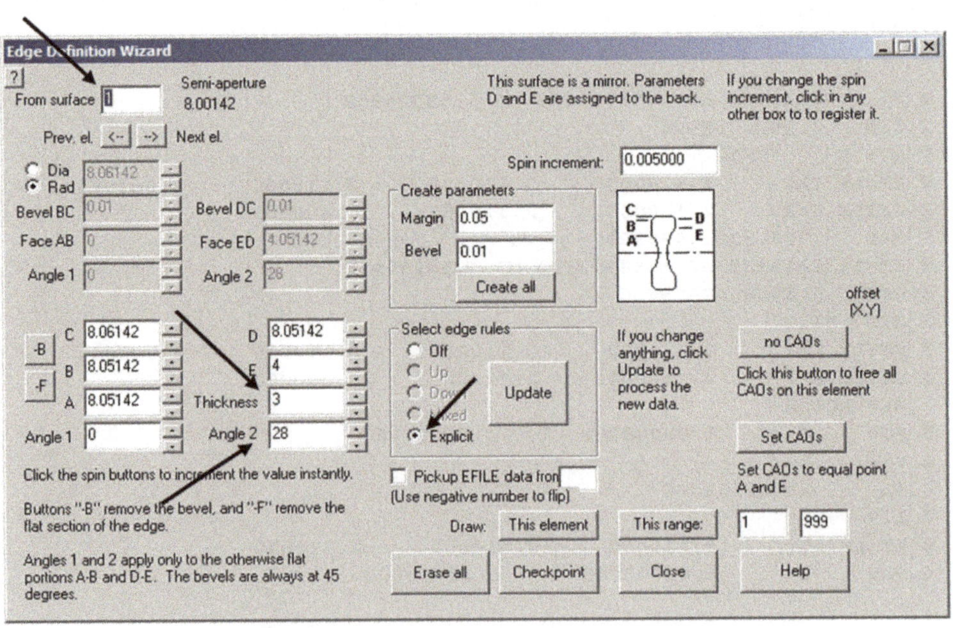

Figure 8.17. The Edge Definition Wizard (MEW) showing data for surface 1.

about: if the peak-to-peak wavefront error is less than 1/4 wave, the image will appear nearly perfect—but there are complications. To obtain this level of performance, a reflective surface has to be twice as accurate, or 1/8 wave, since the OPD error is doubled on reflection. Also, if there are lenses involved, their errors will add to the mix, so the mirror must be even better than that, and some very fussy astronomers will insist that the wavefront be corrected to 1/10 wave instead of 1/4. As noted in chapter 13, tolerances will usually be very tight, and many iterations of fabrication adjustments will be required in such systems.

IOP Publishing

Lens Design
Automatic and quasi-autonomous computational methods and techniques
Donald Dilworth

Chapter 9

Improving a lens designed using a different lens design program

Adding and removing lens elements to improve the design

In this chapter, we start with a lens designed on another program and then apply some of the newer tools to see if we can improve its performance.

Here is the starting lens (**C9L1**), shown in figure 9.1, along with MTF curves at three field points, in figure 9.2.

(Type **MMF**, select the 'Multicolor' option, then click 'Execute', to make the MTF plot.)

This lens operates in the near IR at a speed of F/3.5, and must be telecentric, have low distortion, and be diffraction limited. At first look, this design is not bad, with less than 1/4 wave of aberration. (*Telecentric* means the central ray at all field points should be parallel to the axis, so the exit pupil is at infinity.)

Maximum distortion over the field is just over 0.5 μm, and the maximum departure from telecentricity is about 0.01 radians. Not bad at all—but if we can improve the baseline performance, that will give us more leeway for tolerances, so it is worth a try.

The lens at the moment uses the WAP 3 pupil, which is not a good idea at this stage, so we first make some system changes and then optimize. We will also let the glass types vary, since we do not know if the previous designer used those variables properly—except for the windows at front and back, which we assume the customer wants to keep as is. Here is our MACro (**C9M1**):

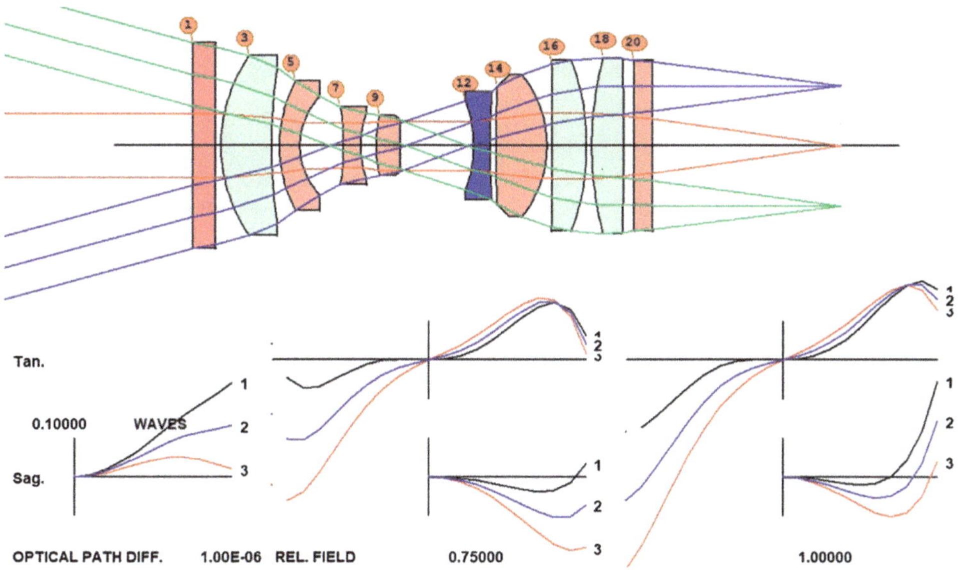

Figure 9.1. Lens to be improved.

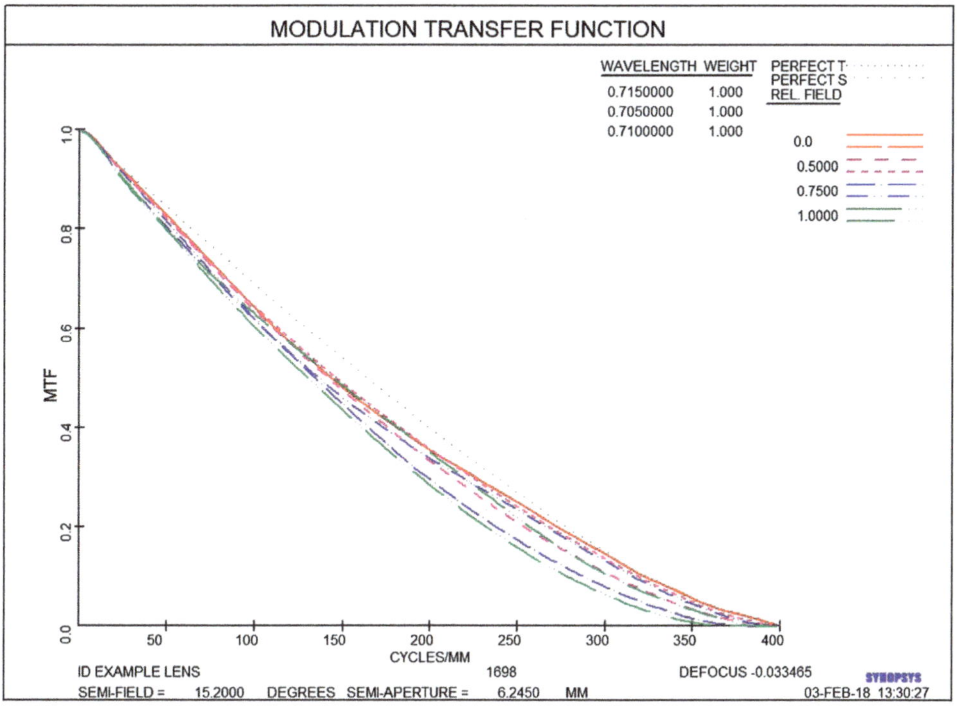

Figure 9.2. MTF of the starting lens.

```
CHG
WAP 1                   ! keep entering beam diameter constant over field
19 UMC -0.14286         ! maintain F/number
CFREE                   ! remove the clear aperture at the stop
END

PANT
VLIST RAD ALL ! all radii will change except 19 and the flat windows
VLIST TH ALL EXCEPT 1 LB2  ! and all thicknesses except 1 and 20
VLIST GLM 3 5 7 9 12 14 16 18
END

AANT
AEC                     ! monitor feathered edges
ACC                     ! and keep thicknesses less than 25.4 mm
M 89.6 1 A TOTL         ! keep total lens length constant
M 0 50 A GIHT           ! control distortion at full field
S P YA 1

M 0 50 A GIHT           ! and at 0.8 FIELD
MUL CONST 0.8
S P YA .8

M 0 50 A GIHT           ! and at half field
DIV CONST 2
S P YA .5

M 0 20 A P HH .7        ! control telecentricity at 0.5 field
M 0 20 A P HH 1         ! and full field

GSO 0 0.1 5 M 0         ! correct OPDs of ray grids at three fields
GNO 0 0.05 4 M .7
GNO 0 0.05 4 M 1
END

SNAP                    ! get snapshot every iteration
SYNO 30                 ! optimize for 30 cycles.
```

The simplest way to create this set of ray-grid aberrations is with the '**Ready-Made Raysets**' button in the MACro editor. In this case, select set number 8, which creates both transverse and OPD targets, and then delete the transverse targets and increase the weighting of the OPD targets at full field. The 'Bare-bones **Rayset**' dialog can also do it, with more options then available ().

Optimize with this file, and then anneal (**55, 2, 50**). The lens is improved, shown in figure 9.3.

Now let us apply some powerful tools. First, run the **Automatic Element Deletion** feature. This finds the element that can be removed with least degradation to the merit function. To run it, first rename the MACro, and then simply add the line

```
AED 3 Q 3 18 ! find which element to delete between surfaces 3 and 19.
```

to the MACro before the **PANT** command and reoptimize. The program reports that the lens at surface 14 can be removed. Allow it to remove that element, and then

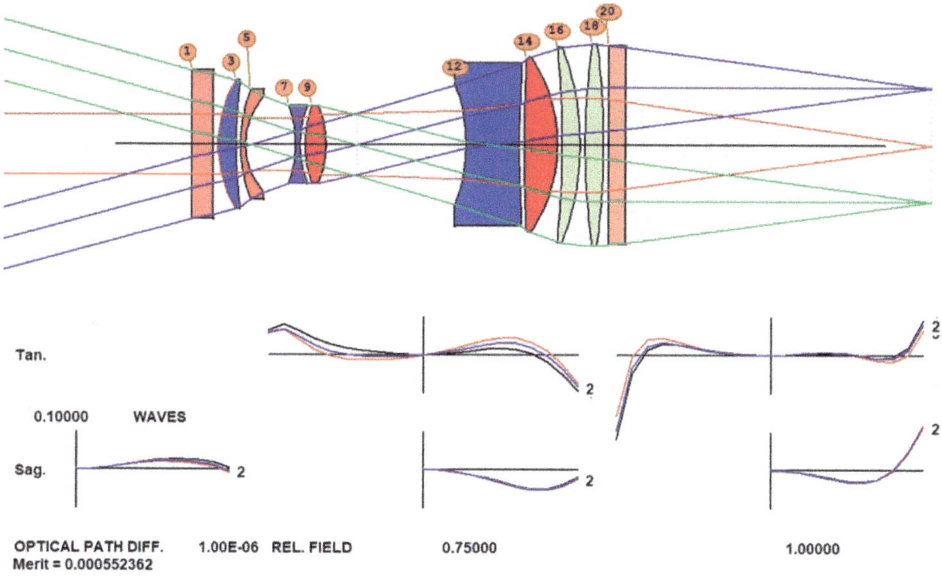

Figure 9.3. Lens as reoptimized.

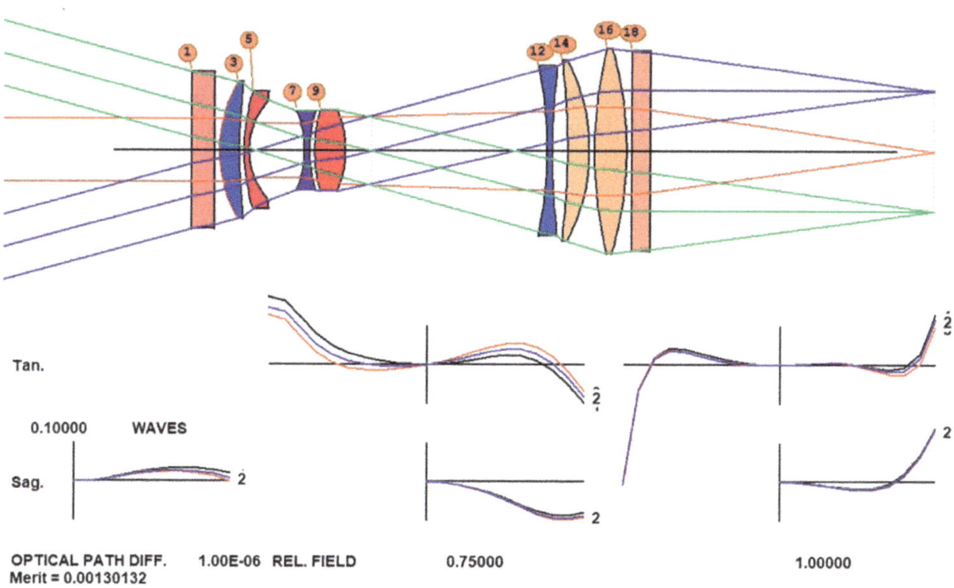

Figure 9.4. Lens with an element removed by AED, optimized, and annealed.

remove the CHG file at the top, comment out the AED line so you do not remove yet another element. Change the GLM variable declaration to `VLIST GLM ALL`. The surfaces are renumbered now, and this entry will vary all elements that are already glass models, so you do not have to keep track of the numbers yourself.

Optimize and anneal again, and you obtain the result shown in figure 9.4.

The lens is not quite as good as before, as expected, but still not bad (and definitely better than the starting lens). Now we will use the **Automatic Element Insertion** feature to see if going back to the previous number of elements gives better results than the original lens.

To do this, change the AED line to

```
AEI 3 3 17 0 0 0 20 1 ! insert one element between 3 and 17.
```

and run the MACro once more. (If you have a multicore PC, you should also add the line

```
CORE nb
```

at the top of your MACro, so AEI runs faster, where **nb** is just under the number of cores you have.)

The program inserts an element at surface 16. Now comment out the AEI line, reoptimize, and anneal. The result is shown in figure 9.5.

A quick run on **MRG** replaces the glass models with a real glass from Ohara and gives the MTF in figure 9.6 (**C9L2**). (MRG is described in more detail in chapter 44; we will use it often.)

This exercise was definitely worth doing. The program has removed the original lens element at 14 and replaced it with a different element at 16. Maximum distortion is now about 0.06 μm, compared with 0.6 before, and the maximum departure from telecentricity is now about 0.000 35, compared with 0.01 before.

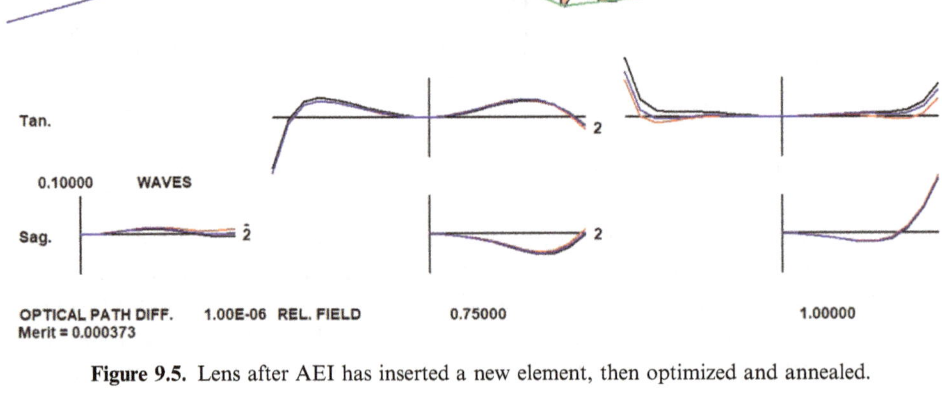

Figure 9.5. Lens after AEI has inserted a new element, then optimized and annealed.

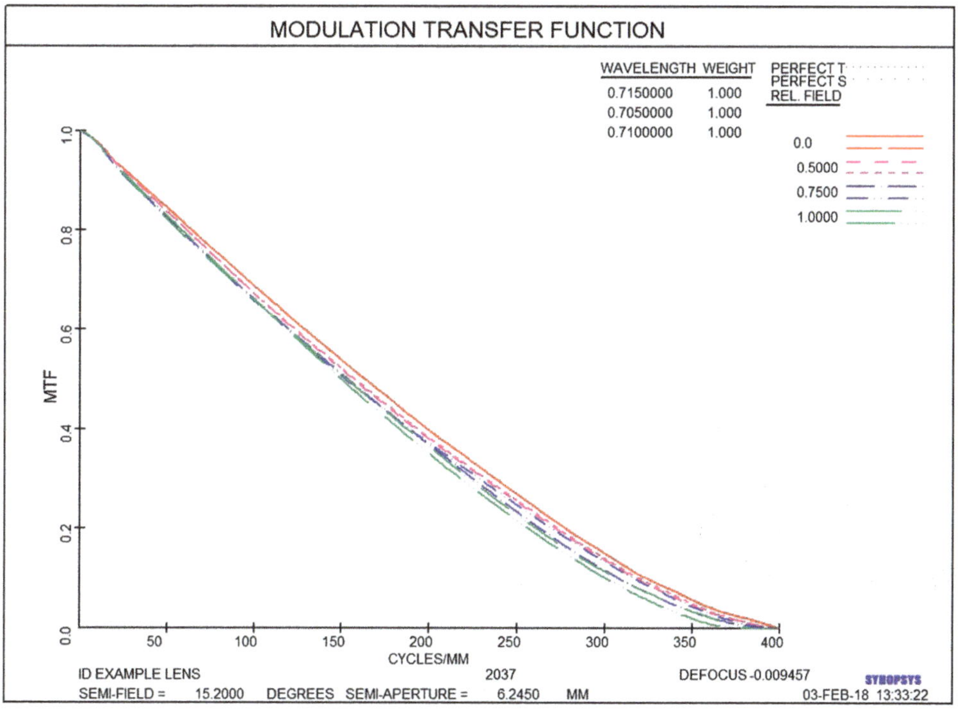

Figure 9.6. MTF of the redesigned lens.

The performance is much improved over the original, and tolerances will most likely be looser as a result.

We have not addressed the question of element thicknesses in this chapter, which is a task we often leave for later. That would be the next step in this design, because some elements are clearly too thin to be practical. Some of the chapters below also carry out that step.

Chapter 10

Third-order aberrations

Use and misuse of third-order aberrations; tolerance desensitization

Many students of lens design, and many managers who hire lens designers, are adamant that aberrations have to be very well controlled. They are partly right—but those requirements invariably refer to third-order aberrations, which should all be zero in the opinion of the manager. This is unwise. The point of this chapter is that third-order aberrations are in fact *not* very important—although they still do have some uses.

The reason they are not important is because most lenses also have higher-order aberrations, and all orders must be properly balanced. People without a deep knowledge of optics do not know that.

Let me illustrate why they are not important. Fetch the lens **C10L1.RLE**. This is a five-element lens with rather good correction, shown in figure 10.1.

Make an optimization MACro (**C10M1**) that will strongly control the third-order aberrations:

```
PANT
VLIST RAD ALL
VLIST TH ALL
VLIST GLM 1 3 6 8 9
END

AANT
M 1 1 A FNUM
M 7.8 1 A BACK
M 0 1 A DELF
M 0 1 A SA3
```

doi:10.1088/978-0-7503-1611-8ch10

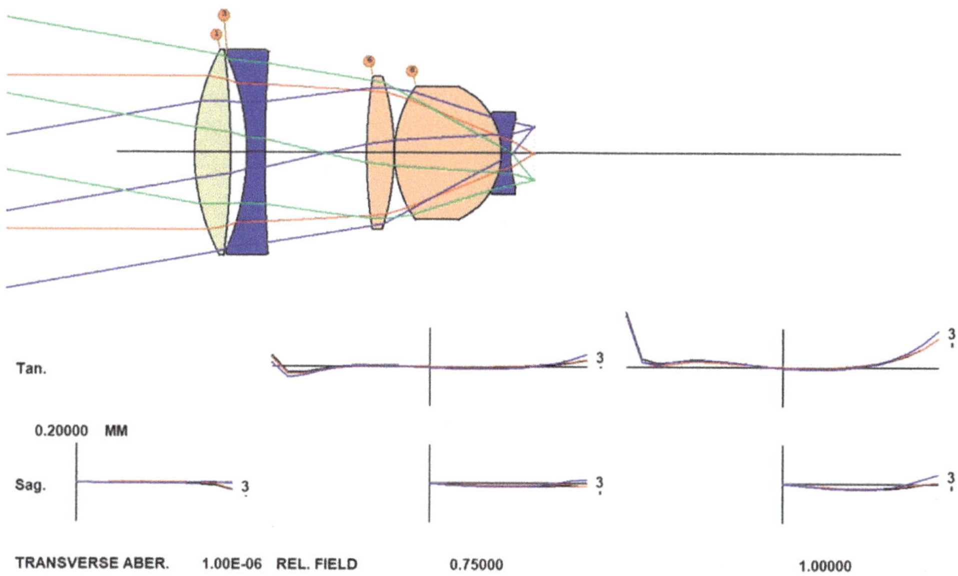

Figure 10.1. Lens with well-balanced aberrations.

```
M 0 1 A CO3
M 0 1 A TI3
M 0 1 A SI3
M 0 1 A PETZ
M 0 1 A DI3
M 0 1 A PAC
M 0 1 A SAC
M 0 1 A PLC
M 0 1 A SLC
END

SNAP
SYNO 30
```

This MACro will vary all the design variables and control the F/number, defocus, and back focus distance, all the while correcting the third-order aberrations to a target of zero. The input **VLIST RAD ALL** will vary all radii, and **VLIST TH ALL** will vary all thicknesses and airspaces, but we could not use the form **VLIST GLM ALL** in this case since that form will only vary those materials that *already* have a glass model, and in this example lens none of them have. So we have to declare the surfaces individually here.

Run this MACro and the lens is horrible, as shown in figure 10.2. The scale of the fan plots is 25 times larger.

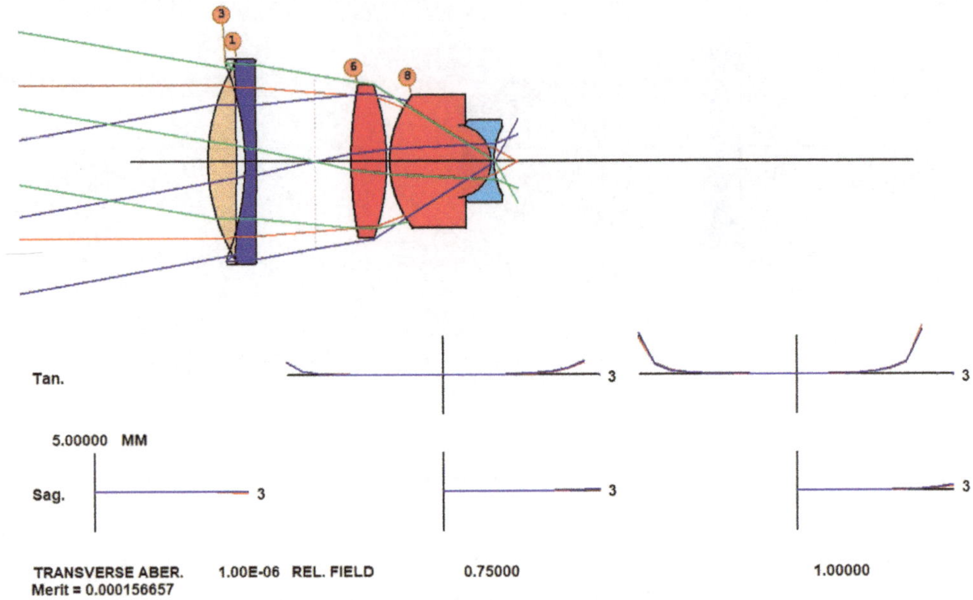

Figure 10.2. Result of correcting third-order aberrations to near zero.

What happened? Did the optimization fail? Ask for the third-order aberrations with the command **THIRD**:

```
SYNOPSYS AI>THIRD

 ID FIVE-ELEMENT LENS                              2146            07-FEB-18    09:05:39

 THIRD-ORDER ABERRATION ANALYSIS


  FOCAL LENGTH   ENT PUP SEMI-APER   GAUSS IMAGE HT
        50.804               25.400            8.958

  THIRD-ORDER ABERRATION SUMS
                 SPH ABERR        COMA  TAN ASTIG  SAG ASTIG     PETZVAL  DISTORTION
                     (SA3)       (CO3)      (TI3)      (SI3)      (PETZ)   (DI3(FR))
                 3.015E-05  -1.732E-05 -1.115E-05 -2.172E-05  -2.701E-05    -0.00040

  PARAXIAL CHROMATIC ABERRATION SUMS
                  AX COLOR   LAT COLOR  SECDRY AX  SECDRY LAT
                     (PAC)       (PLC)      (SAC)       (SLC)
                  -0.00320  -2.982E-05    0.01209     0.00041
  SYNOPSYS AI>
```

Indeed. Those aberrations are very small. How about the starting lens?

```
ID FIVE-ELEMENT LENS

THIRD-ORDER ABERRATION ANALYSIS

FOCAL LENGTH   ENT PUP SEMI-APER   GAUSS IMAGE HT
      50.799              25.400            8.957

THIRD-ORDER ABERRATION SUMS
             SPH ABERR        COMA   TAN ASTIG   SAG ASTIG     PETZVAL DISTORTION
                 (SA3)       (CO3)       (TI3)       (SI3)      (PETZ)   (DI3(FR))
              -0.01812    -0.03732    -0.04232    -0.08743    -0.10998    -0.01754

PARAXIAL CHROMATIC ABERRATION SUMS
              AX COLOR   LAT COLOR  SECDRY AX  SECDRY LAT
                 (PAC)       (PLC)       (SAC)       (SLC)
               0.02653     0.01293     0.01665     0.00407
```

Wow! Those aberrations are much larger, and the larger third-order aberrations gave a better lens. Lesson learned: do not try to out-guess the program when it comes to aberration balancing.

Let me repeat, when you design a lens, you are usually only concerned with two things: **is the image sharp, and is it in the right place?** If people start talking about aberrations, be polite and smile.

10.1 Tolerance desensitization

We mentioned earlier, however, that those aberrations still have a use. The most important of these deals with *tolerance desensitization*. This is because, when lenses are improperly manufactured, it is the third-order aberrations that change the fastest. To help you keep those aberrations small, the program has a set of eight quantities that can be put into the AANT file. If these are small, tolerances tend to be looser. These quantities are SAT, COT, ACD, ACT, ECD, ECT, ESA, and ECO, and are described in table 10.1.

Table 10.1. Definition of eight third-order quantities that can be controlled.

SAT	The sum of the squares of the surface contributions to spherical aberration, SA3.
COT	The sum of the squares of the surface contributions to coma, CO3.
ACD	The sum of the squares of the amount by which CO3 varies as each surface is decentered.
ACT	The sum of the squares of the amount by which CO3 varies as each surface is tilted.
ECD	The sum of the squares of the amount by which CO3 varies as each element is decentered.
ECT	The sum of the squares of the amount by which CO3 varies as each element is tilted.
ESA	The sum of the squares of the element contributions to spherical aberration, SA3.
ECO	The sum of the squares of the element contributions to coma, CO3.

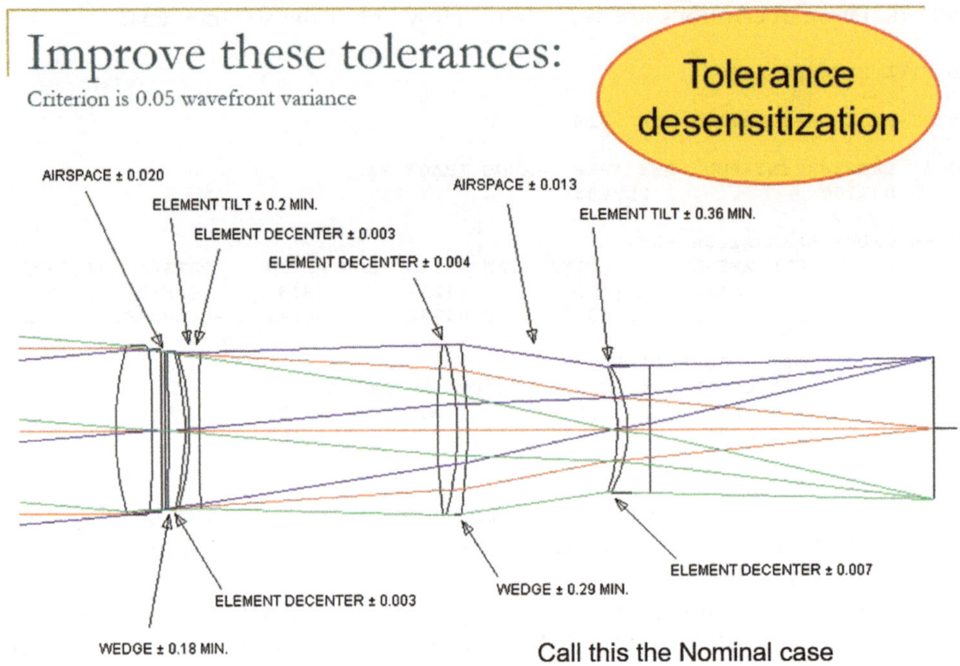

Figure 10.3. Lens with tolerance sensitivity problems.

Table 10.2. Tolerance sensitivity reduction resulting from controlling some third-order quantities.

	3 TH	6 wedge	7 tilt	5 YDC	7 YDC	9 YDC	12 YDC
Nominal	0.034	0.23 min	0.24 min	0.0042	0.0034	0.0053	0.0086
Case A	0.091	0.67	0.42	0.011	0.009	0.011	0.011
Case B	0.112	0.87	0.89	0.015	0.018	0.025	0.014

Here is an example of how you might use these aberrations to loosen the lens tolerances. The lens shown in figure 10.3 (**C10L2**) was optimized, and then a tolerance budget was created by BTOL with a target wavefront quality of 0.05. (BTOL is discussed in more depth in chapter 13.)

Some of the tolerances came back very tight, as shown in table 10.2, in which the nominal data are for this lens. (YDC means Y-decenter.)

It would be expensive indeed to hold lens positions to these tight values. (Look at the centration tolerance on surface 7: 3.4 μm!) So proceed as follows:

1. Run the command **`THIRD SENS`**, to see the current values of these parameters:

```
THIRD SENS
ID 8-ELEMENT TELEPHOTO

NORMALIZED 3RD-ORDER ANALYSIS OF TOLERANCE SENSITIVITY

SS OF SA3 BY SURFACE (SAT) =              85.107903
SS OF CO3 BY SURFACE (COT) =              21.404938
SS OF CO3/YDC BY SURFACE (ACD) =           0.007657
SS OF CO3/TILT BY SURFACE (ACT) =         73.889722
SS OF CO3/YDC BY ELEMENT (ECD)  =          0.003941
SS OF CO3/TILT BY ELEMENT (ECT) =         31.259708
SS OF SA3 BY ELEMENT (ESA) =               1.944190
SS OF CO3 BY ELEMENT (ECO) =               0.492351
```

2. Since we are mainly concerned about centration errors, we might try to reduce the value of ECD, the change in CO3 when an element is decentered. To do so, we add to the AANT file (in **C10M2**) the line

```
M .001 100 A ECD
```

Since ECD is already a small number (compared to the others in the list) we give it a high weight so it makes a difference to the merit function. Keep in mind that you cannot simply target all of these values to zero, since lens elements in general cannot be designed without any aberrations and still have any optical power. Also, these quantities are coupled in obscure ways. If, for example, you reduce the value of SAT, you will probably find that COT also became smaller. You could not give an independent value to both of them and expect that the program could find such a combination. So it is wise to proceed with one at a time until you find the parameter that works best with your lens. In this example, by controlling the value of ECD, then optimizing and annealing, we obtain the lens shown in figure 10.4:

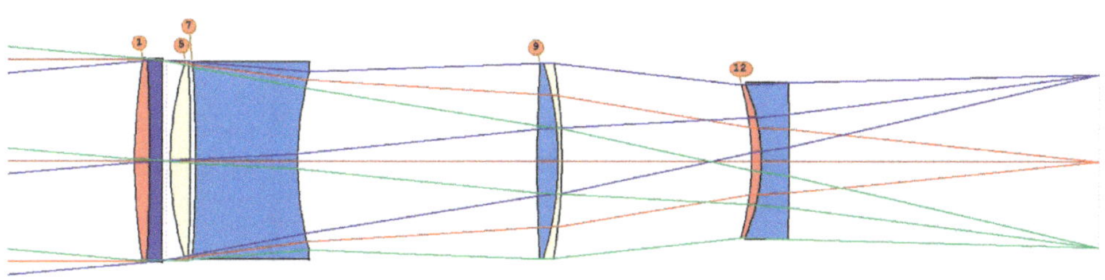

Figure 10.4. Lens with ECD reduced.

```
THIRD SENS

 ID 8-ELEMENT TELEPHOTO                    1166                24-SEP-17
14:38:04

 NORMALIZED 3RD-ORDER ANALYSIS OF TOLERANCE SENSITIVITY

 SS OF SA3 BY SURFACE (SAT) =                  7.039158
 SS OF CO3 BY SURFACE (COT) =                  4.886167
 SS OF CO3/YDC BY SURFACE (ACD) =              0.001647
 SS OF CO3/TILT BY SURFACE (ACT) =            19.595239
 SS OF CO3/YDC BY ELEMENT (ECD)   =            0.001064
 SS OF CO3/TILT BY ELEMENT (ECT) =             8.608645
 SS OF SA3 BY ELEMENT (ESA) =                  0.185937
 SS OF CO3 BY ELEMENT (ECO) =                  0.127815
```

Notice how all of the values returned by **THIRD SENS** have changed, even though we only targeted ECD. Tolerances of this lens are listed as Case A in the above table. Clearly, the tolerances are much looser now, although still a challenge for the shop. Let us experiment some more. This time we will target the value of ACT to the value 7.0, which is 1/10 of the nominal value, and remove the target for ECD:

```
M 7 1 A ACT
```

Optimize and anneal, and the lens now looks as shown in figure 10.5.

The tolerances are listed as Case B, above. This looks like a better budget. (We ignore for this exercise the issue of manufacturability: some elements are much too thin and should be controlled with the ACM or ADT monitors.)

The quantity that you elect to control depends on which tolerances you want to affect. Airspace tolerances, for example, may respond to control over the quantity ESA. Lens thickness tolerances, on the other hand, may respond better to SAT. You will have to understand your lens, and experiment with these tools, to find the best targets and the best BTOL budget.

Sometimes the effect of these quantities is to *increase* the value of the merit function. Normally that would not be a good idea, since if the image becomes worse,

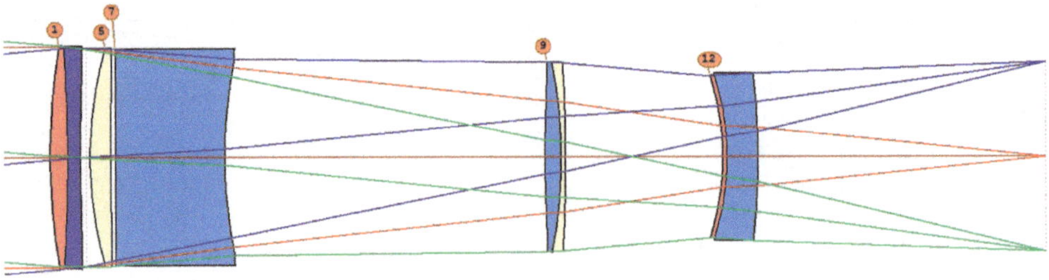

Figure 10.5. Lens with ACT controlled.

tolerances generally become tighter. However, the relaxing effect of the tools in this lesson can sometimes outweigh that effect, giving tolerances that are looser anyway. This only works up to a point, of course, and if the merit function becomes too large, you should require a less-demanding value of that parameter in your AANT file.

There is no guarantee that any of these aberration targets will work in any given case, but experience has shown that they are certainly worth a try. Your tolerances may be relaxed by a factor of from two to ten.

There is yet another way to control the sensitivity of an individual element: with the SECTION aberrations. Whereas the quantities discussed above apply to all surfaces or elements and are therefore very easy to use, the SECTION aberrations apply only to the surface range you specify. If an element is assigned a very tight centration tolerance, even after you try the targets given in this section—which can happen if some tolerances become much looser but the problem element becomes tighter—you might control only the coma or spherical aberration of *that* element. That gives you precise control over the aberrations where you need it, and is sometimes worth the extra step. For example, if the element at surfaces 13 and 14 was very sensitive, you might control the spherical aberration of that element with

```
M 0 .1 A SECTION SA3 13 14
```

and experiment with the target and weights until you obtain the best results. This request will minimize SA3 for the section from surfaces 13 to 14, and its alignment may then be somewhat less sensitive.

Chapter 11

The in and out of vignetting

Pupil definition; reduction due to apertures; adjusting ray targets for reduced pupil

By ‘vignetting’ (pronounced *vin-YETTing*) we mean any property of a lens that blocks some of the rays that would otherwise pass through the stop; it is a topic that different programs handle in different ways. Of course, one would often prefer that the beam size remain constant everywhere in the field, since the effective transmission then does not fall off with field on that account. However, sometimes the best tradeoff is to accept some vignetting in order to avoid the cost and weight of a more complex lens. In such cases one must know how to manage the varying beam size during optimization and how to set up the lens apertures to model that amount of vignetting for image analysis when the design is finished. That is the subject of this chapter.

Here is a sample lens, in figure 11.1, a triplet with substantial vignetting (**C11L1**).

Notice how the size of the beam at the upper and lower field points (in blue and green) is much smaller than the on-axis beam (in red). A look at the RLE file for this lens shows a real stop on surface 3 (to activate ray aiming for the chief ray) and wide-angle pupil option 3 in effect (WAP 3)—type **LE** to see this file:

```
RLE
ID COOKE TRIPLET F/4.5                          747
 WAVL .6562700 .5875600 .4861300
 APS                -3
 WAP                  3
 UNITS MM
 OBB  0.000000      20.00000     5.55500
   0 AIR
   1 CAO       4.69068139       0.00000000        0.00000000
   1 RAD      21.4939500000000    TH       2.00000000
   1 N1 1.61726800 N2 1.62040602 N3 1.62755182
   1 CTE    0.630000E-05
```

doi:10.1088/978-0-7503-1611-8ch11

```
  1 GTB S     'SK16                 '
  2 CAO        4.25560632        0.00000000        0.00000000
  2 RAD    -124.0387000000000    TH        5.25509000 AIR
  3 CAO        3.19251725        0.00000000        0.00000000
  3 RAD     -19.1051800000000    TH        1.25000000
  3 N1 1.61163844 N2 1.61658424 N3 1.62846980
  3 CTE    0.830000E-05
  3 GTB S     'F4                   '
  4 CAO        3.15978037        0.00000000        0.00000000
  4 RAD      21.9794700000000    TH        4.93473000 AIR
  5 CAO        3.48158127        0.00000000        0.00000000
  5 RAD     328.3317499999989    TH        2.25000000
  5 N1 1.61726800 N2 1.62040602 N3 1.62755182
  5 CTE    0.630000E-05
  5 GID 'SK16                 '
  5 PIN     1
  6 CAO        4.00000022        0.00000000        0.00000000
  6 RAD     -16.7537700000000    TH       43.24303731 AIR
  6 TH      43.24303731
  6 YMT        0.00000000
  7 CV       0.0000000000000    TH        0.00000000 AIR
END
```

The WAP 3 option adjusts the entrance pupil dimensions so that the marginal rays at each field point just clear all defined lens apertures. Every surface except the image (at surface 7) has been assigned a hard[1] clear aperture (with CAO data). That is one

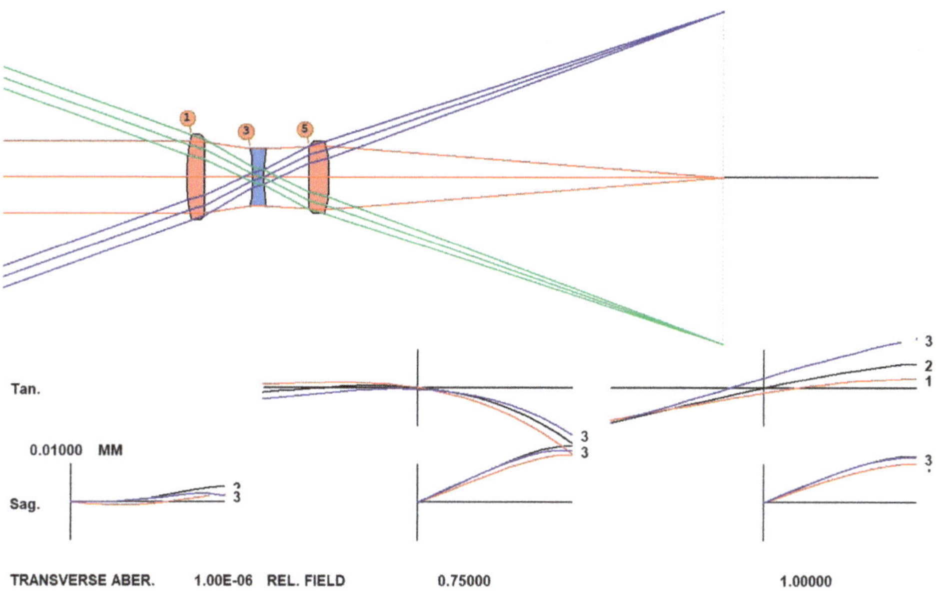

Figure 11.1. Triplet illustrating vignetting.

[1] A 'hard' aperture is one entered by the user, while a soft aperture is calculated by the program. Only the former can vignette rays.

way to implement a desired amount of vignetting. Note how the size of the entering beam changes with field angle. This is most easily shown with the 'PAD Scan' button  as seen in figure 11.2.

However, the WAP 3 option is not the only way, and often is not the best way, to deal with vignetting. During optimization, the size of the beam can change at each

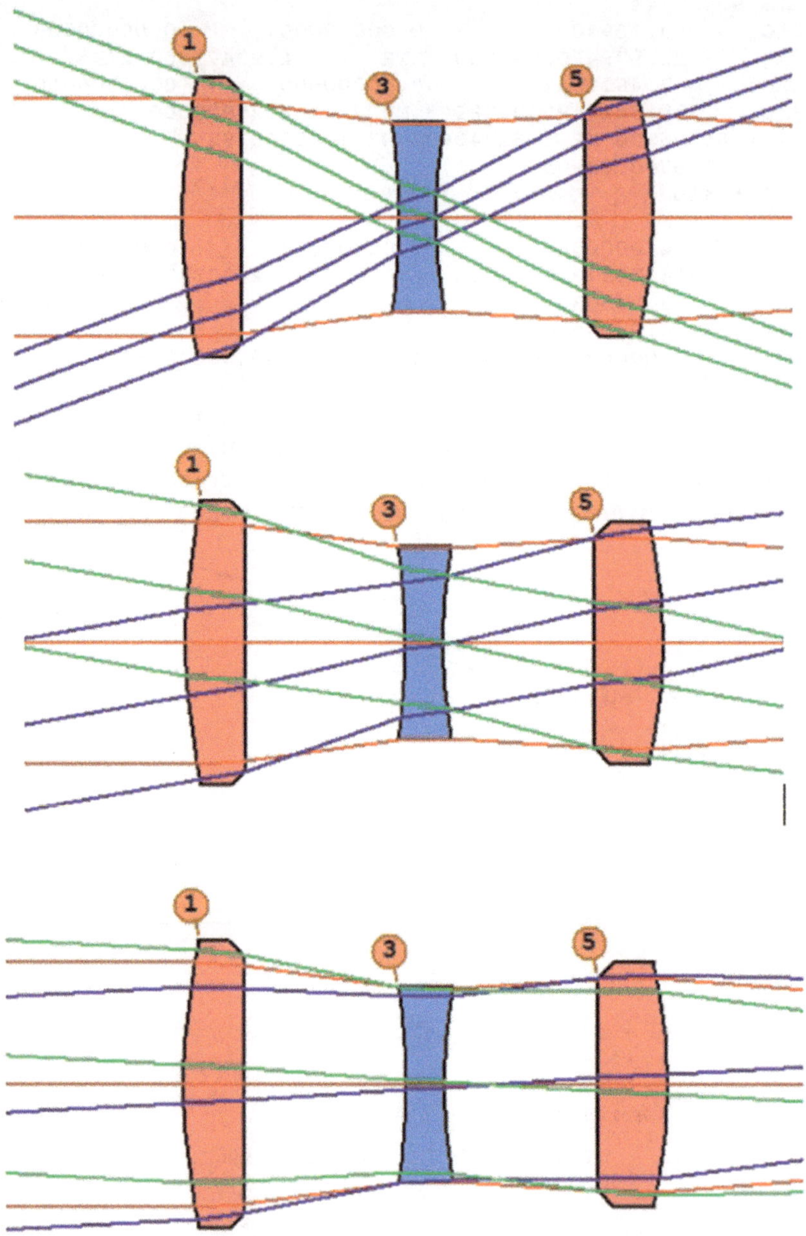

Figure 11.2. The size of the entering beam varies with field, as shown by the 'PAD Scan' button.

surface as the lens changes, and it makes no sense to assign a hard CAO to a surface when you do not even know what that size will be when you are finished. So never use the WAP 3 option during optimization and use it afterwards only when necessary.

Instead, approach vignetting in stages. To illustrate, let us first remove all of the CAOs and the WAP declaration as well:

```
CHG
CFREE
WAP 0
END
```

The image is awful, as shown in figure 11.3. Maybe that is why the starting design used those options. Perhaps it made sense under the original conditions. What can we do with this lens?

A look at the current options (with the command **POP**) shows a YMT solve on 6 but no curvature solve. Let us add one. The lens works at F/4.5 paraxially, and the value of a UMC solve is therefore 0.5/4.5 or –0.1111, the minus sign because the marginal ray is going down at the image. Change the lens and store a copy in the lens library for future reference:

```
CHG
6 UMC -.1111
END
STORE 3
```

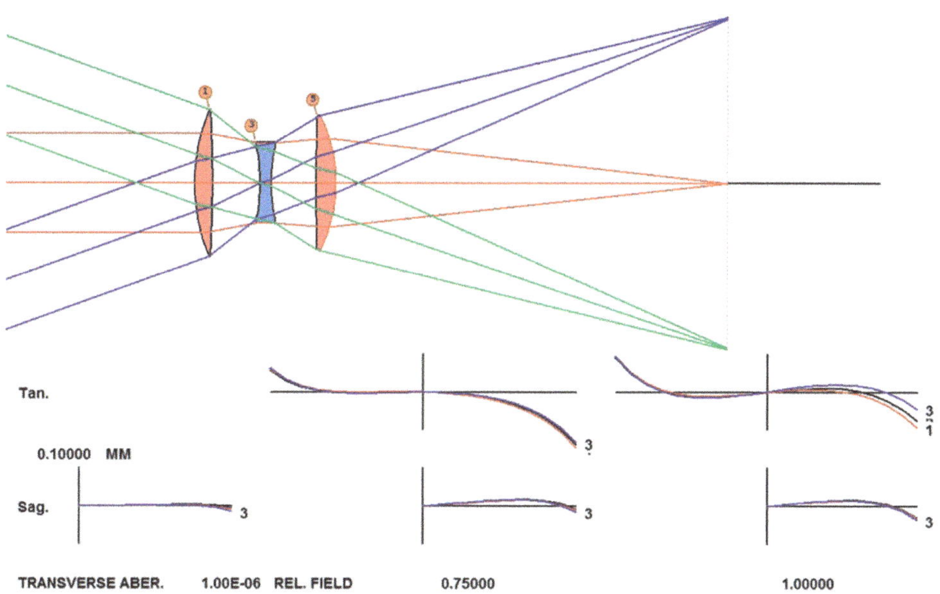

Figure 11.3. Triplet with default apertures and no vignetting, making a poorer image.

Now open a new editor with the command **AEE** and create an optimization MACro:

```
LOG
PANT
VLIST RAD ALL
VLIST TH ALL
END

AANT
AEC
ACC
GSR .5 10 5 M 0
GNR .5 2 3 M .7
GNR .5 1 3 M 1
END

SNAP
SYNO 30
```

Here we have used merit function 6 from the '**Ready-Made Raysets**' button, .

Make a checkpoint and run this MACro. The result is shown in figure 11.4.

The result is not too good. The aberrations are out of control, especially at full field. We have to do better. Tell the optimization program to vignette the beam size

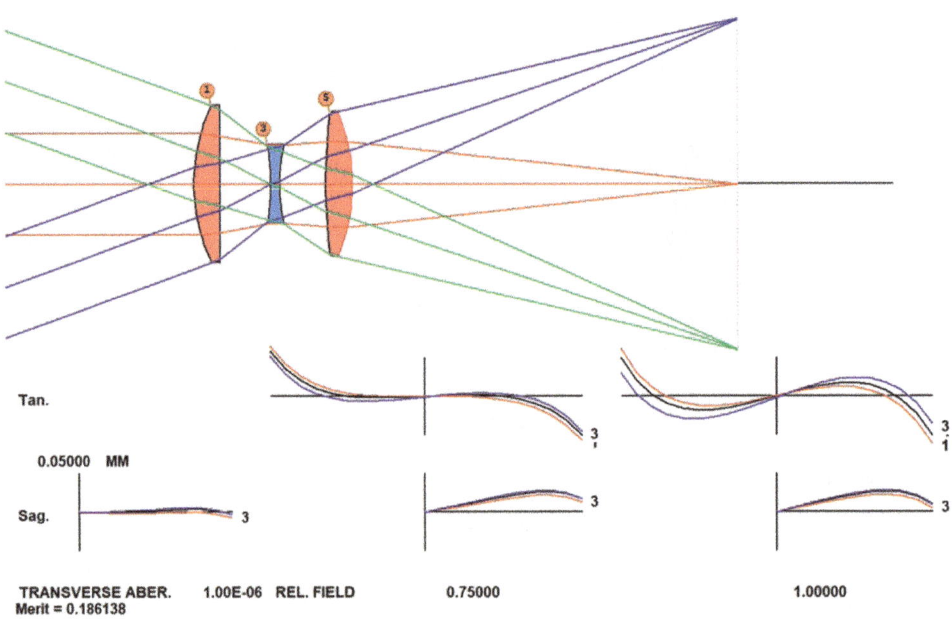

Figure 11.4. Triplet reoptimized to eliminate feathered edges.

to 40% of the on-axis value at full field. This is done by adding to the AANT file a **VSET** directive:

```
AANT
AEC
ACC
VSET .4
GSR .5 10 5 M 0
GNR .5 2 3 M .7
GNR .5 1 3 M 1
END
```

Now run it again. The result is in figure 11.5.

The *edges* of the TFAN became worse, which is not surprising, since they are not corrected anymore. However, if we were to vignette this beam at 40% of the size shown in the PAD display, the image would look much better. Let us assume this is the solution we are after. Now we have to model the elements so that amount of vignetting actually occurs.

That is easy to do. Open the WorkSheet (**WS**), enter into the edit pane the directive **CFIX**, and click 'Update'. All surfaces are now assigned a hard CAO, at the same aperture as the default CAO that was currently in effect. Now, click on surface 6 in the lens drawing. The data for that surface are shown in the edit pane. Select the CAO radius with the mouse and then click the '**SEL**' button. That assigns the top slider to that aperture radius. Move the slider thumb to the left, reducing the

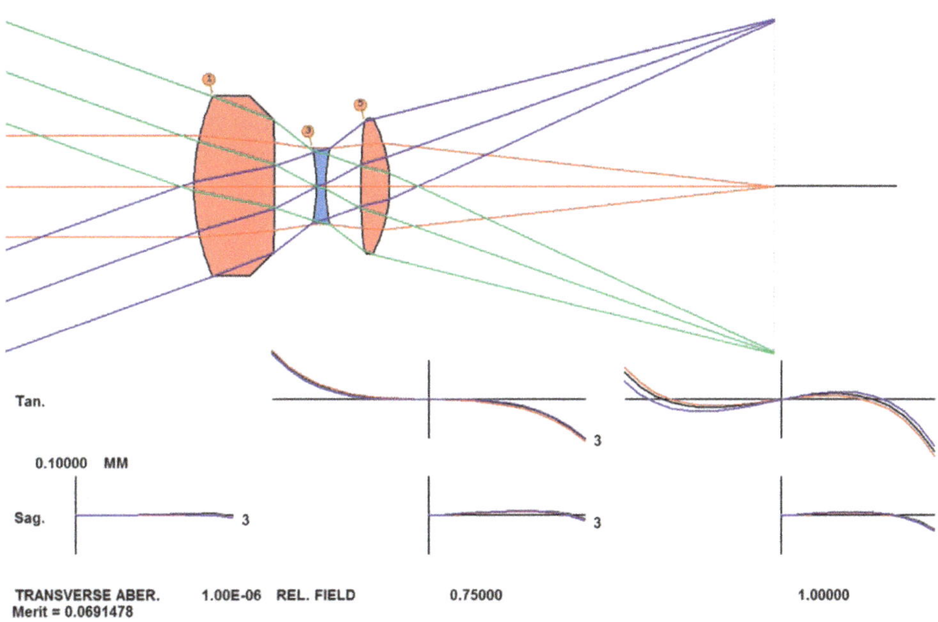

Figure 11.5. Triplet reoptimized with anticipated vignetting to 40% of the aperture.

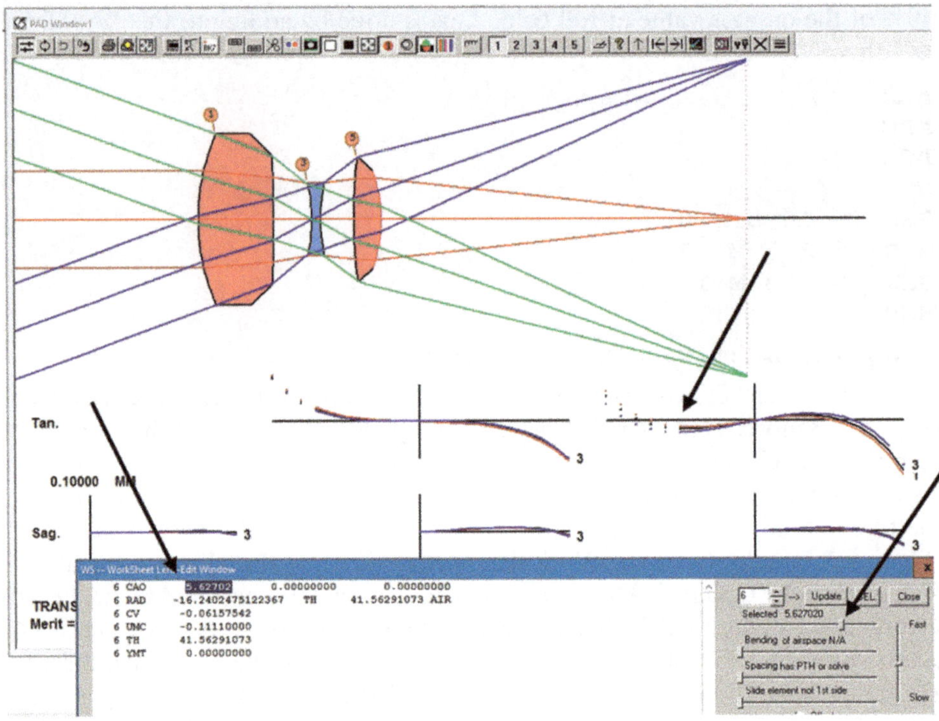

Figure 11.6. Lens with aperture on 6 adjusted to yield the desired vignetting on the left of the TFAN.

aperture. Watch the TFAN at full field and stop when the unvignetted portion on the left side appears to be about at the 40% position, as shown in figure 11.6.

This is about where the beam should be vignetted on that side. Do the same at surface 1. Now the beam is vignetted on both sides, shown in figure 11.7.

Why does the drawing on top of the PAD display still show the original, unvignetted beam? Well, there is an option to change that too, which you can activate by turning off mode switch 65. However, since that makes the picture look like what would happen with the WAP 3 option turned on—and it is *not* turned on at the moment—we prefer to leave this switch turned on so we do not become confused later.

We are almost done. We can activate WAP 3 by adding that directive in the edit pane. That is one way to proceed.

However, here is another: declare a set of **VFIELD** parameters. Close WS and type into the CW:

```
FVF 0 .5 .8 .9 1
```

The program calculates the vignetting factors that just clear the apertures we have entered, at five points in the field. Now the display shows the vignetted beam as it should, shown in figure 11.8.

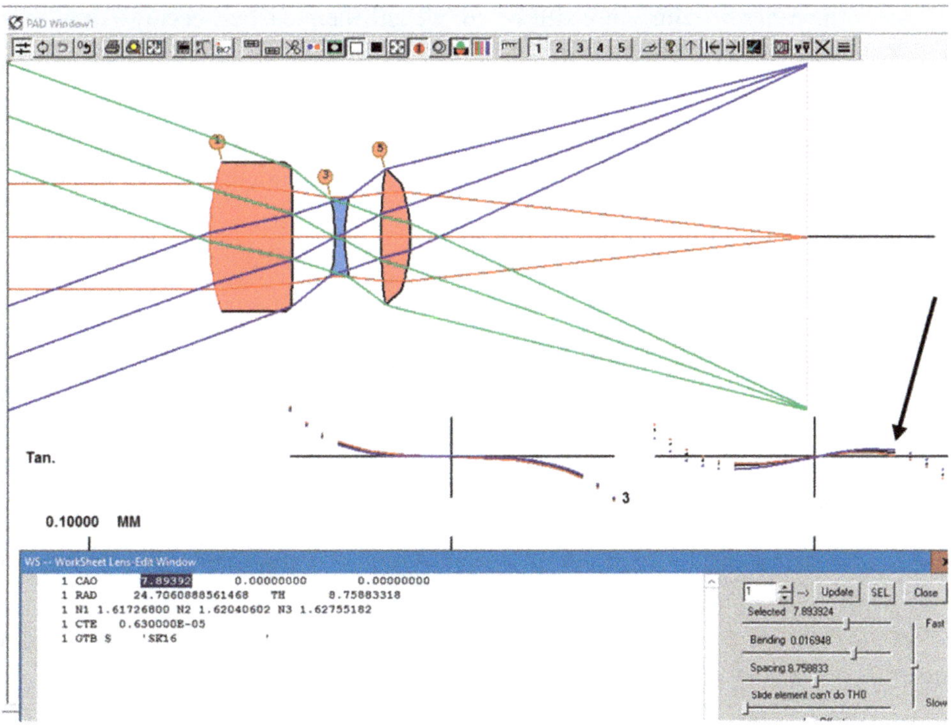

Figure 11.7. Lens with aperture on 1 adjusted for vignetting on the right side of the TFAN.

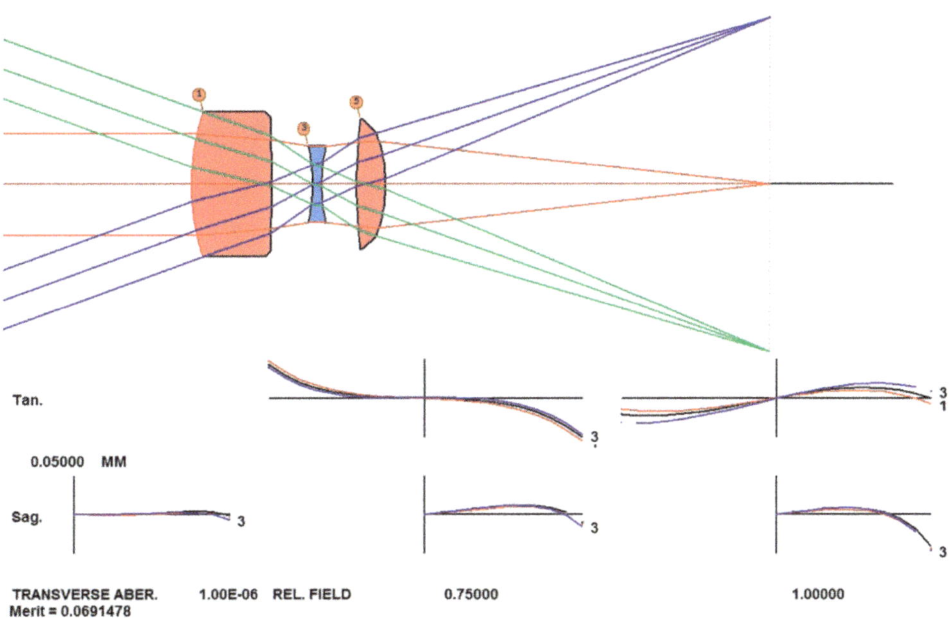

Figure 11.8. Vignetting accounted for via reduced apertures and VFIELD.

Except the other apertures are still where we left them earlier, because they are all hard apertures. In the WS edit pane, type **CFREE** and click 'Update'. Now the lens has default apertures again, this time calculated according to the VFIELD pupil, as shown in figure 11.9.

You can see the pattern. If you now go back to your optimization MACro, remove the VSET directive, and reoptimize, the edge-control monitor will look at

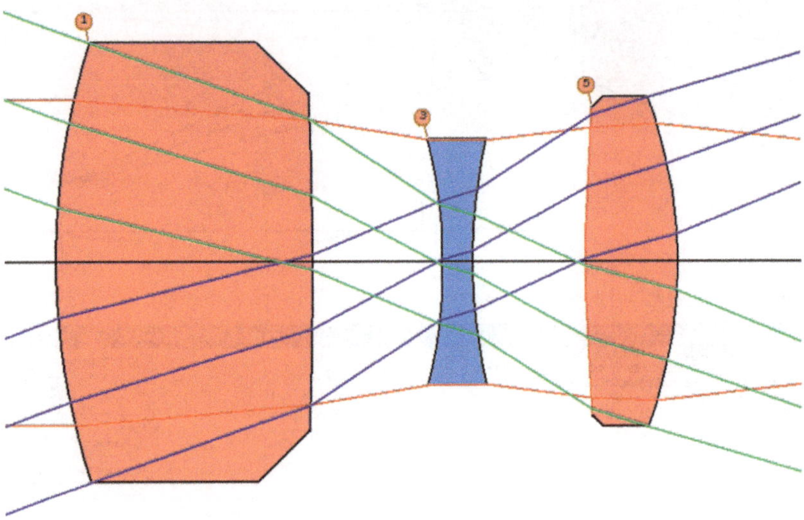

Figure 11.9. Default apertures assigned to honor the vignetting applied with VFIELD.

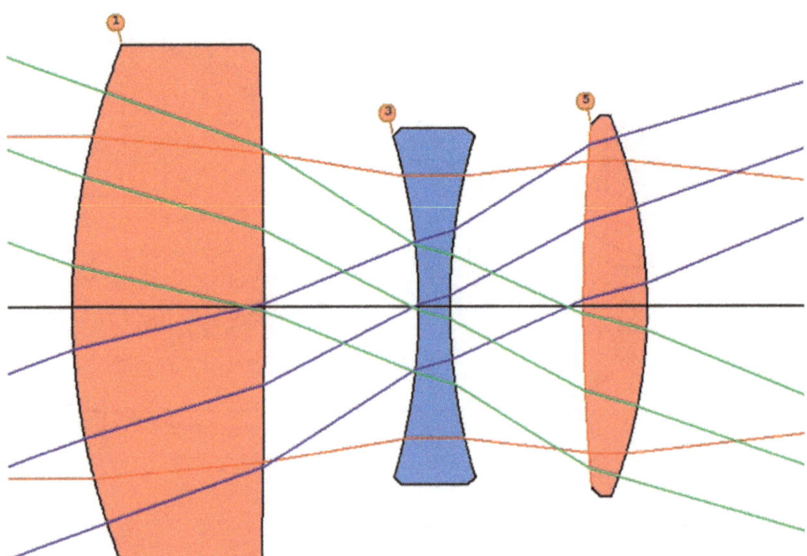

Figure 11.10. Final triplet with vignetting and apertures properly assigned. One must now design the lens cell so that vignetting occurs where the default CAOs were. The elements themselves are larger than this to accommodate the cell and retainer rings, which are usually used with lenses like this one.

the rays in the fans shown above, not the nominal rays. It will therefore reoptimize the lens, reducing the thicknesses if the merit function no longer benefits from the larger values. Then you can adjust the edge geometry with the Edge Wizard (**MEW**) just as you like them, as shown in figure 11.10 (**C11L2**).

What is the difference between the WAP 3 option and VFIELD, since they look much the same? I am glad you asked. WAP 3 requires aiming five rays every time you do something that requires ray tracing. That is a rather slow option. VFIELD, on the other hand, has already done that calculation and only needs to aim the chief ray afterwards, interpolating at the requested field, fast—and that is what vignetting is all about.

Chapter 12

The apochromat

Correcting a lens at three wavelengths

This chapter shows how to design a lens with better color correction than one can obtain with a simple doublet. A concise description of how one can proceed is given in Rutten and van Venrooij's book *Telescope Optics*. The gist of it is, one must use three different kinds of glass that satisfy certain properties. They may easily be selected by inspecting the glass-table display. To illustrate, we will start with a design using glass types N-SK4, N-KZFS4, and N-BALF10 from the Schott catalog. (Those glasses are sometimes recommended for the purpose.) Here is the starting lens file, shown in figure 12.1 (**C12L1**):

```
RLE
ID F10 APO
 WAVL .6500000 .5500000 .4500000
 APS              3
 UNITS INCH
 OBB  0.00    0.5     2.00000    -0.01194     0.00000     0.00000     2.00000
   0 AIR
   1 RAD   -300.4494760791975   TH      0.58187611
   1 N1 1.60978880 N2 1.61494395 N3 1.62386887
   1 GTB S    'N-SK4          '
   2 RAD     -7.4819193194388   TH      0.31629961 AIR
   2 AIR
   3 RAD     -6.8555018049530   TH      0.26355283
   3 N1 1.60953772 N2 1.61628830 N3 1.62823445
   3 GTB S    'N-KZFS4        '
   4 RAD      5.5272935517214   TH      0.04305983 AIR
   4 AIR
   5 RAD      5.6098999521052   TH      0.53300999
   5 N1 1.66610392 N2 1.67304720 N3 1.68543133
   5 GTB S    'N-BAF10        '
   6 RAD    -27.9819596092866   TH     39.24611007 AIR
   6 AIR
   6 CV      -0.03573731
```

doi:10.1088/978-0-7503-1611-8ch12

```
   6 UMC      -0.05000000
   6 TH       39.24611007
   6 YMT       0.00000000
   7 RAD     -11.2104527948015    TH        0.00000000 AIR
 END
```

Open this file, open PAD, and then click the '**GlassTable**' button and select the Schott table from the box that opens. The display is shown in figure 12.2.

This shows the Schott glass map, but it is not what we want for this exercise. Click the 'Graph' button, shown in figure 12.3.

Then select 'Plot P(F,e) vs. Ve', shown in figure 12.4.

The display changes, and now the abscissa is the V-number at the e line (0.546 07 μm) and the ordinate is the quantity $(N_F - N_e)/(N_F - N_C)$.

The theory of the apochromat says that you must select three glasses that *do not* lie on a straight line on this diagram. They must form a triangle, and the larger the area the better. The green circles in figure 12.5 show the present glasses in the triplet. They work quite well—but we can do somewhat better.

Click on the green circle with the number 1 beside it. That is the glass currently on surface 1, N-SK4. Now click the '**Properties**' button, to see the properties of this glass, shown in figure 12.6.

This glass is not all that stable: a humidity rating of 3 and an acid sensitivity of 5. Let us see if we can find a better glass for the first element. (This is exposed to the environment, so it is important.) Close the 'Properties' window and click the 'Graph' button again; then click the radio button for 'Acid Sensitivity' and 'OK'. Zoom in with the mouse wheel as needed near the green circle at N-SK4 so things become bigger, and then click the 'Full Name' button. You will see figure 12.7.

Now you see a red vertical line through the glass locations, showing the acid sensitivity. The line for glass N-SK4 is rather long, since that glass is not very

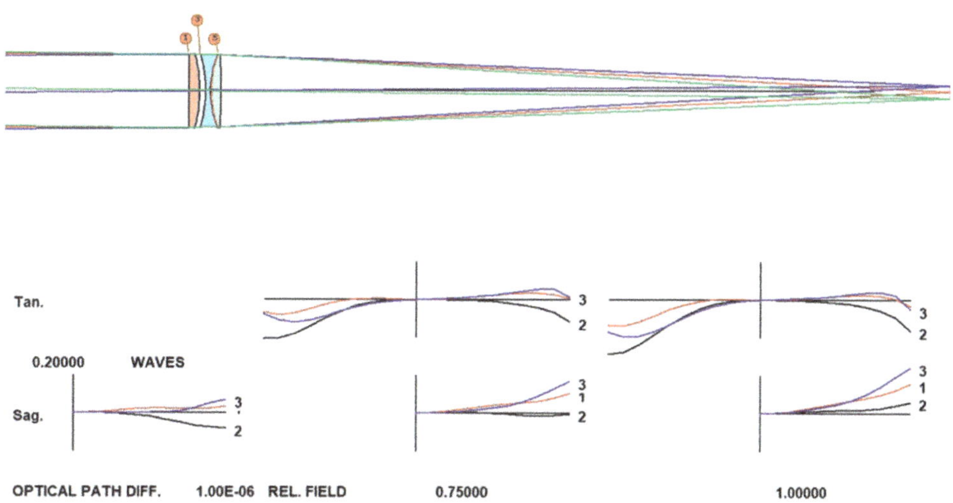

Figure 12.1. Starting point for apochromat design.

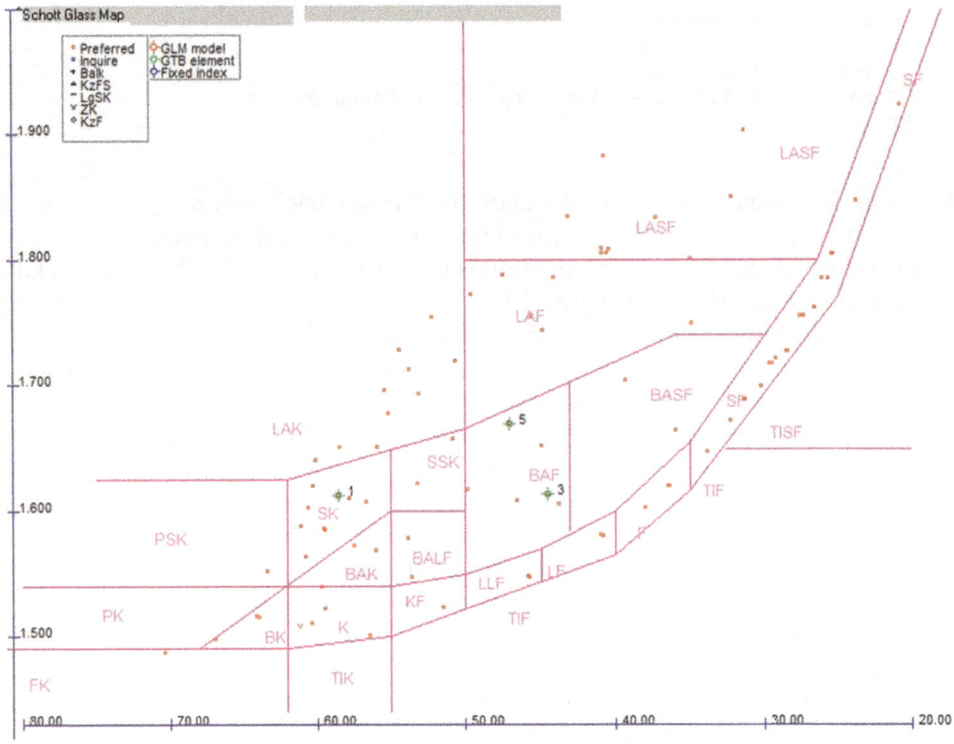

Figure 12.2. The glass map display, showing Nd and Vd.

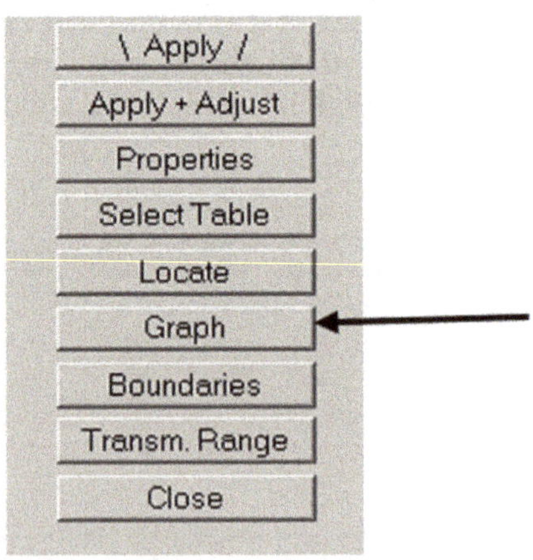

Figure 12.3. Button to show a selected graph in the glass map.

resistant. To the left you see N-BAK2, with no line at all (it is in the best category). Click that glass symbol, and when the name appears in the window on the right, as shown in figure 12.8, click the 'Properties' button again.

This glass has an acid rating of 1, better humidity tolerance, and a lower price as well. There is no reason we cannot use it instead of the previous N-SK4, since it

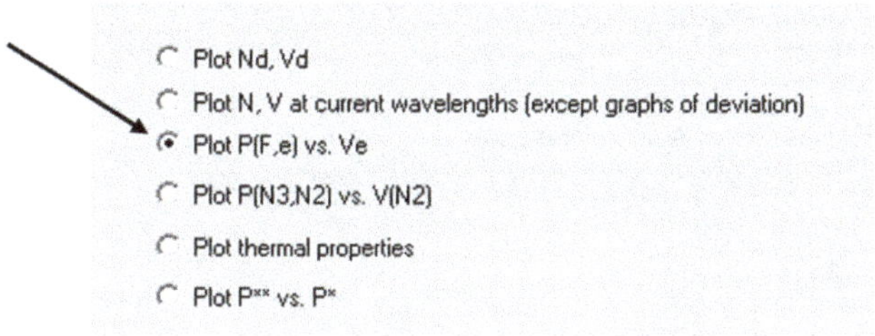

Figure 12.4. Selecting a graph of partial dispersions.

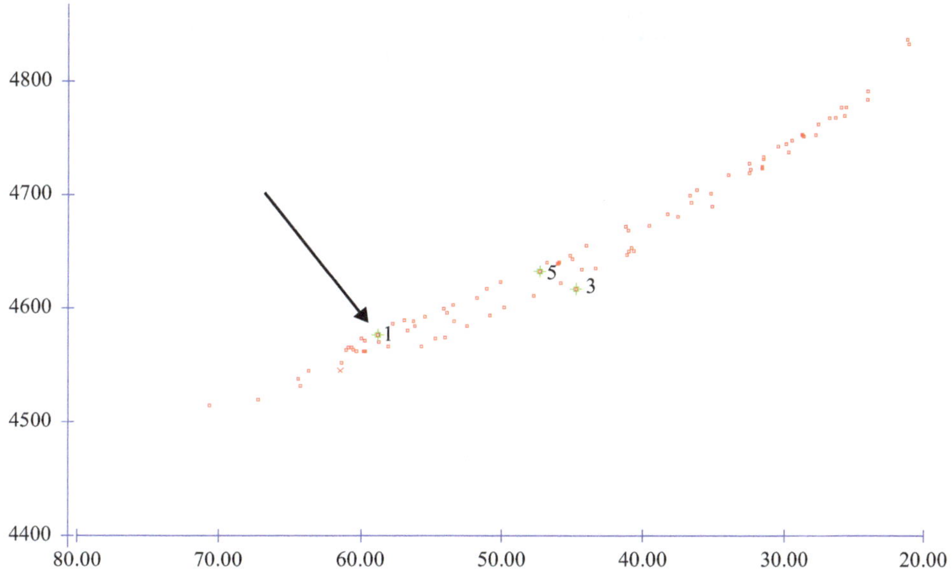

Figure 12.5. The glass map showing a plot of 'P(F,e) vs. Ve'.

makes just as good a triangle with the other glasses. Type the surface number 1 into the 'Surface' box and click '\Apply/'. Glass N-BAK2 is now assigned to surface 1.

Now clean up the display by deleting the names (click 'Spots Only'), and then click 'Graph' and select 'No Graph' and 'OK'. The triangle is just as nice as before.

Of course, the lens is not optimized for this glass, so we have to run the optimization program. Here is a MACro that will do the job (**C12M1**):

```
PANT
VLIST RAD 1 2 3 4 5 7
VLIST TH 2 4
END

AANT
AEC
```

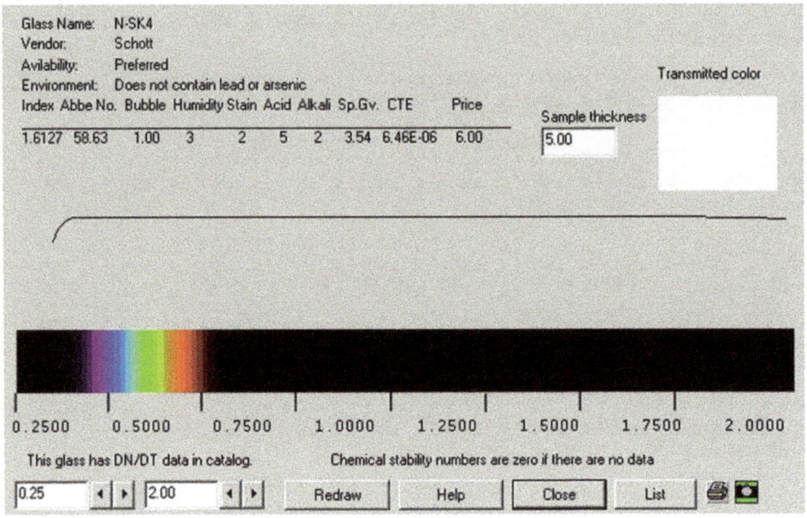

Figure 12.6. Properties of glass N-SK4.

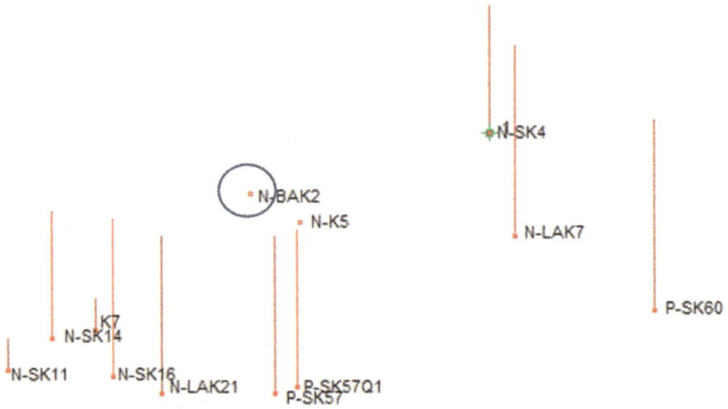

Figure 12.7. Glass map showing acid sensitivity. Glass N-BAK2 is a better choice than N-SK4.

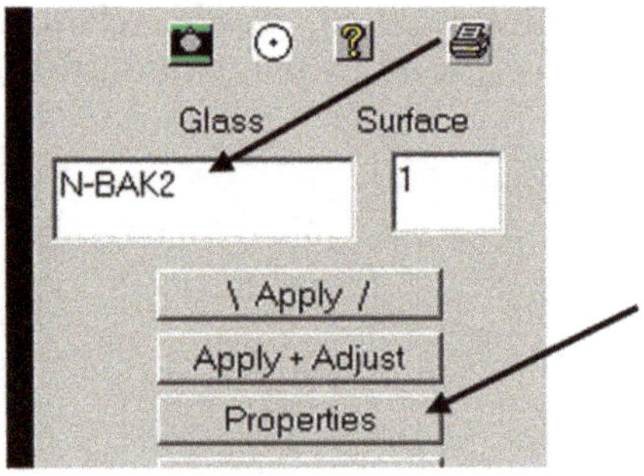

Figure 12.8. Applying a selected glass to surface 1.

```
ACC
GSO 0 1 4 M 0 0
GNO 0 .2 3 M .75 0
GNO 0 .1 3 M 1.0 0
END

SNAP
SYNO 30
```

Run this MACro, and now the correction is better than 1/4 wave on-axis. We have a better design that is cheaper to make, more resistant to the elements, and corrected over the range 0.45–0.65 μm. Here is the RLE file for that design, shown in figure 12.9 (**C12L2**):

```
RLE
ID F10 APO
 WAVL .6500000 .5500000 .4500000
 APS              3
 UNITS INCH
 OBB  0.000000  0.50000  2.00000 -0.00652  0.00000  0.00000  2.00000
   0 AIR
   1 RAD   -167.6807592628928   TH      0.58187611
   1 N1 1.53742490 N2 1.54188880 N3 1.54960358
   1 CTE   0.800000E-05
   1 GTB S    'N-BAK2          '
   2 RAD     -7.0647888938302   TH      0.36076391 AIR
   3 RAD     -6.5538674975636   TH      0.26355283
   3 N1 1.60953772 N2 1.61628830 N3 1.62823445
   3 CTE   0.730000E-05
   3 GTB S    'N-KZFS4         '
   4 RAD      5.3138288434095   TH      0.03937000 AIR
```

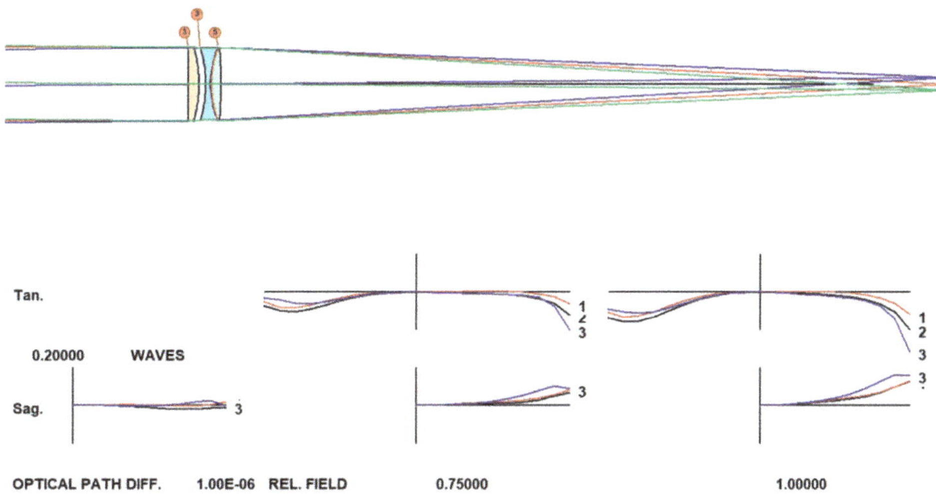

Figure 12.9. Lens reoptimized with improved glass selection.

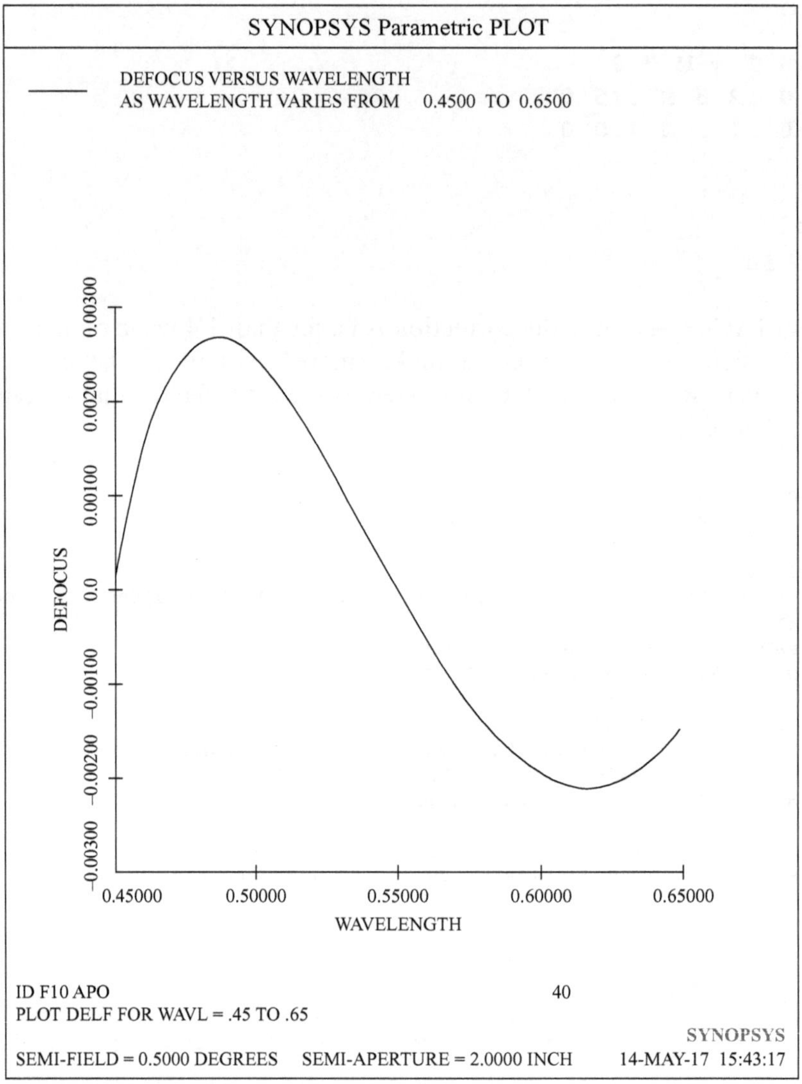

Figure 12.10. Color correction curve for the new design.

```
   5 RAD       5.4083667709938   TH       0.53300999
   5 N1 1.66610392 N2 1.67304720 N3 1.68543133
   5 CTE    0.618000E-05
   5 GTB S     'N-BAF10           '
   6 RAD     -19.4177738787113   TH      39.42904372 AIR
   6 CV       -0.05149921
   6 UMC      -0.05000000
   6 TH       39.42904372
   6 YMT       0.00000000
   7 RAD     -11.1931120564708   TH       0.00000000 AIR
 END
```

Figure 12.11. Diffraction image calculated by the Image Tools (MIT).

Let us see how the defocus varies with color in the new design:

```
CHG
NOP
END
PLOT DELF FOR WAVL = .45 TO .65
```

The result is in figure 12.10.

This analysis shows a defocus of about 0.0026 inches over the design range, and a perfect Airy disk, shown in figure 12.11. (The latter was calculated by Image Tools (**MIT**) with ten wavelengths assigned to the lens to make a good white in the center, and with the coherent effect.) The defocus is not zero because the program has balanced a small shift against the change of spherical aberration with wavelength. Both are small. How did we obtain ten wavelengths? With the Spectrum Wizard (**MSW**).

Note how we plotted **DELF** in this case instead of BACK, as we did previously. Why? Well, we eliminated all solves with the NOP directive, so the back focus is now fixed. However, the paraxial defocus, DELF, now changes with wavelength.

Chapter 13

Tolerancing the apochromatic objective

Tolerance budget calculation; fabrication adjustments

In chapter 12 you designed an apochromatic telescope objective with an extremely good image. In this lesson, you will calculate a tolerance budget for that lens.

Before you can send drawings of the lens elements to the shop, you have to figure out how large the manufacturing errors can be and still yield adequate performance. The result is called a *tolerance budget*. Since this example is a very well-corrected astronomical objective—the fussiest lenses you are ever likely to meet unless you design microlithography objectives—the budget will probably be very tight. That is exactly what we need for this exercise: most lenses are very easy to tolerance with SYNOPSYS, but that would not be an interesting lesson. Use lens **C12L2** from the previous lesson.

Before you can obtain tolerances, however, you have to remove the curvature solve on surface 6:

```
CHG
6 NCOP
END
```

The on-axis image is the most important for this kind of objective, which is often used for planetary observing, and there is some field curvature (which is why we varied the image surface radius in the previous lesson) and some astigmatism, which you cannot correct with a compact lens group at the stop.

The principle behind a tolerance budget is that, if you make all the elements according to that budget, holding radii, thicknesses, wedge, and so on to the tolerances assigned to those parameters, and then assemble the lens while holding all airspaces, and element tilts and decenters to their tolerances, the lens will then yield the requested image quality to the requested statistical confidence level. If that level

doi:10.1088/978-0-7503-1611-8ch13

is, say one sigma, then the quality will be equal or better than requested for 84.27% of a large batch of lenses. If the criterion is two sigma, that confidence rises to 99.53%, and so on.

First, try a simple **BTOL** evaluation. BTOL has many options, but there is a menu for simple cases: **MSB**, for **M**enu, **S**imple **B**TOL. Type **MSB** in the Command Window, and then fill in the boxes as follows. Most of them are already filled in for you; but select the 'TOLERANCE' and 'WAVE' radio buttons (instead of 'DEGRADE SPOT') and click the 'Prepare MC' box to select that option. Everything else can be left as it is, as shown in figure 13.1. Then click the '**GO**' button. (The logic behind BTOL is explained in appendix C.)

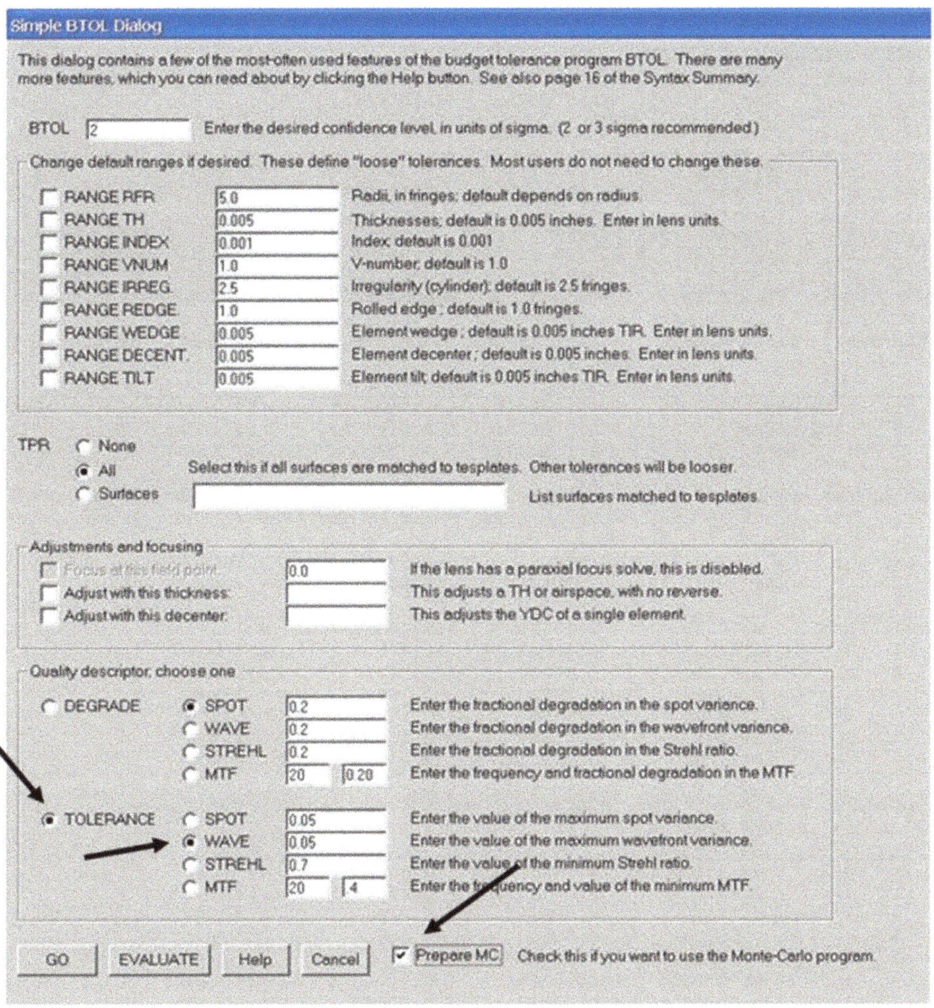

Figure 13.1. Input selection for BTOL via MSB.

When the calculations are finished, look up a little from the bottom line in the CW and you will see the expected performance:

SUMMARY OF OPTICAL PERFORMANCE

REL. Y-HEIGHT (HBAR)	REL. X-HEIGHT (GBAR)	ANTICIPATED STATISTICS OF QUALITY DESCRIPTOR MEAN VALUE	MULTIPLE DEV	EXPECTED	ZOOM
0.000	0.000	0.03179	0.01608	0.04787	1
1.000	0.000	0.03614	0.01423	0.05037	1
0.500	0.000	0.03219	0.01586	0.04805	1

This says that the on-axis image will get a variance of 0.048, as specified. The edge of the field is slightly over that value, but for this lens we do not care much about it. What does the budget look like? Scroll the display up a bit more until you come to the budget results:

BUDGET TOLERANCE ANALYSIS

EL.	SURF	RADIUS	RADIUS TOLERANCE (RADIUS)	(FRINGES)	THICKNESS	-----B----- THICKNESS TOL
1	1	-167.68076	0.08133	0.46749	0.58188	0.00497
1	2	-7.06479	1.37118E-04	0.44382	0.36076	3.11072E-04
2	3	-6.55387	1.14532E-04	0.40527	0.26355	0.00266
2	4	5.31383	6.90004E-05	0.38095	0.03937	1.28392E-04
3	5	5.40837	6.69742E-05	0.36138	0.53301	0.00443
3	6	-19.41777	0.00096	0.40367	39.42904	0.00143
	7	-11.19311	0.00000	0.00000	0.00000	0.00000

ELE	SURF	GLASS NAME	BASE INDEX	INDEX TOL	V-NUMBER	V-NUMBER TOL
1	1	N-BAK2	1.53996d	2.31392E-05	59.70771d	0.09660
2	3	N-KZFS4	1.61336d	1.27816E-05	44.49298d	0.03322
3	5	N-BAF10	1.67003d	1.61258E-05	47.11137d	0.04233

Note: The symbol "d" indicates that the quantity is estimated at 0.58756 uM.
The symbol "F" indicates that the quantity is taken at the primary color.

ELE	SURF	WEDGE TOLERANCE (ARC MIN)	(TIR)	IRREG. TOL (FRINGES)	ROLLED EDGE TOL (FRINGES)
1	1	0.00000	0.00000	0.19958	0.16399
1	2	0.27535	0.00032	0.19756	0.16039
2	3	0.00000	0.00000	0.18412	0.14486
2	4	0.14458	0.00017	0.18073	0.13989
3	5	0.00000	0.00000	0.17287	0.13249
3	6	0.32271	0.00037	0.17751	0.13841
	7	10.97289	0.00223	0.00000	0.00000

ELE	SURF	ELEMENT TILT TOLERANCE (ARC MIN)	(TIR)	Y-DECENT TOL	X-DECENT TOL
1	1	0.34332	0.00040	0.00041	0.00000
1	2	0.00000	0.00000	0.00000	0.00000
2	3	0.18489	0.00021	0.00015	0.00000
2	4	0.00000	0.00000	0.00000	0.00000
3	5	0.15280	0.00018	0.00017	0.00000
3	6	0.00000	0.00000	0.00000	0.00000
	7	0.00000	0.00000	0.00000	0.00000

This is scary. The lens has a 0.0003 inch airspace tolerance between elements 1 and 2, and 0.00013 between 2 and 3. The V-number tolerance is 0.033 on the center element, and you need to hold centration of 0.00015 on that element. *Nobody* can make this lens by just following a budget.

We have to loosen these tolerances. How? Well, one reason tolerances can be tight is because the aberrations of individual elements are large. While third-order aberrations are no longer as useful for the lens designer as they were a generation ago, this is a case where they actually can make a difference, as we saw in chapter 10—but we will not control them directly. Type the command **THIRD SENS**:

```
SYNOPSYS AI>THIRD SENS

 ID F10 APO

 NORMALIZED 3RD-ORDER ANALYSIS OF TOLERANCE SENSITIVITY

 SS OF SA3 BY SURFACE (SAT) =                8.363047
 SS OF CO3 BY SURFACE (COT) =                0.018283
 SS OF CO3/YDC BY SURFACE (ACD) =            0.132904
 SS OF CO3/TILT BY SURFACE (ACT) =           4.158202
 SS OF CO3/YDC BY ELEMENT (ECD)  =           0.038108
 SS OF CO3/TILT BY ELEMENT (ECT) =           1.184945
 SS OF SA3 BY ELEMENT (ESA) =                0.042947
 SS OF CO3 BY ELEMENT (ECO) =                0.000094
```

This listing shows the sum of the squares of the individual surface contributions to various aberrations and their derivatives. The idea is that if those are large, even if they are compensated by other surface contributions, the system becomes sensitive to small errors since then the cancellation will not be as good. The value of SAT, the spherical aberration sum, is 8.363. Let us modify the merit function to bring that sum down. Here is the new MACro (**C13M1**):

```
PANT
VLIST RAD 1 2 3 4 5 7
VLIST TH 2 4
END
AANT
AEC
ACC
M 4 1 A SAT
GSO 0 1 5 M 0 0
GNO 0 .2 4 M .75 0
GNO 0 .1 4 M 1.0 0
END
SNAP
SYNO 30
```

Here we ask for a value of 4 for the quantity SAT. After running this, the lens (**C13L1**) is slightly changed, and SAT now has a value of 4, as requested. Now prepare a new BTOL run (**C13M2**):

```
CHG
NOP
END
BTOL 2

EXACT INDEX 1 3 5
EXACT VNO 1 3 5

TPR ALL
TOL WAVE 0.1
ADJUST 6 TH 100 100

PREPARE MC

GO
STORE 4
```

This increases the wavefront variance tolerance to 0.1 and specifies an adjustment of thickness 6. (The first BTOL run utilized the paraxial thickness solve on 6, but sometimes the tolerances are looser if you let the program depart slightly. The adjustment will take care of that. The **NOP** directive removes all paraxial solves, so this thickness will be free to vary.) Also, specify that the index and Abbe number of the three glasses are exact, which removes them from the budget. In systems as fussy as this, one always requests melt data from the glass vendor, which give the measured index, and then one tweaks up the design with those values, entered as melt data. So errors in those values do not have to be part of the budget anymore.

Run this, and the tolerances are somewhat looser:

BUDGET TOLERANCE ANALYSIS -----B-----

EL.	SURF	RADIUS	RADIUS TOLERANCE (RADIUS)	(FRINGES)	THICKNESS	THICKNESS TOL
1	1	-171.99111	0.89919	4.90822	0.58188	0.00500
1	2	-6.93756	0.00134	4.48085	0.25590	0.00322
2	3	-6.53562	0.00122	4.40878	0.26355	0.00499
2	4	6.24408	0.00093	3.76549	0.03937	0.00144
3	5	6.31033	0.00092	3.68363	0.53301	0.00499
3	6	-19.41777	0.01157	4.91258	39.42904	0.00000
	7	-13.12585	0.00000	0.00000	0.00000	0.00000

ELE	SURF	GLASS NAME	BASE INDEX	INDEX TOL	V-NUMBER	V-NUMBER TOL

Note: The symbol "d" indicates that the quantity is estimated at 0.58756 uM.
The symbol "F" indicates that the quantity is taken at the primary color.

ELE	SURF	WEDGE TOLERANCE (ARC MIN)	(TIR)	IRREG. TOL (FRINGES)	ROLLED EDGE TOL (FRINGES)
1	1	0.00000	0.00000	0.94816	0.25930
1	2	0.76881	0.00090	0.93774	0.25442
2	3	0.00000	0.00000	0.86792	0.23442
2	4	0.51242	0.00059	0.85807	0.23132
3	5	0.00000	0.00000	0.81569	0.22006
3	6	0.88272	0.00103	0.83286	0.22681
	7	23.42596	0.00473	0.00000	0.00000

ELE	SURF	ELEMENT TILT TOLERANCE (ARC MIN)	(TIR)	Y-DECENT TOL	X-DECENT TOL
1	1	0.92368	0.00108	0.00113	0.00000
1	2	0.00000	0.00000	0.00000	0.00000
2	3	0.89070	0.00102	0.00051	0.00000
2	4	0.00000	0.00000	0.00000	0.00000
3	5	0.57124	0.00066	0.00064	0.00000
3	6	0.00000	0.00000	0.00000	0.00000
	7	0.00000	0.00000	0.00000	0.00000

Now the lens will obtain a variance of 0.1 everywhere in the field, at the two-sigma level. Is that too large? To find out, run the Monte-Carlo routine to see what the as-made lenses will look like. The starting lens is now in the library at location 4, and we want a worst-case example to be placed into location 5.

Type in the CW:

MC 50 4 QUIET -1 ALL 5.

This will run a batch of 50 lenses, made according to the budget above, compile statistics for the batch, and then save the worst-case example. If you do not know the arguments of the command, just type the letters **MC** and look at the tray. The format of the command is displayed for you, as shown in figure 13.2, and if you need more

MC [nsamples {libloc/MULTI} [QUIET] [qtol {qnum/ALL} qlib]]

Figure 13.2. The Trayprompt display for MC.

information, just press the <F2> key while the command is shown in the tray to open the Help File at that topic (or type **HELP MC**). You have to run BTOL before MC will work, since it uses the budget from BTOL.

MC runs its 50 cases and prints some statistics. You can look at a histogram of the results by typing **MC PLOT** after the run is finished. The on-axis image now has a variance of 0.1 at a two-sigma confidence.

However, this looks too easy. We need to examine the worst-case example. Switch to ACON 2 (type **ACON 2** or click the button [2]) and type **GET 5**. This is where MC put that example. Now look at the PAD display, shown in figure 13.3.

Select the 'OPD Fan Plots' option for the bottom display, and you see that the lens has more than a quarter-wave of aberration at the on-axis image—we told you this would not be an easy problem. This budget is still very tight, with centration tolerances of less than a micrometer.

Anyone with experience with astronomical optics could have predicted all of this, of course. One can rarely just build such a lens according to the print and expect it to work. We have a great budget that nobody can hold, and a slightly more practical budget that makes a bad lens. What is next?

13.1 Fabrication adjustment

Clearly, we need some fabrication adjustments. In this scenario, one makes an element, measures it, puts those data into the lens prescription, and reoptimizes, varying the other elements. Then one makes another element to the new design, measures and adjusts all over again, continuing in this way until everything is made. At assembly, one then adjusts centration and possibly tilts to obtain the best possible image. Go back to ACON 1 again and store the lens in library location 4. We will

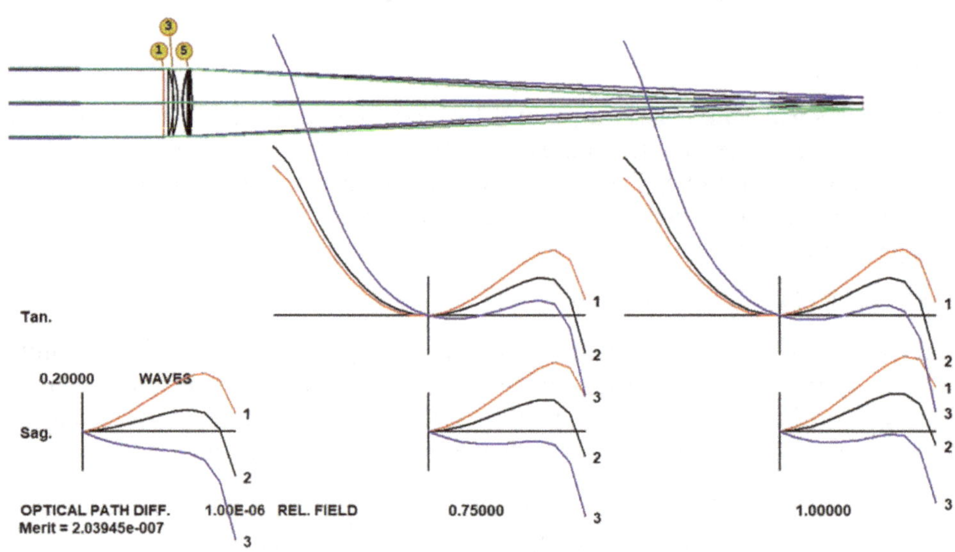

Figure 13.3. Worst-case example from MC. Fabrication adjustment is needed.

now analyze the statistics with FAMC. (This analysis will use the BTOL budget you prepared above, with a tolerance of 0.1 on the wavefront.)

You really should read about this topic before you use it. Type **HELP FAMC** (FAMC is Fab Adjust MC). Here is our MACro (**C13M3**):

```
FAMC 50 4 QUIET -1 ALL 5
PASSES 20
FAORDER 5 3 1

PHASE 1
PANT
VLIST RAD 1 2 3 4 5 6
VLIST TH 2 4 6
END

AANT
GSO 0 1 5 M 0
GNO 0 1 5 M 1
END
SNAP
EVAL

PHASE 2
PANT
VY 3 YDC 2 100 -100
VY 3 XDC 2 100 -100
VY 5 YDC 2 100 -100
VY 5 XDC 2 100 -100
VY 6 TH
END
AANT
GNO 0 1 4 M 0 0 0 F
GNO 0 1 4 M 1 0 0 F
END
SNAP
SYNO 30

PHASE 3
```

Here is what is going on in this MACro:

1. Request **FAMC**, with arguments the same as for the MC analysis that was run above.
2. In phase 1, the program will alter the elements in the order given in the **FAORDER** line, changing the parameters by a random amount within the BTOL budget. This simulates making the hardest element first, and so on. It

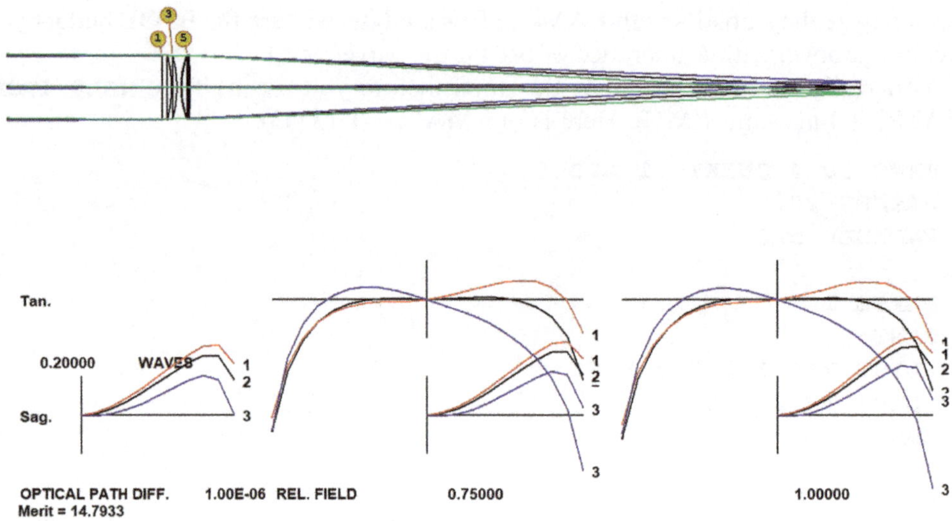

Figure 13.4. Worst-case example with fabrication adjustments.

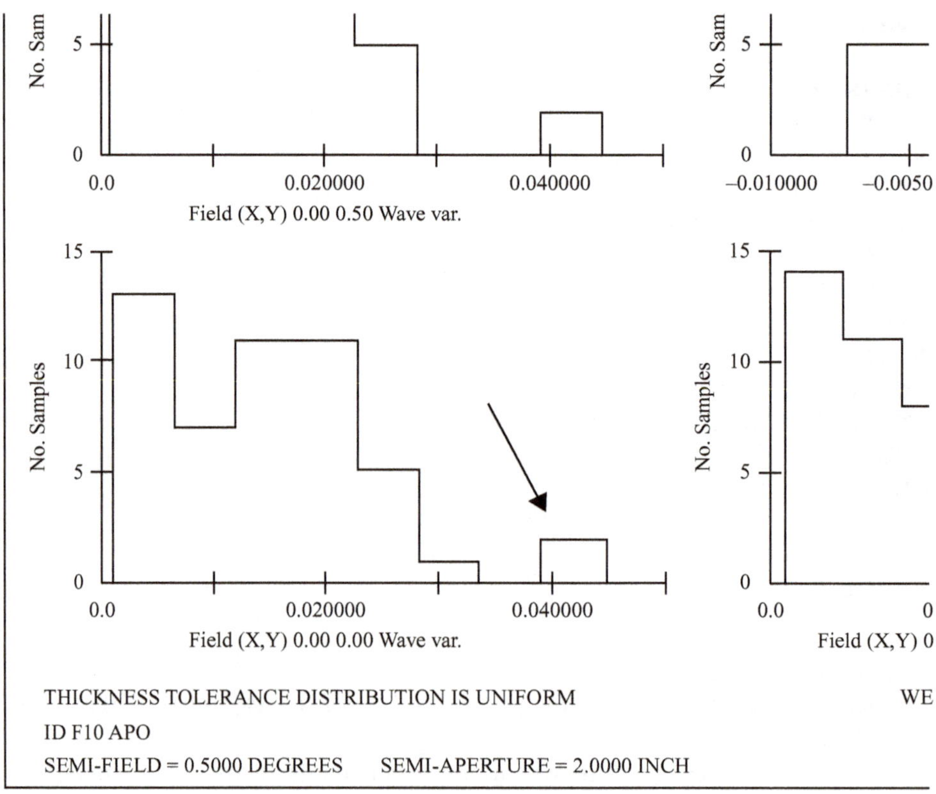

Figure 13.5. Statistics of the MC analysis with fabrication adjustments.

will optimize the lens with the variables and merit function listed in the **PHASE 1** portion, as each element is manufactured, deleting those variables that apply to elements that are already finished.

3. When the simulated elements are all manufactured, it simulates mounting in the cell according to the tilt and decenter tolerances. Then it optimizes once again, varying the decenter of elements two and three in both x and y (errors are simulated in both directions, so the compensation must be as well) in accordance with the **PHASE 2** parameters. It also varies thickness 6 once again, since a large centration change produces a small focus shift as well. The merit function also corrects rays over both halves of the pupil (via the **F** in the GNO lines) because there is no bilateral symmetry anymore once the errors are simulated.

Run this MACro and look at the worst case again, shown in figure 13.4.

This lens has wavefront errors of just over a quarter wave on-axis—but remember this is the worst case. Most of the 50 in this batch are much better. Type `MC PLOT` and look at the histogram that describes the on-axis image, which is on the lower left of the plot, shown in figure 13.5.

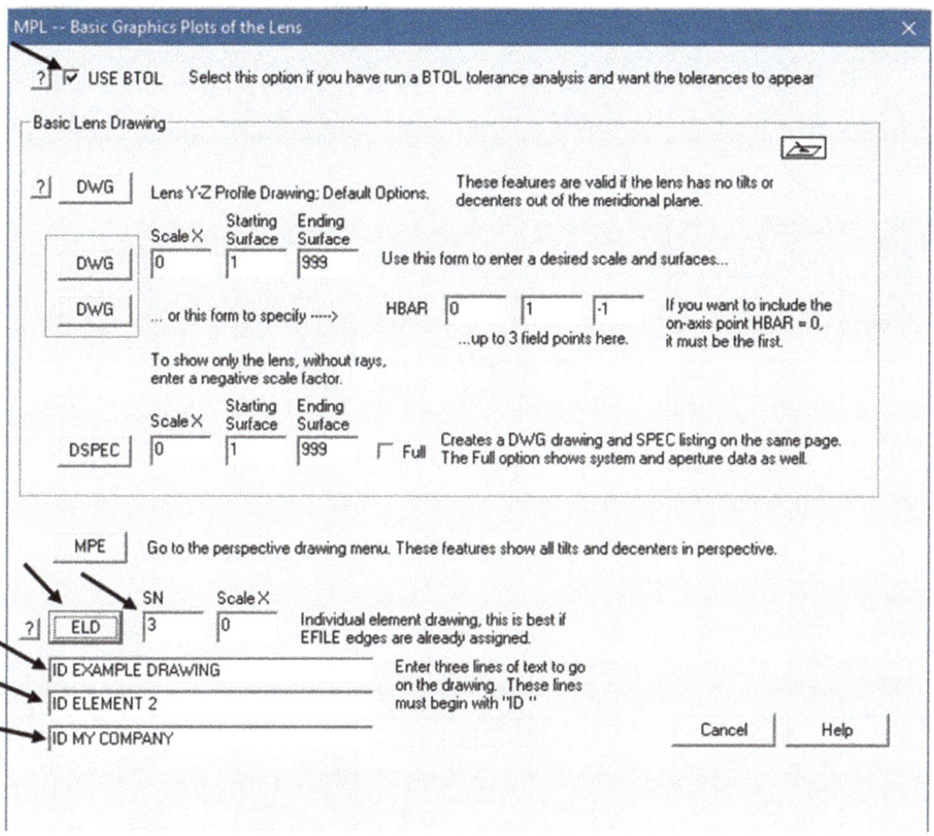

Figure 13.6. The MPL dialog for preparing input for an element drawing.

The worst-case example is that lonely box way out at the end, and it is much worse than the others. There is always a small chance of getting a lemon, after all, and this is it. Nonetheless, there is a high probability that the lens will work just fine if built according to this rather tight budget.

Note that the budget itself was not altered, or recalculated, when we decided to use FAMC. What we did was to employ a budget that did not work very well and made it work much better. As a fringe benefit, we do not have to worry about the tight centration errors anymore—which are expensive to hold—since the elements are to be adjusted at assembly anyway. So things have become much easier.

However, there is a price to pay: you have to get melt data when the glass is delivered, tweak up the design with those data, the shop must make the elements in the order given, measure them carefully, and then send those data back to the designer, who will reoptimize everything at each step. Provision must be made for adjusting the centration of elements two and three on the test bench, and then locking everything down once the image is tuned up. However, that is what precision optics is all about.

About those adjustments: it is easier to adjust the centration of only one element, not two as in this example. What happens if you delete the adjustments on, say surface 3 and redo the whole procedure? If it works, the assembly technician will thank you. Try it and find out!

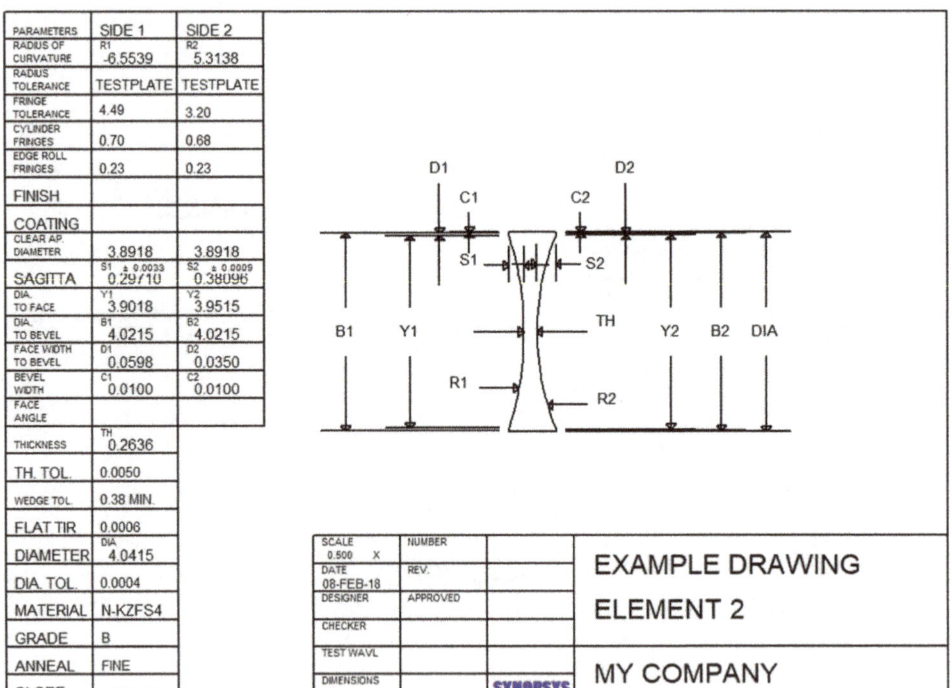

Figure 13.7. Element drawing, annotated with BTOL tolerances.

13.2 Transferring tolerances to element drawings

Now that you have a tolerance budget, you can prepare lens drawings to send to the shop. That can be done automatically too. Open the MPL dialog (**MPL**) and enter the data as shown in figure 13.6.

The '**USE BTOL**' checkbox tells the program to pick up the tolerances from the BTOL budget and add them to the element drawing, as shown in figure 13.7.

These tolerances are all added as *annotations*, rather than as plain text, which means you can edit them with the Annotation Editor and change any values you wish. If you select the first **DWG** button, the program creates an assembly drawing, also with tolerances added. Figure 13.8 shows an example, where we entered a scale factor of –0.8. (The minus sign tells the program to draw just the elements and not add rays to the drawing.) This drawing shows the decenter tolerances as calculated by BTOL, and in this case you will want to edit those annotations and add a note that the centration is to be adjusted at assembly.

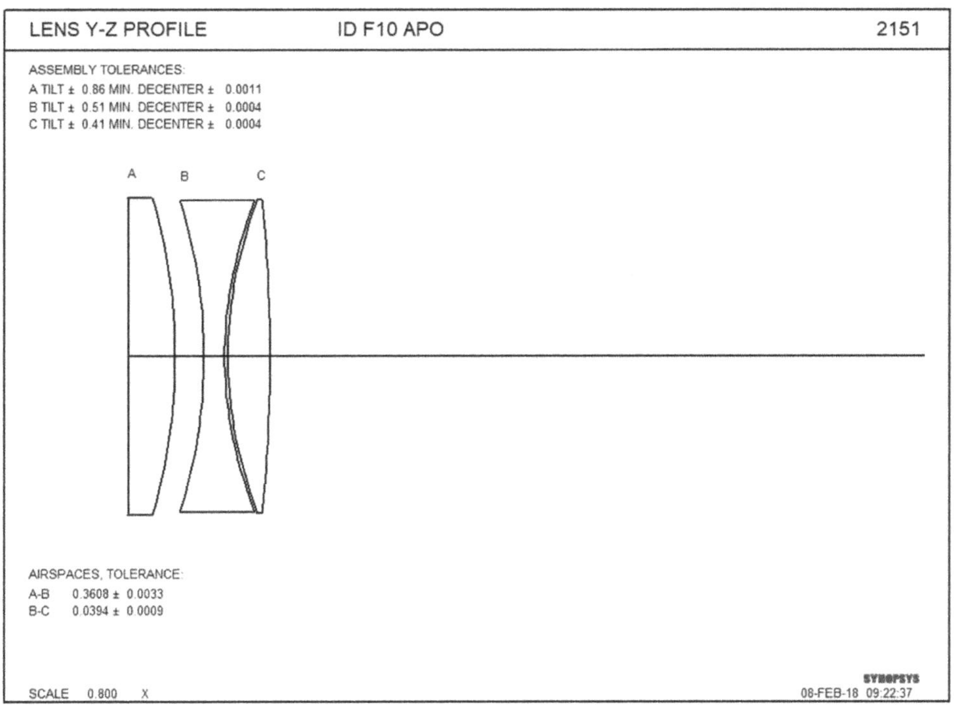

Figure 13.8. Assembly drawing prepared by DWG with airspace, tilt, and decenter tolerances added.

IOP Publishing

Lens Design

Automatic and quasi-autonomous computational methods and techniques

Donald Dilworth

Chapter 14

A near-infrared lens example

Glass selection for the near IR spectrum

Chapter 12 showed how to design an apochromatic objective for use in the visible spectrum. Now we will design one over the wavelength range from 1.06 to 1.97 μm, in the near-infrared (NIR), which is more difficult.

The challenge when designing a lens for the infrared is finding optical materials that are useful over the spectral range and whose cost and chemical properties are attractive. The task in this lesson is to redesign an existing lens, replacing some undesirable materials with ordinary optical glass. The reference system is bundled as 1.RLE, with the `ID MIT 1 TO 2 UM LENS`. You can `FETCH` that lens and examine its performance. Set the scale of the OPD curves to 0.5 waves, as shown in figure 14.1.

This lens has three elements of ZNS and one of AS2S3, making four in all. Those names refer to zinc sulphide and arsenic trisulphide glass, and we would like to avoid those materials if possible. The first-order properties we need to match are as follows:

- Entering beam radius 17.5 mm.
- Chief-ray angle 0.935 degrees.
- Back focus distance 16.3 mm.
- Cell length 50 mm.
- Wavelengths 1.9701, 1.5296, and 1.06 μm.

14.1 Design approach

Rather than try to change the materials in the present lens, all of which have an index greater than 2.0, let us start from scratch. For this we will use the design search program—but first we have to decide which glass types to use: if we just run **DSEARCH** and let it find model glasses, it will not come back with any of the unusual glasses that make a big difference in the NIR. (The model represents an *average* of selected glasses in the visible region.) So we have to steer it.

Open the glass table display (**MGT**), select the 'Guangming' table, and then click the 'Graph' button and select the option shown in figure 14.2.

The data are now off the screen, so click in the display and zoom out with the mouse wheel until you see a collection of red dots. Then pan with the right mouse button to center things and zoom in again. You should see the display shown in figure 14.3. Click the 'Full Name' button, then click on each of the spots circled and write down the name of the glass.

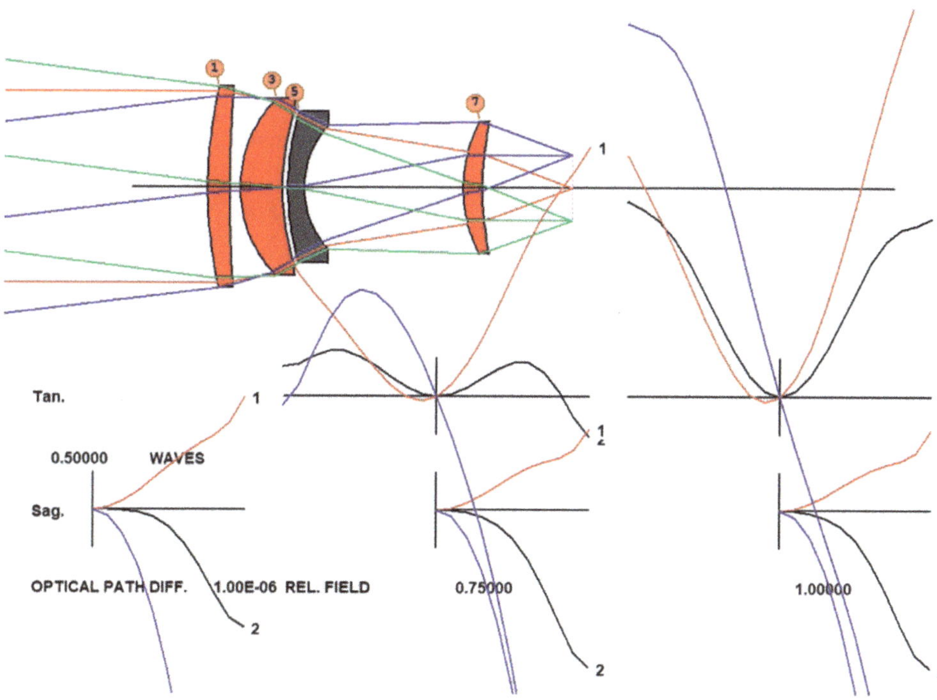

Figure 14.1. An NIR lens, to be redesigned.

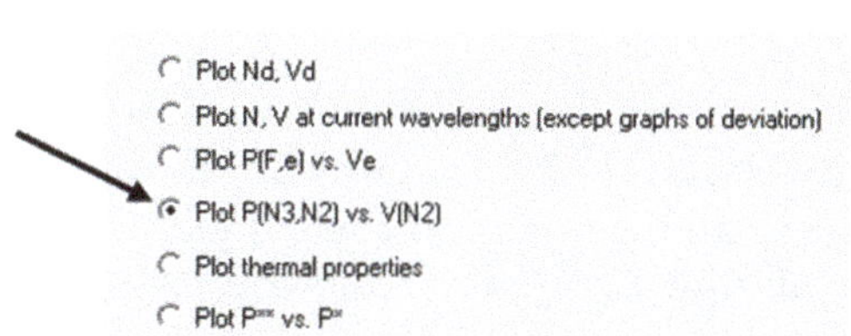

Figure 14.2. Selecting a graph option in the glass map utility.

Figure 14.3. Selecting four promising glass types for the NIR design.

The four glass names circled, D-FK61, G-ZF52, H-ZH88, and H-F51, definitely do not behave the same as all the others. We will direct DSEARCH to only use two of them, and later use all four for a comprehensive glass search.

Here is the DSEARCH input (**C14M1**):

```
CORE 14
TIME
DSEARCH 3  QUIET ! the best lens will show up in library 3 (and also in PAD)
 SYSTEM                 ! system requirements follow
 ID NIR EXAMPLE         ! lens identification
 OBB 0 .935 17.5        ! specify the object
 WAVL 1.97 1.53 1.06    ! and the wavelength range
 UNITS MM
 END

 GOALS             ! here we set the goals
 ELEMENTS 5        ! since glass has a lower index, we'll ask for 5.
 FNUM 1.428
 BACK 16 .1
 TOTL 50 .1
 STOP FIRST        ! there seems to be no reason to let the stop position vary
 STOP FIX          ! so we put it in front and keep it there
 NPASS 100
 ANNEAL 200 20 100
 RSTART 300       ! a useful starting radius,
 TSTART 1         ! and this thickness on each element to start with
 QUICK 60 90
FOV 0 .5 1
FWT 2 1 1
GLASS POS       ! positive elements will use this glass type
G D-FK61
GLASS NEG       ! and negative this type.
G H-ZF88
END

 SPECIAL        ! here we give requirements that are not defaults
ACM 3 .1 1      ! auto edge control (AEC) and center thickness control (ACC) are
                ! defaults,
 ACC 10 .1 1    ! maximum thickness 10 mm
 ACA           ! but we add to these ACM, so thicknesses do not get too thin, ACA,
 ASC   ! so rays do not approach the critical angle, and ASC so surfaces do not
 END           ! get too close to the hemisphere point.

 GO            ! this starts the process.
TIME
```

In 20 s, DSEARCH produces a picture of the ten best configurations it found, shown in figure 14.4.

We now have a very good five-element lens, shown in figure 14.5, but it has only the two glass types we specified. It is time to do a more comprehensive search.

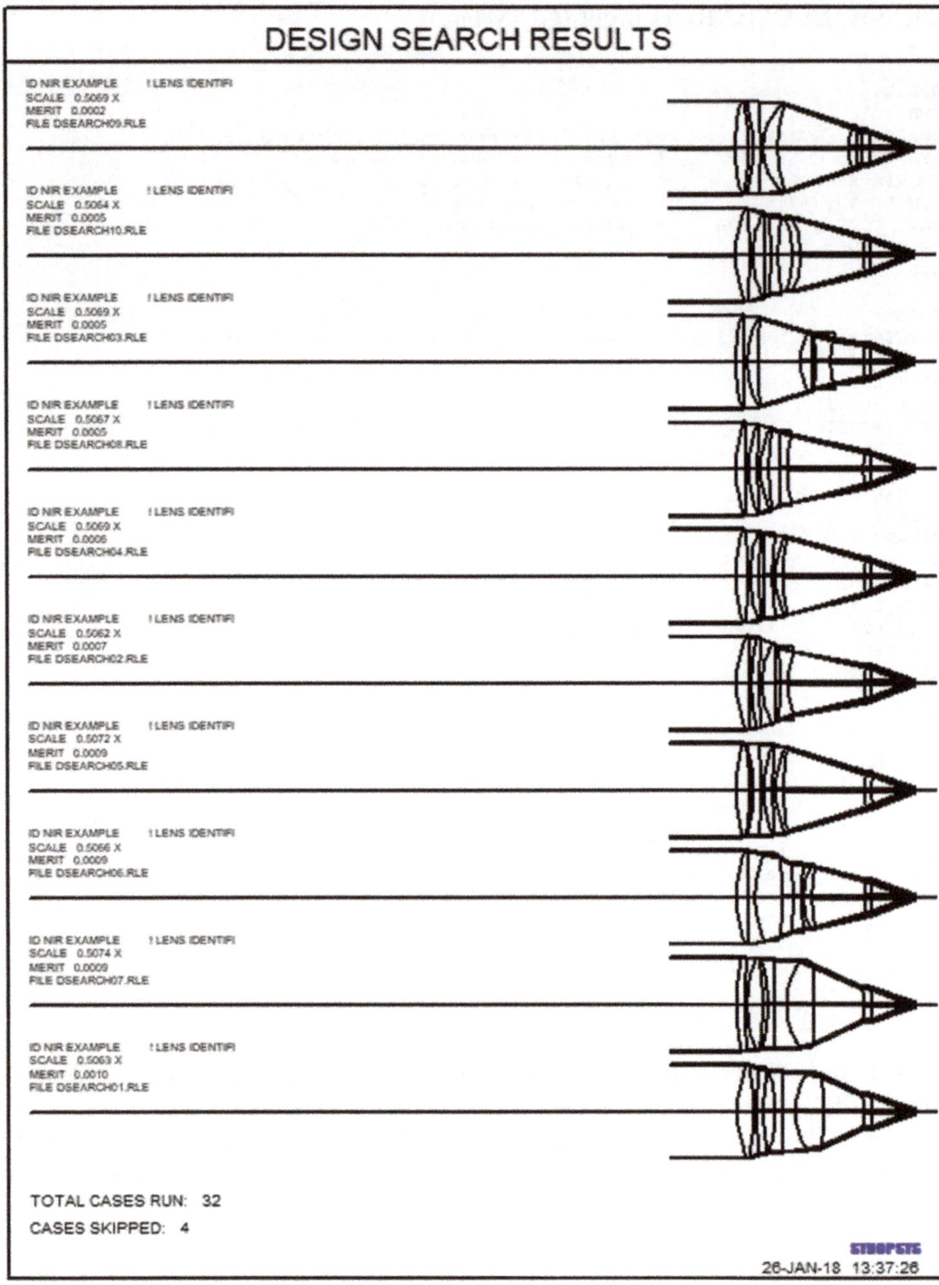

Figure 14.4. Display of the ten best lenses returned by DSEARCH.

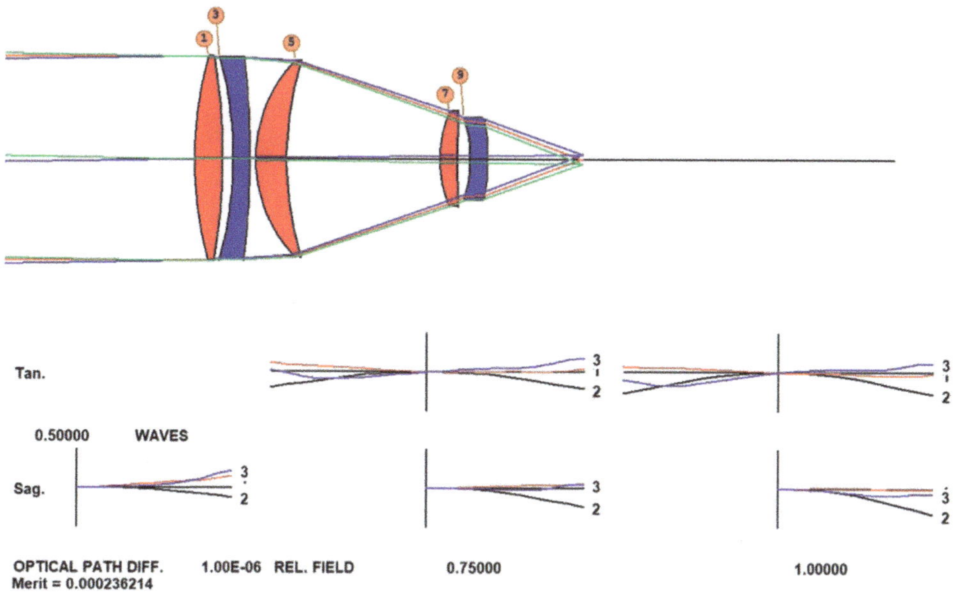

Figure 14.5. Best lens returned by DSEARCH for an NIR design.

Look at the MACro DSEARCH_OPT .MAC, which DSEARCH has constructed and should be open in a new editor window:

```
PANT
VLIST RD ALL
VLIST TH ALL
END
AANT P
AEC
ACC
GSR      0.000000      2.000000      4  M      0.000000
GNR      0.000000      1.000000      4  M      0.500000
GNR      0.000000      1.000000      4  M      1.000000
M   0.160000E+02  0.100000E+00 A BACK
M   0.500000E+02  0.100000E+00 A TOTL
 ACC 10 .1 1
 ACM 3 .1 1    ! AUTO EDGE CONTROL (AEC,ACC) ARE DEFAULTS
 ACA           ! BUT WE ADD TO THESE ACM, SO THICKNESSES DO NOT GET TOO THIN, ACA,
 ASC          ! TO AVOID THE CRITICAL ANGLE, AND ASC SO CURVES DO NOT GET TOO STEEP
END
SNAP/DAMP 1
SYNOPSYS  100
```

Save this MACro with the name NIR_OPT.MAC. This is the optimization MACro that will be run over and over when we execute **GSEARCH**. That program will determine which glass should go on which elements.

Now make a new MACro (type **AEE** to open a new editor and type the data (**C14M2**) below):

```
CORE 14
GSEARCH 3 QUIET LOG

SURF
1 3 5 7 9
END

OFILE 'NIR_OPT.MAC'
NAMES
G G-ZF52
G D-FK61
G H-ZF88
G H-F51

END
USE 2
GO
```

Run this MACro, and the lens has improved even more, as shown in figure 14.6. The performance is now just over 0.25 waves of aberration, in color 3 at full field.

It looks like we have a solution **(C14L1)**. There is almost no primary or secondary chromatic aberration. We have succeeded in replacing the undesirable materials with ordinary glass, and the performance became much better than the original at the same time:

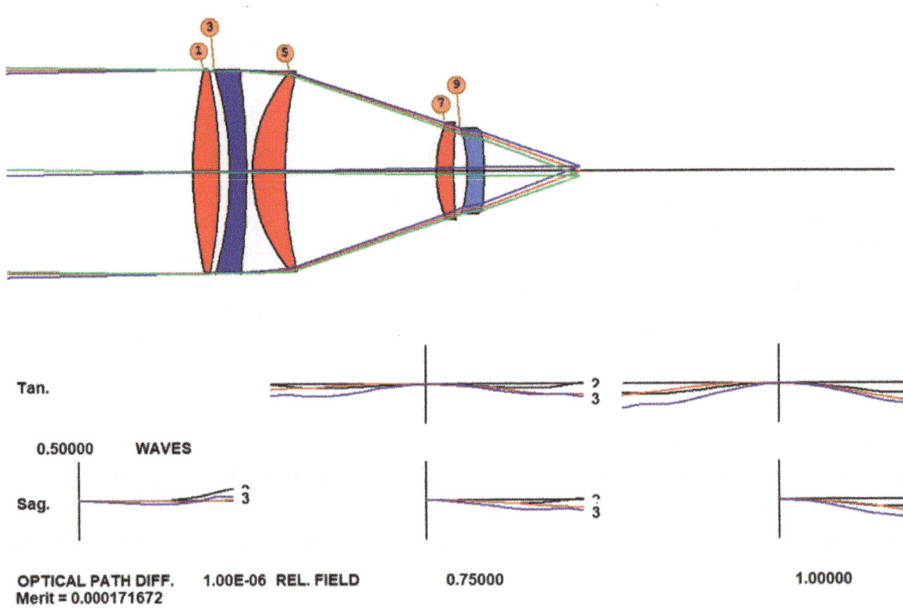

Figure 14.6. Lens returned by GSEARCH.

```
SYNOPSYS AI>SPEC

ID NIR EXAMPLE          ! LENS IDENTIFI    2044          24-JAN-18   10:03:24
ID1 DSEARCH CASE WAS 0000000000000000000010110      22
LENS SPECIFICATIONS:

SYSTEM SPECIFICATIONS
---------------------------------------------------------------------------------
OBJECT DISTANCE     (THO)        INFINITE  FOCAL LENGTH      (FOCL)        49.9800
OBJECT HEIGHT      (YPP0)        INFINITE  PARAXIAL FOCAL POINT            15.9991
MARG RAY HEIGHT    (YMP1)         17.5000  IMAGE DISTANCE    (BACK)        15.9991
MARG RAY ANGLE     (UMP0)          0.0000  CELL LENGTH       (TOTL)        50.0024
CHIEF RAY HEIGHT   (YPP1)          0.0000  F/NUMBER          (FNUM)         1.4280
CHIEF RAY ANGLE    (UPP0)          0.9350  GAUSSIAN IMAGE HT(GIHT)          0.8157
ENTR PUPIL SEMI-APERTURE          17.5000  EXIT PUPIL SEMI-APERTURE        24.7859
ENTR PUPIL LOCATION                0.0000  EXIT PUPIL LOCATION            -54.7895

WAVL (uM) 1.970000 1.530000 1.060000
WEIGHTS   1.000000 1.000000 1.000000
COLOR ORDER    2    1    3
UNITS                              MM
APERTURE STOP SURFACE (APS)         1     SEMI-APERTURE    17.53132
FOCAL MODE                         ON
MAGNIFICATION              -4.99800E-11
POLARIZATION AND COATINGS ARE IGNORED.
SURFACE DATA
----------------------------------------------------------------------------------
SURF         RADIUS       THICKNESS   MEDIUM             INDEX       V-NUMBER
----------------------------------------------------------------------------------
  0        INFINITE       INFINITE   AIR
  1        81.04292        4.57832   D-FK61             1.48647     78.02 GUANGMIN
  2       -91.61021        1.75391   AIR
  3       -62.04337        2.89159   H-ZF88             1.87811     26.89 GUANGMIN
  4      -139.85794        1.00000   AIR
  5        25.65212        5.65231   D-FK61             1.48647     78.02 GUANGMIN
  6        76.43347       25.93375   AIR
  7        23.23659        2.91149   D-FK61             1.48647     78.02 GUANGMIN
  8       103.76135        2.40473   AIR
  9       -24.09198        2.87631   H-F51              1.60755     25.46 GUANGMIN
 10       -40.39312S      15.99911S  AIR
IMG        INFINITE

KEY TO SYMBOLS
--------------
A  SURFACE HAS TILTS AND DECENTERS      B  TAG ON SURFACE
G  SURFACE IS IN GLOBAL COORDINATES     L  SURFACE IS IN LOCAL COORDINATES
O  SPECIAL SURFACE TYPE                 P  ITEM IS SUBJECT TO PICKUP
S  ITEM IS SUBJECT TO SOLVE             M  SURFACE HAS MELT INDEX DATA
T  ITEM IS TARGET OF A PICKUP
THIS LENS HAS NO SPECIAL SURFACE TYPES
THIS LENS HAS NO TILTS OR DECENTERS
SYNOPSYS AI>
```

If these lenses are acceptable mechanically, the problem is solved.

Except, what is the transmission at 1.97 μm? Type **FIND TRANS IN COLOR 1**. It comes back 98.18%—very good (coatings and reflection losses are ignored here because the lens is not in polarization mode).

But what if the value had come back too low? Well, then go back to the glass map and display the absorption at 1.97 μm—and select glasses with shorter data bars. Lens design is all about tradeoffs, after all, and with these tools you can obtain the best one rather easily.

IOP Publishing

Lens Design

Automatic and quasi-autonomous computational methods and techniques

Donald Dilworth

Chapter 15

A laser beam shaper, all spherical

Reshaping a laser beam from Gaussian to uniform with spherical elements

The output from a laser has a Gaussian intensity profile, which is nonuniform, and for some applications one would like to make it uniform. That is the job of a laser beam shaper.

The job can be done in several ways. With simple lenses, with spherical surfaces, it requires balancing rather large amounts of spherical aberration in a way that redistributes the light, reducing the energy density at the center of the beam while increasing it near the edges, at the same time keeping the wavefront aberrations under control. It is easier to do with aspheric surfaces, where one has more control over the amount of aberration to be introduced, and also easier if one uses diffractive optical elements. The problem is that the latter two are more expensive than spherical lenses, so we want to see what we can do with those first.

With this in mind, we will try an all-spherical approach to start with, to determine how uniform we can make the beam and how many elements we will need. Let us start with a very simple setup, one that we will modify as we go along.

The problem is to convert a HeNe laser with a waist radius of 0.35 mm to a beam that is 10 mm in diameter and uniform to within 10%. (There is no reason to use DSEARCH for this problem, since the only practical starting design is very simple.)

Here is the input file for our very primitive starting point (**C15M1**), shown in figure 15.1:

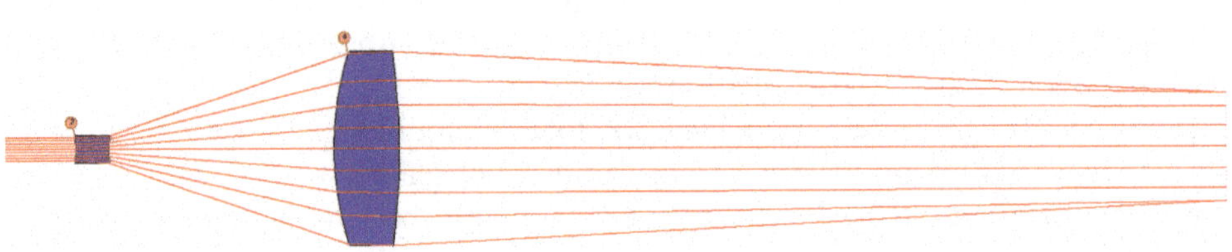

Figure 15.1. Rough-guess initial system for a laser beam shaper.


```
RLE                      ! Beginning of lens input file.
ID LASER BEAM SHAPER
WA1 .6328                ! Single wavelength
UNI MM                   ! Lens is in millimeters
OBG .35 2                ! Gaussian object; waist radius -.35 mm; define full
                         ! aperture as twice the 1/e**2 point.
1 TH 22                  ! Surface 2 is 22 mm from the waist .
2 RD -5 TH 2 GTB S       ! Guess some reasonable lens parameters; use glass type SF6
                         ! from Schott catalog
SF6
3 UMC 0.3 YMT 5          ! Solve for the curvature of surface 3 so the marginal ray
                         ! has an angle of 0.3;
                         ! find spacing so ray height is 5 mm on next surface
4 RD 20 TH 4 PIN 2       ! Guesses for surface 4
5 UMC 0 TH 50            ! Solve for curvature of 5 so beam is collimated.
7                        ! Surfaces 6 and 7 exist
AFOCAL                   ! because they are required for AFOCAL output.
END
```

(This system is in **AFOCAL** mode, which means that the output will be collimated. Since transverse aberrations make no sense if the image is at infinity, the program translates them into *angular* aberrations, no 'perfect lens' required. The program needs two dummy surfaces at the end, where it does the translation, here surfaces 6 and 7, which are coincident at the end.)

First, let us examine how the energy density falls off from the center to the edge of the aperture. There are three ways to do this. The simplest is with the `FLUX` command:

```
FLUX 100 P 3
```

This input will show the flux on surface 3, which shows the expected falloff, as illustrated in figure 15.2.

Another way to visualize this uses the FLUX aberration. This form gives you more flexibility, since you can specify the aperture and field points yourself. Type **`STEPS = 100`**, and then make a MACro (**C15M2**) with the following lines (note the symbol definition for DD):

```
DD: DO MACRO FOR AIP = -1 TO 1
COMPOSITE               ! Ready a composite definition.
CD1 P FLUX 0 0 AIP 0 3 ! Composite data number 1 is flux at Y- coordinate of AIP
                        ! (defined later) on surface 3.
= CD1
Z1 = FILE 1
= 1 + Z1
ORD = FILE 1
```

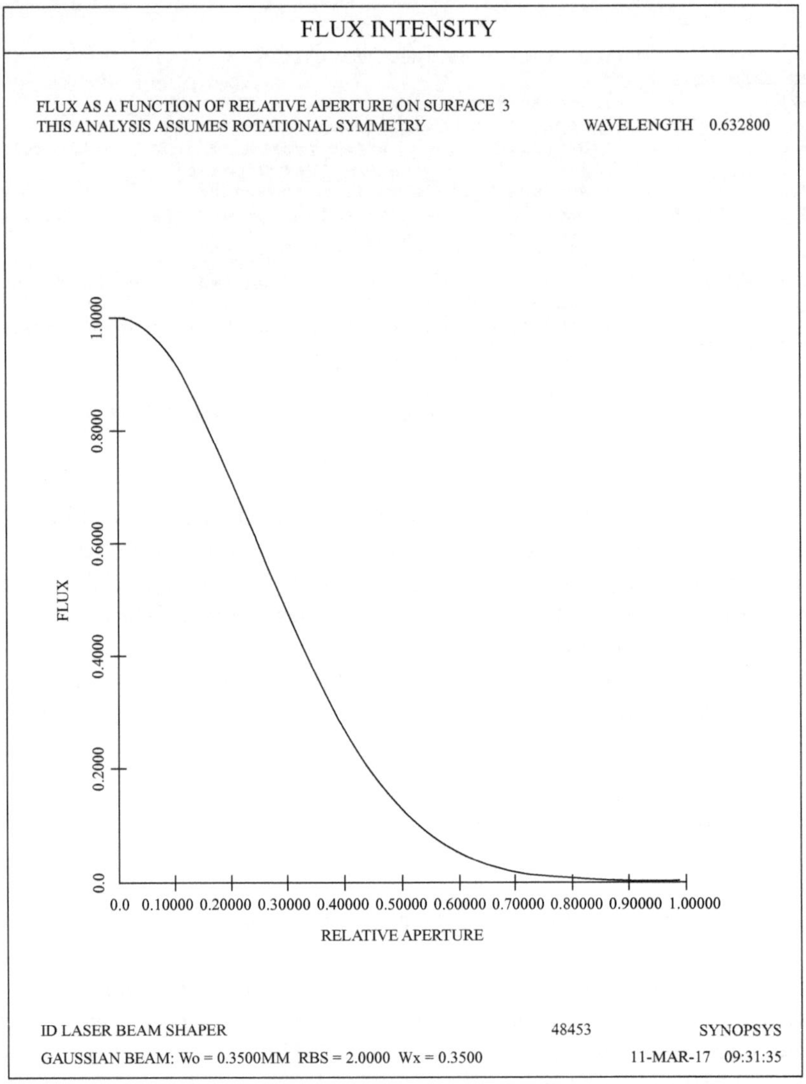

Figure 15.2. Flux falloff due to Gaussian intensity profile, calculated with the `FLUX` command.

Run this MACro once, and then type **`DD`**. The program loops over values of AIP from −1 to 1 and plots the flux density. Let me explain the logic:

- **`CD1 P FLUX`** ... calculates the flux falloff at zone AIP (which is a loop variable) at surface 3.
- **`= CD1`** is an equation, the result of which is automatically placed in FILE location 1.
- **`Z1 = FILE 1`** picks up that value and puts it into variable Z1.
- **`= 1 + Z1`** adds 1.0 to the result. This is then the total flux, since Z1 is the falloff.

- **ORD = FILE 1** picks up this value and uses it for the ordinate of a plot, the abscissa of which is the loop variable, AIP.

The plot is shown in figure 15.3.

Again, you see the Gaussian flux profile, evaluated out to twice the $1/e^2$ point according to the OBG definition. (The third way uses the **DPROP** diffraction propagation feature. That is more complicated to set up and run, but can account for diffraction along the beam, which the other two do not. An example will be shown later.)

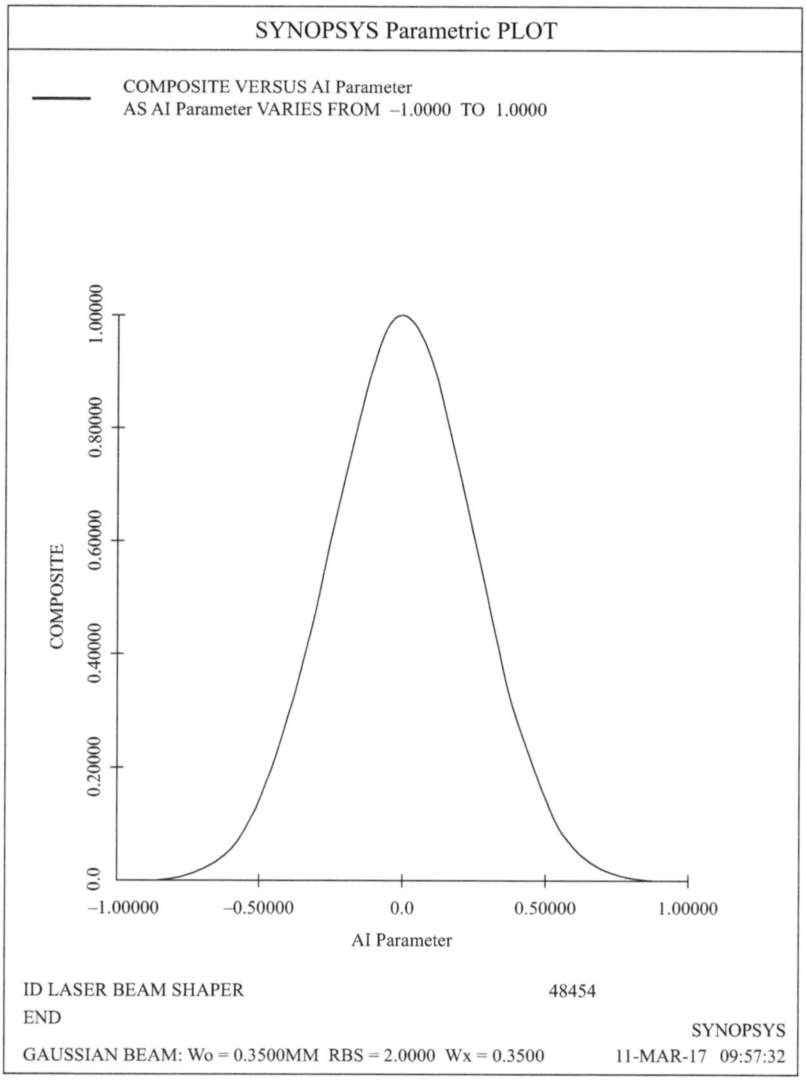

Figure 15.3. Flux intensity as calculated by an AI loop.

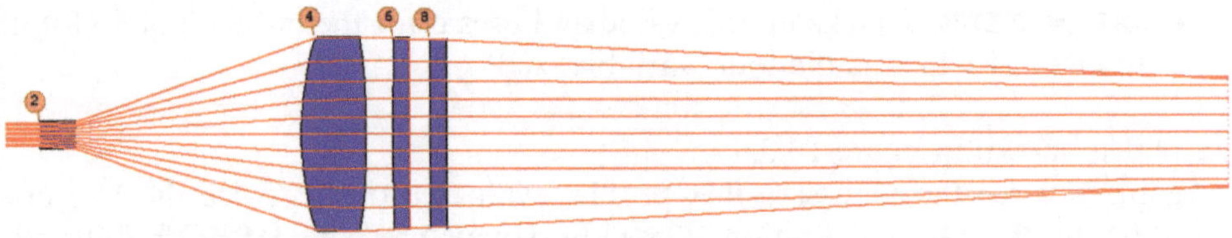

Figure 15.4. Lens with two added elements.

The object of this lesson is to make the flux as uniform as possible, with a target of 10% change over the aperture.

Here, we have simply guessed some starting lens dimensions, but already we see a hint of a solution: notice how the marginal rays converge toward the axis while the center rays are more collimated. The energy will indeed be more concentrated at the edges than before, which is a step in the right direction—but we also want the entire beam to be collimated, so we need a way to straighten out the rays afterwards. That is more difficult.

We suspect that we need more than just two elements, so let us add two more. Using the WorkSheet, as you did in chapter 6, click the button, and click the 'Insert Element' button in the WorkSheet toolbar. Then click on the axis in the PAD display, to add an element to the right of surface 5. Do the same thing again, a little further to the right. Now the system should appear as shown in figure 15.4.

We will try to optimize this system, but first let us make a checkpoint, so we can instantly go back if things do not work as we hope. Click the 'Checkpoint 'button.

Now we need to set up an optimization MACro to see if we can even things out. Here is a start (**C15M3**):

```
CHG
NOP
9 UMC
END

PANT                    ! Start of variable parameter definitions.
VLIST RAD ALL           ! Vary all radii.
VLIST TH 3 5 6 7 8      ! Vary the airspaces and five thicknesses (so AEC !works on
                        ! those elements).
END

AANT                    ! Start of merit function definition
AEC 1 1 1
ACC 4 1 1
ACA 60 10 1             ! stay away from the critical angle
LUL 100 1 1 A TOTL      ! Prevent the system from growing too large; assign upper-
                        ! limit of 150 on TOTL.
M 5 10 A P YA 0 0 1 0 9    ! Ask for a beam radius of 5 mm on surfaces 9 and 10
M 5 10 A P YA 0 0 1 0 10
M 0 1 A P FLUX 0 0 1 0 10 ! Ask for a flux falloff of zero at several zones on 10
M 0 1 A P FLUX 0 0 .99 0 10
```

```
M 0 1 A P FLUX 0 0 .985 0 10
M 0 1 A P FLUX 0 0 .98 0 10
M 0 1 A P FLUX 0 0 .97 0 10
M 0 1 A P FLUX 0 0 .96 0 10
M 0 1 A P FLUX 0 0 .95 0 10
M 0 1 A P FLUX 0 0 .94 0 10
M 0 1 A P FLUX 0 0 .93 0 10
M 0 1 A P FLUX 0 0 .92 0 10
M 0 1 A P FLUX 0 0 .91 0 10
M 0 1 A P FLUX 0 0 .85 0 10
M 0 1 A P FLUX 0 0 .8 0 10
M 0 1 A P FLUX 0 0 .7 0 10
M 0 1 A P FLUX 0 0 .5 0 10
M 0 1 A P FLUX 0 0 .3 0 10
GSO 0 1 5 P               ! Control the output ray OPD over an SFAN of 5 rays.
END                       ! End of merit function definition.
SNAP
SYNO 100
```

Run this MACro, and the new lens is shown in figure 15.5. (Your results may differ, since the exact place you clicked to insert the elements is unpredictable.)

To see if things improved, we need to evaluate the flux uniformity again. Type

FLUX 100 P 10.

The flux is not improved, as shown in figure 15.6.

It looks like we are not there yet—what should we do?

Remember that we defined our Gaussian object so that 'full aperture' is twice the $1/e^2$ point. That means we are trying to redistribute the energy so the very faint outer edge is as bright as the center—maybe that is asking too much. Let us restore the checkpoint and edit the object specification. In the WorkSheet, enter the number 0 in the surface box and click 'Update'. This shows the current system specifications in the edit pane, including the object selection, as seen in figure 15.7.

The object is currently defined as **OBG 0.35 2.000000**. Change this to **OBG .35 1** and click 'Update'. Now, 'full aperture' is at the $1/e^2$ point, not twice as far out as before. Save this as a new checkpoint, run the last MACro again, and anneal for a few cycles (**22, 1, 50**). We obtain the results in figure 15.8 and the flux uniformity shown in figure 15.9.

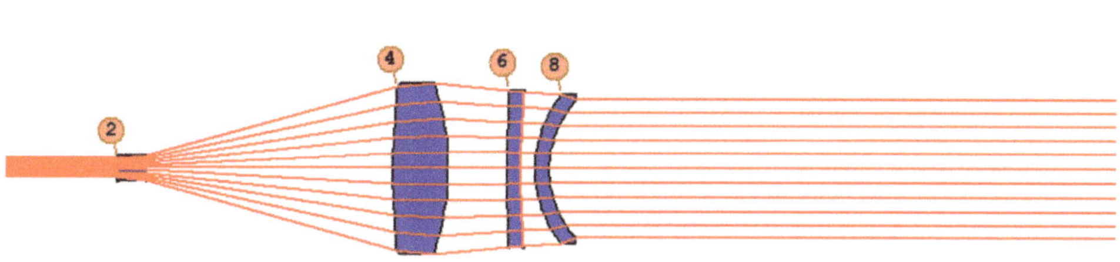

Figure 15.5. Lens optimized with flux aberrations.

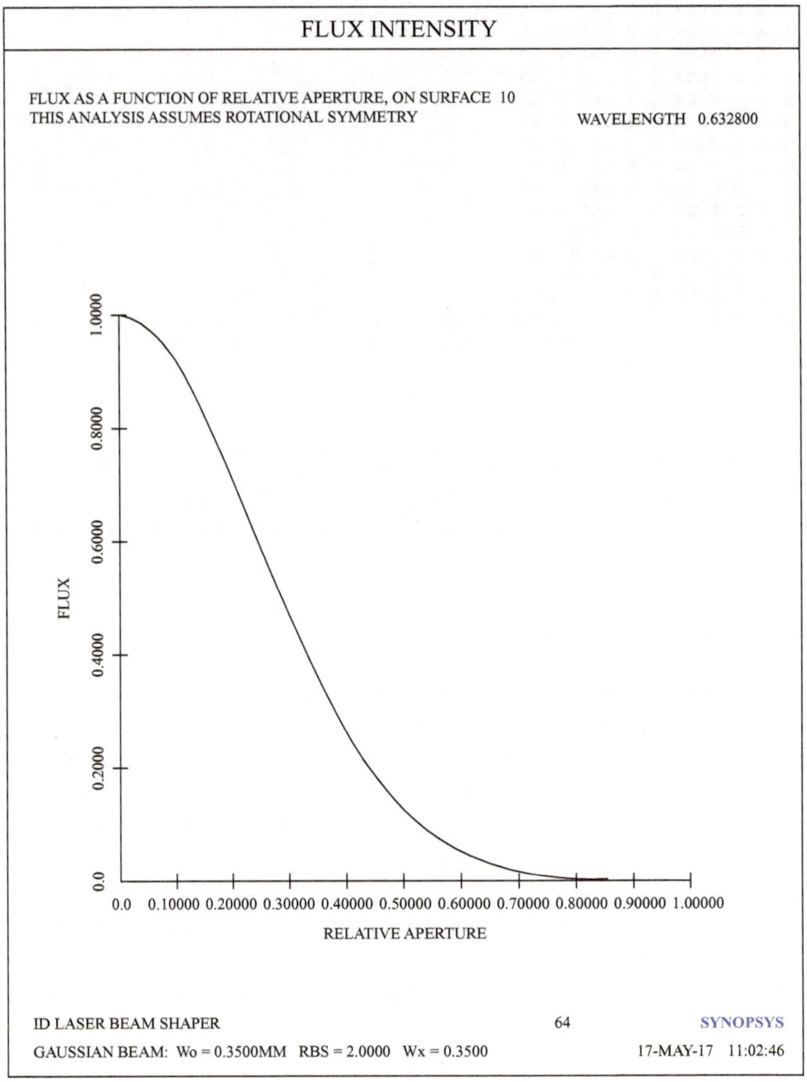

Figure 15.6. Flux falloff after reoptimizing.

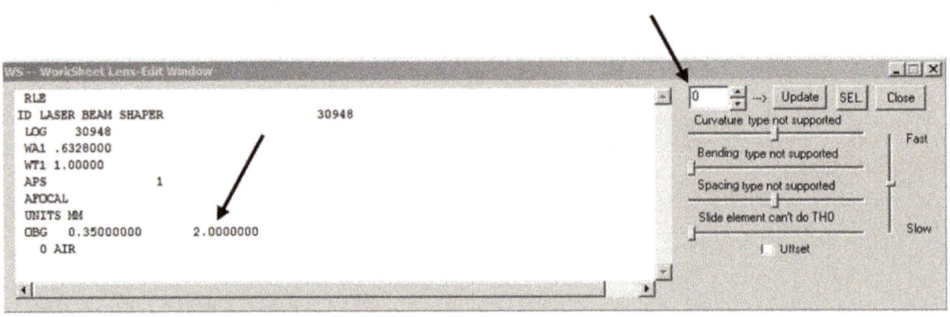

Figure 15.7. Edit pane in WS, showing the object specifications.

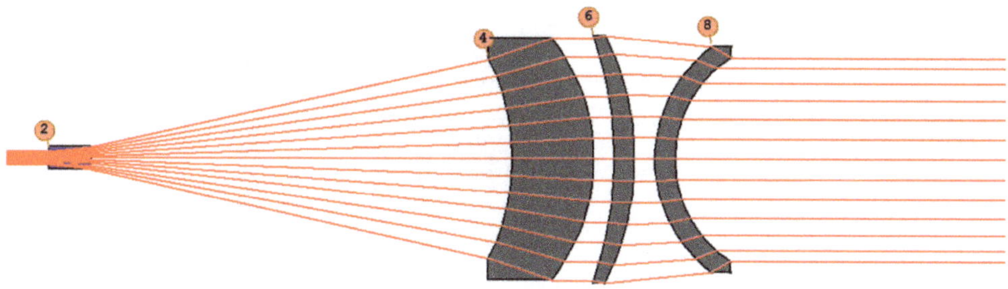

Figure 15.8. Lens reoptimized.

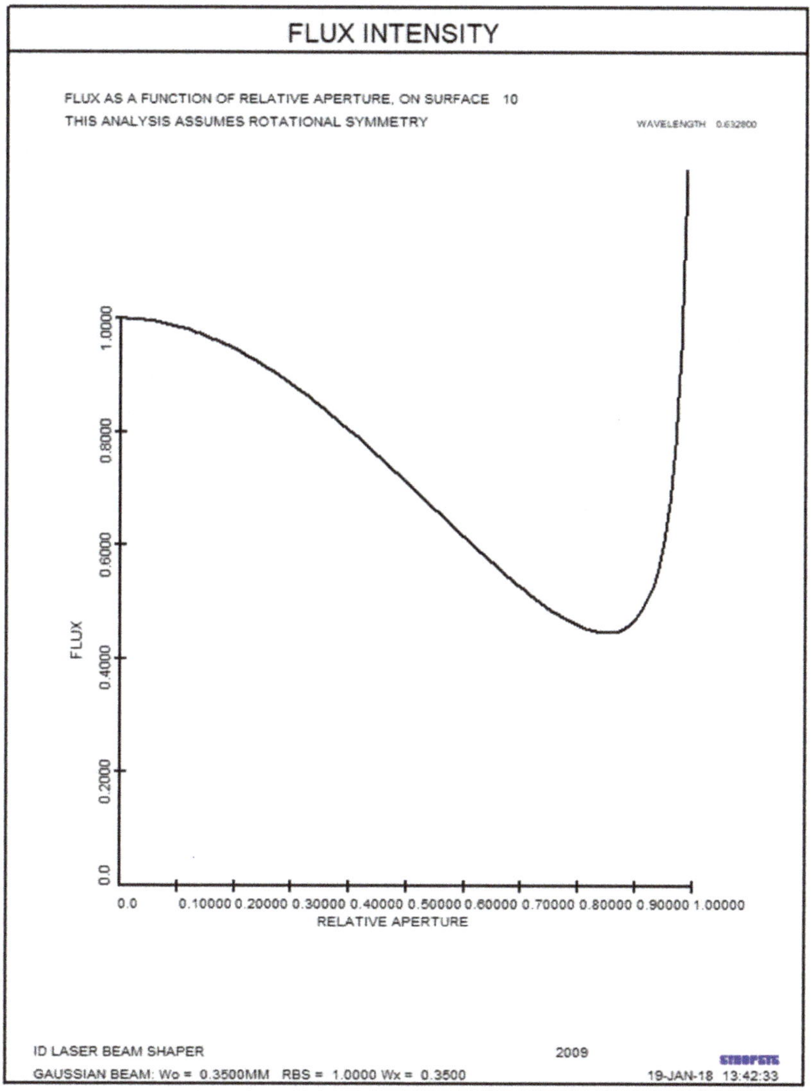

Figure 15.9. Flux uniformity after optimizing with added elements.

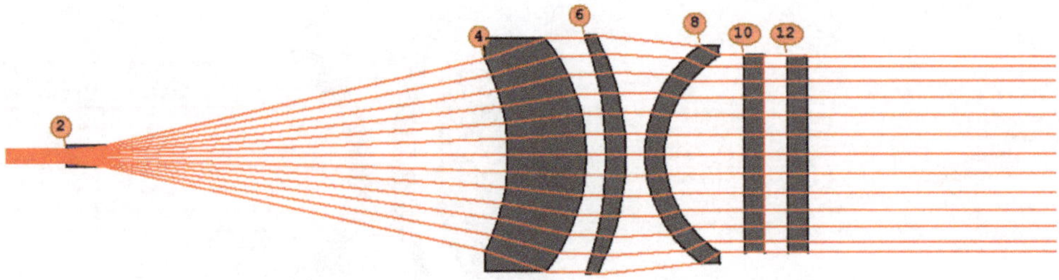

Figure 15.10. Lens with six elements, before optimizing.

Now the flux is somewhat better, but still not uniform enough. We mentioned that it was not so easy to keep the ray angles under control at the same time one flattens the intensity profile. The OPDs are about 0.25 waves.

This seems to be about as good a balance as we can achieve with four elements. What happens if we add some more? Let us start with this design and add two more elements, shown in figure 15.10.

Now we need to add the new variables to the PANT file; also, since the merit function specifies the surface number where certain quantities are to be evaluated, and this number changes whenever we add an element, we can make things easier by changing the surface number to a special symbol, LB1, which means ‘last but one’. Since the highest surface in the lens is currently number 15, this symbol automatically turns into the number 14. Now, if we decide to add or delete elements as we search for a solution, we will not have to edit that number every time. We also add a GSR directive after the GSO, to better keep the ray angles under control, and we lower the weight on the OPD fan to better balance things. We also assign the UMC solve to surface 13 instead of 9 and declare all thicknesses variables (**C15M4**):

```
CHG
NOP
13 UMC
END

PANT            ! Start of variable parameter definitions.
VLIST RAD ALL   ! Vary all radii.
VLIST TH ALL    ! Vary the airspaces and all thicknesses (so AEC works on those
                ! elements).
END

AANT                    ! Start of merit function definition
AEC 1 1 1
ACC 4 1 1
LUL 100 1 1 A TOTL      ! Prevent the system from growing too large; assign upper-
                        ! limit of 150 on TOTL.
M 5 10 A P YA 0 0 1 0 LB1 ! Ask for a beam radius of 5 mm on surfaces 9 and 10
M 5 10 A P YA 0 0 1 0 LB1
M 0 1 A P FLUX 0 0 1 0 LB1 ! Ask for a flux falloff of zero at several zones on 10
M 0 1 A P FLUX 0 0 .99 0 LB1
M 0 1 A P FLUX 0 0 .98 0 LB1
```

```
M 0 1 A P FLUX 0 0 .97 0 LB1
M 0 1 A P FLUX 0 0 .96 0 LB1
M 0 1 A P FLUX 0 0 .95 0 LB1
M 0 1 A P FLUX 0 0 .94 0 LB1
M 0 1 A P FLUX 0 0 .93 0 LB1
M 0 1 A P FLUX 0 0 .92 0 LB1
M 0 1 A P FLUX 0 0 .91 0 LB1
M 0 1 A P FLUX 0 0 .85 0 LB1
M 0 1 A P FLUX 0 0 .8 0 LB1
M 0 1 A P FLUX 0 0 .7 0 LB1
M 0 1 A P FLUX 0 0 .5 0 LB1
M 0 1 A P FLUX 0 0 .3 0 LB1
GSO 0 .1 5 P              ! Control the output ray OPD over an SFAN of 5 rays.
GSR 0 1 5 P
END                       ! End of merit function definition.
SNAP
SYNO 100
```

Run this MACro, and then anneal (22, 1, 50). After many iterations of annealing, the lens (C15L2) looks like figure 15.11.

The flux is well within the target of 10% uniformity, as shown in figure 15.12:

```
FLUX 100 P 14
```

The OPD errors are now under 0.09 waves. It appears that doing this job with all-spherical lenses is possible but requires six elements.

Now we understand why people generally use aspherics or diffractive elements for this kind of job. Then there are fewer elements, so it can be worth the extra manufacturing trouble.

To finish this lesson, let us look at the ray pattern coming out of the system. Go to the footprint dialog (**MFP**), turn off switch 27, select surface 13, 10× scale, and ask for 600 rays. Here is the ray distribution, shown in figure 15.13.

This is pretty good for a beam shaper; rays are more spread out near the center and compressed near the edge, which is exactly the right way to make the beam more uniform.

Now we will use **DPROP** to evaluate the final envelope. In this design, the beam is expanded early on, from surface 3, and diffraction will not play a significant role thereafter.

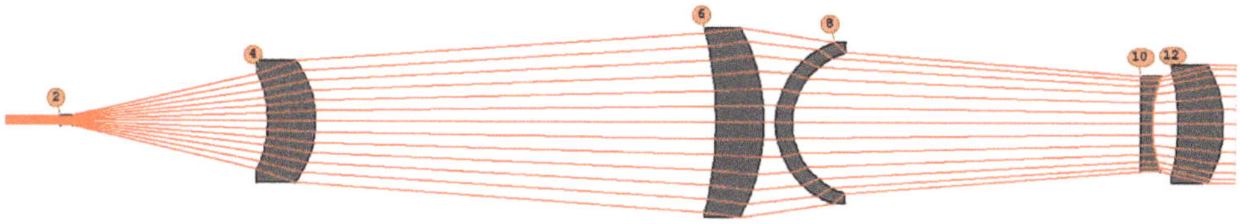

Figure 15.11. Lens optimized with six elements.

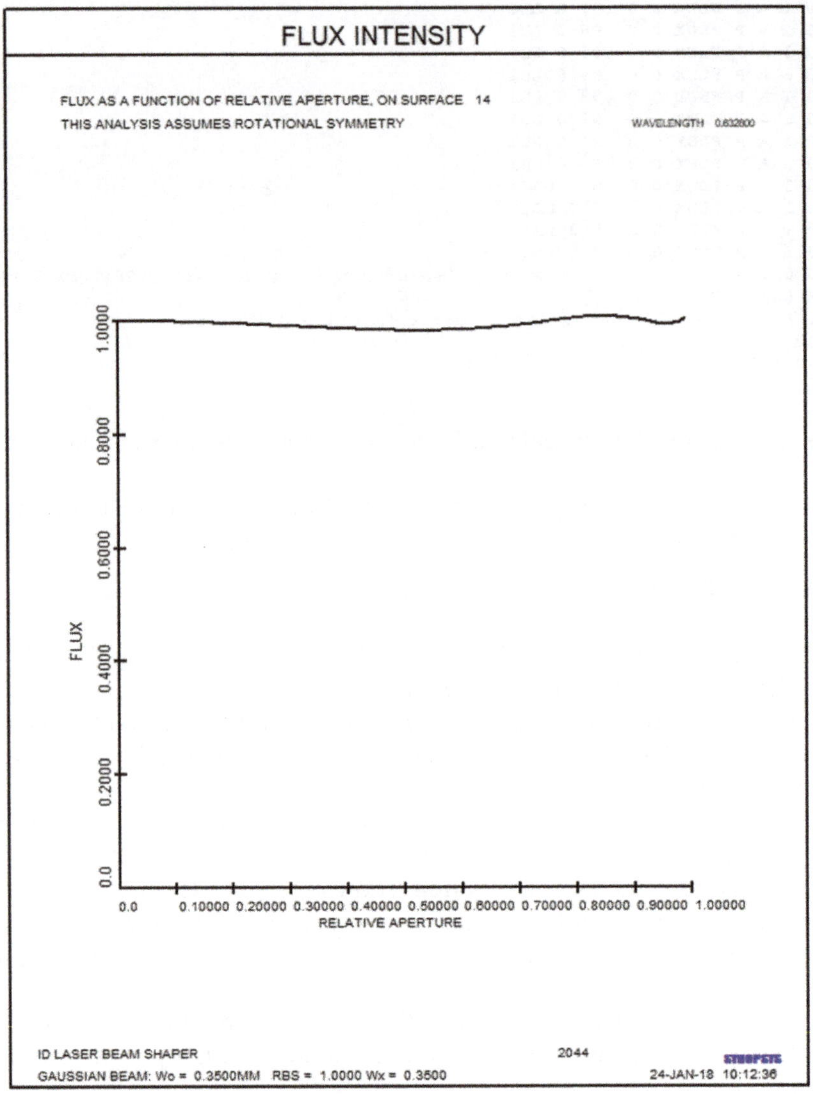

Figure 15.12. Flux uniformity with six-element design.

Here is the input needed to run DPROP:

```
CHG
CFIX
1 TH 0
END
DPROP P 0 0 13 SURF 3 R RESAMPLE
```

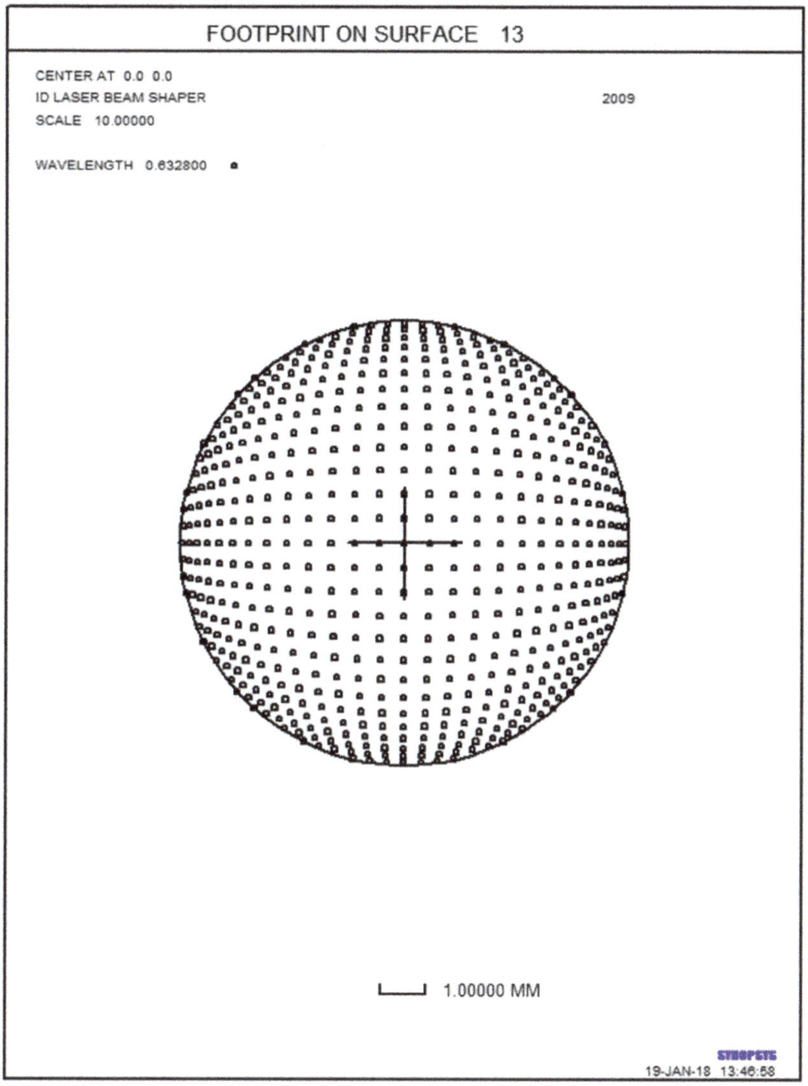

Figure 15.13. Footprint of final lens; rays are concentrated near the edges, compensating for the Gaussian falloff of the entering beam.

Here, we changed all apertures to fixed values, which is always recommended when running DPROP, since that program will often examine a larger area than the lenses allow, if diffraction can send even a small amount of energy there. The fixed apertures prevent this. The wavefront on surface 13 now looks like figure 15.14. It is not too far from flat.

In the next chapter we will develop a beam shaper with aspheric surfaces to illustrate the effect of those very useful variables.

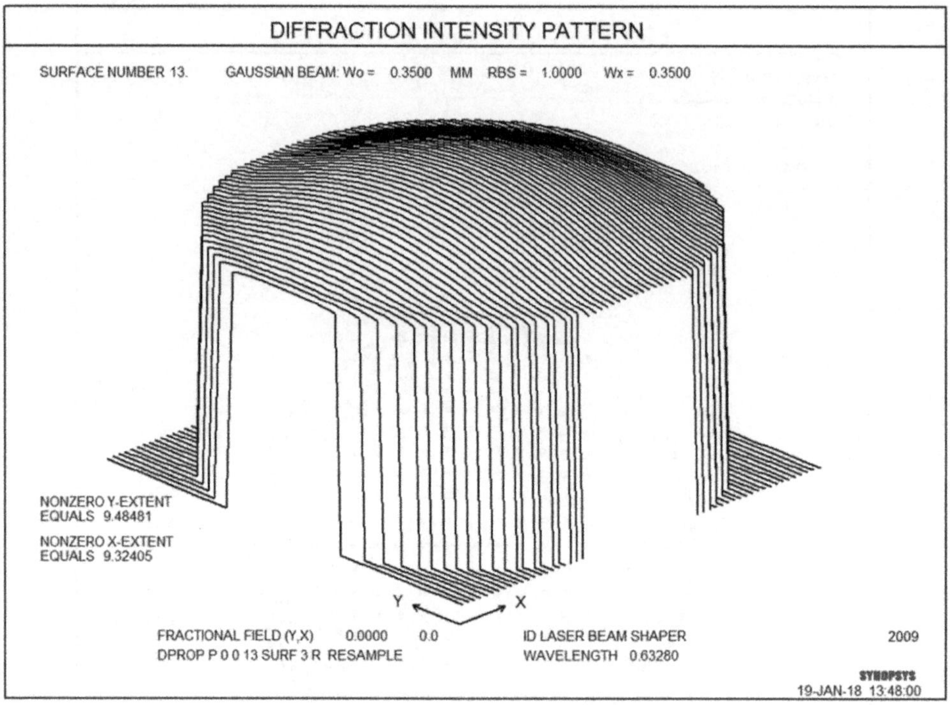

Figure 15.14. DPROP analysis of the final design.

This exercise was intended to show you some of the tools you have available and illustrate what to look out for and how to deal with the inevitable surprises and tradeoffs that turn up when you face a difficult and unfamiliar challenge.

Chapter 16

A laser beam shaper, with aspherics

With aspherics one can design a laser beam shaper with only two elements

In chapter 15 you designed a laser beam shaper to flatten out the Gaussian beam profile of a small HeNe laser. To keep manufacturing costs down, you tried in that lesson to do the job with spherical surfaces, since those are much easier to make than aspherics. You came up with a six-element design that seemed to meet the specs. Perhaps that design could be further improved, but we also must ask whether six spherical lenses will be cheaper to make than two aspherics. If not, then an aspheric design looks more attractive.

Let us start with the same two-element configuration that we used in chapter 15, modified so that we only flatten the flux out to the $1/e^2$ point. Going out to twice that aperture appeared to be impractical before, and we suspect that it will again. Here is our starting point (**C16M1**), in figure 16.1:

```
RLE                          ! Beginning of lens input file.
ID LASER BEAM SHAPER
WA1 .6328                    ! Single wavelength
UNI MM               ! Lens is in millimeters
OBG .35 1            ! Gaussian object; waist radius -.35 mm; define full aperture at the 1/e**2 point.
1 TH 22              ! Surface 2 is 22 mm from the waist .
2 RD -5 TH 2 GTB S ! Guess some reasonable lens parameters; use glass type SF6 from Schott catalog
SF6
3 UMC 0.3 YMT 5    ! Solve for the curvature of surface 3 so the marginal ray has an angle of 0.3; find
                   ! spacing so ray height is 5 mm on next surface
4 RD 20 TH 4 PIN 2           ! Guesses for surface 4
5 UMC 0 TH 50                ! Solve for curvature of 5 so beam is collimated.
7                            ! Surfaces 6 and 7 exist
AFOCAL                       ! because they are required for AFOCAL output.
END                          ! End of lens input file.
```

doi:10.1088/978-0-7503-1611-8ch16

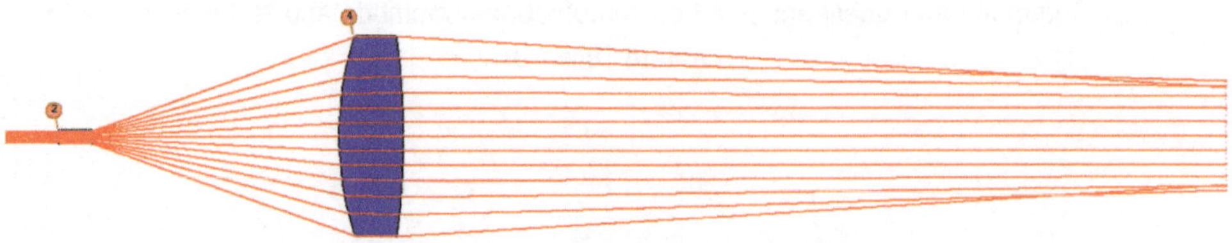

Figure 16.1. Starting system for beam shaper.

Since we learned a great deal in chapter 15, we can start this problem with this merit function (**C16M2**):

```
CHG
NOP              ! Be sure there are no pickups or solves.
4 PIN 2
5 TH 10 UMC 0 ! move surface 6 before the caustic
END

PANT                     ! Start of variable parameter definition.
VLIST RAD 2 3 4 5        ! Vary four radii.
VLIST TH 3               ! Vary the central airspace.
VY 3 CC          ! Vary the conic constant on surface 3.
VY 4 CC          ! And on surface 4.
VY 3 G 3         ! Add three aspheric terms to surface 3.
VY 3 G 6
VY 3 G 10
VY 4 G 3         ! And three to surface 4.
VY 4 G 6
VY 4 G 10
END

AANT             ! Start of merit function definition.
AEC 1 1 1        ! Enable automatic edge feathering control.
ACC 4 1 1        ! Enable automatic center thickness monitoring
ASC              ! Enable automatic slope control, so curves don't get too steep.
LUL 100 1 1 A TOTL              ! Limit the paraxial total length to no more than
                                ! 150 mm.
M 5 100 A P YA 0 0 1 0 LB1
M 5 100 A P YA 0 0 1 0 LB2      ! Assign a target of 5 mm to the marginal ray on
                                ! surfaces 5, 6.

M 0 1 A P FLUX 0 0 1 0 LB1      ! Target the flux difference between the marginal
                                ! ray point and the on-!axis point to 0 on 6.
M 0 1 A P FLUX 0 0 .99 0 LB1    ! Target the flux at the 0.99 aperture point.
M 0 1 A P FLUX 0 0 .98 0 LB1    ! And so on, for a set of zones.
M 0 1 A P FLUX 0 0 .97 0 LB1
M 0 1 A P FLUX 0 0 .96 0 LB1
M 0 1 A P FLUX 0 0 .95 0 LB1
M 0 1 A P FLUX 0 0 .94 0 LB1
M 0 1 A P FLUX 0 0 .93 0 LB1
M 0 1 A P FLUX 0 0 .92 0 LB1
M 0 1 A P FLUX 0 0 .91 0 LB1
M 0 1 A P FLUX 0 0 .9 0 LB1
M 0 1 A P FLUX 0 0 .89 0 LB1
M 0 1 A P FLUX 0 0 .88 0 LB1
```

```
M 0 1 A P FLUX 0 0 .86 0 LB1
M 0 1 A P FLUX 0 0 .84 0 LB1
M 0 1 A P FLUX 0 0 .82 0 LB1
M 0 1 A P FLUX 0 0 .8 0 LB1
M 0 1 A P FLUX 0 0 .7 0 LB1
M 0 1 A P FLUX 0 0 .5 0 LB1
M 0 1 A P FLUX 0 0 .3 0 LB1
GSO 0 .01 10 P            ! Target the OPD of an SFAN of 10 rays to zero, with a
                          ! weight of .01
GSR 0 50 10 P             ! And also target the ray angles to zero.
END

SNAP
SYNO 50
```

While this is pretty straightforward, we should point out some things: why is GSR used to target the ray angles? Usually, GSR controls the actual x-coordinate of each ray relative to the chief ray—but since this system is in AFOCAL mode, where the output is collimated, this entry targets the output *angles* instead.

Where did we specify that the ray and flux targets are to be taken on surface 6? Well, this system has a total of seven surfaces, counting the two dummies at the end that are required for AFOCAL angle conversion. In the previous chapter you learned that the mnemonic LB1 means 'last but one', and here it is replaced by the number 6 when the input is processed. This form of input is valid within PANT and AANT files and is a real time-saver when you want to use the same number in several places.

In this exercise, we have chosen to vary the conic constant and three aspheric terms on two surfaces. There are also higher-order terms available, using a variety of surface specifications—but this is a simple task and we want to see how close we can come with just those terms. This form of aspheric has 22 terms available, but only terms G 3, 6, 10, 16, 18, 19, 20, 21, and 22 are rotationally symmetric. They vary the fourth-, sixth-, eight-, and tenth- through twentieth-order aspheric terms, and here we did not even use the last six.

Now let us run this MACro. Things might still get better, so anneal (**22, 1, 50**).

That brings the merit function down to 4.5E−5, which is a sign that we have reached a good solution, shown in figure 16.2.

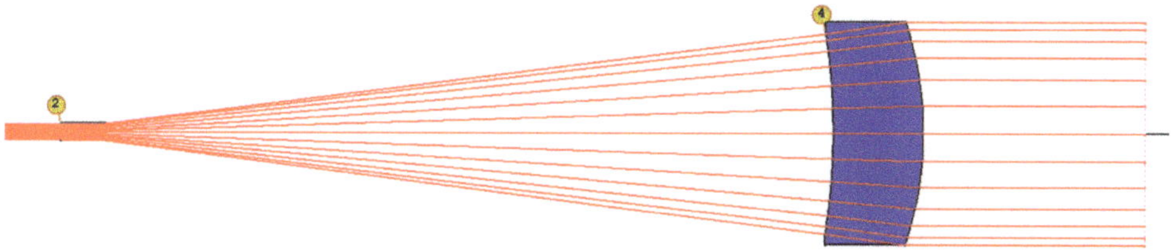

Figure 16.2. Two-lens design with aspherics.

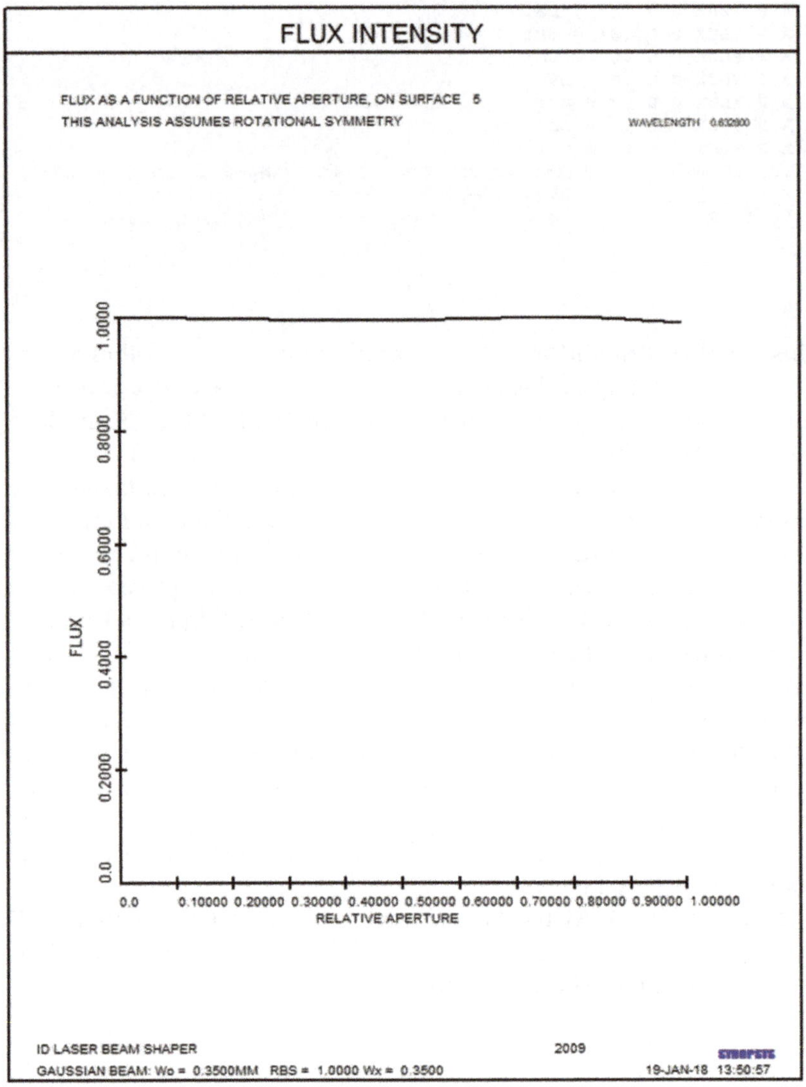

Figure 16.3. Flux uniformity of the aspheric design.

Here is the flux plot for our final design (**C16L1**), in figure 16.3:

```
FLUX 100 P 5
```

The flux is almost perfectly uniform. How about the OPD errors?

```
SYNOPSYS AI>OPD

SYNOPSYS AI>TFA 5 P

 ID LASER BEAM SHAPER                              2055            26-JAN-18   13:44:48
 TANGENTIAL RAY FAN ANALYSIS

 FRACT. OBJECT HEIGHT                       HBAR       0.000000    GBAR       0.000000
 COLOR NUMBER                                1
  REL ENT PUPIL    WAVEFRONT ABERR
       YEN              OPD (WAVES)
 ________________________________
     -1.000              -0.000711
     -0.800              -0.001623
     -0.600               0.001277
     -0.400              -0.000757
     -0.200              -0.000929
      0.200              -0.000929
      0.400              -0.000757
      0.600               0.001277
      0.800              -0.001623
      1.000              -0.000711
 SYNOPSYS AI>
```

Now this design is essentially perfect, with just over 1/1000 wave of error, and it only required two elements. We can be pleased at that. It looks like we do not need the six-element design of chapter 15 after all.

Just to be sure, let us also examine the output wavefront with DPROP, shown in figure 16.4:

```
STORE 9
CHG
CFIX
END
DPROP P 0 0 5 SURF 2.5 R RESAMPLE
GET 9
```

This is very close to what we are after. Now the only question is how difficult the aspherics will be to manufacture.

To see how far the aspherics are from the closest fitting sphere (CFS), enter

```
ADEF 3 PLOT
ADEF 4 PLOT
```

and obtain the results in figure 16.5.

Both of these aspherics are only a few micrometers away from the CFS. It appears that this is manageable. Let us look at the fringe pattern, relative to the CFS, shown in figures 16.6 and 16.7:

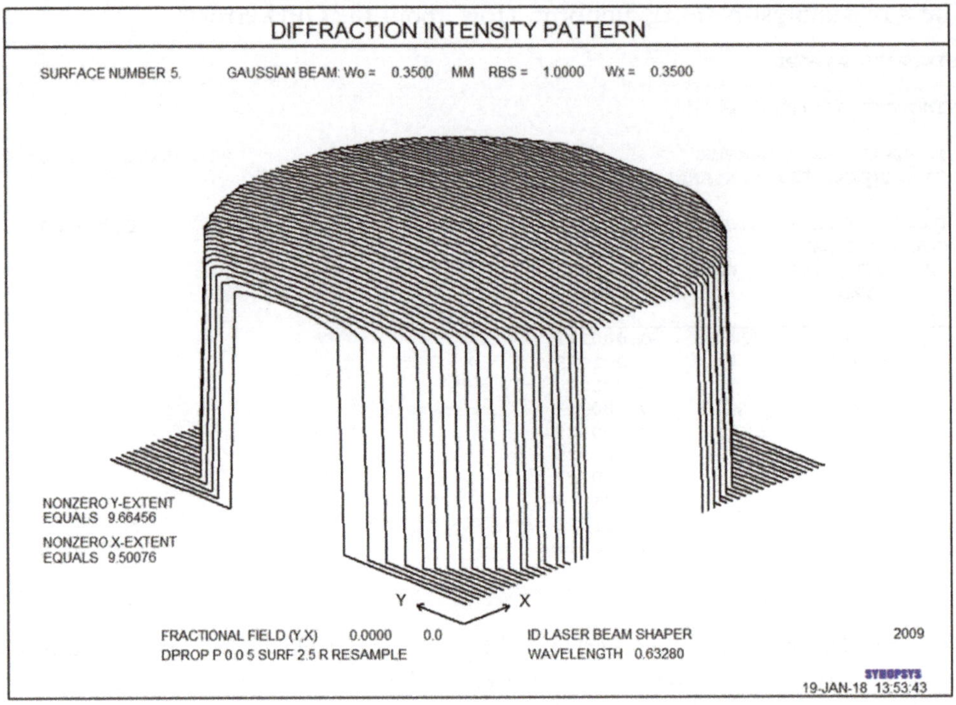

Figure 16.4. Output wavefront of aspheric design.

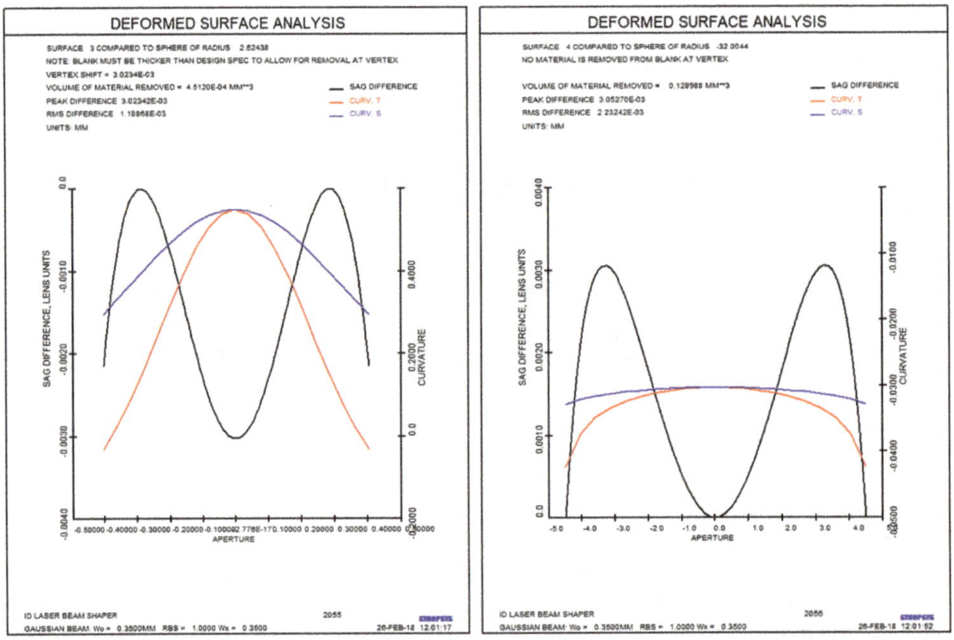

Figure 16.5. Analysis of aspheric surfaces.

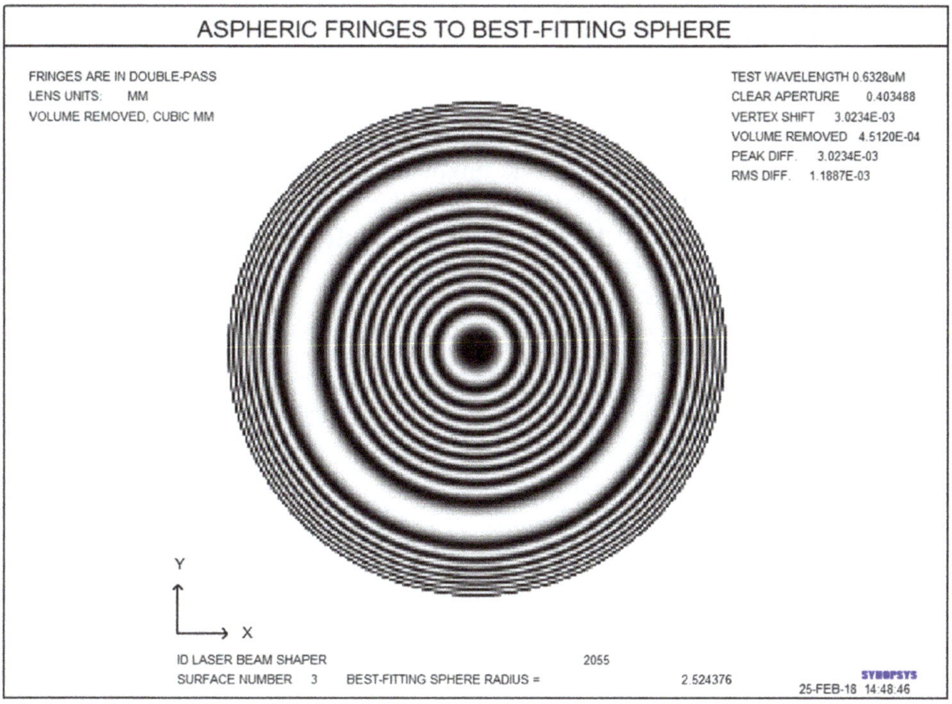

Figure 16.6. Fringe pattern relative to the closest-fitting sphere to surface 3.

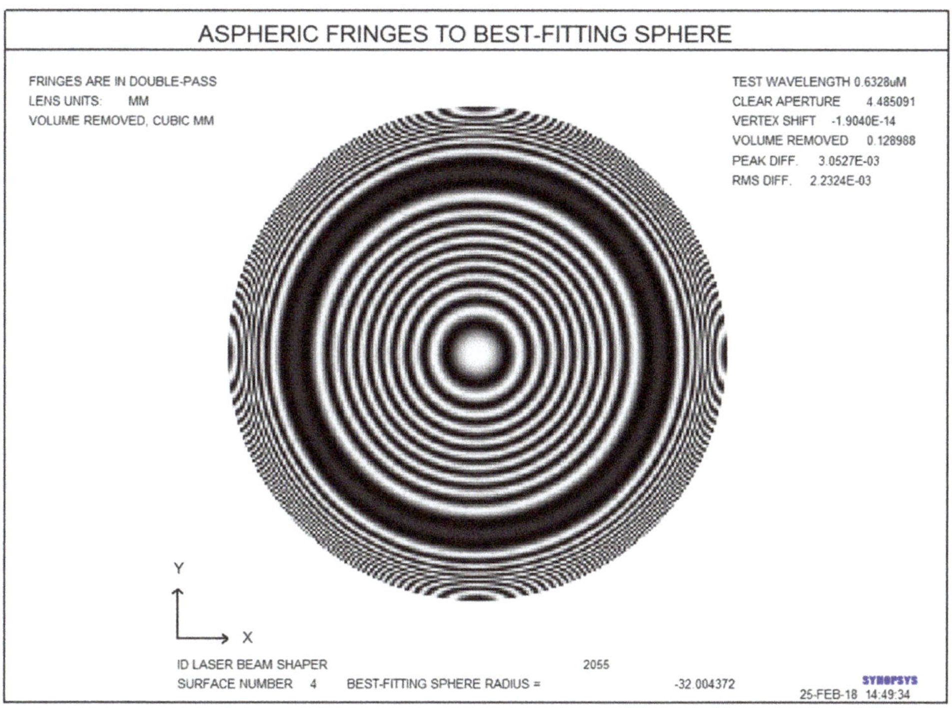

Figure 16.7. Fringe pattern for surface 4.

```
ADEF 3 FRINGES
ADEF 4 FRINGES
```

This may or may not be a challenge to the folks in the shop. Depending on how the aspherics are made and measured, one might try to reduce the aspheric departure somewhat, trading off that departure against the performance. See chapter 24 to learn how one can use the CLINK optimization feature to do just that.

IOP Publishing

Lens Design

Automatic and quasi-autonomous computational methods and techniques
Donald Dilworth

Chapter 17

A laser beam expander with kinoform lenses

A laser beam shaper with kinoform lenses requires only two elements

In chapter 15 you saw how a laser beam expander could be designed with ordinary spherical lenses and learned that one would need six elements to achieve good performance that way. Chapter 16 did the same thing with only two aspheric elements, with excellent results. This lesson will show that you can do as well with diffractive optical elements (DOEs), also known as kinoform lenses.

The problem is to convert a HeNe laser with a waist radius of 0.35 mm to a beam that is 10 mm in diameter and uniform to within 10%.

Here is the input file for our starting point (**C17M1**):

```
RLE                          ! Beginning of lens input file.
ID KINOFORM BEAM SHAPER
WA1 .6328                    ! Single wavelength
UNI MM                ! Lens is in millimeters
OBG .35 1             ! Gaussian object; waist radius -.35 mm; define full
                      ! aperture = 1/e**2 point.
1 TH 22               ! Surface 2 is 22 mm from the waist.
2 RD -2 TH 2 GTB S    ! Guess some reasonable lens parameters; use glass
                      ! type SF6 from Schott catalog
SF6

3 TH 20               ! Surface 3 is a kinoform on side 2 of the first
                      ! element
3 USS 16              ! Defined as Unusual Surface Shape 16 (simple DOE)
CWAV .6328            ! Zones are defined as one wave phase change at this
                      ! wavelength
HIN 1.7988 55         ! Assume the zones are machined into the lens.  You
                      ! can also apply a film of a different index.
RNORM 1

4 TH 2 GTB S          ! The first side of the second element is also a DOE
SF6
```


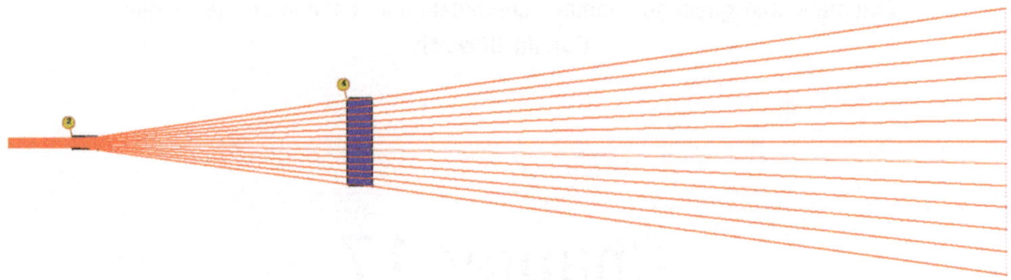

Figure 17.1. Starting point for DOE beam shaper.

```
4 USS 16
CWAV .6328
HIN 1.7988 55
RNORM 1

5 CV 0 TH 50          ! Start with a flat surface
7                     ! Surfaces 6 and 7 exist
AFOCAL                ! because they are required for AFOCAL output.
END                   ! End of lens input file.
```

We guessed a value for RD number 2, and we came close enough to start with. Here is the system at this stage, with no aspheric DOE terms yet, shown in figure 17.1. If you are not yet familiar with the USS surface shapes, open the help file by typing

```
HELP USS
```

in the Command Window and selecting type USS 16, a simple DOE.

The beam is expanded but not collimated, and the intensity profile is still that of the Gaussian input beam. The task is to find the DOE OPD terms that will accomplish both of our goals. To start with, we will keep both sides of the second element flat but add aspheric terms to it to define the DOE. Here is an optimization MACro that might do the job (**C17M2**):

```
PANT           ! Start of variable parameter definitions.
RDR .001       ! This is a very small beam, so use smaller derivative increments
               ! to start with
VY 2 RAD
VLIST TH 3     ! Vary the airspace
VY 3 G 26      ! Vary term Y**2,
VY 3 G 27      ! Y**4,
VY 3 G 28      ! and Y**6
VY 3 G 29      ! and Y**8

VY 4 G 26      ! Do the same at surface 4
VY 4 G 27
VY 4 G 28
VY 4 G 29

END
```

```
AANT            ! Start of merit function definition
AEC 1 1 1
ACC 4 1 1
LUL 150 1 1 A TOTL          ! Prevent the system from growing too large
M 5 1 A P YA 0 0 1 0 5      ! Ask for a beam radius of 5 mm on surface 5

M 0 1 A P FLUX 0 0 1 0 6    ! Ask for a flux falloff of zero at several zones
M 0 1 A P FLUX 0 0 .98 0 6
M 0 1 A P FLUX 0 0 .97 0 6
M 0 1 A P FLUX 0 0 .96 0 6
M 0 1 A P FLUX 0 0 .95 0 6
M 0 1 A P FLUX 0 0 .94 0 6
M 0 1 A P FLUX 0 0 .93 0 6
M 0 1 A P FLUX 0 0 .92 0 6
M 0 1 A P FLUX 0 0 .91 0 6
M 0 1 A P FLUX 0 0 .85 0 6
M 0 1 A P FLUX 0 0 .8 0 6
M 0 1 A P FLUX 0 0 .7 0 6
M 0 1 A P FLUX 0 0 .5 0 6
M 0 1 A P FLUX 0 0 .3 0 6
GSO 0 .1 10 P          ! Control the output ray OPD over an SFAN of 10 rays,
GSR 0 100 10 P         ! and some transverse aberrations too.
END                    ! End of merit function definition.

SNAP
SYNO 40
```

This PANT file varies some of the general-purpose G variables, which we used in the previous lesson to vary the aspheric terms on the lens elements. However, in this case, surfaces 3 and 4 are already defined as USS type 16, which is a simple DOE surface, and those terms therefore alter selected coefficients defining that shape. (The help file describes how the G terms are applied to each of the USS surface types.)

Run this, and the lens looks promising. So run it again, and then anneal (**22, 1, 50**). You obtain the lens in figure 17.2.

This is not bad at all. Let us try varying some of the higher-order coefficients. Add new terms on both DOEs in the PANT file, up to G 31, which is the Y**12 term. After reoptimizing, the lens (**C17L1**) looks much the same, but the merit function drops to 3.4E−7—it looks like the run converged.

How does the flux vary over the aperture now? Type the command

```
FLUX 100 P 6
```

and you obtain a beautiful curve, almost straight, shown on the left in figure 17.3.

This is indeed an excellent design. The question now is, can anyone make it? What is the spatial frequency on surface 4? If it is too high, fabrication technology

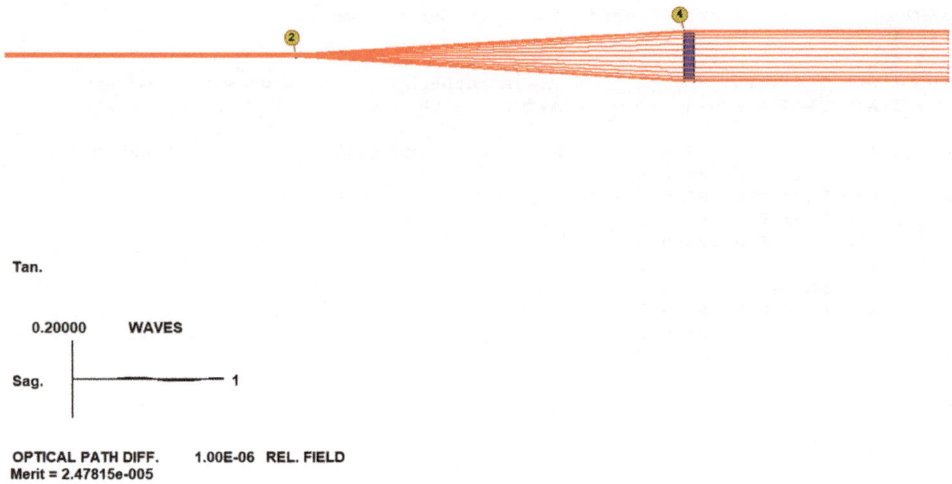

Figure 17.2. DOE-based beam shaper, after optimization.

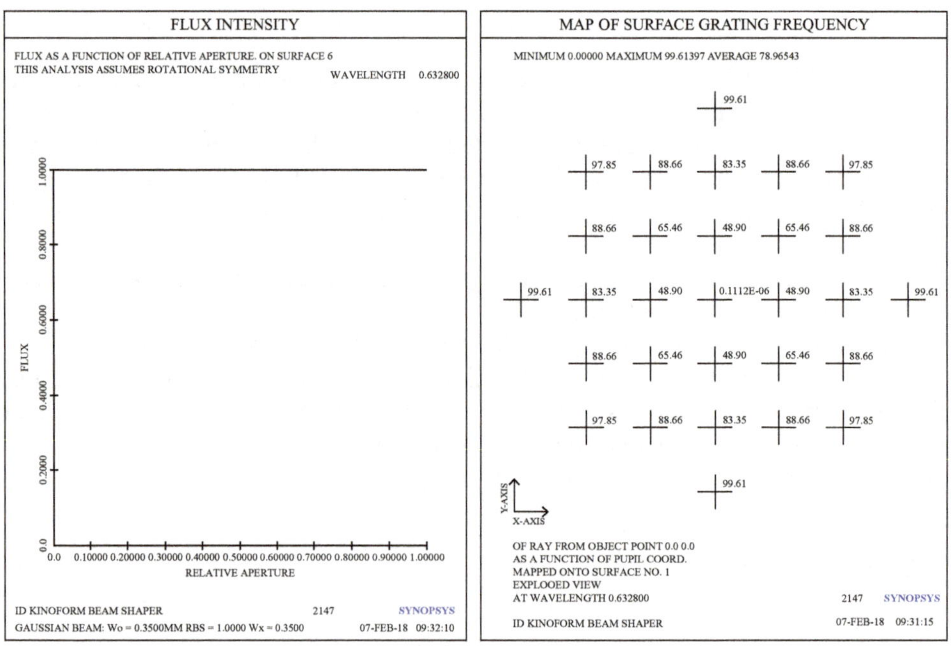

Figure 17.3. Flux uniformity of DOE-based beamshaper (left) and MAP of surface spatial frequency on surface 4.

may have trouble with it. Open the **MMA** (MMA) dialog to select the input for a **MAP** command, select a map of 'HSFREQ' over 'PUPIL' with object 'POINT 0' and raygrid 'CREC' with a grid of 7, 'DIGITAL' output, and 'PLOT'. The result shows a frequency of 99.61 c mm^{-1} at the edge of the lens, on the right in figure 17.3.

That works out to just 10 µm/cycle, which is possible but perhaps not easy. Can we reduce that to, say, 50 c mm^{-1}? Add the variable 5 RAD to the variable list and add a new aberration to the AANT file:

```
M 50 .01 A P HSFREQ 0 0 1 0 4
```

The program now controls the spatial frequency at the marginal ray intercept on surface 4. Reoptimize, and now surface 5 is slightly convex and the spatial frequency on 4 is right at 50 c mm^{-1}. The flux uniformity is as good as before—mission accomplished!

How well did we do? Run the **DPROP** command, asking for the profile at surface 3, before the beam has been restructured. This shows the Gaussian profile of the beam at that point, as in figure 17.4:

```
DPROP P 0 0 3 SURF 3 L RESAMPLE
```

Now we do the same on surface 6, with the results shown in figure 17.5—essentially perfect!

```
DPROP P 0 0 6 SURF 3 L RESAMPLE
```

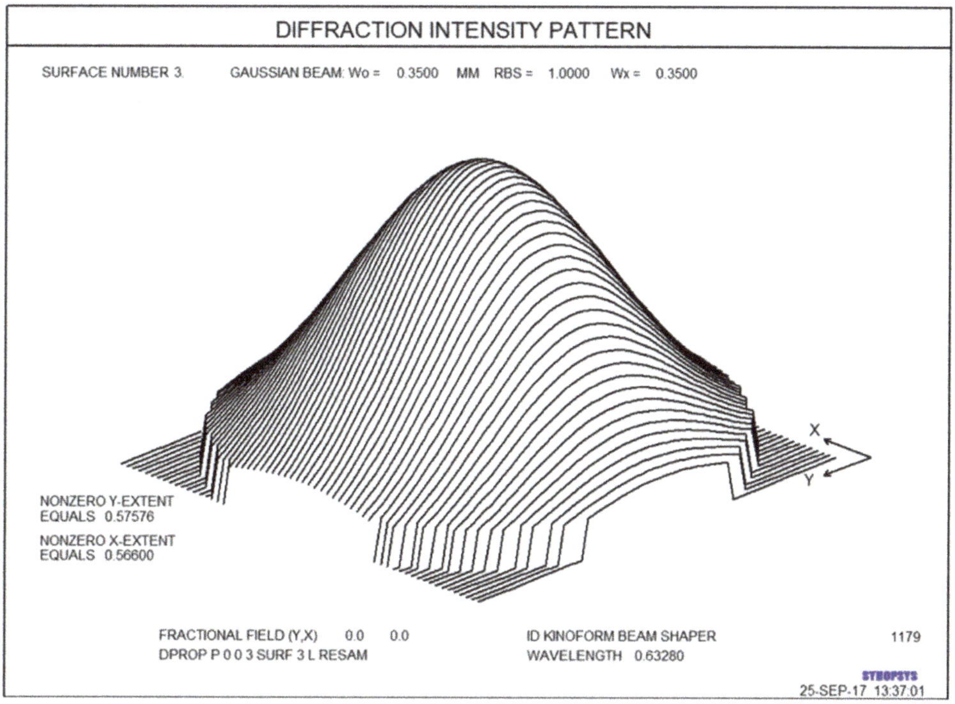

Figure 17.4. Beam intensity plot at surface 3.

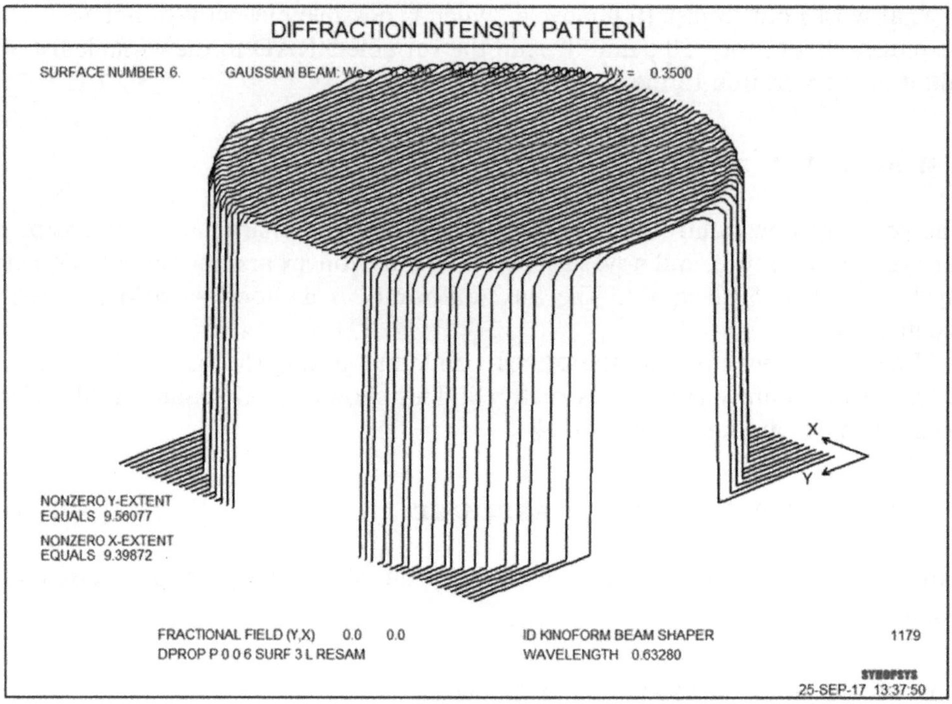

Figure 17.5. Beam intensity plot at surface 6.

IOP Publishing

Lens Design

Automatic and quasi-autonomous computational methods and techniques

Donald Dilworth

Chapter 18

A more challenging optimization challenge

Designing a lens with plane-parallel plates; inserting real glass types for the glass models

In chapter 7 you designed a seven-element lens starting from plane-parallel surfaces, which is about as close to starting from scratch as you can get in this business. That lesson was intended to demonstrate the speed of the PSD III optimization algorithm, which is one of the factors that make modern number crunching so effective.

In this lesson, you will start with the same system—but in this case, you want to achieve a high MTF at four field points and substitute catalog glass types for the glass models of the earlier lesson. To do the latter, you will use the automatic real-glass insertion program, **ARGLASS**[1].

Here is the input and optimization MACro (**C18M1**):

```
RLE                    ! The starting system.
ID TEST PSD III
OBB 0 20 12.7
WAVL CDF
UNITS MM
1 TH 5 GLM 1.6 50
2 TH 5
3 TH 5 GLM 1.6 50
4 TH 5
5 TH 5 GLM 1.6 50
6 TH 5
7 TH 5 GLM 1.6 50
8 TH 5
9 TH 5 GLM 1.6 50
10 TH 5
11 TH 5 GLM 1.6 50
12 TH 5
13 TH 5 GLM 1.6 50
14 TH 50
```

[1] ARGLASS™ is a trademark of Optical Systems Design, Inc., a Maine, USA corporation.

doi:10.1088/978-0-7503-1611-8ch18

```
15
APS 7
END
PAD/U          ! Show the initial system.
TIME              ! Start a timer, then define a symbol, AWT, for the ap. weight

AWT: 0.5       ! almost equal weight over aperture
QUIET          ! not showing everything on the monitor speeds things up

PANT           ! Define variables.
VY 1 YP1        ! Vary the paraxial stop position.
VLIST RAD 1 2 3 4 5 6 7 8 9 10 11 12 13 14
VLIST TH ALL
VLIST GLM ALL
END

AANT              ! Start of merit function definition.
AEC
ACC
M 33 2 A GIHT
GSR AWT 5 5 M 0     ! Note how weights are assigned to the several fields.
GNR AWT 5 4 M .3    ! This creates a ray grid at the .3 field point
GNR AWT 5 4 M .5    ! These for the 0.6 field point
GNR AWT 5 4 M .65   ! These for the 0.75 field point
GNR AWT 4 4 M .8    ! These for the 0.8 field point
GNR AWT 4 4 M 1     ! Full field
END

SNAP 100
DAMP 1
SYNOPSYS 10
SYNOPSYS 50
SYNOPSYS 50
SYNOPSYS 100
ANNEAL 50 10
ANNEAL 50 10

LOUD              ! Restore output to the monitor
MERIT?

STORE 3           ! Store the results in the library.

TIME                           ! See how long the job took
MOF M 0 40 80 0 Q 30 20 10     ! Calculate the MTF over field.
```

The job runs for about 35 s and produces the lens in figure 18.1 (**C18L1**) and the MTF in figure 18.2.

Optically, this lens is superb—but some elements are too thin. We have to fix that. Select the lines in the MACro from PANT to the first SYNOPSYS command. Type <Ctrl>+C to copy them to the clipboard. Then click the 'NewMACro' Window button and type <Ctrl>+V to paste those lines into the new editor window. Now add a line to the AANT section:

```
ADT 6 1 10
```

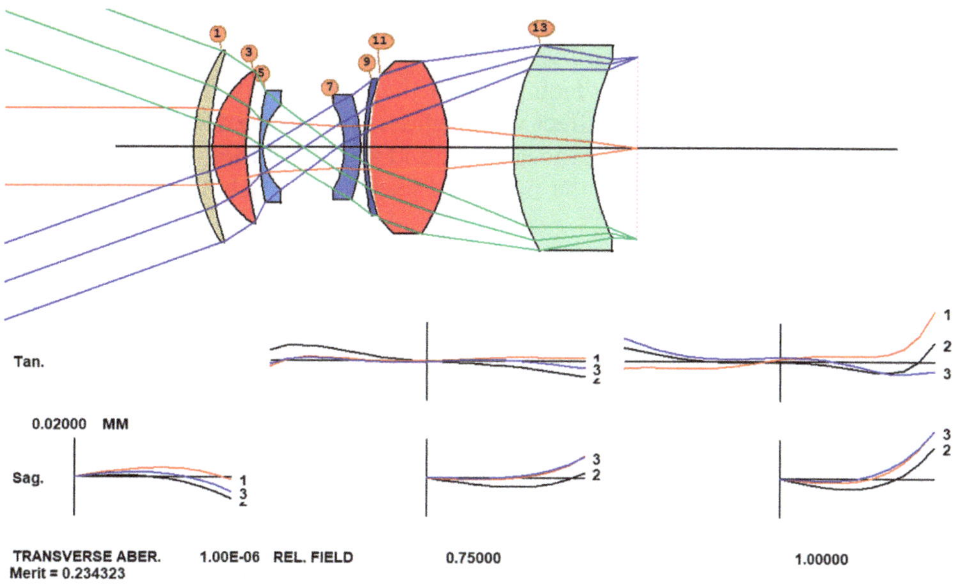

Figure 18.1. Results of first optimization.

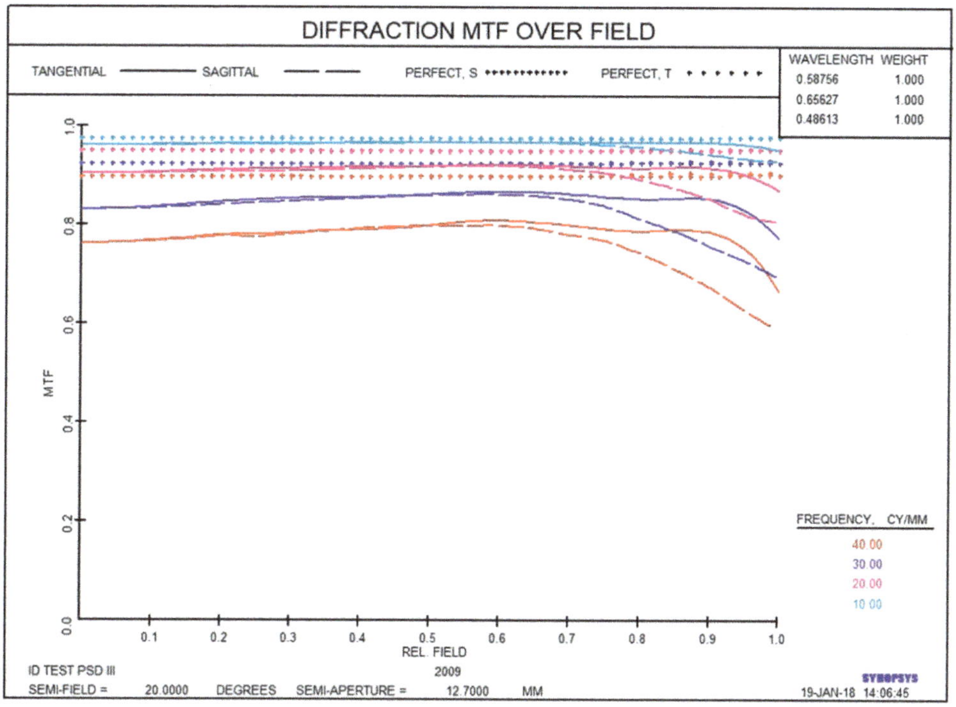

Figure 18.2. MTF over field of the optimized lens.

and run this version. The elements are thicker, and the MF goes up to 0.255. Anneal **(55, 2, 50)**, and the MTF is again very good.

Now open the dialog (**MRG**), which prepares the input for **ARGLASS** and lets you specify a number of filters that affect which glasses the program selects. You might only want inexpensive glasses, or those with good acid resistance, for example. Here is what is selected when we run the program, selecting the Schott catalog and Sort **(C18L2)**:

```
– ARGLASS 6 QUIET
Lens number      6 ID TEST PSD III
 GLASS N-SF66            HAS BEEN ASSIGNED TO SURFACE    9; MERIT =    0.241992
 GLASS N-LASF31A         HAS BEEN ASSIGNED TO SURFACE    1; MERIT =    0.250926
 GLASS N-LASF31A         HAS BEEN ASSIGNED TO SURFACE   13; MERIT =    0.243235
 GLASS N-LAK21           HAS BEEN ASSIGNED TO SURFACE    3; MERIT =    0.266684
 GLASS N-SF15            HAS BEEN ASSIGNED TO SURFACE    7; MERIT =    0.244131
 GLASS N-LAK21           HAS BEEN ASSIGNED TO SURFACE   11; MERIT =    0.263512
 GLASS F2                HAS BEEN ASSIGNED TO SURFACE    5; MERIT =    0.272759
 Type <ENTER> to return to dialog.
```

To examine the properties of these glasses, enter the command

```
PGA ALL                  !Print Glass Attributes, all glasses
```

and you obtain a table, part of which is shown here:

```
***********************************************************
 GLASS ATTRIBUTE FOR SURFACE NO.   11
 SCHOTT          N-LAK21
 GLASS IS A PREFERRED TYPE.
 GLASS IS ENVIRONMENTALLY SAFE (NO Pb OR As).

 PRICE   BUBBLE   HUMIDITY   STAIN   ACID RESIST   ALKALI RESIST   SP GRAVITY
   3.5        1          4       3             6               4         3.74
 THIS GLASS HAS A LIST OF TRANSMISSION VALUES ATTACHED
 VALID RANGE OF TRANSMISSION DATA:
 LOW     HIGH
  0.320   2.500
 GLASS HAS SELLMEIER INDEX COEFFICIENTS:
    0.1227181E+01  0.4207837E+00  0.1012848E+01  0.6020757E-02  0.1968629E-01
0.8843701E+02
 GLASS HAS 6 DNDT VALUES FROM GLASS TABLE:
 -2.3600E-06  1.1500E-08  1.1100E-11  3.1000E-07  2.7800E-10  2.3400E-01
 THERMAL COEFFICIENT (ALPHA) =   0.680E-05

 ***********************************************************
```

This looks like what you are after. This lens is shown in figure 18.3. To finish up, you can insert a dummy surface where the stop wants to be and then assign a real stop there and reoptimize.

We recommend you run this exercise yourself (you will need a license, since the read-only mode will not allow you to save the lens, and the 12-surface mode will not allow seven elements). Try changing some of the field weights or the aperture weight and running it again. The results are rather sensitive to those changes, and you will

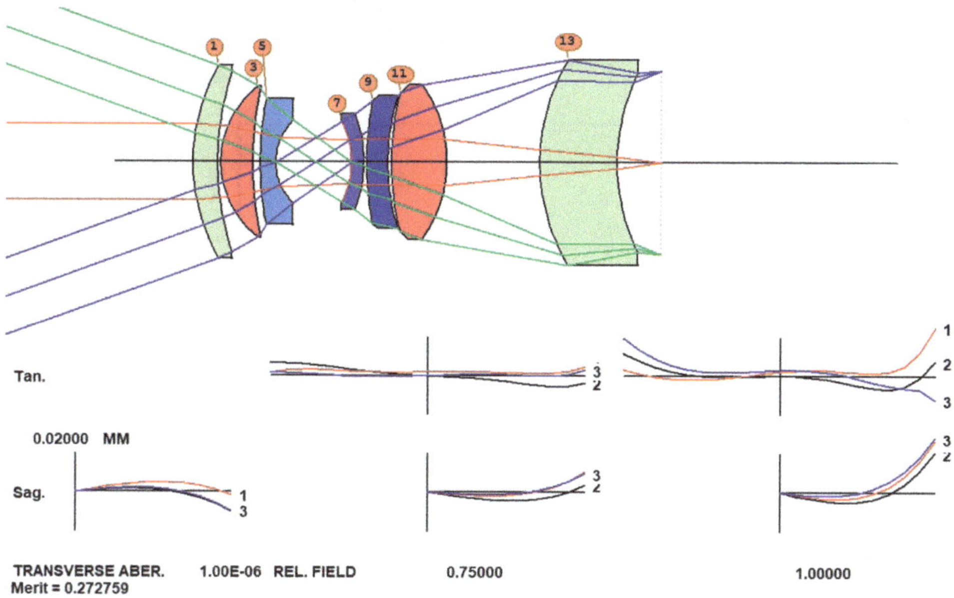

Figure 18.3. The final lens, with real glass.

need to get a feel for what works and what does not as you develop your own lens design skills.

This example started with plane-parallel plates and produced a rather good lens. The final form strongly resembles the classic double Gauss, with a central stop and elements on each side bent inward. It is always pleasant when DSEARCH rediscovers one of the classical lens forms.

What happens if you run this job on DSEARCH? (That program starts with nonzero powers, assigned according to its rules, and finds many more designs.) I tried it on this problem and obtained an even better solution. Try it yourself and see! Adjust the input variables to see what happens. This is your most powerful tool, so it makes sense to learn how to use it.

18.1 Glass absorption

A final question: this lens has two dense flint elements, at surfaces 7 and 9. Glasses in that region of the glass map tend to absorb at the shorter wavelengths—which can affect transmission in the blue. It is a good idea to check. Type the command **XCOLOR**, and you obtain the plot in figure 18.4. There you see that the transmitted color indeed has a slight yellow tint. If that is a problem for the application, you have to find a glass type with better blue transmission than N-SF66 from the Schott catalog, used on surface 9 in this example. The glassmap display (**MGT**) can clarify the situation, as illustrated in figure 18.5, where the length of the red line is a function of the absorption at 0.4 μm. On the left you see that glass N-SF66 has a long red line, because its absorption at that wavelength is rather high, and there are no glasses nearby that are much better. However, sometimes a different glass

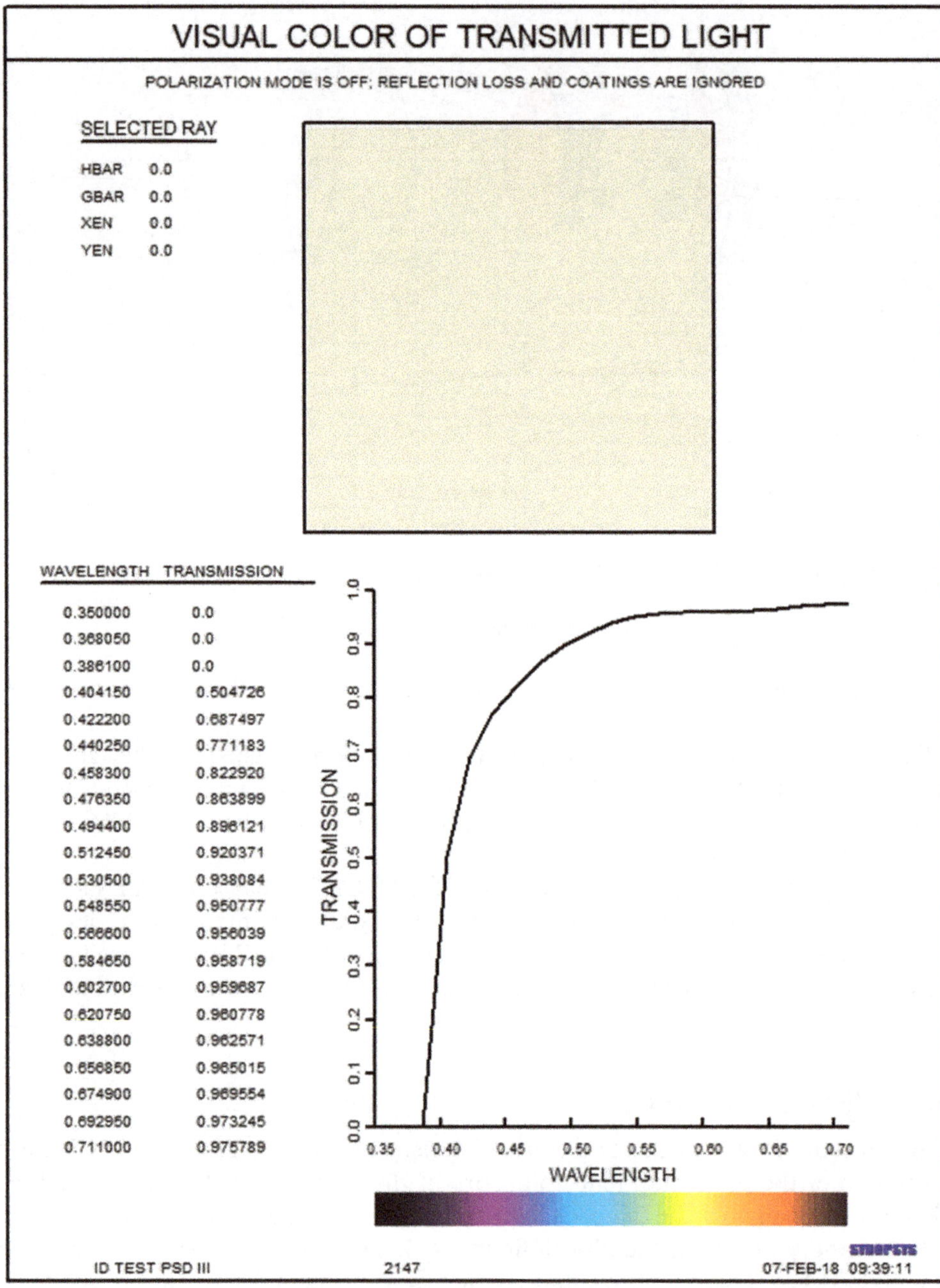

WAVELENGTH	TRANSMISSION
0.350000	0.0
0.368050	0.0
0.386100	0.0
0.404150	0.504726
0.422200	0.687497
0.440250	0.771183
0.458300	0.822920
0.476350	0.863899
0.494400	0.896121
0.512450	0.920371
0.530500	0.938084
0.548550	0.950777
0.566600	0.956039
0.584650	0.958719
0.602700	0.959687
0.620750	0.960778
0.638800	0.962571
0.656850	0.965015
0.674900	0.969554
0.692950	0.973245
0.711000	0.975789

Figure 18.4. The transmitted color through the lens. Absorption in the blue region of the spectrum causes the transmitted beam to appear somewhat yellow, which may be a problem. A change of glass type may be in order.

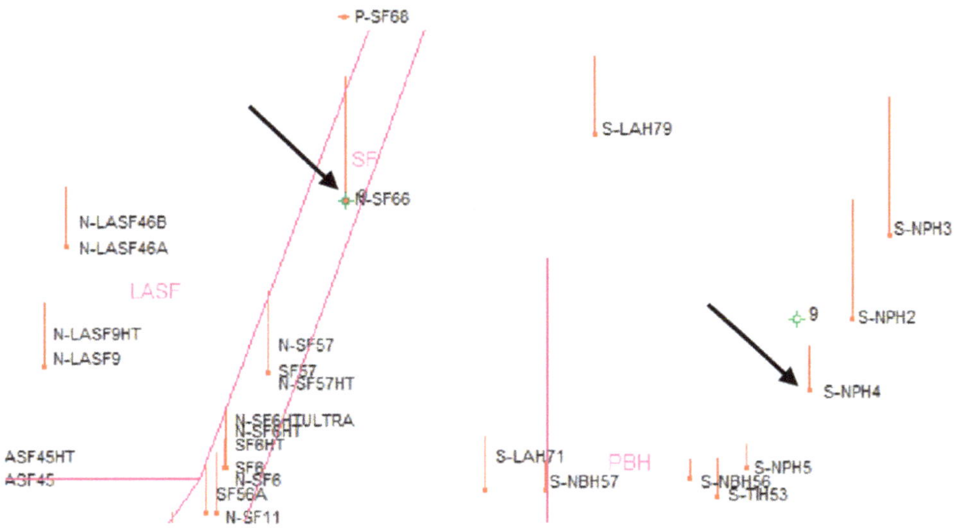

Figure 18.5. Absorption at a wavelength of 4 μm; the Schott catalog on the left, Ohara on the right. For this lens, type S-NPH4 may be a better choice than N-SF66.

company can provide a more suitable glass, as shown on the right, where glass S-NPH4 from Ohara looks like a better choice for this lens.

In the event you need a certain glass type and cannot find a suitable substitute, the only other option is to reduce the thickness of the offending element as much as physically practical. If the lens is intended for aerial reconnaissance, the yellow tint is of no concern, since those systems generally use a yellow filter to cut through atmospheric haze anyway.

IOP Publishing

Lens Design

Automatic and quasi-autonomous computational methods and techniques

Donald Dilworth

Chapter 19

Real-world development of a lens

Global search for a seven-element lens; correcting for two object distances

In chapter 18 you designed a seven-element lens starting with nothing but plane-parallel surfaces and had the program fit the design to catalog glass types automatically with the ARGLASS feature. This lesson will develop the lens further, as you would do if you wanted to manufacture it, and describes some additional procedures that would then be appropriate. To make it a real 'real-world' lesson, we will show how a designer will follow various clues in order to arrive at a solution, and how not all clues lead to success. That is important too: it is instructive to see how sometimes one wanders into blind alleys. As you develop your skills as a lens designer, you will encounter many of them, and should not be discouraged since it happens to all of us.

You will do this lesson in two ways; first with DSEARCH along with a number of other tools. Then, in chapter 21, we show another approach that is actually quicker and easier. You should know about the tools used in both of those approaches.

First, run DSEARCH to find a good starting point. Here is the input (**C19M1.MAC**):

```
CORE 14
DSEARCH 6  QUIET
SYSTEM
ID DSEARCH SAMPLE
OBB 0 20 12.7
WAVL 0.6563 0.5876 0.4861

UNITS MM
END
GOALS
ELEMENTS 7
```



```
FNUM 3.575
BACK 50 SET
STOP MIDDLE
STOP FREE
RSTART 600
RT 0.5
FOV 0.0 .5 .7 .9 1
FWT 2 1 1 1 1
NPASS 55
ANNEAL 200 20 Q 44
COLORS 3
SNAPSHOT 10
QUICK 55 55
END
SPECIAL PANT

END
SPECIAL AANT
LUL 150 1 1 A TOTL
END
GO
```

Notice the **RT** parameter in this file. That controls how individual rays are weighted in the merit function, as you saw in chapter 7. A value of zero gives all the rays in a given grid the same weight, while a higher value will weight rays near the center of the pupil more than rays near the edge. That is a useful way to increase the resolution of a lens; the ray fans may fly away strongly right near the edge, but if the central portion is very flat the resolution will be high anyway. This is a parameter you will often want to experiment with. A value of 0.5 is a good initial guess. The initial radius assigned to each surface is also a useful parameter. Here, we set it about six times the desired focal length, which is often a good choice too.

We also set the back focus distance to exactly 50 mm in this case, which means the program will not use a YMT solve on the last airspace. This, too, is a parameter that can sometimes explore different branches. We will fix up element thicknesses later, when we have a good configuration, so there is no monitor applied at this stage.

Run this, and then optimize and anneal (**50, 2, 50**), using the file DSEARCH_OPT, which is in a new editor window. The lens, shown in figure 19.1, is quite good.

Suppose you want the lens to work over a range of object distances from one meter to infinity. There are two ways to implement that requirement: with multi-configurations, which is very flexible but complicated, or by declaring it a zoom lens in which the object distance zooms. The second approach is better here, since it is simpler, does what you want, and you can examine intermediate object distances very easily. You have to set up this lens as a ZFILE zoom lens. However, first you have to assign a real stop where it looks like it wants to be, at surface 9:

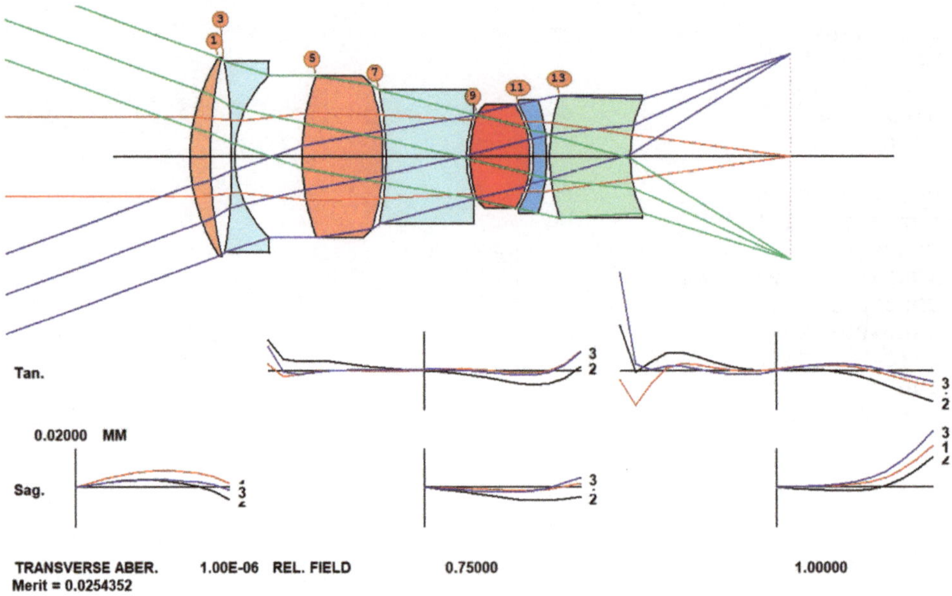

Figure 19.1. Lens as returned by DSEARCH, then optimized and annealed.

```
CHG
APS -9
END
```

Then optimize and anneal again, after deleting the variable YP1. Now the lens is ready to be turned into a zoom lens.

Enter the following in a new editor:

```
CHG
15 CAO 32        ! fix the CAO on the image (so FFIELD works)
FFIELD           ! adjust the object height so the image fills the CAO there
14 YMT           ! assign a paraxial focus solve to surface 14
ZFILE 1          ! start of the ZFILE section
14 14            ! there is one zooming group, the last thickness
ZOOM 2          ! ZOOM 1 is default; ZOOM 2 gets OBA object on the next line
OBA 1000 -366.554 12.7       ! the object description at this zoom
END              ! end of changes
```

This input sets a hard aperture at the image so the FFIELD directive has a target, puts a thickness solve on 14 so all zooms refocus automatically, and declares a single zooming group, surface 14. Then it defines the object distance for ZOOM 2 at 1000 mm distance, with a negative YPP0 because the value in ZOOM 1 is also negative, and they have to have the same sign.

Run this MACro, and the lens changes to a zoom lens, with only a single airspace zooming in this case. Now you see a new toolbar on the right side of the monitor. What does the image look like in ZOOM 2? Click on button 2, and you see the lens at that zoom setting, shown in figure 19.2.

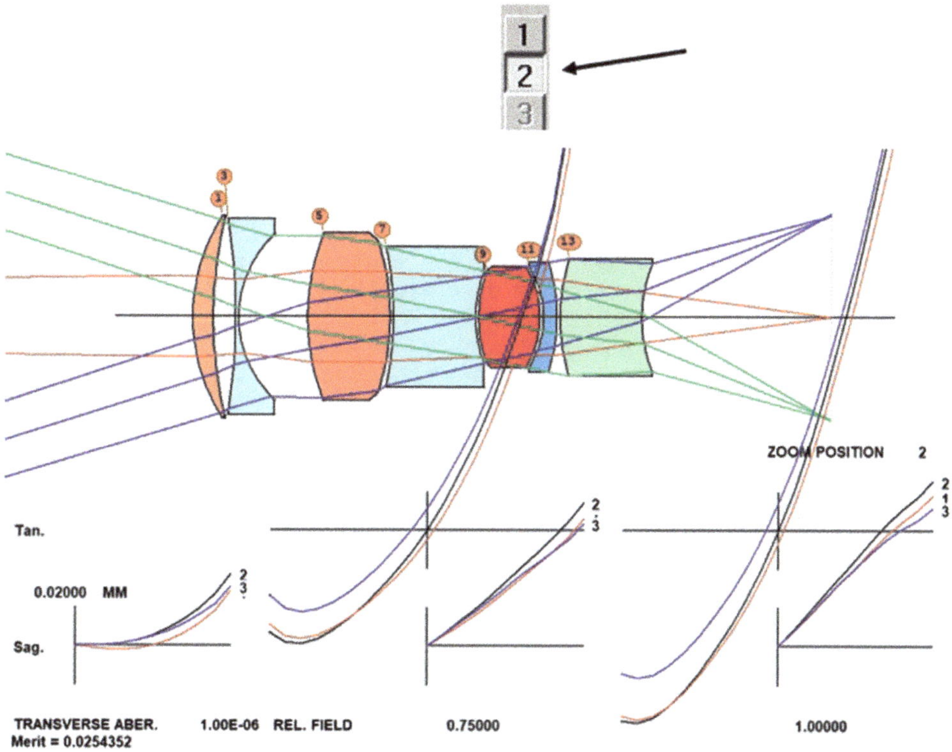

Figure 19.2. Lens with modified object distance.

Pretty awful! The aberrations changed markedly when we changed the object distance. We have to correct the image at both conjugates. Here is a MACro that may do the job (**C19M2**):

```
AWT: 0.5
PANT           ! Define variables.
CUL 1.9        ! Set upper limit of 1.9 on index variables.
FUL 1.9
!VY 1 YP1      ! Don't vary YP1
VLIST RAD ALL ! Varies all radii that are not flat.
VLIST TH ALL  ! varies all thicknesses and airspaces except for the
! back focus, thickness 14, which has a solve in effect
VLIST GLM ALL
END

AANT                    ! Start of merit function definition.
AEC                     ! Activate automatic edge-feathering monitor
ACC                     ! and maximum center thickness monitor.
ADT 6 .1 10             ! Keep diameter/thickness ratio 6 or more
!M 33 2 A GIHT ! Comment this out, since the FFIELD will control scale
LUL 150 1 1 A TOTL
M 50 .1 A BACK          ! Since the back focus will vary, keep it reasonable
M 90.61 1 A FOCL        ! Add this requirement so the focal length doesn't
                        ! change
```

```
GSR AWT 10 5 M 0      ! Note how weights are assigned to the several field
                      ! points,
       ! and the symbol AWT controls the aperture weighting.
GNR AWT 5.5 4 M .5    ! This creates a ray grid at the ½ field point
GNR AWT 5.5 4 M .7    ! These for the 0.7 field point
GNR AWT 3 4 M 1       ! Full field gets the lowest weight.

ZOOM 2          ! Targets for zoom 2 (with the object at one meter)
GSR AWT 10 5 M 0      ! Note how weights are assigned to field points.
GNR AWT 5.5 4 M .5    ! This creates a ray grid at the ½ field point
GNR AWT 5.5 4 M .7    ! These for the 0.7 field point
GNR AWT 3 4 M 1       ! Full field gets the lowest weight.
END

SNAP
SYNO 50
```

Run this and anneal, and the lens is better but still not very good, with about equal and opposite errors at the ends of the zoom range, as shown in figure 19.3.

Some subtleties deserve mention: the **GLM ALL** variable will vary all glass models currently in the lens, which means all elements, since DSEARCH uses the glass model unless told otherwise. We have to control the focal length, since the object height will be continuously adjusted so the image CAO is filled at full field, and the image height will therefore not be available as a target to control it.

This is better than zoom 2 was before, but there is still a loss of resolution. What to do? We need more variables. What should we add?

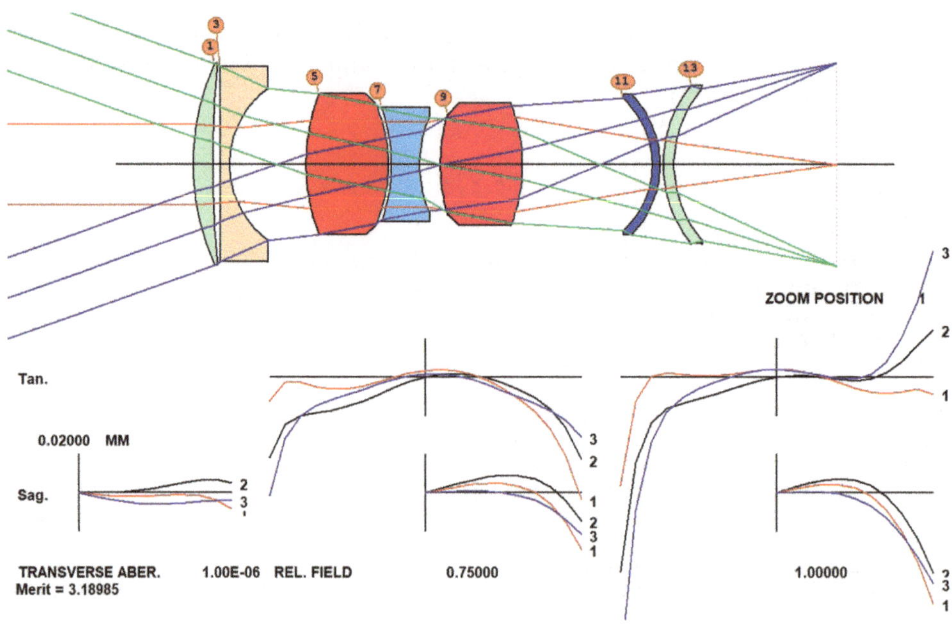

Figure 19.3. Lens reoptimized for two object conjugates; the quality is not good.

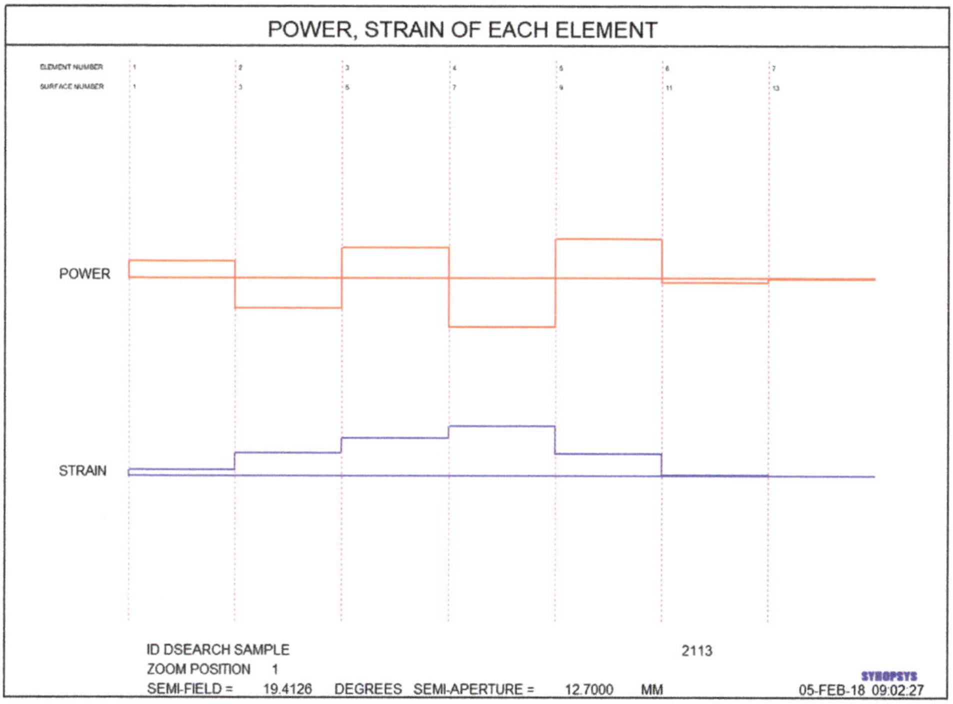

Figure 19.4. Plot produced by the `STRAIN P` command.

A classic tool for cases like this is the **STRAIN** calculation. The idea is that the elements with the largest strain are contributing most of the aberrations, and splitting an element there might relieve it. (Strain is defined here as the sum of the squares of the third- and fifth-order aberrations of that element.)

Type **`STRAIN P`** in the CW. You see the results in figure 19.4.

Indeed, element 4 has the largest strain. Now you can do one of two things: you can split that element and reoptimize, or you can use a different tool that can figure out the best place to add an element. We will try it both ways. First, save this version, so you can go back if things do not work out. Type

`STORE 1.`

Then go to the WorkSheet (type **`WS`** or click on the button). Then click the button , which lets you split an element by clicking in the PAD display on the axis inside that element. Click between surfaces 7 and 8, splitting the element. Your lens now looks like figure 19.5.

When the program splits (or adds) an element, it assigns an index pickup, because at that moment it has no other index data. Change the index pickup on surface 9 to a glass model by typing

`9 GLM`

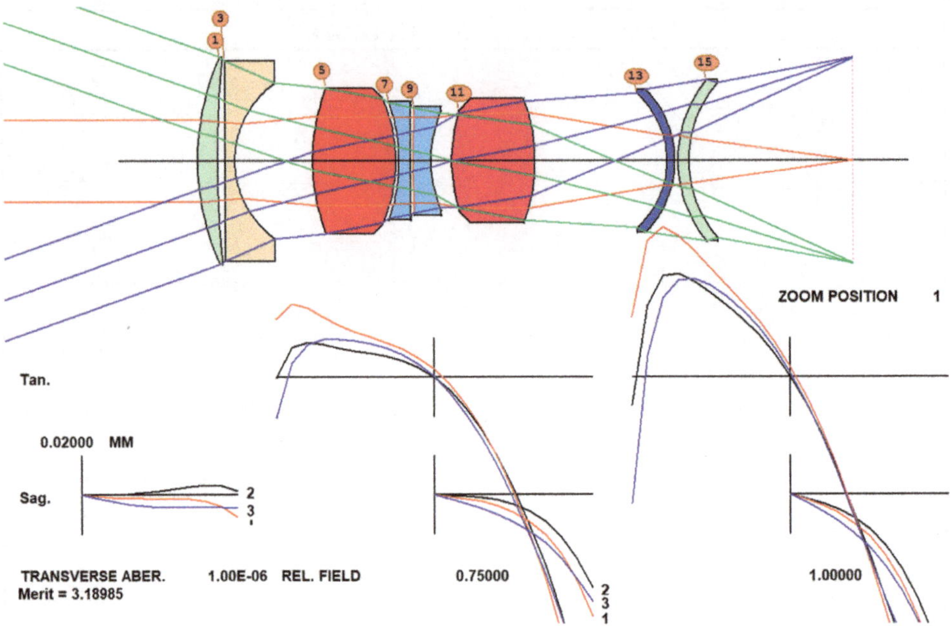

Figure 19.5. Lens with split element, before optimizing.

in the WS edit pane, and click 'Update'. That assigns the element a model glass with properties similar to what were there before.

Make a new checkpoint, close WS, and run the optimization again and anneal. The MF goes to 3.03. This is not working. Now what?

This is the way lens design has long been done, using classic tools, and it was a slow and arduous process. One can modify the lens and then try optimizing again, and again, and again....

Today we have better tools. Go back to the version you stored before splitting the element:

```
GET 1
```

and then add a line before the PANT file,

```
AEI 2 1 14 0 0 0 10 2
```

This will run the Automatic Element Insertion tool (**AEI**). Now the program will search for the best place to insert a new element. Run this, and the lens is better. Comment out the AEI line and run the MACro again, then anneal. The result (**C19L1**) is shown in figure 19.6.

The program has inserted a new element at surface 13, and the merit function came down to 1.75. There is a lesson here: the program can usually figure out how to improve a lens better than you can, so it is better to let AEI do it than to try things

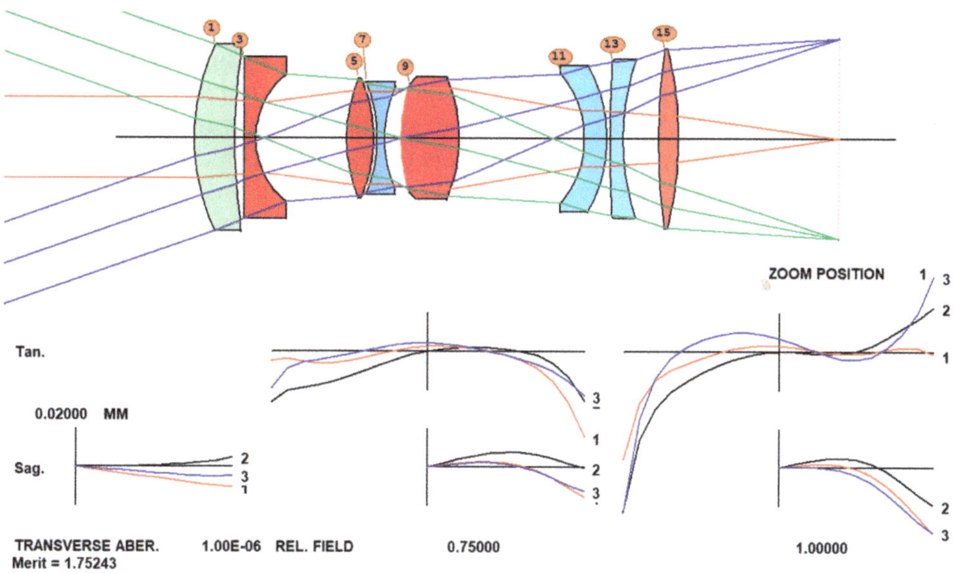

Figure 19.6. Lens reoptimized with new element inserted by AEI.

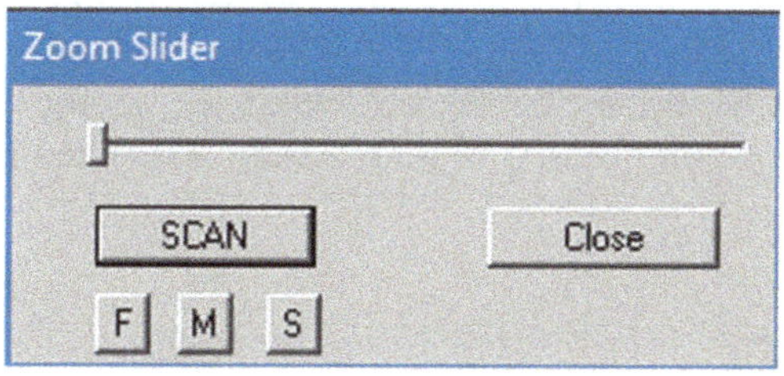

Figure 19.7. The Zoom Slider.

yourself, even though you have ideas that seem to make sense. Those things sometimes work, but AEI is better.

Here you see an improvement, and the MTF is also better, as you can check yourself. Now we have a lens that is somewhat corrected for both infinity conjugate and at one meter. But what about in-between distances? It would be a rude surprise if you built the lens and found that intermediate distances yielded a poor image. We have to check.

That is one of the reasons we chose to use the ZFILE zoom feature for this job. One can easily scan over the zoom range and spot any points that perhaps need attention. Click the button at the bottom of the zoom-selection bar: . This opens the Zoom Slider[1], shown in figure 19.7, which is fun to watch.

Click the 'SCAN' button and watch the PAD display. The image plane slowly moves back, from the infinity focus to the 1 m focus position, and then forward

[1] Zoom Slider™ is a trademark of Optical Systems Design, Inc., a Maine, USA, corporation.

again. The good news is, the image quality shows little change over the entire range. (If it had changed, you could use the CAM command to create an intermediate focus position, making a total of three zooms, and then add some more targets for the new ZOOM 3 position in the AANT file. You can create and target up to 20 zooms this way, as you will learn if you type **HELP CAM** to read about that feature.)

This lesson shows how asking a lens to do something it was not selected for is generally a bad idea. We obtained a lens that compromises many of the goals we set, and balanced things as best it could. Chapter 21 shows how a different approach works better in cases like this.

Now we need to assign real glasses again.

But, wait a minute. The sixth element shown in figure 19.6 bothers us. What is it doing? Can we remove it? You have to try. Remove the AEI directive and replace it with

```
AED 6 QUIET 1 15
```

and run it again—the program says the *seventh* element can be removed! Allow it to do this, then comment out the AED directive and optimize and anneal some more (**C19L2**). The merit function goes to 2.01, and you have eliminated an element, as shown in figure 19.8. See how AED can make better decisions than you can? This design is not quite as good as before, however, so we would probably go back to that version.

So that is how it is done: figure out what is wrong and use the tools in SYNOPSYS to fix it. Sometimes it is quick and sometimes not. That is what lens

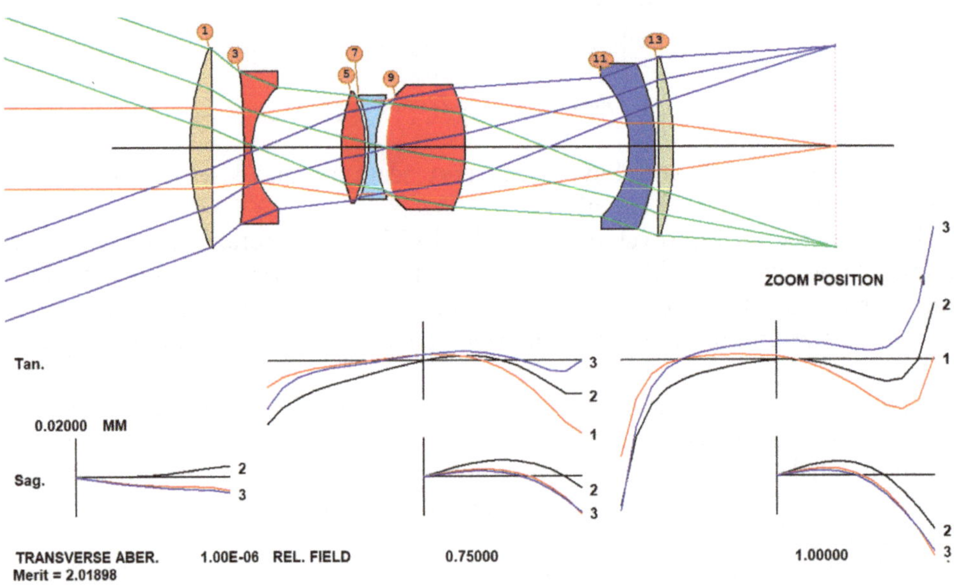

Figure 19.8. Lens with element removed by AED.

design is all about, blind alleys and all. If this lens is still not good enough, another run on AEI might do the trick. Chapter 21 shows a better way to approach this kind of problem, letting the software handle the whole job.

That is enough for this lesson.

I almost forgot: why did we enter the surface number (14) for the zooming group, since the YMT solve will override it anyway? Well, the program requires a group definition, and it will not work otherwise. That is to save you from a serious mistake if you ever leave those data out for a real zoom lens.

Chapter 20

A practical camera lens

Global search for a camera lens design

Here are the goals for this chapter:

1. Focal length 90 mm.
2. Semi-field angle 20 degrees.
3. Semi-aperture 25.4 mm.
4. Cell length approximately 100 mm.
5. Back focus distance 50 mm or greater.

In this chapter we will let **DSEARCH** find a starting point. Type **MDS** in the Command Window, to open the Design Search Menu, shown in figure 20.1.

Enter the data shown by the arrows and click 'OK'. We will modify this input later, when we see the results, and we guess the lens will need seven elements. The program asks you for a file name, so type a name of your choice. An editor window opens, containing the input required to run DSEARCH (**C20M1**):

```
CORE 14
 TIME
 DSEARCH 1  QUIET
 SYSTEM
 ID DSEARCH SAMPLE
 OBB 0 20 12.7
 WAVL 0.6563 0.5876 0.4861

 UNITS MM
 END
 GOALS
 ELEMENTS 7
 FNUM 3.54
 BACK 0 0
 TOTL 100 0.1
 STOP MIDDLE
```


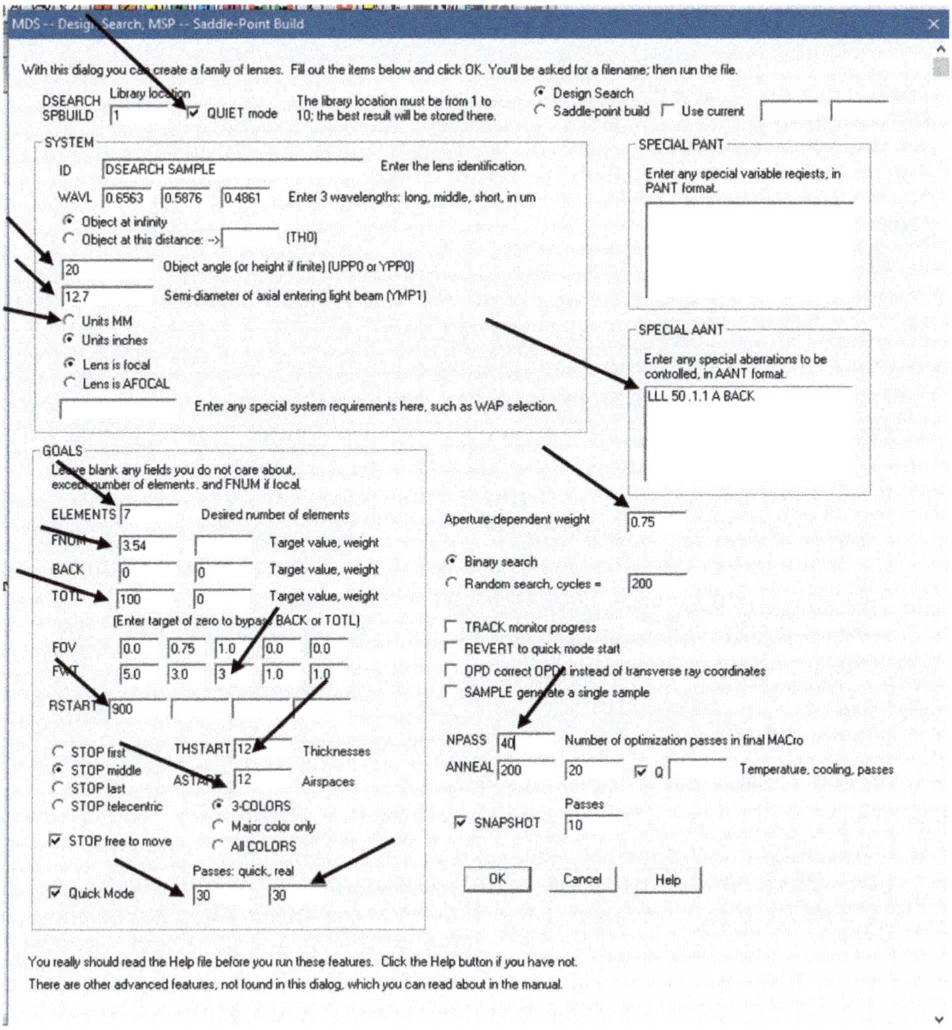

Figure 20.1. The DSEARCH dialog, opened with the command **MDS**.

```
STOP FREE
RSTART 900
THSTART 12
ASTART 12
RT 0.75
FOV 0.0 0.75 1.0 0.0 0.0
FWT 5.0 3.0 3
NPASS 50          ! this gives the number of passes in the final MACro
ANNEAL 200 20 Q
COLORS 3
SNAPSHOT 10
QUICK 50 50       ! 30 passes in quick mode, 30 in real mode
```

```
END
SPECIAL PANT

END
SPECIAL AANT
LLL 50 .1 1 A BACK
END
GO
TIME
```

We elected not to assign a weight to the back focus distance in the dialog, preferring to put that requirement in the SPECIAL section, where we gave it a one-sided requirement. **LLL** means Limit, Lower Limit, and this lets the back focus become larger than 50 without any penalty—but not smaller.

Run this MACro, and you obtain a collection of potential starting points, shown in figure 20.2. Some of the lenses returned by DSEARCH have a negative element in front. We term such lenses *inverse telephoto*. A telephoto lens has a negative group at the *image* end, which serves to increase the size of the image, thereby obtaining the focal length of a longer lens in a small package. The inverse telephoto configuration, with a negative group in front, is often used for wide-angle lenses, where one wants just the opposite: the focal length is shorter than the physical size of the lens.

Let us examine these lenses more closely. Type, in the CW,

```
EM DSS
```

This command loads and runs the file DSS.MAC, which DSEARCH has created. That MACro will open each of the lenses returned by DSEARCH, show it on the PAD display, and then wait for you to press the <Enter> key before opening the next. You will see that many of them have similar quality. When you see one you like, just press the <Esc> key to stop the MACro at that point. We like the one saved by DSEARCH with the name DSEARCH07.RLE, which was also the top one in this example. (The names in your results may be different, since the order depends on which cores finished when.) The lens is shown in figure 20.3.

Run the optimization MACro DSEARCH_OPT, which DSEARCH has opened in a new window; then anneal **(50, 2, 50)**. The lens changes slightly.

How good is this lens? Open the **MOP** dialog (`MOP`) and enter the data shown in figure 20.4. Click the 'MOF' button, and you obtain the MTF curves in figure 20.5. Call this lens version 1.

Not too good—and we learned some things. The MTF varies with field, so we might need more field points in the DSEARCH input. We also decide that, since this lens is not far from the diffraction limit and we want to improve the MTF, we should try the **TOSHEAR** directive instead of the default transverse aberrations.

To understand this input, you have to understand how the MTF is calculated. A common method is to evaluate a *convolution integral*, which combines two copies of the exit pupil, one of them sheared in x or y by an amount that depends on the frequency to be calculated. If the OPD errors at a given point are the same as those at the sheared point, the MTF is perfect, as far as those points go. Calculated over

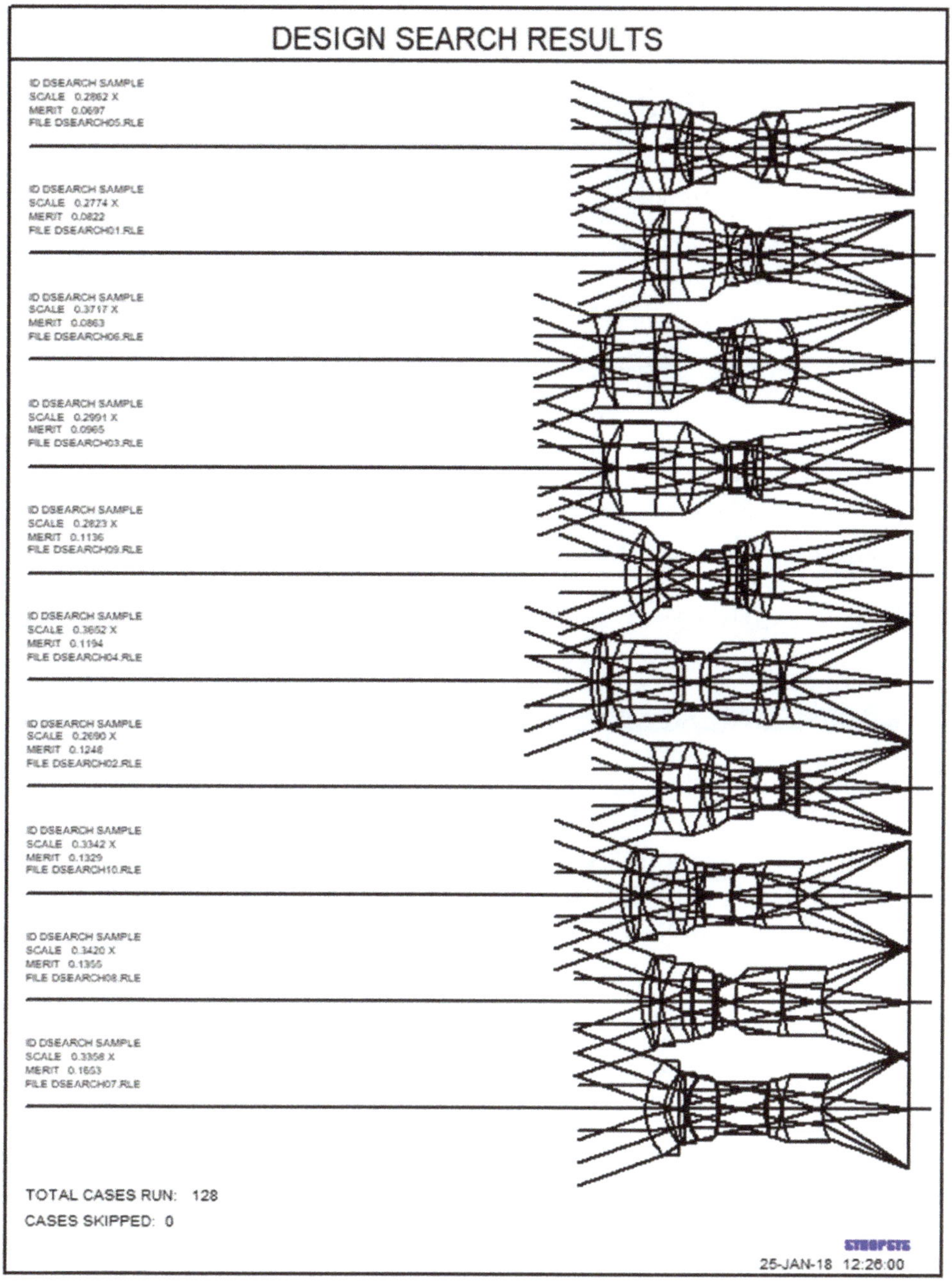

Figure 20.2. Ten lenses returned by DSEARCH.

the whole pupil, the result is the MTF of the lens, in the scalar approximation. The requests below combine a set of transverse ray targets with a set of GSHEAR requests when it makes the MF. The latter target the difference in OPD over a set of sheared pupil locations.

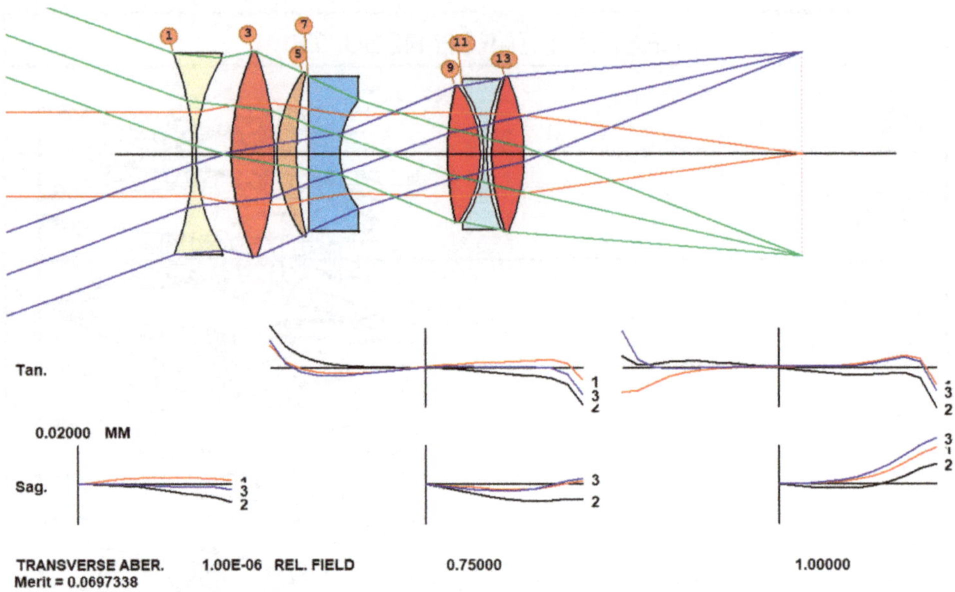

Figure 20.3. A good candidate from the DSEARCH results.

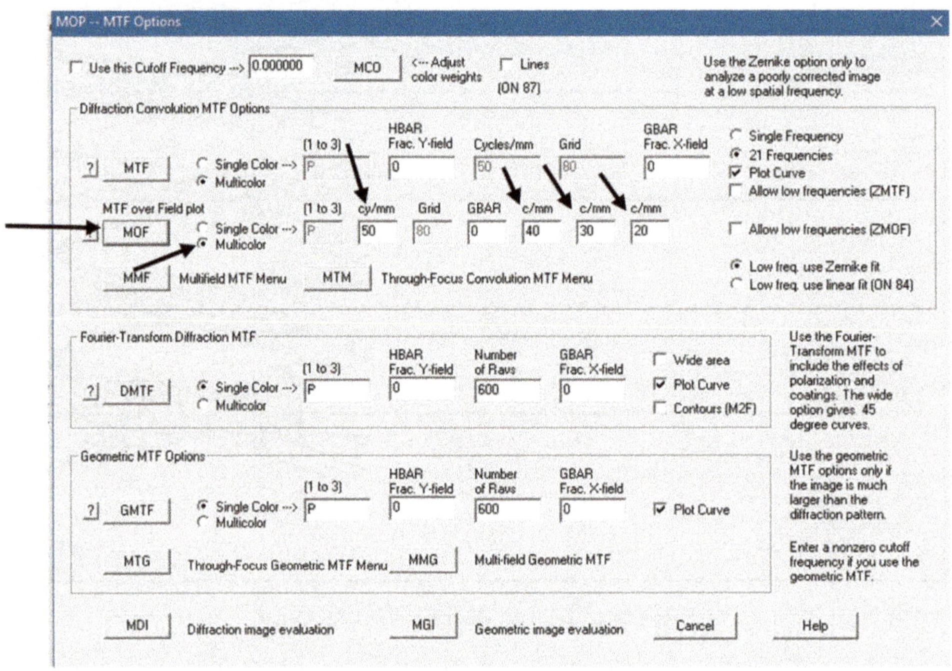

Figure 20.4. Dialog to run MOF, with four spatial frequencies requested.

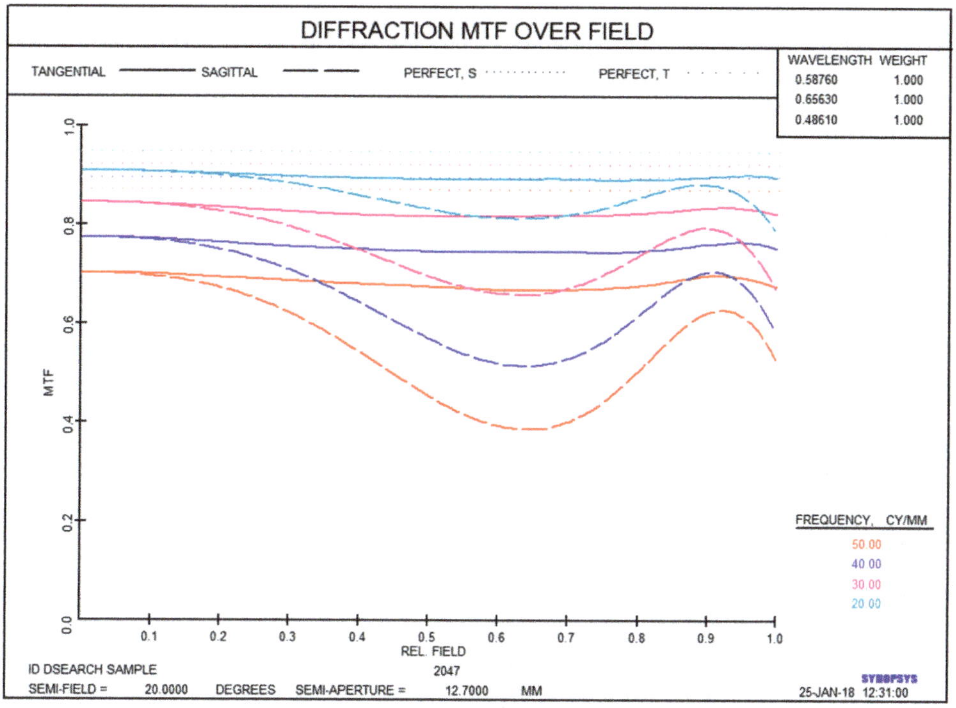

Figure 20.5. MTF curves for version 1.

Here is the revised input for the GOALS section:

```
GOALS
 ELEMENTS 7
 FNUM 3.54
 BACK 0 0
 TOTL 100 0.1
 STOP MIDDLE
 STOP FREE
 RSTART 900
 THSTART 12
 ASTART 12
 RT 0.75
 FOV 0.0 .5 .7 .9 1
 FWT 1 1 1 1 1
 TOSHEAR
 NGRID 6
 NPASS 40
 ANNEAL 200 20 Q
 COLORS 3
 SNAPSHOT 10
 QUICK 80 80
 END
```

Run this DSEARCH file, then optimize and anneal. You obtain a very different lens (**C20L1**), shown in figure 20.6, and higher MTF curves, shown in figure 20.7.

Perhaps we cannot do better with only seven elements. Add a line before the `PANT` command,

```
AEI 4 1 123 0 0 0 50 10
```

and run the MACro again. The program adds an element at surface 5. Then comment out the AEI line, optimize, and anneal once more. The MF comes down to 0.11.

This is version 3, shown in figure 20.8, with the MTF in figure 20.9. This is an excellent lens.

This lens works very well, but some elements are too thin. It is time to gently push thicknesses up.

Add a new monitor to the AANT file:

```
ADT 7 .1 1
```

Then run it again and anneal. The elements are more reasonable. This is the version 4 design (**C20L2**) in figure 20.10; MTF curves are in figure 20.11. Can we do even better? Perhaps we can by relaxing the requirements somewhat.

We ran the DSEARCH input again, but with a target of 150 mm for the total length instead of 100. The lens form that came back was very different, and MTFs were higher (after running AEI, optimizing, and annealing), as seen in figures 20.12 and 20.13 (**C20L3**).

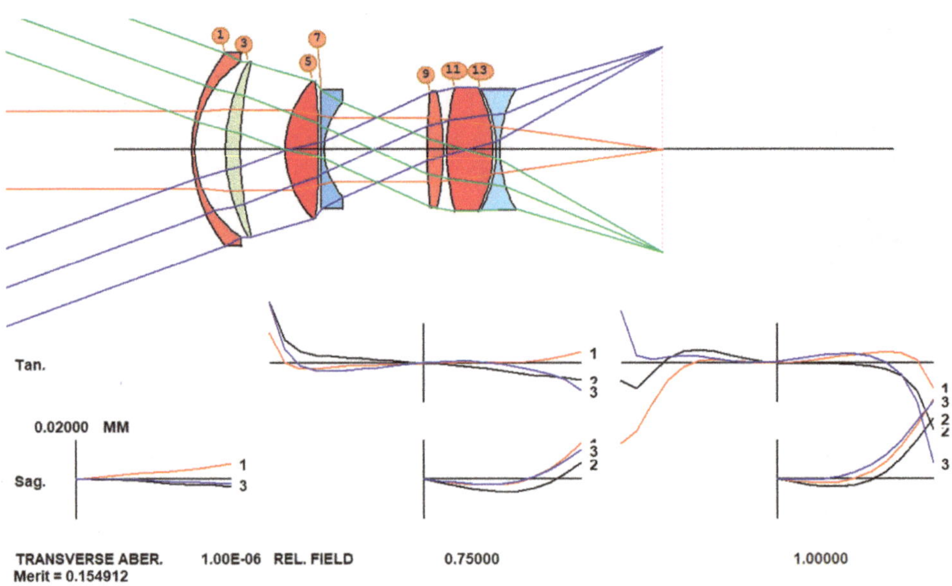

Figure 20.6. Version 2 lens, optimized and annealed.

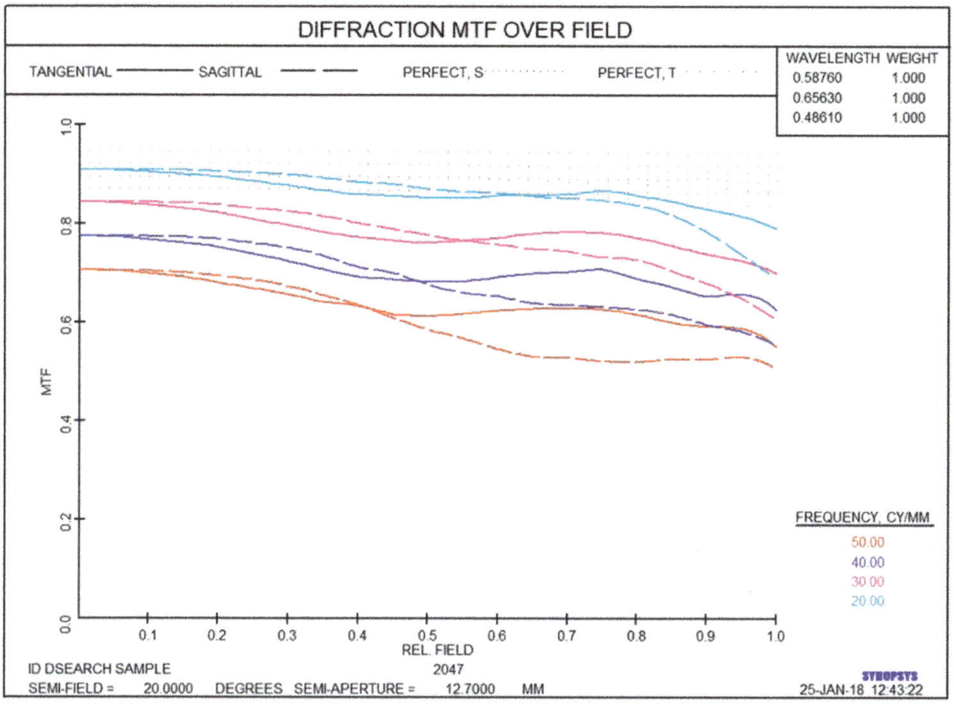

Figure 20.7. MTF of version 2 lens.

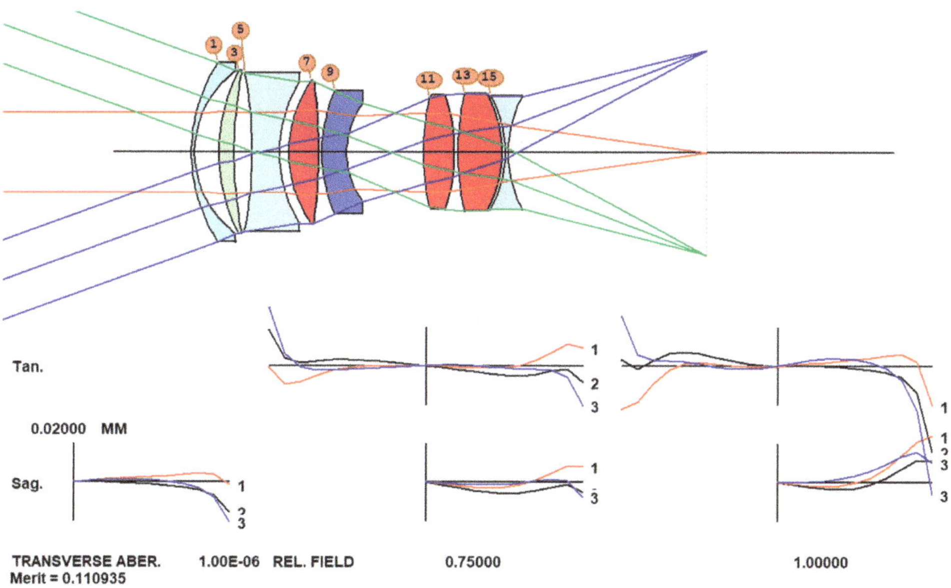

Figure 20.8. Version 3 lens, with element 3 inserted by AEI, optimized and annealed.

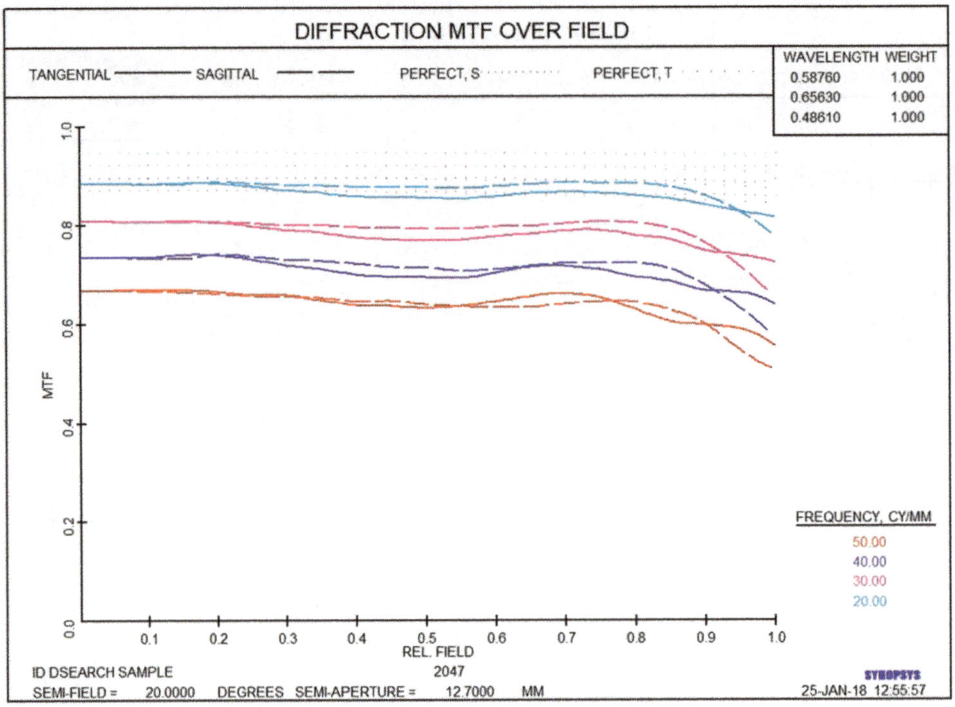

Figure 20.9. MTF of version 3 lens.

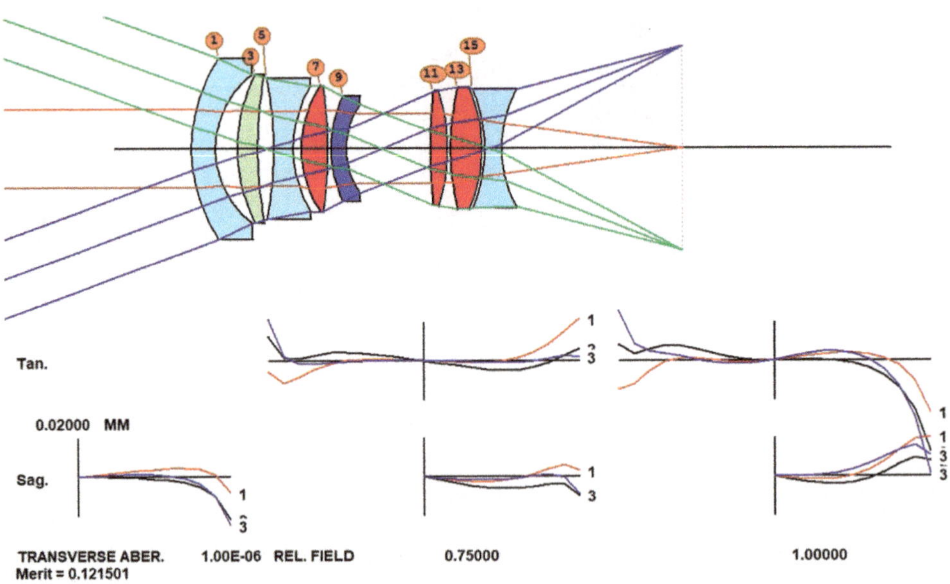

Figure 20.10. The version 4 lens.

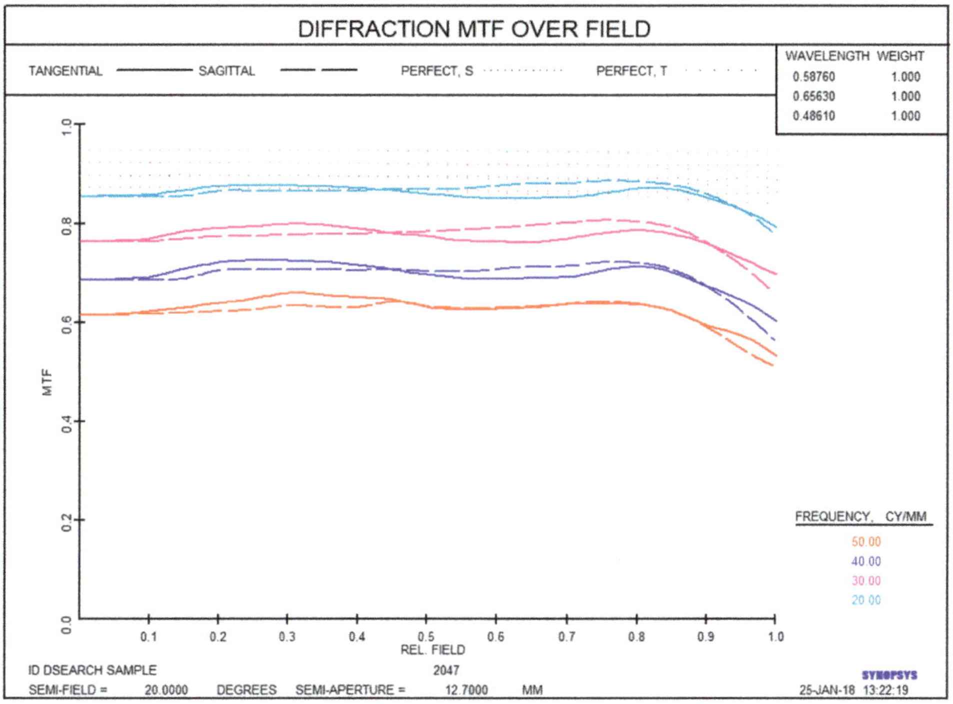

Figure 20.11. MTF of the version 4 lens.

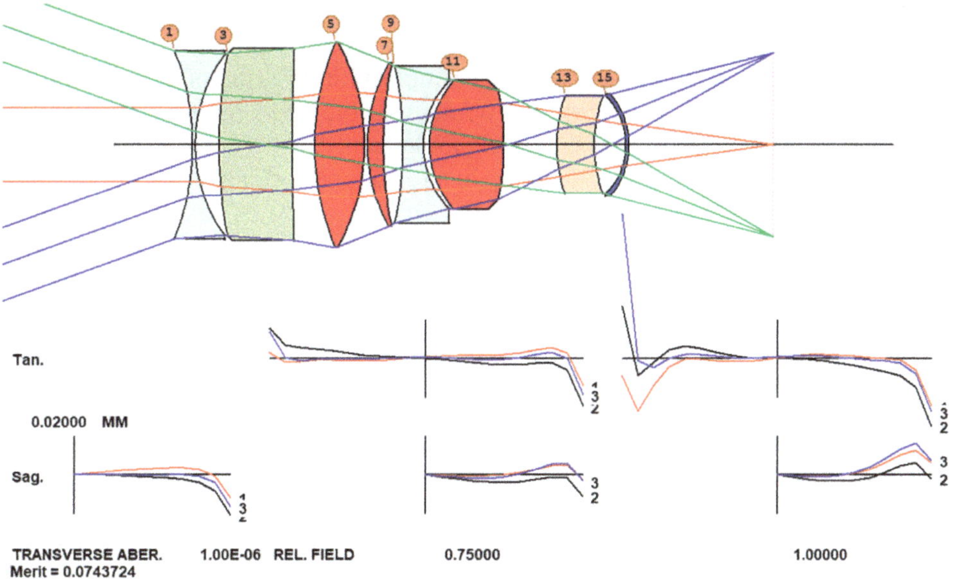

Figure 20.12. Lens form returned by DSEARCH when a longer lens was permitted, then improved with AEI.

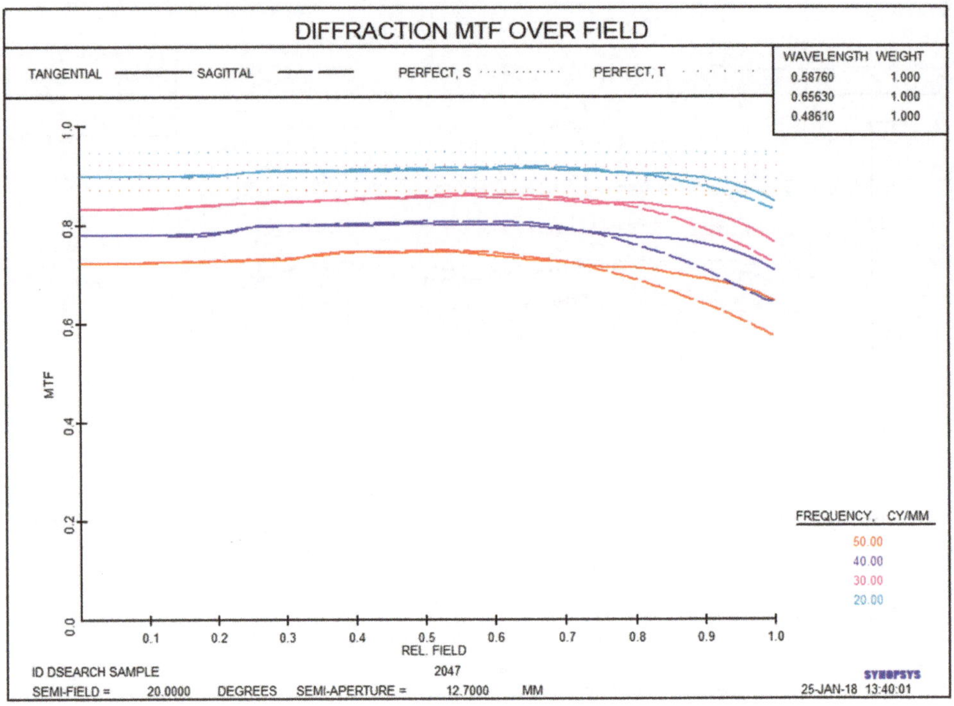

Figure 20.13. MTF of the lens in figure 20.12.

Now it is easy to tell a customer that, with *this* length you can obtain *this* MTF, and with *that* length you obtain *that*. What happens if you allow the back focus distance to be shorter? Again, tradeoffs are simple with this kind of autonomous lens design tool. Asking for a lower limit of 10 mm, we obtained a design with only seven elements that was even better than the eight-element lens we obtained in version 4. What would have taken weeks to develop in a former period can now be evaluated in a few minutes with these powerful tools.

With so many tools available, one naturally wants to know which ones to use. If you want an eight-element lens, for example, should you ask DSEARCH for eight, or ask it for, say, six—and then use AEI twice to obtain to eight elements? Can we predict what Nature will favor?

Experience alone will tell. We ran this job again doing just that, asking DSEARCH for six elements, and then running AEI twice. Testing all ten of the DSEARCH results in this way, we found that nine of them had quality similar to the previous designs. Figure 20.14 shows the best of the lot (**C20L4**), with MTF curves in figure 20.15. It seems this might be a good strategy, but owing to the chaotic nature of the process, one cannot draw a firm conclusion; a different problem might work best the other way. So put this idea in your toolbox and use it when it makes sense.

This lens does not resemble any of the classic forms derived from a triplet, which have the stop in the middle near a negative lens. One of the other cases, with essentially identical performance, is shown in figure 20.16; that lens does somewhat

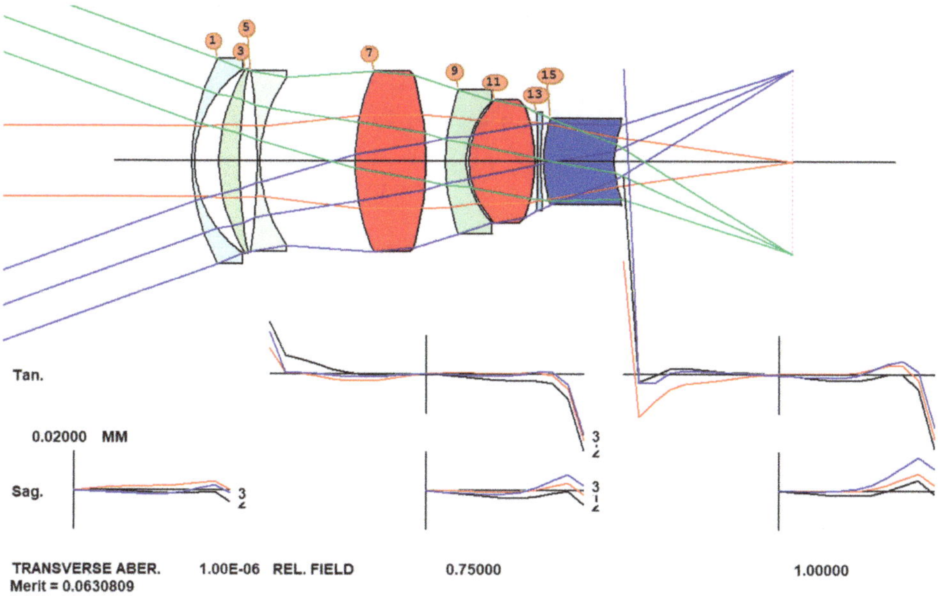

Figure 20.14. Lens found by asking DSEARCH for six elements, running AEI twice on all ten cases, then optimizing the best one.

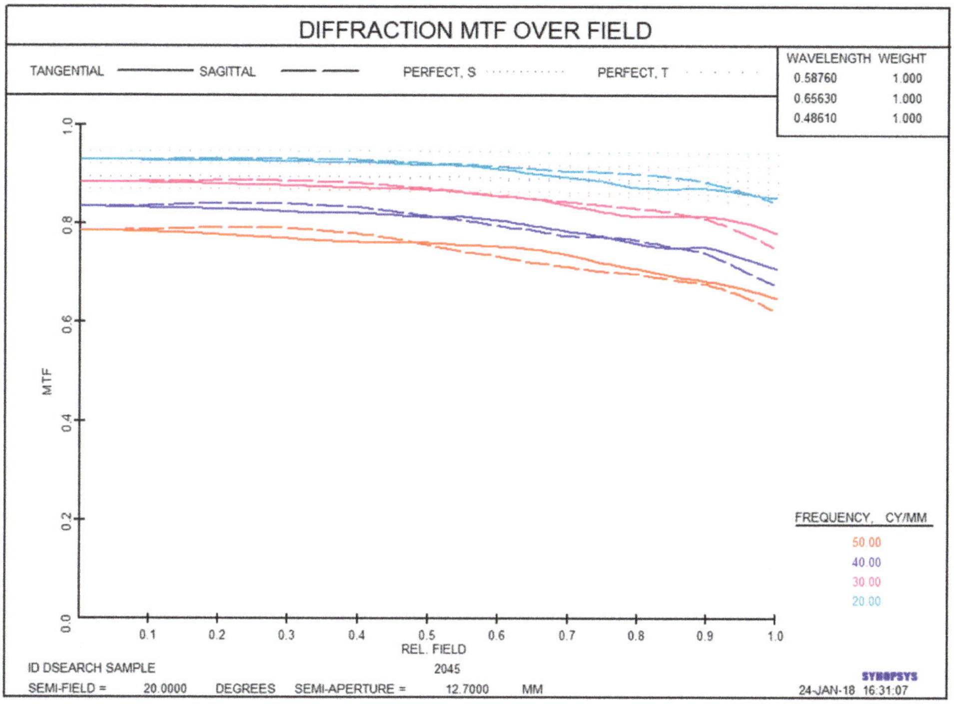

Figure 20.15. MTF of the lens in figure 20.15. This was found when a DSEARCH lens of six elements was increased to eight with AEI.

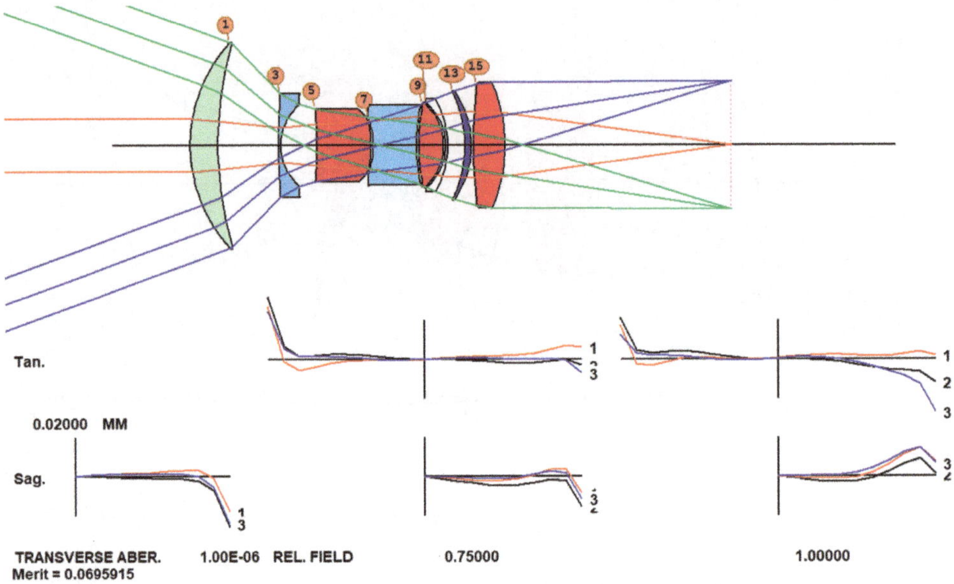

Figure 20.16. A different configuration found by increasing a six-element lens to eight elements.

resemble a classic form. The ancient designers found constructions such as the double Gauss that worked well and did not explore many other configurations thereafter, giving us another case when automatic methods prove superior. The search tools are not biased and can find solutions a human would likely never have imagined.

Back to our question; we ran DSEARCH again, asking for eight elements this time instead of using AEI to get to that number. The lens that came back was, in this example, not quite as good as the above, but close. It seems that AEI is a powerful tool you will want to use often.

This lesson illustrates quite graphically that, in lens design, you are dealing with a landscape with a great many solutions. Unless you already have a very good starting lens, your chance of finding a great one improves as you experiment with the parameters of DSEARCH.

The art of lens design has changed significantly with the advent of new tools like DSEARCH. Whereas in former ages an expert designer would work for days or weeks on a single design, using sophisticated knowledge to steer the process, today one can produce a great many designs in a matter of minutes and then select the most promising for further work. Some of those designs often turn out to be superior to what an expert was able to come up with in days past.

In a practical situation, I would next finalize the lens, see how the image correction changes with conjugate (if that is a requirement, and reoptimize the lens as shown in chapter 19 if so), insert real glass types with ARGLASS or GSEARCH, assign a real stop where implied, decrease the thickness of some elements and increase others, reoptimize, define edge geometry with the Edge Wizard, match the curves to a vendor's testplate list with TPM, prepare tolerances with BTOL, make element drawings with ELD, system drawings with DWG, and so on. There is much to do when designing a lens, and this has been a practical lesson that shows how to use a few of the many tools that are available. We started with nothing more than a list of requirements and in a fairly brief time came up with some rather good lenses.

Feel free to experiment with this problem, and try various values for the starting radius, thickness, airspaces, and other parameters for DSEARCH, with switch 98 turned off.

20.1 Reusing dialog commands

Before leaving this chapter, I would like to point out some very handy features. You have run the **MOF** (**M**TF **O**ver **F**ield) analysis several times from the **MOP** dialog (**M**TF **OP**tions) when analyzing these lenses, and of course you can always type **MOP** to go back there when you want. However, after you have run a command from a dialog such as this, notice the prompt in the CW:

```
Type <ENTER> to return to dialog
```

If you now just press the <Enter> key, you go right back there, which is simpler. If you have run some other command in the meantime, and that prompt is no longer there, just click the '**Last Menu**' button on the top toolbar and you go back to the most recent dialog, which also saves some typing.

Here is another handy tool: after you run a command from a dialog, if you type **LMM** (**L**oad **M**enu **MA**Cro) in the CW, the program opens a new editor window with the command form already filled in, in this case

```
MOF M 0 50 80 0 Q 40 30 20
```

Now you can save this as a new file, giving it a name like MOF.MAC, and you can run it with the command

```
EM MOF
```

which will save the labor of typing the data next time if you want to perform the same analysis. There is more. If you define a symbol, say

```
U9: EM MOF
```

then just clicking the U9 button on the left toolbar will execute that MACro. Put that symbol definition in your CUSTOM.MAC MACro, and it comes back whenever you restart SYNOPSYS. You have just defined a whole new command. Just hit the U9 button and you obtain that MTF analysis—no typing needed.

Chapter 21

An automatic real-world lens

Using ZSEARCH to design a lens from scratch with two conjugates

In chapter 19 you developed a seven-element lens with DSEARCH and then changed it into a zoom lens so you could correct at two different object conjugates. You found you needed to add an additional element with AEI to improve performance and ended up with eight elements. That exercise was not completely successful and was tedious besides—and illustrates an important concept: ask for *exactly* what you want, if possible. In that case, you did not.

Since DSEARCH does not know about zoom lenses, what you got was a very good *fixed-focus* lens—and then you added a new requirement to focus at a shorter distance. So the lenses that came back from DSEARCH were not selected for that property. That is always a risky process. Perhaps they were not well suited to the job.

Would it not be nice if the search routines could monitor the performance at two different conjugates? Then the lenses that come back would naturally tend to perform well on that score. Well, they can. We have said nothing so far about **ZSEARCH**[1], which is similar to DSEARCH except it works on zoom lenses. Let me summarize the requirements for this same job by listing the input for ZSEARCH (**C21M1**)—you should also review ZSEARCH in the User's Manual at this point, to better follow along:

```
CORE 14
ZSEARCH 4 QUIET
SYSTEM
ID LESSON 21 ZOOM
OBB 0 20 12.7        ! OBJECT AT ZOOM 1, AT INFINITY
WAVL CDF
UNI MM
END
```

[1] ZSEARCH™ is a trademark of Optical Systems Design, Inc., a Maine, USA corporation.

doi:10.1088/978-0-7503-1611-8ch21

```
GOALS
ZOOMS 3                 ! CORRECT AT THREE ZOOMS
GROUPS 8 0 0            ! ONE GROUP, EIGHT ELEMENTS
ZGROUPS Z 0             ! ZOOM THE FIRST GROUP
GIHT 32 32              ! KEEP THE IMAGE HEIGHT CONSTANT
BACK 50 .01             ! BACK FOCUS DISTANCE IN ZOOM 1
RSTART 100
THSTART 10
ASTART 10
RT 0.25
FINAL
OBA 1000 -366.5 12.7! OBJECT DEFINITION FOR ZOOM 3
APS 10                  ! PUT THE STOP NEAR THE MIDDLE OF THE LENS
NPASS 50
COLOR M

FOV 0 .5 .7 1
FWT 2 1 1 1
NGRID 5
SNAP 10
ANNEAL 30 10 Q 50
QUICK 30 50
END

SPECIAL AANT
ADT 6 .01 10
LUL 350 .1 1 A TOTL
M 32 .1 A GIHT
END

GO
```

This input declares two different object distances, infinity and 1000 mm, specifies that the lens will have three zoom positions (the minimum allowed by ZSEARCH), declares a single zooming group of eight elements (the entire lens), and gives a maximum allowed total length of 350 mm.

Run this job, and then optimize and anneal (**55, 2, 50**) the top lens that comes back. The result (**C21L1**) is shown in figures 21.1 and 21.2.

Compare this to the results from chapter 19 and you see the importance of our observation. When you ask ZSEARCH for what you *really* want, it has no trouble finding a configuration that meets those requirements. The lens looks nothing like the previous results and works better besides. This kind of tool is a real time-saver, for sure.

However, like all tools, it has limitations. ZSEARCH does not support a curvature or thickness solve, since too many things might be zooming, and it does not support the variable YP1, which we have used before to find the best place for the stop. So we assigned the stop to surface 10. We also specified the desired GIHT in the SPECIAL AANT section of the ZSEARCH file to maintain the desired focal length. Then we put an upper limit on the total cell length.

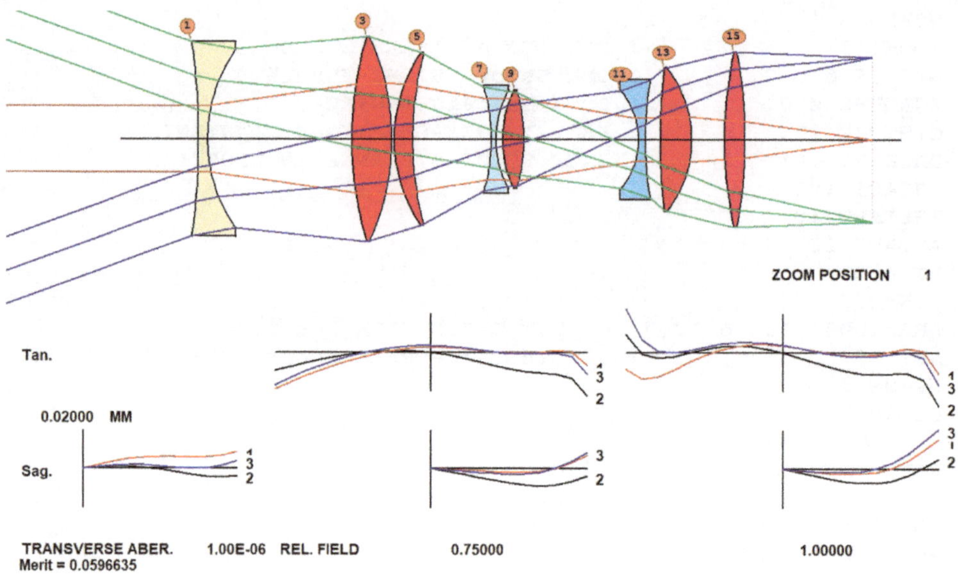

Figure 21.1. Lens corrected at two conjugates returned by ZSEARCH, at infinity conjugate, optimized and annealed.

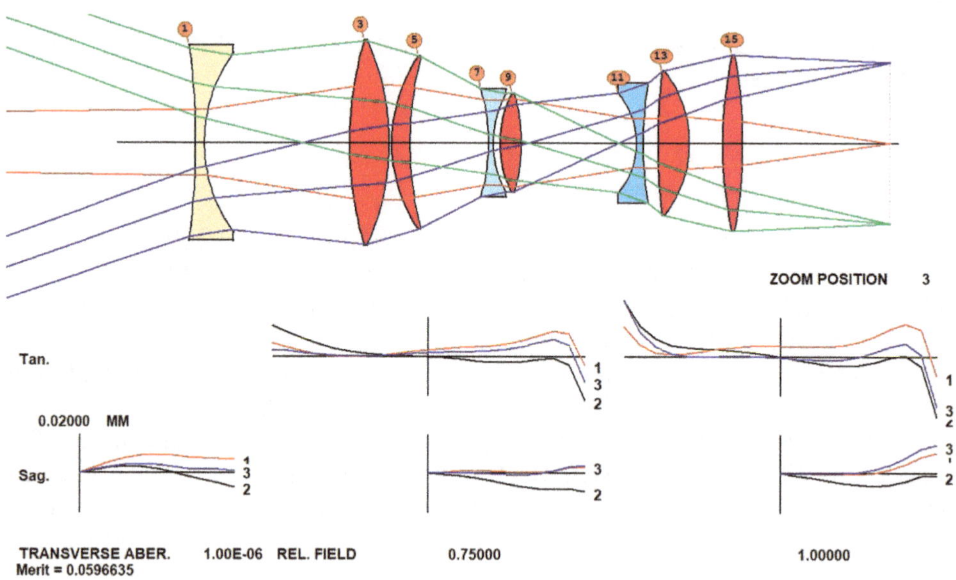

Figure 21.2. Lens at near conjugate.

If you look closely at the lens at both conjugates, you will notice that the size of the beam at the stop, on surface 10, changes slightly with the conjugate. It is a fine point, but easy to address. Open WS, and type in the edit pane:

```
APS -10
CSTOP
WAP 2
```

and click ‘Update’. Then optimize and anneal. This will fix the size of the clear aperture on 10 to whatever the paraxial raytrace needs for the marginal ray, and then adjust the size of the entrance pupil at each zoom and field point so the real beam just clears that aperture. All of this is discussed more fully in the next chapter.

The input to ZSEARCH included the monitor

```
ADT 6 .01 10
```

to prevent lenses from becoming too thin. This monitor must be used with caution, since you do not want to miss a great configuration because the program penalized it early on, never mind it could be fixed up later. However, here we have specified a low weight, 0.01, which means that it will not have much influence in the early stages when the other aberrations are large, but will steer the design when it approaches optimum and those become much smaller. We often try our searches both ways.

Now, we ask: do we really need all those elements? Add the line

```
AED 4 QUIET 1 123
```

before the PANT file and optimize again. The program says element number 5 can come out. Accept that change; then delete the AED line, reoptimize and anneal again. The MF now is 0.089, the lens (**C21L2**) is not quite as good as before, but now it has only seven elements instead of the previous eight. This version is shown in figure 21.3. It looks like we really need eight.

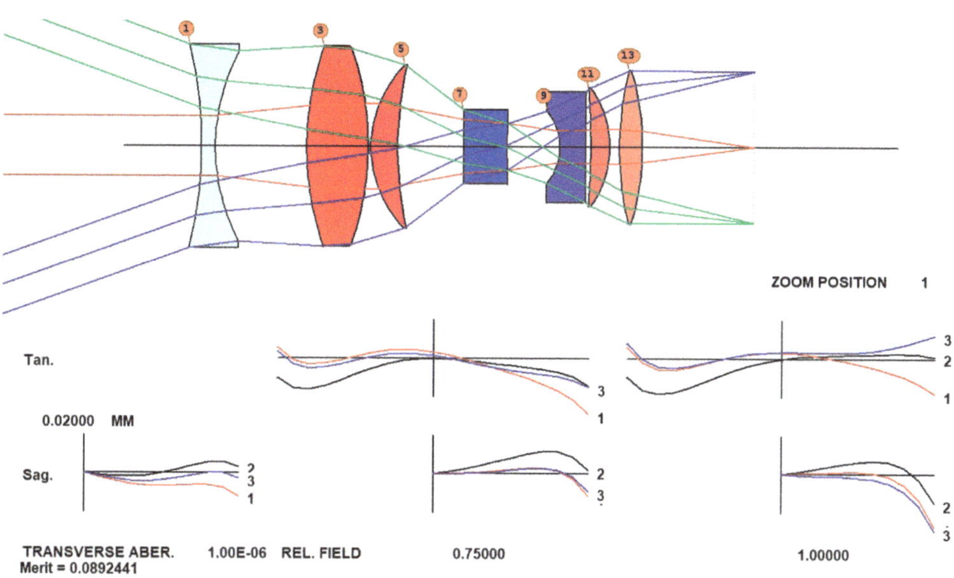

Figure 21.3. Zoom lens after AED has removed an unnecessary element.

What would happen if you assigned the stop to a different surface? Try it and see. Also, experiment with the value of RSTART. Every change you make sends the program up a different branch of the tree. You can quickly identify a great many promising configurations with these tools, and you can always run AEI and AED to improve the lens further or to reduce the number of elements.

Chapter 22

What is a good pupil?

Pupil definition; vignetting; ray aiming; wide-angle pupil options

Those familiar with other lens design codes already know about two common pupil definitions: a paraxial pupil that only applies to simple systems or where the stop is out in front, and, for more complex systems, 'ray aiming', which is intended to model a real stop somewhere inside the system.

For those not familiar with these concepts, let me explain. Start this lesson with the lens (**C22L1**) shown in figure 22.1. Surface 7 has been declared the stop surface—but it is a *paraxial* stop at the moment.

This lens has two problems: the chief ray does not go through the center of surface 7, and the upper and lower rim rays do not hit that surface at the edge, as they should. Figure 22.2 shows the lens after those problems have been corrected. Rays from every field point now fill the aperture of surface 7.

When you tell the program to trace a ray, it first has to know where to aim the ray so it hits the stop at the desired point. For example, the ray at HBAR = 1 and YEN = 1 (the full-field marginal ray) should hit surface 7 (the stop) at the edge of the aperture. How does it know where to aim? That is the whole issue with pupil definitions.

There are two possible stop definitions: **paraxial** and **real**. The definition in the first example was *paraxial*, declared with

```
RLE
...
APS 7
...
END
```

doi:10.1088/978-0-7503-1611-8ch22

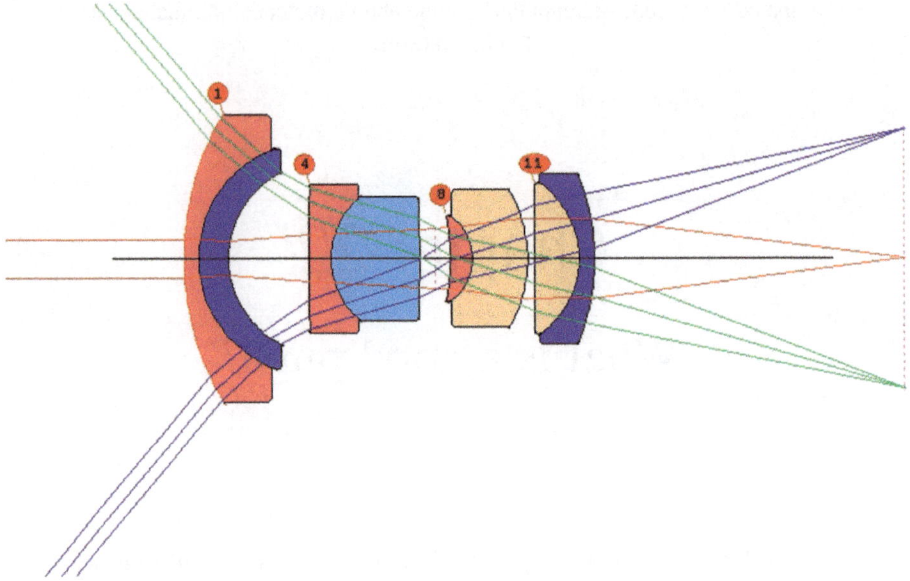

Figure 22.1. Wide-angle lens with paraxial pupil, which is not adequate in this case.

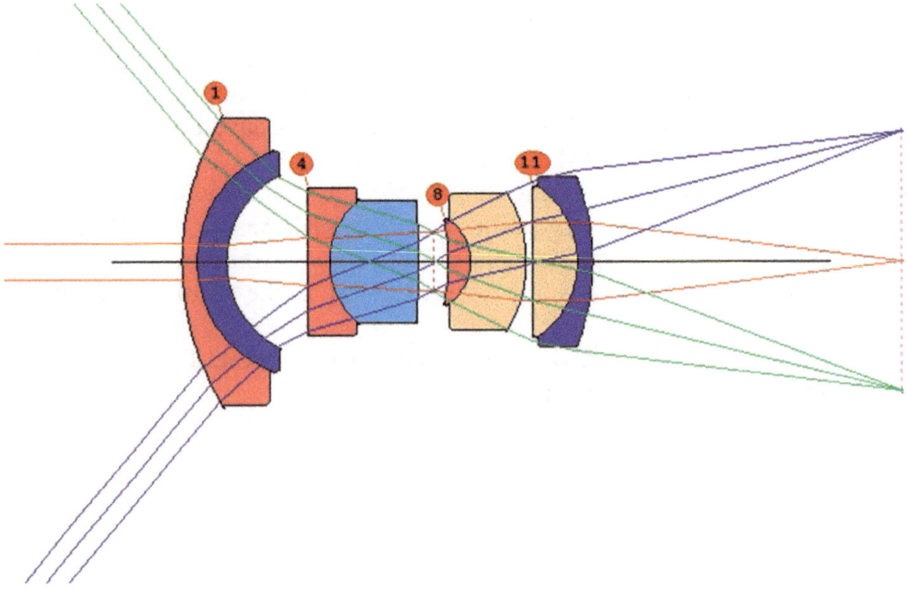

Figure 22.2. Wide-angle lens with correct pupil definition.

To fix these problems, we first declare surface 7 to be a *real* stop, with

```
CHG
APS -7
END
```

Now the chief ray is traced properly, as shown in figure 22.3. The minus sign indicates that this is a real stop and the chief ray will be found by iteration.

Although the chief ray is now acceptable, the marginal rays are not. You need another common declaration, which will adjust the size of the pupil so the stop is nicely filled. This is the **WAP 2** option (there are three wide-angle-pupil possibilities, illustrated in figure 4.7). It finds the shape of the entrance pupil by iterating a few rays at the edge of the stop. However, this option requires a hard aperture on the stop surface so it knows where to aim. Let us assume for the moment that the aperture is not defined. You can do a **CAP** listing—to see the values of all current apertures—and then assign a 'hard aperture' to surface 7. The value in this case turns out to be 3.9937, so you could enter that value either in a CHG file or with the WorkSheet. Here is how to do it with a CHG file (CAO means Clear Aperture Outside):

```
CHG
7 CAO 3.9937
END
```

An easier way is simply to type, in a CHG file or in the WorkSheet edit pane, **7 CFIX**. That fixes the current value, whatever it is at the moment, so you do not need

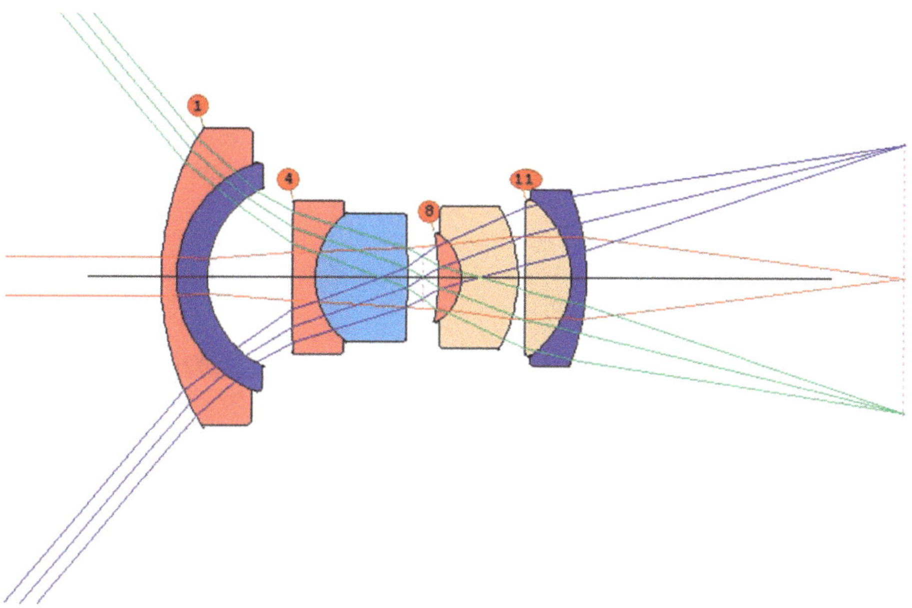

Figure 22.3. Lens with real pupil assigned but without aperture aiming.

to type it yourself. Now change to WAP 2 with the WorkSheet as shown in figure 22.4 and click the ‘Update’ button.

You obtain the lens in figure 22.2. Now both the chief ray and the marginal rays go to the right place on surface 7. You have turned on ray aiming for a total of five rays.

So far, this is not too complicated, and many users will not need anything else. However, suppose you are optimizing the lens and the required aperture on surface 7 keeps changing. In that case, the hard aperture you assigned will be incorrect almost immediately.

No problem. You can activate an option to recalculate that aperture every time the lens is changed. This is done by adding the directive **CSTOP** to the lens input file. Then the program will alter the CAO on 7 so it always equals the paraxial marginal ray height there.

(If the lens has pupil aberrations so large that a *real* axial marginal ray requires a different aperture than the *paraxial* ray, change this to **CSTOP REAL**. You can even specify *which* real ray is used to define this aperture, as explained in the User’s Manual.)

But what is the point of all this? Is it not easier just to use the kind of ‘ray aiming’ that some other codes do?

Yes, it is easier—but it is much slower. As usually implemented, when those programs trace a grid of rays for any kind of image analysis, they create a square grid at the stop and then iterate *every ray* so it goes through that grid point. All that iteration takes time, a lot of time. Here is an example of a very wide-angle design (**C22L2**), in figure 22.5.

The stop is on surface 9, and it is nicely filled by the WAP 2 option. Look at a footprint on that surface showing rays from the full-field point, in figure 22.6.

This is certainly *not* a uniform square grid. Those programs that employ ‘ray aiming’ fill this aperture with the *wrong* distribution, and then, to account for this wrong distribution, they alter the effective energy of each ray according to the actual ray density at that point. While this can indeed produce a correct evaluation of the image, one has to ask why they spend so much time with all the ray iterations in the first place.

Instead, SYNOPSYS finds the size and shape of the *entrance pupil*, and then fills *that* with a uniform grid. Here is that pupil, on surface 1, for the above lens. It is regular, as it should be, shown in figure 22.7.

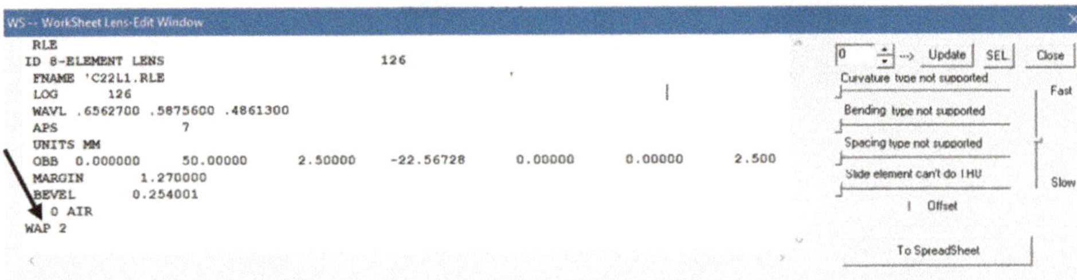

Figure 22.4. Worksheet edit panel, where a wide-angle pupil (WAP) number 2 has been declared.

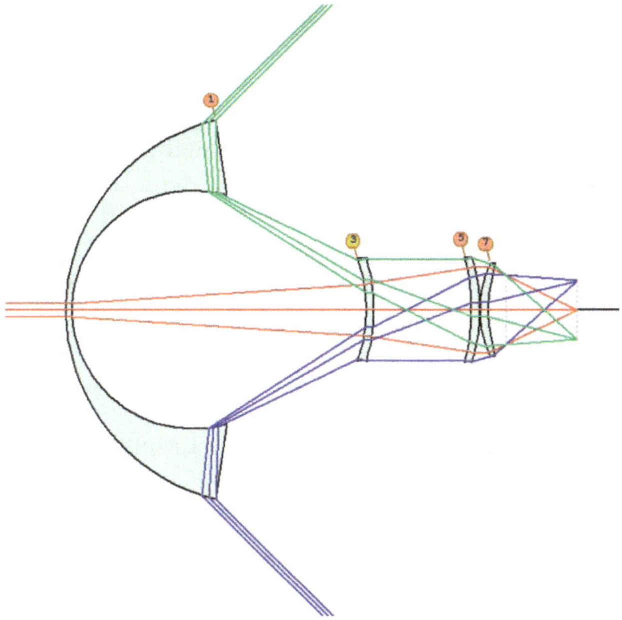

Figure 22.5. A very wide-angle lens using the WAP 2 option.

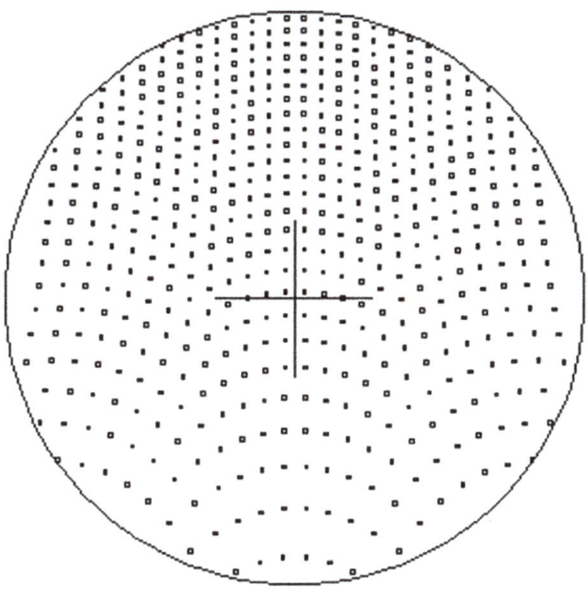

Figure 22.6. Ray distribution at the stop for the lens in figure 22.5. This is not a uniform distribution.

The WAP pupil options in SYNOPSYS model the *outline* of this distribution, so a regular grid can fill it as it should. It is not necessary to iterate each ray, so it is much faster, and the distribution at the stop is correctly modeled. For this extreme example, a simple outline is a bit too small, however (normally it is modeled by an elliptical shape). A better pupil, in this case, is found by declaring **RPUPIL** in the lens file. Now it starts with a rectangle that encloses that ellipse and deletes any rays that fall outside the stop aperture. Figure 22.8 shows that shape as it enters the lens, and figure 22.9 shows an outline of what actually gets through.

We like this better than the slow 'ray aiming' method.

Do not forget to check out the dialogs **MPW** (Menu, Pupil Wizard) and **MOW** (Menu, Object Wizard), where you can define the kind of pupil you need by checking boxes and selecting from various options. Both of those dialogs do much the same thing, but they are organized differently, so you can choose which you like best.

22.1 Which way is up?

The unique pupil definition in SYNOPSYS offers an interesting possibility—which is handy but takes some getting used to.

Let me illustrate. First, I will show you some rays that do not go where you expect them too, and then I will describe a simple way to keep things straight. Fetch the lens bundled as 1.RLE:

```
FETCH 1
```

Now look at it in PAD, as shown in figure 22.10.

The lens at the moment has been assigned a *paraxial* stop on surface 4. In PAD, click the 'PAD Top' button, and select the option to draw a single ray. Click

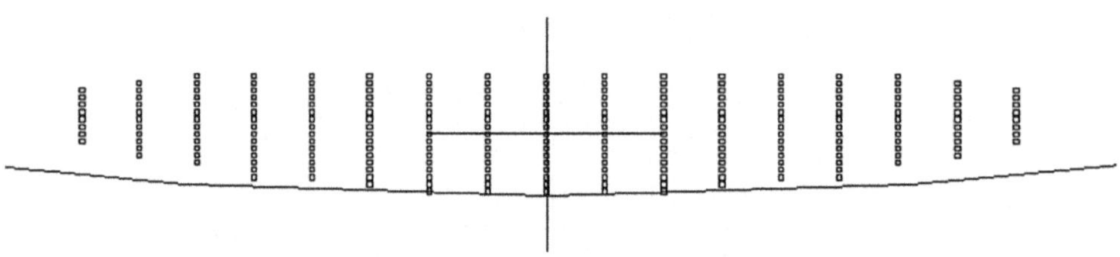

Figure 22.7. Ray grid at entrance pupil for the wide-angle lens.

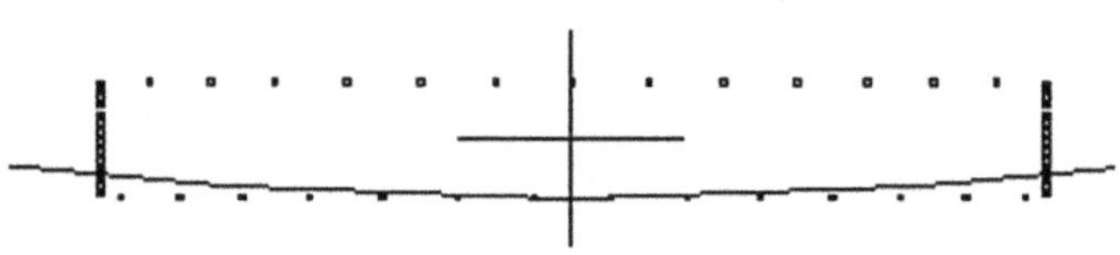

Figure 22.8. Outline of input ray pattern.

Figure 22.9. Outline of the pattern that gets through the lens.

'OK', and a small box opens where you can select which ray to draw with two sliders. Move the top slider to full field (HBAR = 1) and the bottom slider to full aperture (YEN = 1). This object has been defined with a positive *angle* coming in, which means that the 'full field' rays start from an object *below* the axis, as shown in figure 22.11.

You see the full-field marginal ray, as expected. Now move the top slider to the bottom of the field (HBAR = −1), as shown in figure 22.12.

Again, the ray enters at the top of the pupil. This is the basic idea of the paraxial pupil. Simple—but not always adequate. Close the ray display dialog and change the stop designation to

```
APS -4
```

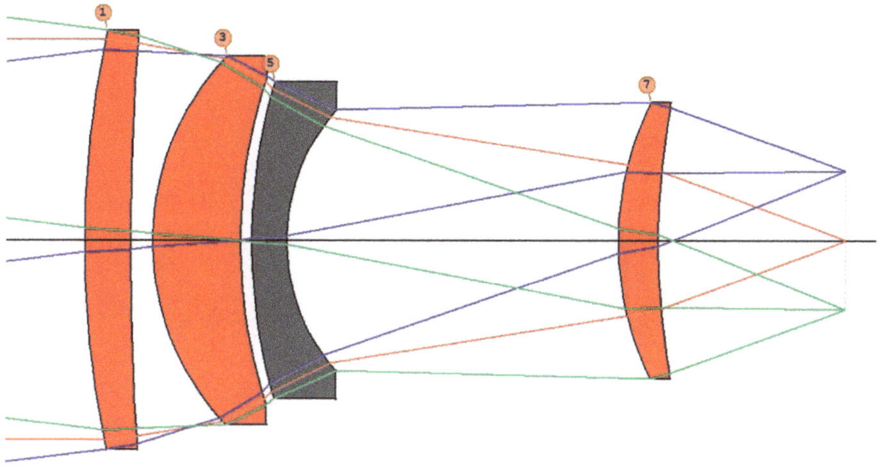

Figure 22.10. PAD display of lens for demonstrating pupil options.

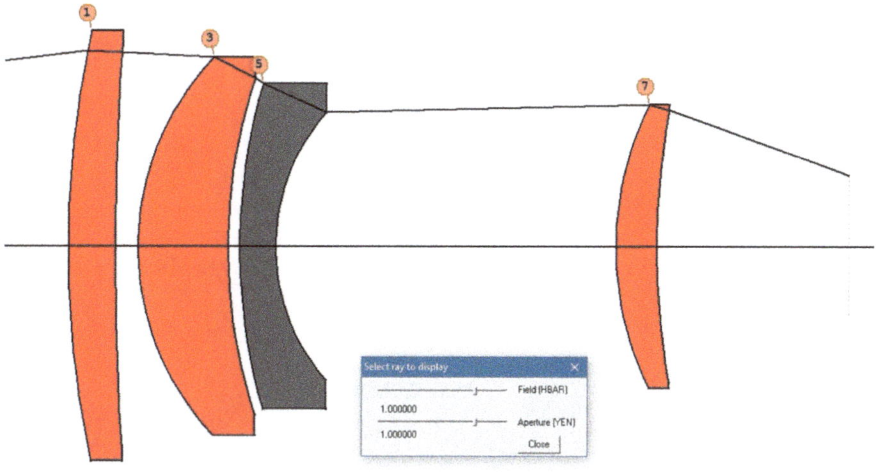

Figure 22.11. PAD display showing a single user-selected ray at the top of the field and aperture.

in the WorkSheet. Remember, the full-field object is located at a *negative y*-coordinate, far to the left of the lens.

Now it is time to get interesting. Open the single-ray dialog again and set it to full aperture and full field once more. Now you see the ray in figure 22.13.

What happened? The 'full aperture' ray is now at the *bottom* of the pupil. What is going on?

It is simple: this feature is designed to make it easy to correct feathered edges, no matter where in the field you are looking. (We have used the AEC monitor in many of these lessons to control edge feathering, a feature that usually works very well. However, there is also a provision in the program to control feathering along a selected ray at a given surface, and to use that effectively, you have to know which ray to trace. Look up ECP and ECN in the help file to read about that feature.)

In the lens in figure 22.13, you would correct along the 'upper' rim ray (the ray shown) if feathering were a problem. Now go to the lower field point, HBAR = −1, as in figure 22.14.

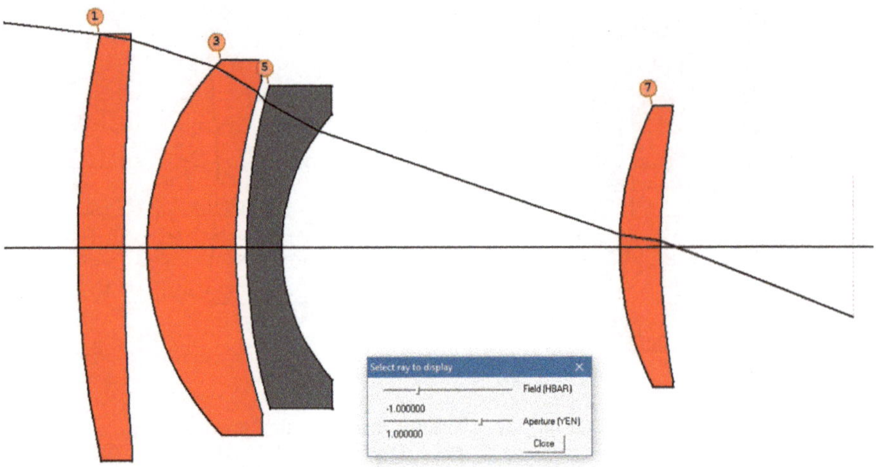

Figure 22.12. PAD display showing a single ray at the bottom of the field.

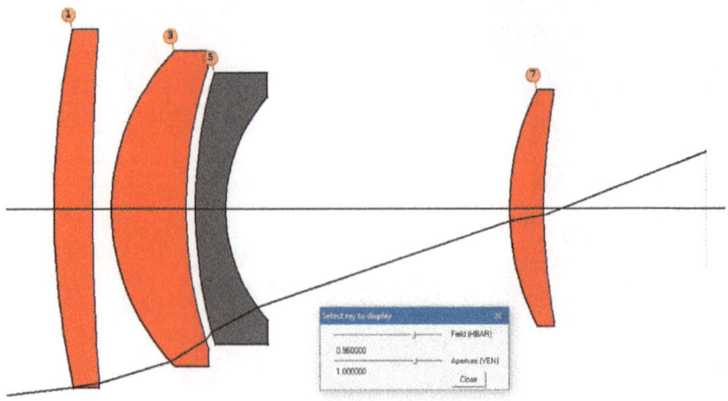

Figure 22.13. Example lens with a real pupil defined, at the top of the field.

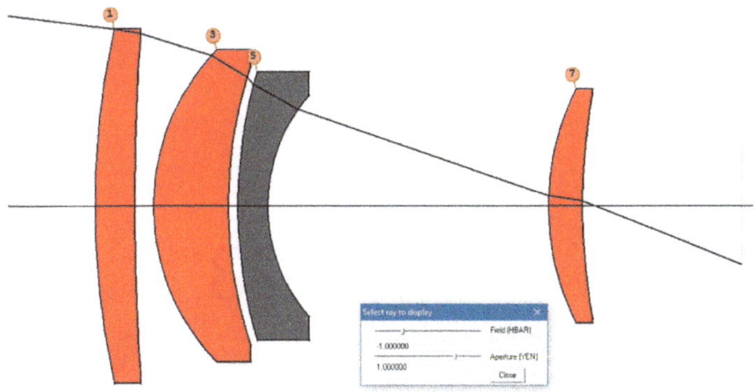

Figure 22.14. Example lens at the bottom of the field with a real pupil.

Aha! The ray to correct is *still* the upper rim ray! The program *rotates* the entire entrance pupil according to the orientation of the field point from which you are tracing. If you trace a point in the skew field, the 'upper marginal ray' would turn out to be the extreme *skew* ray, so again you can easily control feathered edges. If the program left the definition of the upper and lower marginal rays the same for all field points (as it is with a paraxial pupil), it would not be so easy to do; you would have to figure out which skew ray to fix and then create an aberration for it. This is much simpler—once you get used to it.

So how can you easily figure out which ray to examine or correct for feathered edges? It is simple: while the PAD display is open, press the <**F7**> key. Only the 'lower' rim ray at full field shows up. The <**F8**> key shows only the 'upper'. Presto! You can tell which ray is where with a single keystroke.

There is another advantage to this kind of pupil definition: the entrance pupil is usually modeled as an ellipse, as illustrated in the first part of this chapter, and it turns out that the ellipse also rotates with field point. So it can model the vignetted pupil at all points in the field.

Take a look at section 2.6.1 in the User's Manual for an example of a rotated pupil.

One more thing: the program decides which ray to call the 'upper' ray based on the sign of the height of the full-field object. Since that was negative in this example, it flipped the marginal rays for positive HBAR. At negative HBAR, the object comes from a positive y-coordinate, and the opposite happens.

What about at HBAR = 0? Well, to avoid confusion from being neither positive or negative, the program displays a very small but nonzero field point there.

Confused? Try the <F7> and <F8> keys—simple.

IOP Publishing

Lens Design
Automatic and quasi-autonomous computational methods and techniques
Donald Dilworth

Chapter 23

Using DOEs in modern lens design

Improving a lens with a kinoform surface

In this lesson, we will start from scratch, design a five-element lens, and then see if adding a diffractive optical element (DOE) to the lens can improve its performance.

Here is the problem, as defined by the entries in the Design Search dialog (**MDS**). This dialog will create a MACro that runs the DSEARCH command, and is shown in figure 23.1 with all the data filled in.

This input will design a lens at F/3.5 with a semi-field angle of 25 degrees and an aperture radius of 12 mm. We elect to control the back focus with the 'SPECIAL AANT' entry, which lets the distance grow but will not let it become less than about 22 mm. We also ask for the angles of the upper and lower rim rays at full field with respect to the surface normal at every surface to be no more than 60 degrees, with a low weight, with the ACA request, so we do not obtain solutions with wild angles at the image and to avoid steep ray refraction, which would introduce high-order aberrations that are better avoided.

When you click the 'OK' button, the program loads the MACro. The **CORE 14** directive at the top will speed things up, and we specify a grid number of 6, because aspherics and DOEs can cause high-order aperture aberrations and we will likely need more than the default grid of 4 (**C23M1.MAC**):

```
CORE 14
DSEARCH 1  QUIET
SYSTEM
ID DSEARCH SAMPLE
OBB 0 25 12
WAVL 0.6563 0.5876 0.4861
```


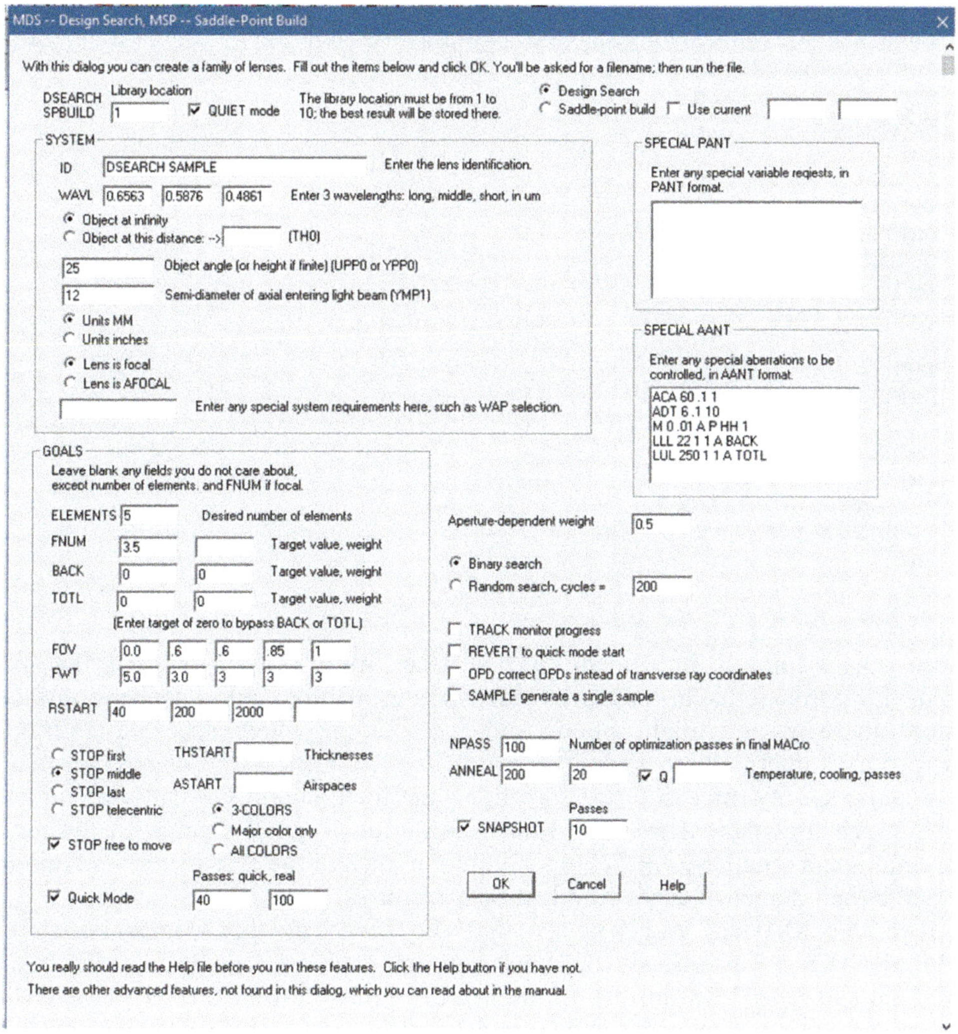

Figure 23.1. MDS dialog with selections for an example lens.

```
UNITS MM
END
GOALS
ELEMENTS 5
FNUM 3.5
BACK 0 0
TOTL 0 0
STOP MIDDLE
STOP FREE
```

```
RSTART 40 200 2000
RT 0.5
FOV 0.0 .4 .6 .85 1
FWT 5.0 3.0 3 3 3
NPASS 100
NGRID 6
ANNEAL 200 20 Q
COLORS 3
SNAPSHOT 10
QUICK 40 100
END
SPECIAL PANT

END
SPECIAL AANT
 ACA 60 .1 1
 ADT 6 .1 10
 M 0 .01 A P HH 1
 LLL 22 1 1 A BACK
 LUL 250 1 1 A TOTL
END
GO
```

Since we are going to implement DOE surfaces, we also elect to specify five field points for correction. This is a good idea when using any kind of aspheric surfaces, since otherwise one might obtain great correction where specified and poor correction at intermediate fields.

We also specify three different starting values for the radius of curvature of each case, to be investigated in turn. Remember, even a small change to the initial conditions can send DSEARCH to a different branch of the design tree, and this will increase the number of cases searched by a factor of three.

Run this MACro, and you see that the best lens that comes back from DSEARCH is not too good—but what can you expect with only five elements at this field and speed? Optimize with the MACro prepared by DSEARCH, and then anneal (**50, 2, 50**). The lens is shown in figure 23.2, a classic inverse telephoto.

We can probably obtain better results by requesting a greater number of elements—but here we want instead to see how much improvement we can obtain by changing one of the lenses to a DOE. Let us try it. Add to the optimization MACro another line at the top (**ADA** means **A**utomatic **D**OE **A**ssignment):

```
ADA 5 QUIET
PANT
VY 0 YP1
VLIST RD ALL
```

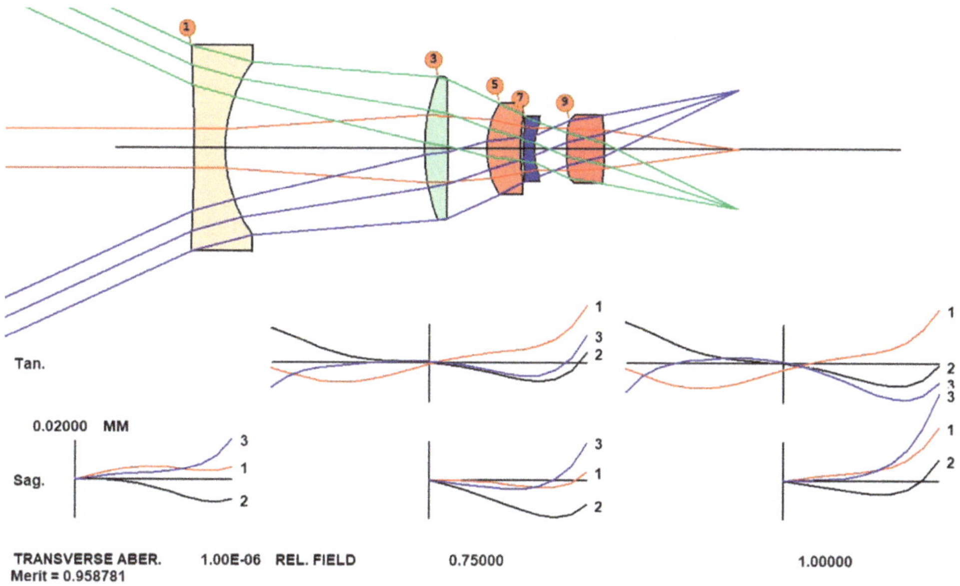

Figure 23.2. Five-element lens, all spherical, returned by DSEARCH, optimized, and annealed.

```
VLIST TH ALL
VLIST GLM ALL
END
AANT P
AEC
ACC
GSR      0.500000     5.000000     6  2     0.000000
GSR      0.500000     5.000000     6  1     0.000000
GSR      0.500000     5.000000     6  3     0.000000
GNR      0.500000     3.000000     6  2     0.400000
GNR      0.500000     3.000000     6  1     0.400000
GNR      0.500000     3.000000     6  3     0.400000
GNR      0.500000     3.000000     6  2     0.600000
GNR      0.500000     3.000000     6  1     0.600000
GNR      0.500000     3.000000     6  3     0.600000
GNR      0.500000     3.000000     6  2     0.850000
GNR      0.500000     3.000000     6  1     0.850000
GNR      0.500000     3.000000     6  3     0.850000
GNR      0.500000     3.000000     6  2     1.000000
GNR      0.500000     3.000000     6  1     1.000000
GNR      0.500000     3.000000     6  3     1.000000
  ACA 60 .1 1
  ADT 6 .1 1
```

```
   M 0 .01 A P HH 1
   LLL 22 1 1 A BACK
   LUL 250 1 1 A TOTL
END
SNAP/DAMP 1
SYNOPSYS  100
```

Run this version, and the program finds that a DOE at surface 9 works best, as shown in figure 23.3. The lens is much better.

The command **ASY** shows the data for this DOE:

```
SYNOPSYS AI>ASY

SPECIAL SURFACE DATA

 SURFACE NO.    9 -- UNUSUAL SURF TYPE 16 (SIMPLE DOE)
WAVELENGTH OF OPD DEFINITION:          0.587600
INDEX OF DOE MATERIAL: PICKUP SUBSTRATE
NORMALIZING RADIUS:           17.404900
DIFFRACTION ORDER:            -1
XD  1     0.007901 (CV)       XD 11     77.600306 (R**2)   XD 12   -12.495727 (R**4)
XD 13     5.748107 (R**6)     XD 14    -1.117877 (R**8)

THIS LENS HAS NO TILTS OR DECENTERS
```

We are curious what would happen if we added a *second* DOE. That is simple to test. Add variables to the PANT file for the DOE terms that ADA just added,

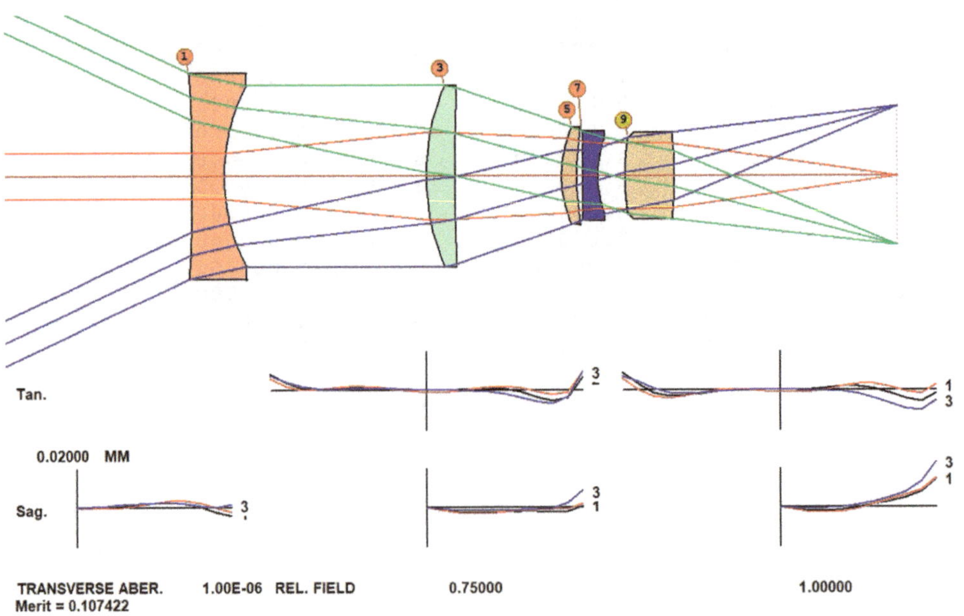

Figure 23.3. Lens with a DOE added by ADA.

```
VY 9 G 16
VY 9 G 26
VY 9 G 27
VY 9 G 28
VY 9 G 29
```

and run the MACro again. How can you tell which G-variables to vary? Look in the User's Manual under **USS** (Unusual Surface Shapes), select type 16, and you see that those terms will vary the base curvature and OPD terms from second to eighth order.

XD number	Function	XDD index	G-variable
1	Base curvature	1	16
2	Base conic constant	1	17
3	Rho**4 term of base curve	1	18
4	Rho**6	1	19
5	Rho**8	1	20
6	Rho**10	2	21
7	Rho**12	2	22
8	Rho**14	2	23
9	Rho**16	2	24
10	Rho**18	2	25
11	Rho**2 term of OPD expansion	3	26
12	Rho**4	3	27
13	Rho**6	3	28
14	Rho**8	3	29
15	Rho**10	3	30
16	Rho**12	4	31
17	Rho**14	4	32
18	Rho**16	4	33
19	Rho**18	4	34
—	D0: blaze depth constant term		54
—	D1: linear term		55
—	D2: quadratic term		56

This time it wants a DOE at surface 4, shown in figure 23.4, and the merit function was reduced to 0.0445.

It is time to assign a real stop to surface 7. Then modify the PANT file so it will vary the coefficients on both DOEs, and include some higher-order terms as well. Term `G 32` is the twelfth-power coefficient, while the default from ADA only goes

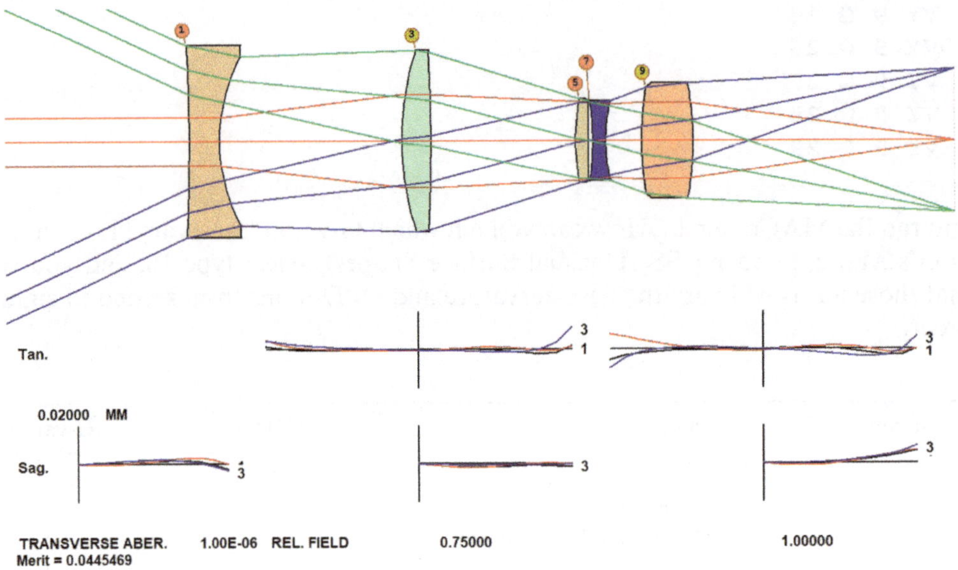

Figure 23.4. Lens with DOEs on two surfaces, assigned by ADA.

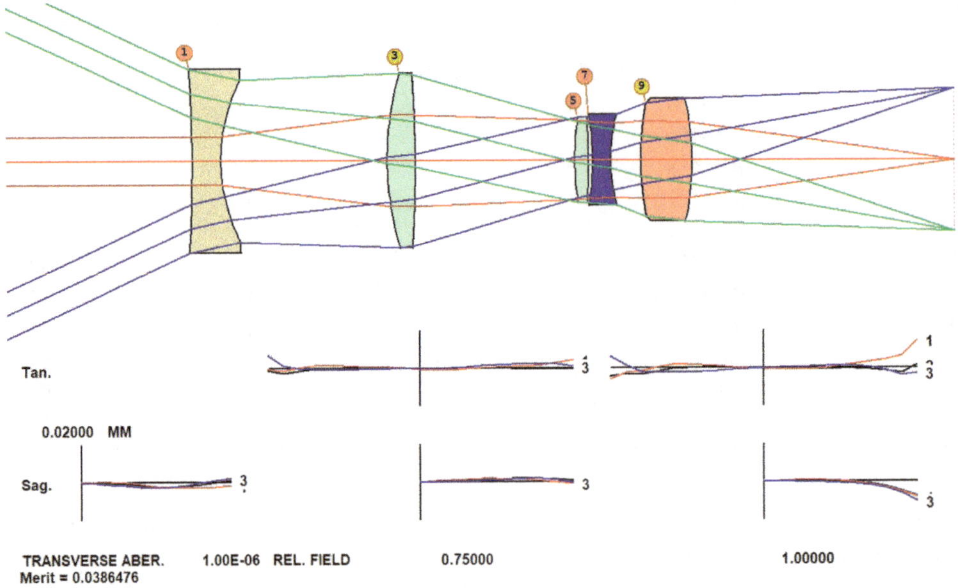

Figure 23.5. Lens optimized with two DOEs.

to the eighth power. (Be sure to comment out the ADA command, so you do not obtain a third DOE!) Also, comment out the YP1 variable so the stop stays on 7:

```
!ADA 5 QUIET

PANT
!VY 0 YP1
VLIST RD ALL
VLIST TH ALL
VLIST GLM ALL

VY 4 G 16
VY 4 G 26
VY 4 G 27
VY 4 G 28
VY 4 G 29
VY 4 G 30
VY 4 G 31
VY 4 G 32

VY 9 G 16
VY 9 G 26
VY 9 G 27
VY 9 G 28
VY 9 G 29
VY 9 G 30
VY 9 G 31
VY 9 G 32

END
...
```

Now run this again, and then anneal (**50, 2, 50**). You obtain the design in figure 23.5 (**C23L1**).

This lens is rather close to the diffraction limit, and it might make sense to do this exercise again, this time with OPD targets, but we will continue as it is.

One naturally asks, how much light gets through? A DOE has to *diffract* light into a desired direction, and of course diffraction involves multiple orders. It would be wise to check. The program assumes that the zones on the DOEs are *blazed* in such a way that the diffracted direction coincides with the *refracted* direction, but that assumption cannot be exact for all rays. Open the **MMA** dialog (`MMA`), select the transmission of a ray (not the transmission of a *beam*), 'Map' over 'PUPIL', 'Y-point' 1.0 in the field, 'Ray Pattern' 'CREC' with a grid number of 51, 'EXPLODED', and 'Show color scale'. Click 'Execute', and you obtain a picture showing the lens transmission mapped over the aperture, shown in figure 23.6. The assumption mentioned above, also called the *Bragg condition*, seems to be valid enough for practical purposes. Transmission is very high. (The system is not in polarization mode, so reflection losses are ignored, and the materials at the moment are all glass models, which do not come with absorption coefficients. Those effects can be analyzed later.)

It would be interesting to see how many spherical elements one would need to obtain this kind of quality; it will certainly be more than five. The next step would be to replace the glass models with real glasses—but we will leave that exercise for the student.

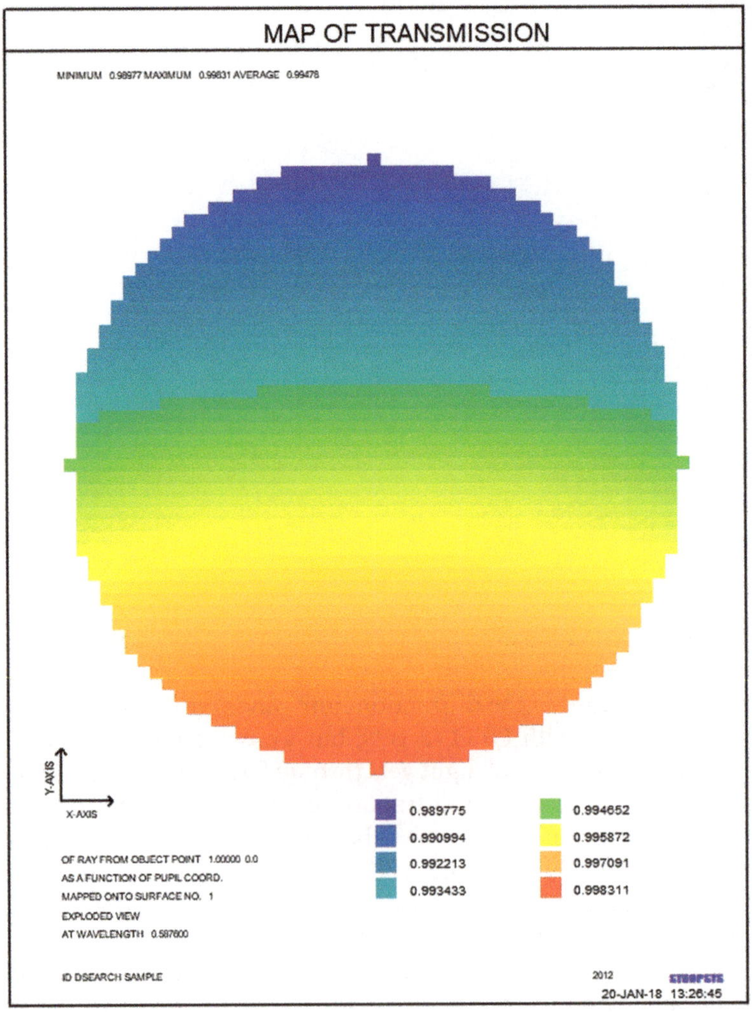

Figure 23.6. MAP of ray transmission over the pupil at the edge of the field.

This lesson has shown how converting a lens surface to a DOE can significantly improve image quality—or let you obtain the quality you need with fewer elements. Of course, it all depends now on whether the lens vendor can *make* the DOEs. These may not be too easy. Here is the DMASK profile at surfaces 3, on the left, and 9, on the right, shown in figure 23.7:

```
DMASK 4 PROFILE
DMASK 9 PROFILE
```

The second might be a challenge for the shop. Let us examine the spatial frequency. Open the MAP dialog again with **MMA**, select a map of 'HSFREQ' over 'PUPIL', on surface 9, 'Object point' 0, 'Ray Pattern' as 'CREC 9', 'DIGITAL', and 'Execute'. The highest frequency is over 7 c mm^{-1} at the edge. This is looking pretty good, but that of course depends on the capability and technology of the shop that will make them. This plot is in figure 23.8.

We expect that, as this technology matures, the designs presented here will become more and more practical. In any event, it is better to be ahead of the

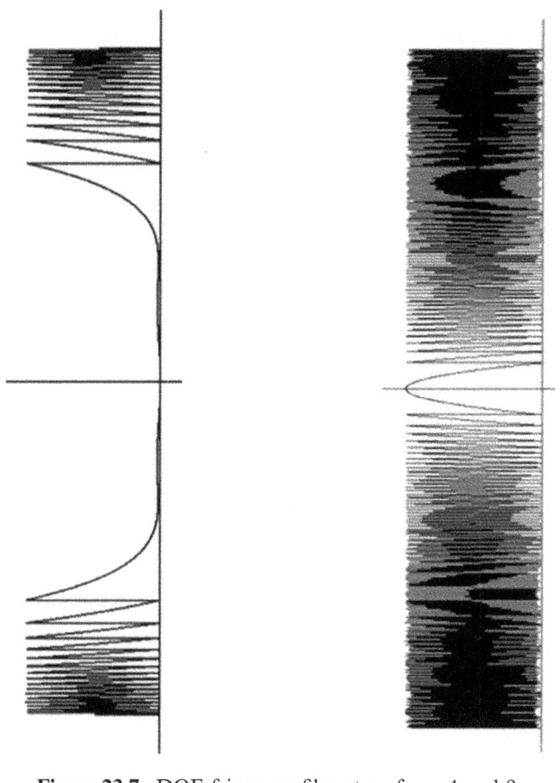

Figure 23.7. DOE fringe profiles at surfaces 4 and 9.

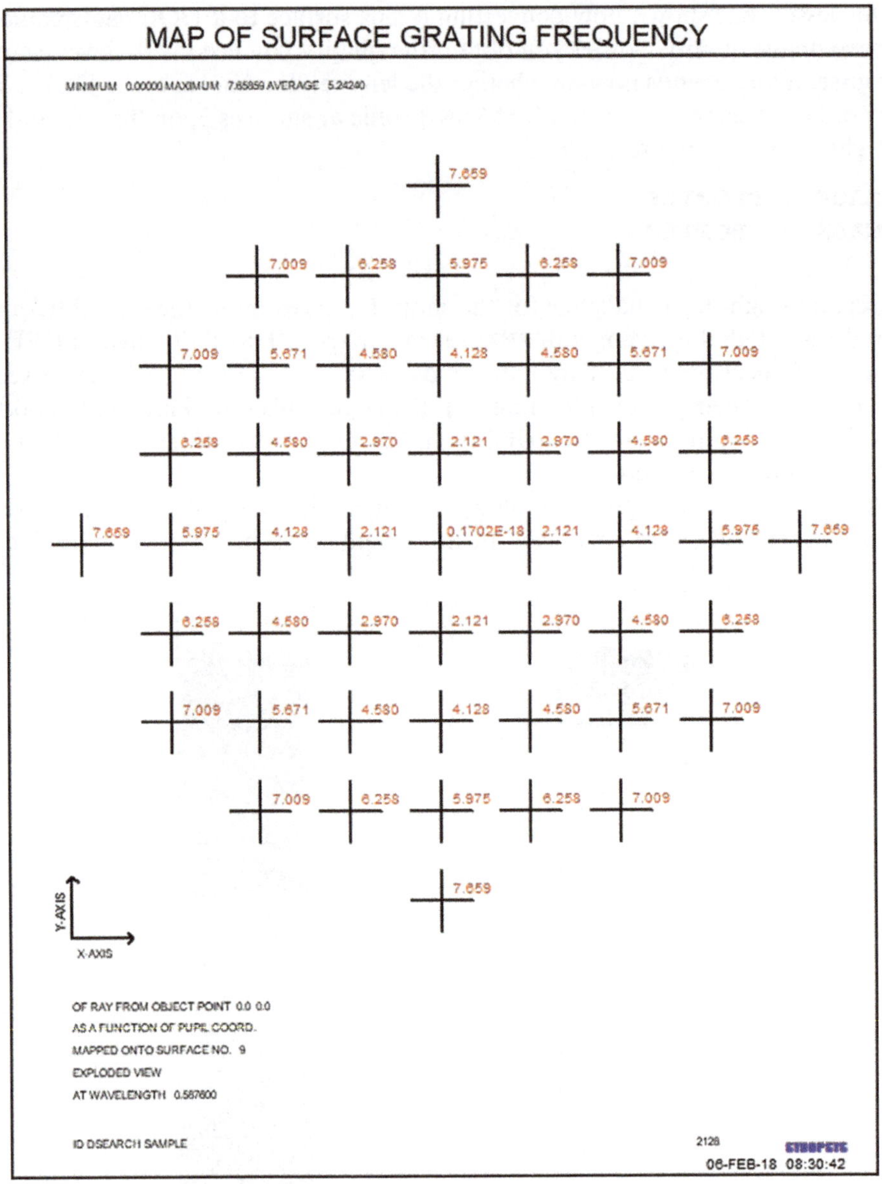

Figure 23.8. Surface grating frequency of the DOE on surface 9.

technology rocket than running behind, trying to keep up. We invite lens vendors with DOE capability to comment on this lesson and perhaps offer insights and design tradeoffs as they understand them today.

IOP Publishing

Lens Design
Automatic and quasi-autonomous computational methods and techniques
Donald Dilworth

Chapter 24

Designing aspheres for manufacturing

Improving a lens with aspherics; controlling departure from best-fitting sphere; CLINK

This lesson shows how to add aspheric terms to a spherical surface, thereby improving the image. Then the optimization controls the RMS departure of the aspheric surface from the closest fitting sphere (CFS) in order to make it easier to manufacture.

This is the starting lens (**6.RLE**), a poorly corrected triplet, shown in figure 24.1. Type `FETCH 6` to open this lens.

Let us perform a simple optimization run, to see how things improve. First, we will use only spherical surfaces. Here is the MACro (**C24M1**):

```
PANT
VLIST RAD 1 2 3 4 6
VY 1 TH 20 3
VY 2 TH
VY 3 TH 20 3
VY 5 TH
VLIST GLM 1 3
END

AANT
AEC              ; AUTOMATIC EDGE CORRECTION
ACCACC           ; AUTOMATIC CENTER THICKNESS CONTROL

GNR .5 1 3 2 0
GNR .5 1 3 2 .5
GNR .5 1 3 2 .7
GNR .5 1 3 2 1.
```

doi:10.1088/978-0-7503-1611-8ch24

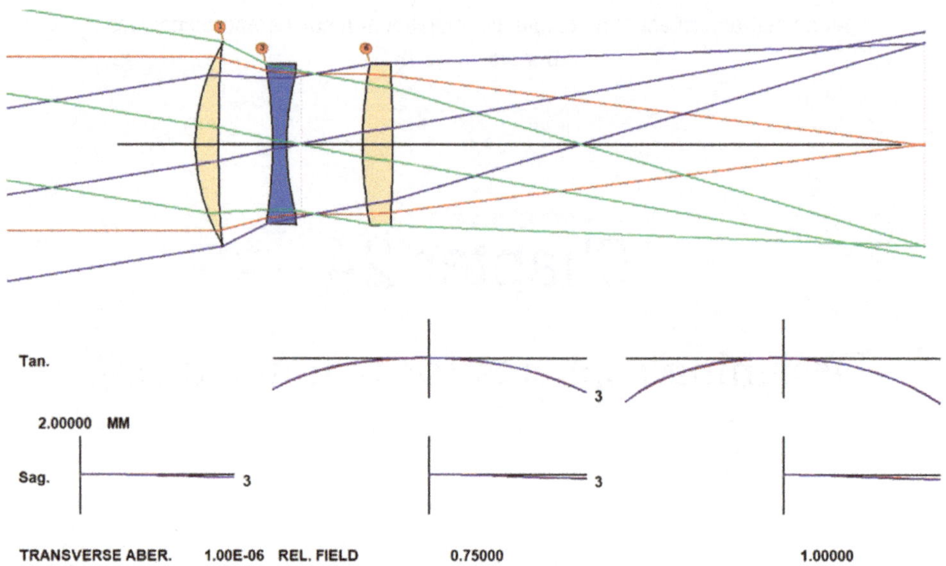

Figure 24.1. Starting triplet lens.

```
GNR .5 1 2 1 0
GNR .5 1 2 3 0
GNR .5 1 2 1 1
GNR .5 1 2 3 1
END

SNAP            ; REQUEST SNAPSHOTS AS OPTIMIZATION RUNS
SYNO 50         ; REQUEST OPTIMIZATION FOR 50 PASSES
```

Run this and the lens is improved; but the fifth-order spherical aberration is balanced by under-corrected third-order spherical, as shown in figure 24.2.

Now we will use the Automatic G-term Testing feature **AGT** to see if adding some general aspheric terms will improve things. Add a line before the PANT file in your MACro:

```
AGT 5 QUIET 1 .01 3 6 10 16
```

Here, we ask the program to test terms G 3, 6, 10, and 16 on surface 1, and keep any that reduce the merit function by 1% or more. Since surface 1 is not currently aspheric, those terms will apply to the default power-series asphere and will vary terms of power 4, 6, 8, and 10 in the aperture.

Run this, and the lens is better, as one would expect, as shown in figure 24.3. The program reports that only term G 3 was useful.

What is the value of those aspheric terms? The **ASY** listing gives the coefficients:

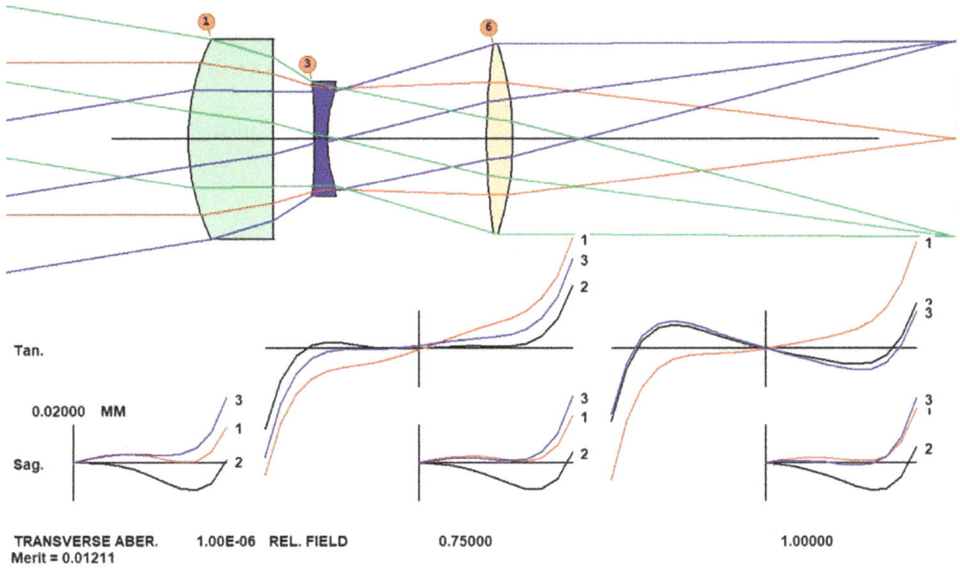

Figure 24.2. PAD display of the improved lens.

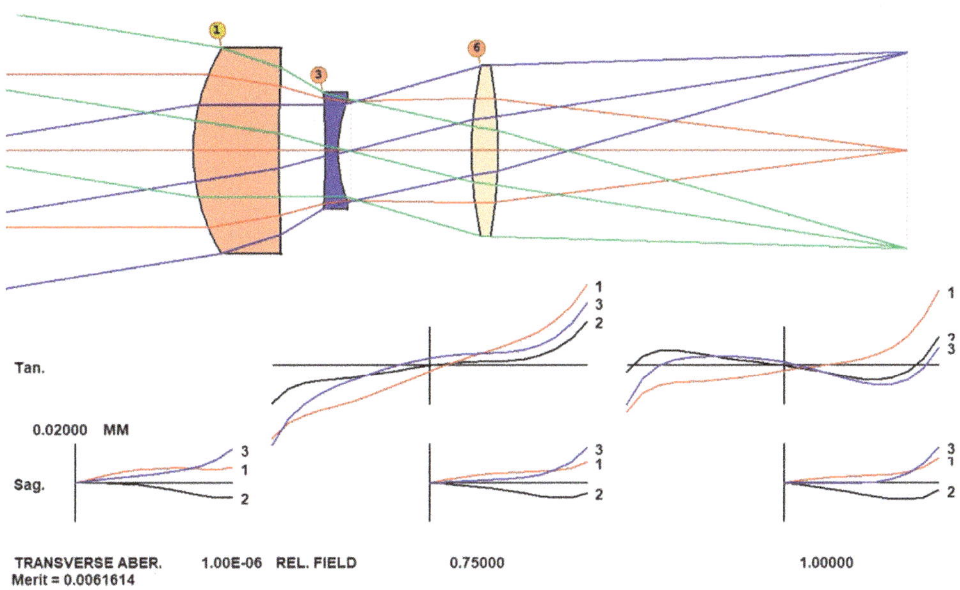

Figure 24.3. Lens optimized with one aspheric term.

```
SYNOPSYS AI>ASY

SPECIAL SURFACE DATA

 SURFACE NO.    1 -- RD + POWER-SERIES ASPHERE
G 3    -8.493897E-08 (R**4)

THIS LENS HAS NO TILTS OR DECENTERS
```

We have used only one of the 22 possible G-series aspheric terms so far. How close is this surface to the closest-fitting sphere?

Type the command **ADEF 1 PLOT**, and you obtain the plot in figure 24.4. The maximum sag difference is about 5.8 μm.

24.1 Adding unusual requirements to the merit function with CLINK

The plot in figure 24.4 shows a large departure from the CFS, almost 6 μm. That might be hard to control accurately. Let us see if we can obtain a similar

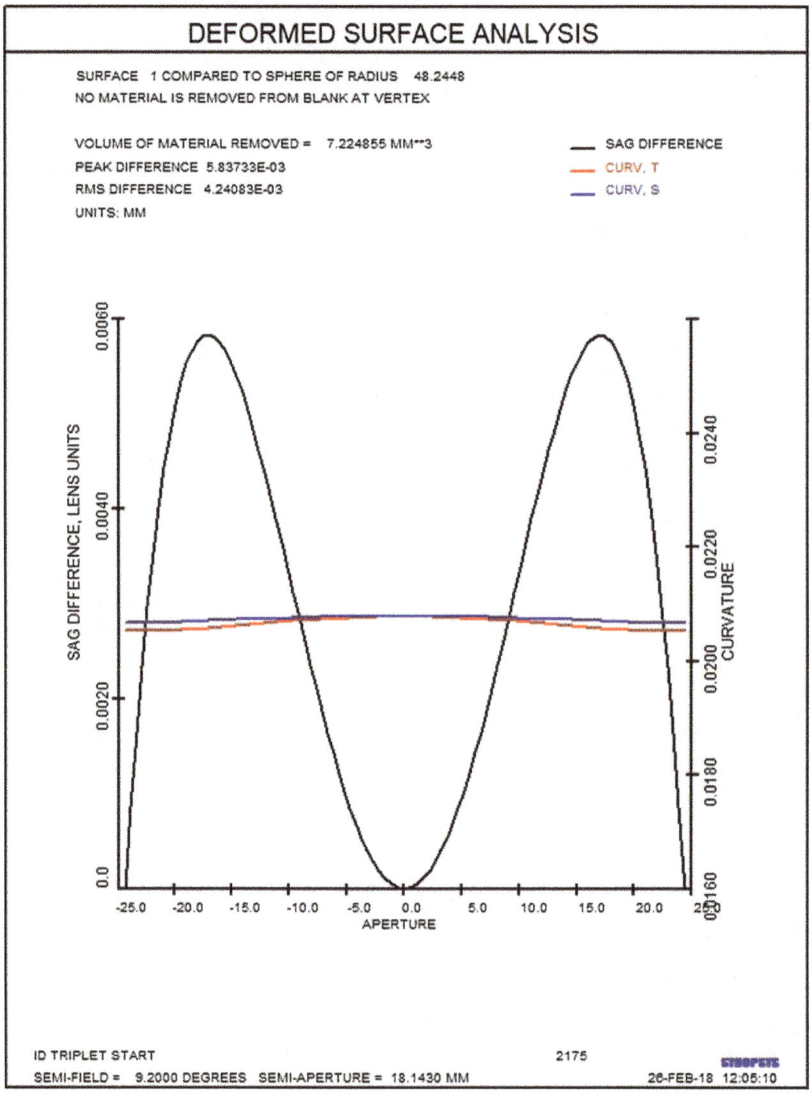

Figure 24.4. Analysis of aspheric surface 1.

performance with an asphere that has a smaller departure. Add variables to the PANT file as shown below, to see if we can obtain a similar performance with a smaller RMS departure:

```
VY 1 G 3
VY 1 G 6
VY 1 G 10
VY 1 G 16
```

Comment out the AGT line, and add to the AANT file the lines

```
M 0 5 CLINK
ADEF 1
CD1 FILE 6
= CD1
```

Here we use the CLINK option, which causes the optimization program to run the next command (**ADEF** in this case) and then pick up the desired quantity from the FILE BUFFER. (To learn about this useful feature, type **HELP CLINK** in the CW.)

SYNOPSYS already has a command for this kind of analysis, described in section 10.3.3 of the User's Manual, but what would you do if it did not? This lesson shows how other features can be used to do the same thing, and it is a good idea to know how to use these other features in case you want to do something for which there is no command. (To use the intrinsic form, add to the AANT file the line M 0 50 A ADIFF sn, instead of the CLINK section above, where sn is the surface number.)

Where did the weighting factor of 5 come from? The RMS difference between the asphere and the CFS is 0.0037 mm, and the largest of the other aberrations is 0.015. To see those values, you can use the handy FINAL nb command. Enter FINAL 5, to see the five largest aberrations, and you obtain the following table:

```
FINAL 5
 ABERRATION LIST
        NAME              TARGET          WEIGHT               RAW VAL.  FINAL ERROR
R. EFFECT

   80                   0.0000000     0.9362033 SR      0.0144    0.134860E-01 0.029585
     A     2  YC     0.70000    0.50000    0.83333    0.00000       ACON 2

   90                   0.0000000     0.9362033 SR     -0.0131   -0.122242E-01 0.024307
     A     2  YC     0.70000    0.50000   -0.83333    0.00000       ACON 2

  102                   0.0000000     1.7262735 SR     -0.0068   -0.118239E-01 0.022742
     A     2  YC     1.00000    0.16667    0.50000    0.00000       ACON 2

  156                   0.0000000     1.1508490 SR      0.0133    0.152882E-01 0.038020
     A     1  YC     1.00000    0.25000    0.75000    0.00000       ACON 2

  160                   0.0000000     2.5733766 SR     -0.0057   -0.147917E-01 0.035591
     A     1  YC     1.00000    0.25000   -0.25000    0.00000       ACON 2
 SYNOPSYS AI>
```

So this weighting will give the RMS departure an error comparable to the error of the largest ray aberration. Of course we can adjust it later, when we see how things work out.

How can one figure out the other lines to be added to the AANT file? It is simple. When the ADEF command runs, it puts a copy of some of its output into the AI buffer. Run the command **ADEF 1**, and then ask the AI question

BUFFER?

to see what got put where.

```
SYNOPSYS AI>BUFF?

 The current FILE BUFFER contains
   1         0.02057423  BEST FIT CV
   2        48.60449871  BEST FIT RD
   3        -0.00000000  VERTEX SHIFT
   4         0.71748646  VOL. REMOVED
   5         0.00143769  PEAK DIFF.
   6         0.00049916  RMS DIFF.
```

Location number 6 has the desired RMS difference—the quantity we want to reduce.

Run this new optimization, and the ray fans are almost unchanged. What does the aspheric look like now? The result is in figure 24.5:

```
SYNOPSYS AI>ASY

 SPECIAL SURFACE DATA

  SURFACE NO.    1 -- RD + POWER-SERIES ASPHERE
 G 3    -3.494551E-08 (R**4)   G 6   5.662387E-12 (R**6) G 10   6.833226E-14 (R**8)
 G 16   -2.933776E-17 (R**10)
```

Now the RMS difference is only 0.0005. Here we have a lens with only 12% as much aspheric departure as before, and essentially identical performance (**C24L1**).

Now it might make sense to adjust that weighting factor, increasing it in stages until the performance starts to degrade. That might yield a lens that is even easier to make than the one above.

One can also control the *peak* difference instead of the RMS, by simply picking up `FILE 5` instead of 6. Let us try that idea. Change the extra AANT entries you added above to

```
M 0 5 CLINK
ADEF 1
CD1 FILE 5
= CD1
```

and reoptimize. Now the peak departure is 0.000 77 mm, shown in figure 24.6.

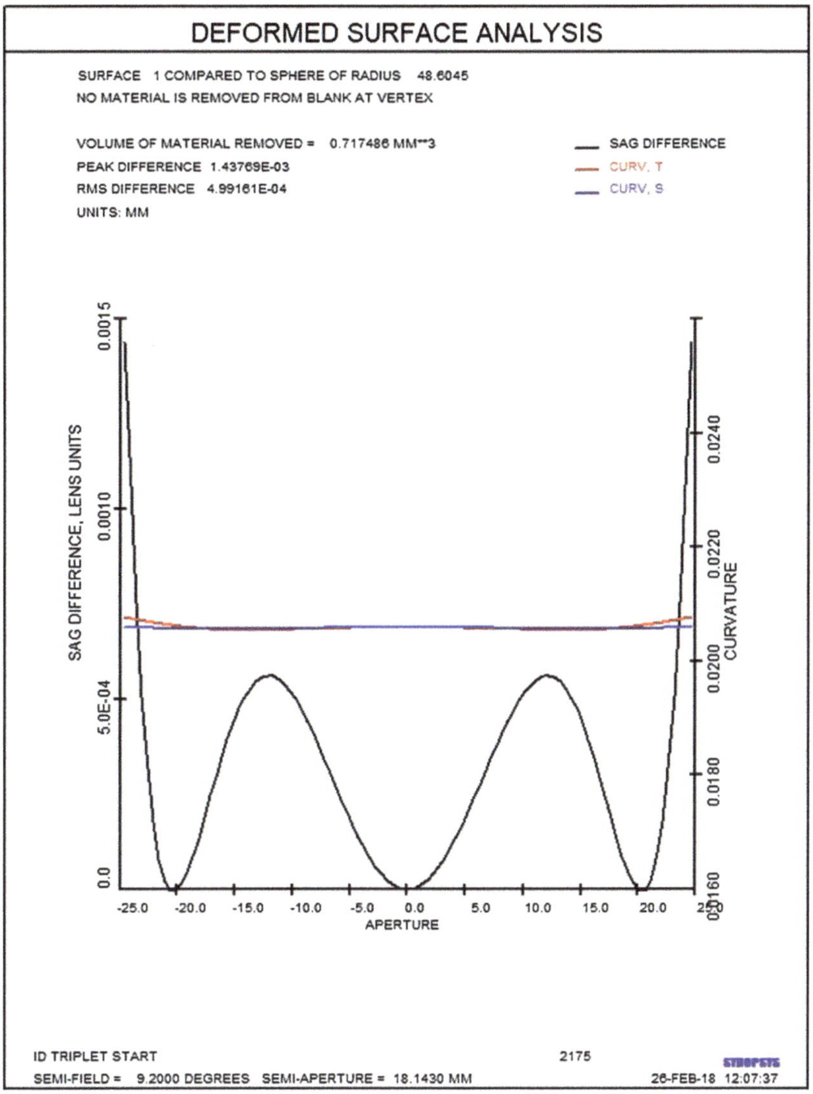

Figure 24.5. Aspheric analysis shows a smaller departure from the CFS.

How hard would it be to make this aspheric? Let us see:

```
ADEF 1 FRINGES
```

The result is in figure 24.7.

This is a very nice gentle asphere, with only a few fringes to the best-fitting spherical surface. The fringes are seen in double-pass; the actual difference is only half that shown.

With these tools, you can utilize aspheric elements while taking care that they are easily manufactured.

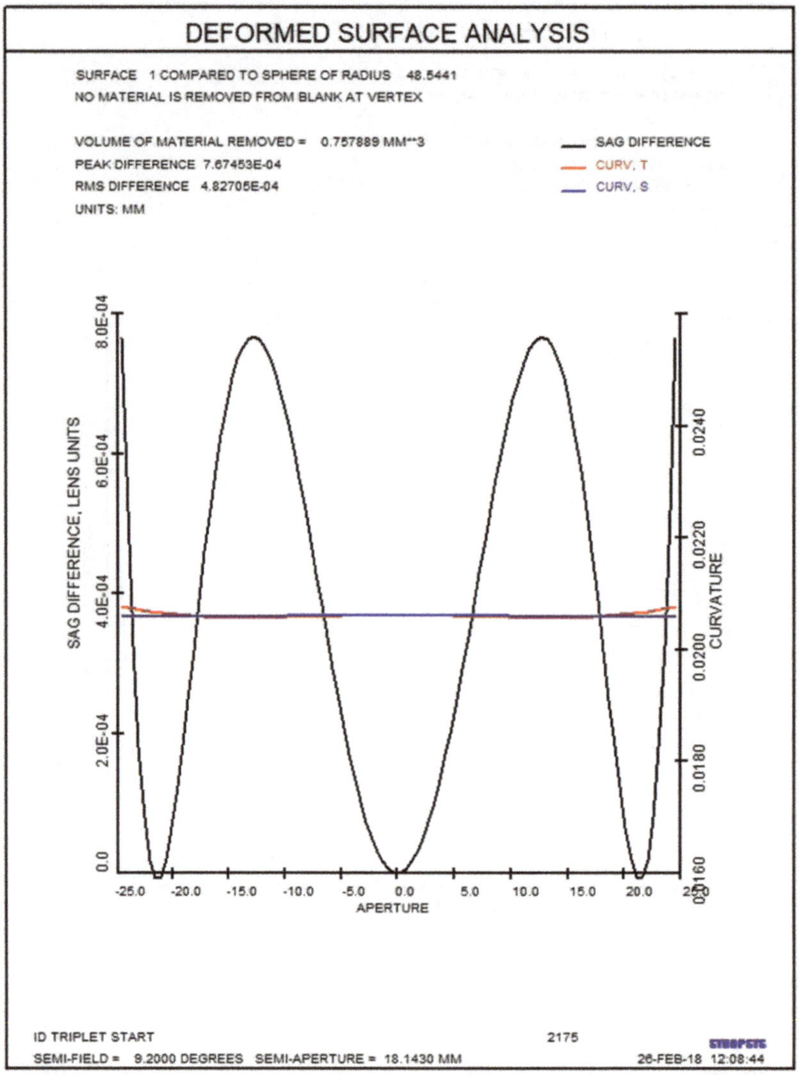

Figure 24.6. Analysis of aspheric surface after controlling the peak difference with the CFS.

24.2 Defining an aberration with COMPOSITE

Yet another way to create a custom aberration definition is with the COMPOSITE format, which lets you combine quantities with an algebraic equation. For example, consider the following AANT file (which is not related to the lens above):

```
AANT
M 0 1 COMPOSITE
CD1 PYA 10
CD2 PYB 10
```

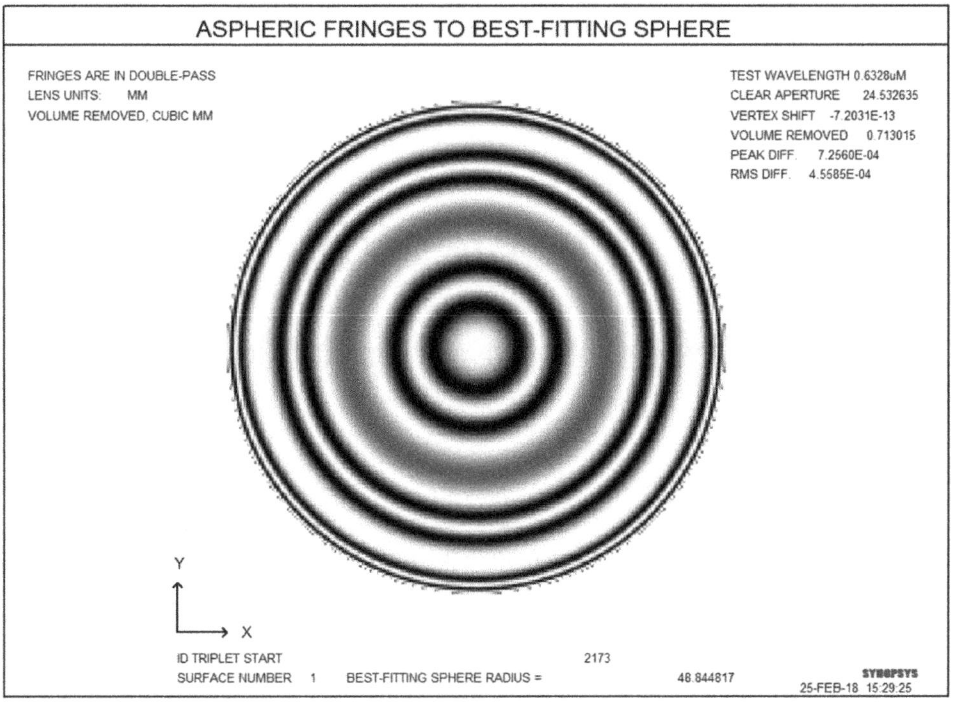

Figure 24.7. Analysis of aspheric surface with reduced peak difference.

```
CD3 GBD
CD4 RAD 10
= ATAN(CD3) + ASIN((SQRT(CD1**2 + CD2**2))/CD4)
END

SYNO 5
```

Here we use four of the nine CD*n* variables, which are defined in the AANT file as shown above. Variable CD1 is set equal to the paraxial *y*-coordinate of the marginal ray on surface 10, CD2 is the *y*-coordinate of the chief ray there, CD3 is the Gaussian beam divergence at the last surface, and CD4 is the radius of curvature of surface 10. When these are all defined, the aberration consists of the entered equation, which involves the arctan, arcsine, and square root. The result of that calculation is then targeted to the value 0 with a weight of 1.0. Chapter 10 of the User's Manual describes this feature in more detail. (This example is shown only as an illustration; we would likely never actually use that definition for a real lens.)

Chapter 25

Designing an athermal lens

Selecting glass types and housing material to correct both chromatic aberration and thermal effects

This chapter will show how to design a lens that must stay in focus over a wide temperature range.

First, we must discuss the interesting concept of the 'achrotherm', which applies to lenses that are corrected for both chromatic aberration and temperature changes at the same time. The theory is actually quite simple[1].

To design this kind of system, you select two glass types that obey a special requirement. They may be found by using the glass table display (**MGT**), where you click on the '**Graph**' button and then select '**Thermal properties**'.

The graph in this case shows the quantity $1/V_d$ on the abscissa (V_d is the Abbe number) and the quantity β on the ordinate, defined as

$$\beta = a_{\mathrm{g}} - \frac{1}{n-1}\frac{dn}{dt},$$

where α_{g} is the coefficient of thermal expansion of the glass, n is the index of refraction, and dn/dt is taken from the glass table. The idea is to select two glasses on the diagram and draw a line connecting them. Extend this line to the right to where $1/V$ equals zero. The height of the intercept, which can be read from the ordinate axis, is the CTE (coefficient of thermal expansion) of the housing required to athermalize the lens.

This feature needs to know the thermal coefficient of the housing material. Open any lens file and specify aluminum 6061:

[1] The original paper is Tamagawa Y, Wakabayashi S, Tajime T and Hashimoto T 1994 Multilens system design with an athermal chart *Appl. Opt.* **33** 8009–13.

doi:10.1088/978-0-7503-1611-8ch25

```
CHG
ALPHA A6061
END
```

Now you can use the glass table to select some likely glass pairs. In **MGT**, select the Ohara catalog and look at the thermal properties, as explained above. The program draws a green symbol over to the right, which is a function of the **ALPHA** of the lens housing that you just entered. Click on the left-most glass, and it draws a line between that glass and the symbol, as shown in figure 25.1.

The line starts at glass S-NPH3. Now you need another glass that falls close to the same line. Type S-FPM3 is quite close, so those two look like a good pair. It might be useful to have a second set, so click the glass S-TIM35 and then see glass type S-PHM53 close to the new line, as shown in figure 25.2.

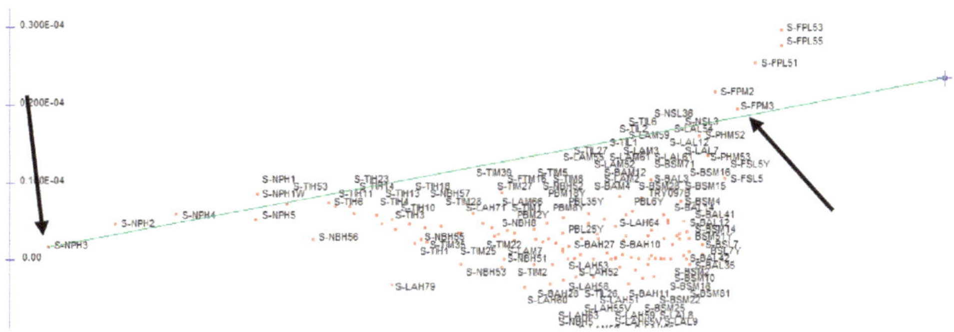

Figure 25.1. Glass-table display showing thermal properties.

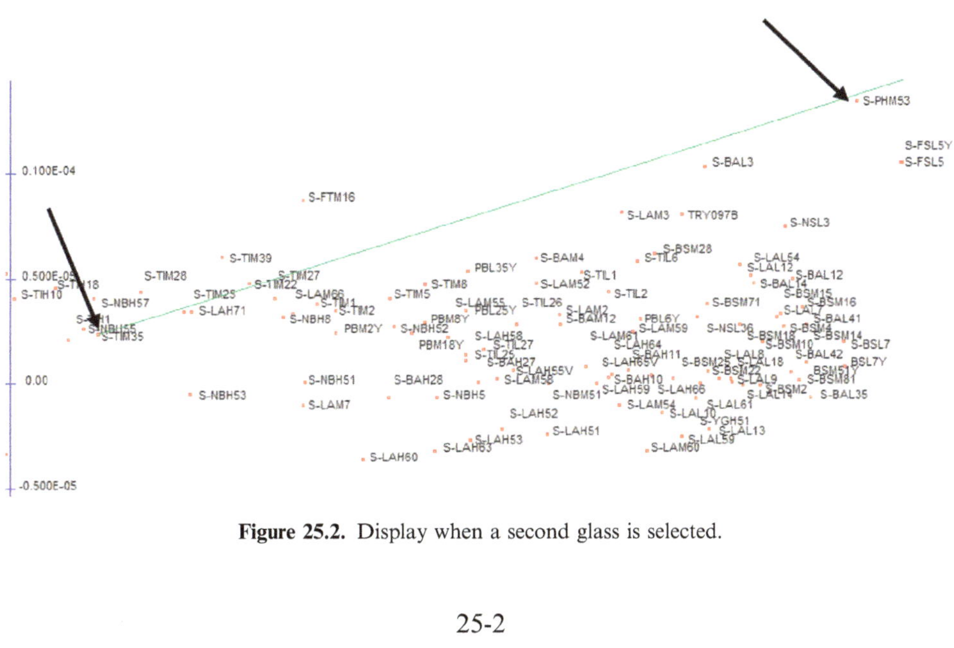

Figure 25.2. Display when a second glass is selected.

We have selected four glass types. Now let us design a lens with those glasses and hope the thermal properties can be controlled.

Create a DSEARCH MACro, specifying only the first two glasses found above. The **MDS** dialog (`MDS`) is a good place to enter the requirements. Give it a file name when you click OK, and it creates a MACro, which you can further edit as you need. Here is a good example (**C25M1**):

```
CORE 14
TIME
DSEARCH 1 QUIET
SYSTEM
ID DSEARCH ATHERMAL
OBB 0 25 2.5
WAVL 0.6563 0.5876 0.4861

UNITS MM
END
GOALS
ELEMENTS 5
FNUM 2
BACK 0 0
TOTL 10 0.1
STOP MIDDLE
STOP FREE
RT 0.5
FOV 0.0 0.75 1.0 0.0 0.0
FWT 5.0 3.0 1.0 1.0 1.0
NPASS 40
GLASS POSITIVE
O S-NPH3
GLASS NEGATIVE
O S-FPM3
ANNEAL 200 20 Q
COLORS 3
SNAPSHOT 10
QUICK 33 40
END
SPECIAL PANT

END
SPECIAL AANT
END
GO
TIME
```

Run this, and in a minute you see the best lens it found. Optimize the lens some more, using the MACro that DSEARCH created. The image is pretty good, at least at a temperature of 20 °C (**C25L1**).

Now you have to check the thermal characteristics. In the WorkSheet, declare the housing to be aluminum 6061, as you did above. Then delete the thickness solve, so the lens will not automatically become refocused as the temperature changes. In WS, type **NTOP** to remove the solve. (The shadowed lenses will automatically remove the curvature solve.)

Now prepare another MACro to initiate *thermal shadowing* of this lens:

```
THERM
ATS 50 2
ATS 100 3
END
```

Run this, and the program puts a copy of the lens in ACON 2, recalculated for 50 degrees, and in ACON 3 at 100 degrees. The ACON 1 display is in figure 25.3.

Now the critical test: click the button for ACON 2,

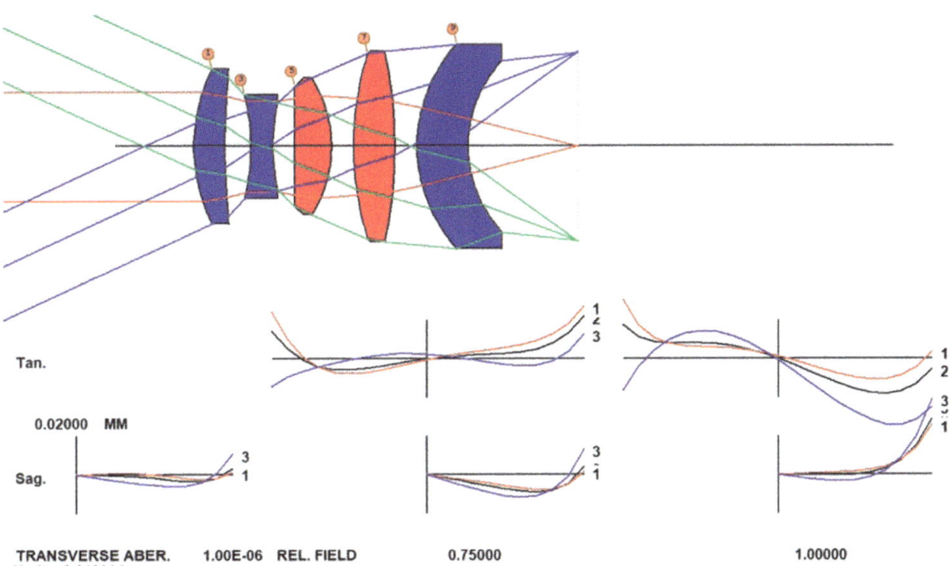

Figure 25.3. Athermalized lens at 20 degrees.

And nothing changes! The rayfan curves look almost identical to those in ACON 1. How about ACON 3, at 100 degrees? Again, almost identical. Our lens is satisfactorily athermal.

What if it had not been so good? Well, then you could correct the image in all three ACONs in the optimization file. That should tweak up any remaining problems.

You did not even need the second pair of glasses we found. What happens if you use those instead of the first pair? Try it and see.

This exercise shows how designing an achrotherm lens is not difficult, using the proper tools. When you have finished, be sure to type

```
THERM OFF
```

in the Command Window. Then you can work on other projects and they will not be shadowed—which you very likely would not want.

IOP Publishing

Lens Design
Automatic and quasi-autonomous computational methods and techniques
Donald Dilworth

Chapter 26

Using the SYNOPSYS glass model

Glass variables; boundary conditions

When you vary the properties of an optical glass in SYNOPSYS, you are asking the program to find values of the index N_d and Abbe number V_d that will correct aberrations and that are within the boundaries of the commercial glass maps. That is pretty straightforward—but you need more than just those two parameters. The program must calculate the index of refraction at each wavelength in the lens, and you hope the value so found will closely resemble the behavior of real glasses in that section of the map. That is the purpose of the glass model. The area on the glass map over which glass models can roam is shown in figure 26.1, which also shows the preferred glasses from the Schott glass company. This display is shown in response to the command **MGT**. The boundaries are given by the lines on the left and right side of the glass display.

A model glass can be inserted into the lens via the SpreadSheet, or more quickly with the keyboard or WorkSheet. For example, the input

```
CHG
1 GLM 1.6 55
END
```

assigns a model glass to surface 1 with the N_d and V_d values specified. You can declare glass variables in the PANT file with input such as

```
VY 1 GLM
VY 3 GBC
VY 5 GBF
VLIST GLM 1 5 8
VLIST GLM ALL
```


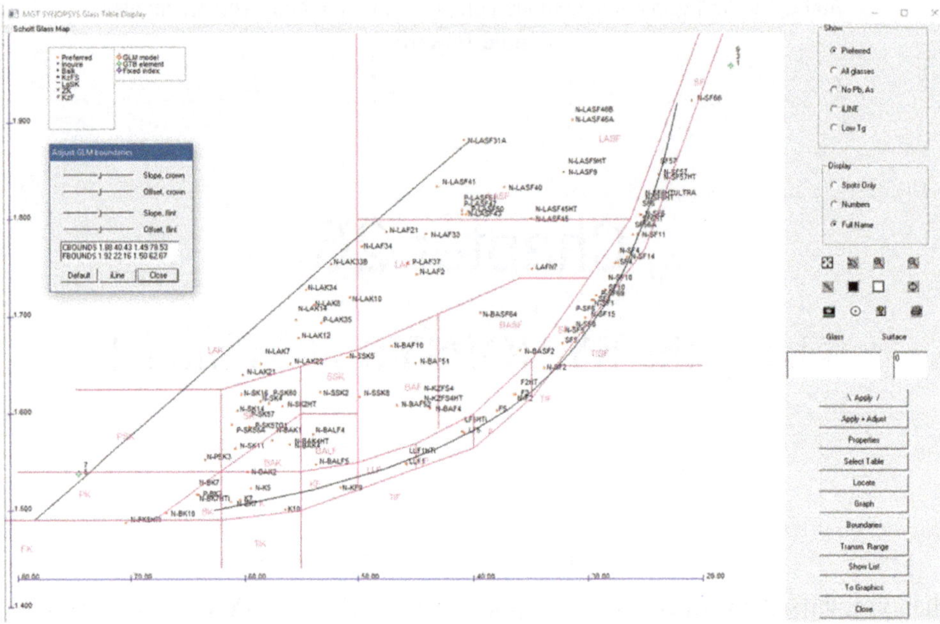

Figure 26.1. MGT display showing glasses from Schott and the boundaries of GLM variables.

The **`VLIST GLM ALL`** form varies all glasses that are already declared model glasses, while the form **`VY sn GLM`** forces the material to be a glass model, if it is not already, as does the **`VLIST`** form with surface numbers added. In that case, the program first finds a model that closely resembles the current glass and assigns it to the lens. **`GBC`** and **`GBF`** are used to vary a glass along the crown or flint boundary.

Glass boundaries are tricky to implement. During optimization, the index often wants to become very high, and of course the V_d of many elements would like to be infinite. That would be great mathematically, but such materials do not exist—so the program has to constrain the glass model to the useable part of the glass map. To do so, it does something clever: when any of the models tries to go over the boundary on the left or right, the program first restricts the change so it goes exactly to that boundary, and then redefines that variable, changing the **GLM** variable to a **GBC** (glass bounded, crown) or **GBF** (glass bounded, flint) instead. Then the glass model will move up or down along that boundary. As a result, the glass model remains within the glass map, and you are left with a single variable where previously there were two. If the glass tries to go over the upper or lower limit on the index, the program reduces the change so it goes exactly to that boundary. In this way, the glass model variable always remains inside the region of the glass map where actual glasses are to be found.

Once a glass has become pinned to the crown or flint boundary, however, it remains there for the duration of that run. It sometimes happens that after a design has been much improved, one of the glasses would work better if it left the boundary. This is simple to test: just run the optimization again. The glasses will start out free to move anywhere, and they can immediately leave the boundary if it improves the lens. The simulated annealing program also frees up all glass models that have become pinned to a boundary before it reoptimizes the lens the first time.

Of course, you do not expect the model glass found during optimization to coincide exactly with any real glass in a selected vendor's catalog, but that is not a problem since you usually can find one whose properties resemble the model closely enough. Then you just substitute that glass and reoptimize. ARGLASS, which you have used in previous chapters, makes that job easy. There is a search program for glasses, GSEARCH, that can often find just the right combination. Chapter 35 gives an example.

Many high-quality designs have to compensate for secondary color to some degree, and in order for the program to optimize glasses while accounting for that aberration, the partial dispersion of the model must be reasonably close to that of nearby real glasses so the correction is maintained when a real glass is substituted later. 'Partial dispersion' refers to the curvature of the index curve, which is different at different wavelengths.

However, now it gets tricky. SYNOPSYS uses a polynomial expression for its glass models that yields the index at any wavelength in the visible region, given only the glass-map coordinates (N_d, V_d), the coefficients having been found via a least-squares fit to selected glasses from the Schott table. Figure 26.2 shows the Schott glass map, where the 'Graph' option has been selected to show the partials 'P(F,e) vs. Ve' (open the glass map with MGT or the PAD button BK7, select 'Schott', click the button Graph and select that option, as shown in figure 26.3).

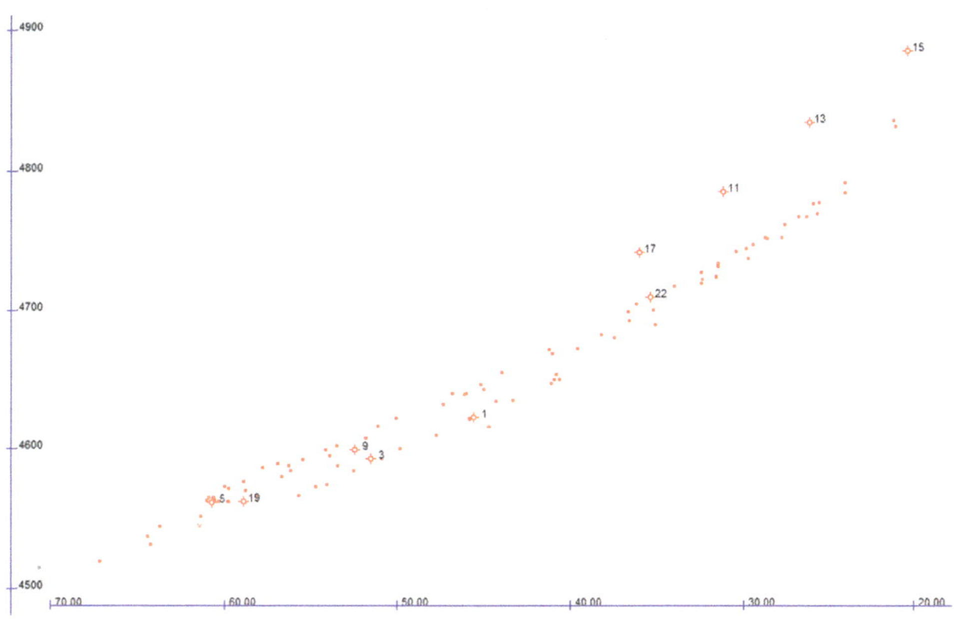

Figure 26.2. Glass map display showing selected partials along with glass model equivalents.

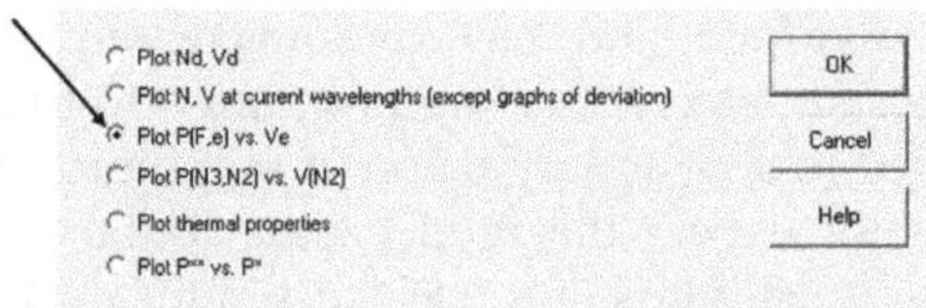

Figure 26.3. Selecting the partials map.

For the above example, we prepared a lens of eight elements with glass models assigned as shown by the red circles in figure 26.2. The goal is for the model to approximate the same distribution as the real glasses, which it does well enough to be useful.

How close is the glass model to a real glass? We prepared a lens where element 1 uses glass SK6 and surface 3 is assigned a GLM with the same N_d and V_d as that glass. Then we prepared an AI plot showing how the two compare, as shown in figure 26.4:

```
STEPS = 50
MULTI PLOT INDEX OF 1 FOR WAVL = .3 TO 2
ADD PLOT INDEX OF 3 FOR WAVL = .3 TO 2
END
```

The fit is close enough for design purposes, especially over the visual range of about 0.4–0.7 µm, and is useful over a wider range, although slightly less accurate, as shown on the plot. (The model coefficients were calculated over the range 0.35–0.9 µm.) For wavelengths far outside this range, such as for the NIR design in chapter 14, it is best to proceed as outlined in that chapter.

Now we will show how to adapt the glass model for special conditions. A good example is when designing for the UV spectrum, where one is restricted to iLine glasses from the Ohara glass company. How can one vary the glass model while staying in the region where those glasses can be found? It is simple. Here is the glass map, in figure 26.5, showing just the iLine glasses, selected by that radio button. The vertical lines show the relative prices; the glasses in red are preferred, while black denotes an inquiry glass.

If we just varied the GLM variables as usual, we would probably wind up with very high-index materials that are not very close to one of the iLine glasses. We can prevent that by changing the boundaries. Click the button 

and the program displays the current (in this case the default) boundaries, as in figure 26.6.

Now, click the 'iLine' button on the boundaries dialog. You see the region where iLine glasses are to be found, shown in figure 26.7. You can also adjust the boundary lines with the sliders in this dialog, should you want to.

There are four parameters that can be specified in the PANT file for controlling the glass boundaries, and the edit box shown in figure 26.6 gives data for the `CBOUNDS` and `FBOUNDS` directives. Select those lines and then copy–paste them into

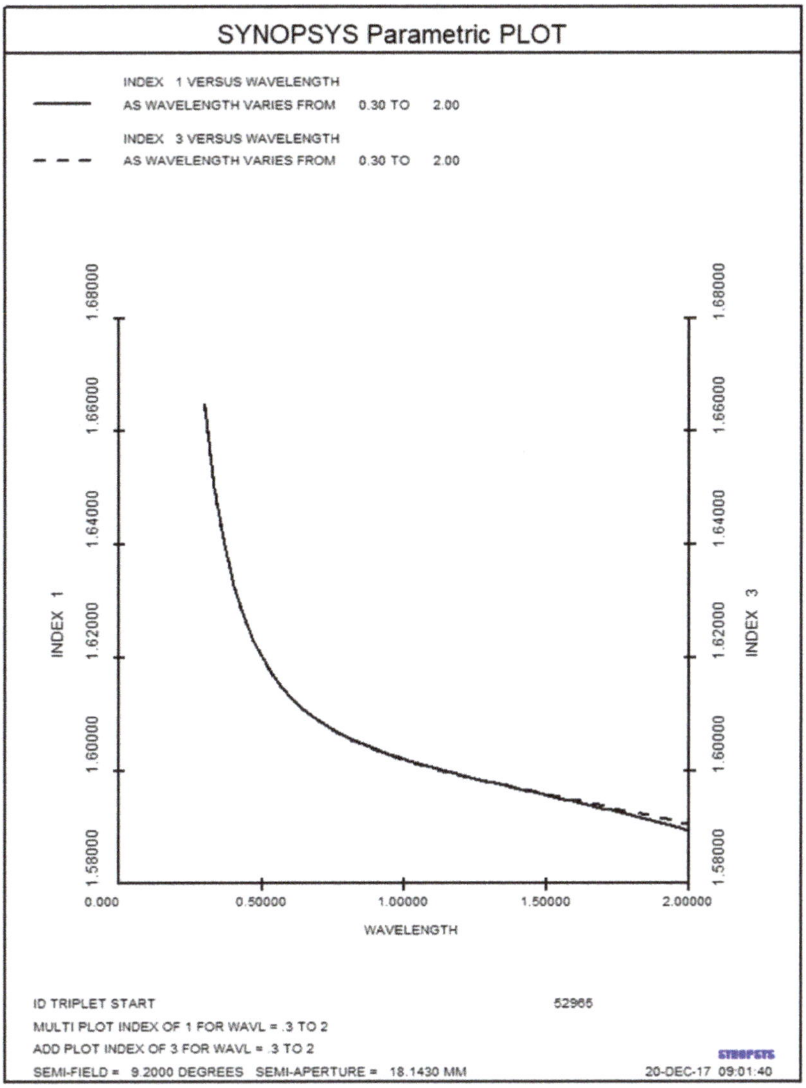

Figure 26.4. Comparison of real and model glass index over the spectral range of 0.3–2 μm.

the PANT file, near the top. Then add another line, giving an upper limit to the GLM index variables of 1.6, with the `CUL` (crown, upper limit) line. The PANT file now starts with

```
PANT
CBOUNDS 1.88 9.43 1.49 82.55
FBOUNDS 1.92 22.16 1.50 62.67
CUL 1.6
...
```

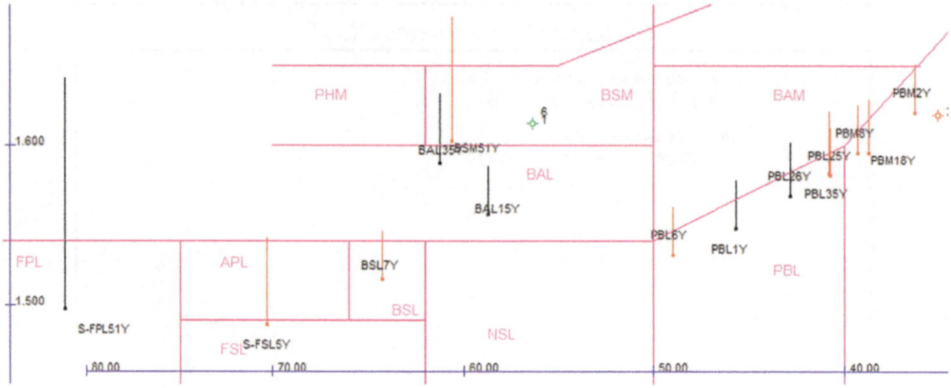

Figure 26.5. Glass map display showing only iLine glasses from Ohara.

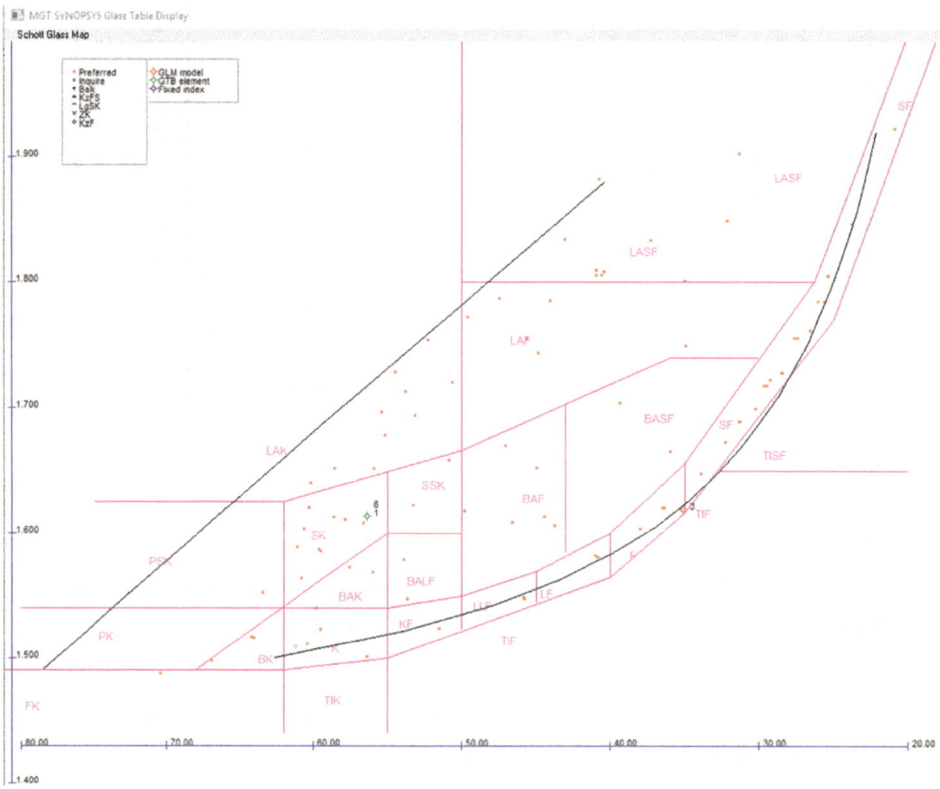

Figure 26.6. Glass map display showing default model boundaries.

Now, when the glasses are varied, they will remain in the area shown above, and one would expect no trouble finding an iLine glass to match the model.

One final note: when you give the program a glass model, you are specifying the *input* to the polynomial. The actual index that comes back at each wavelength is

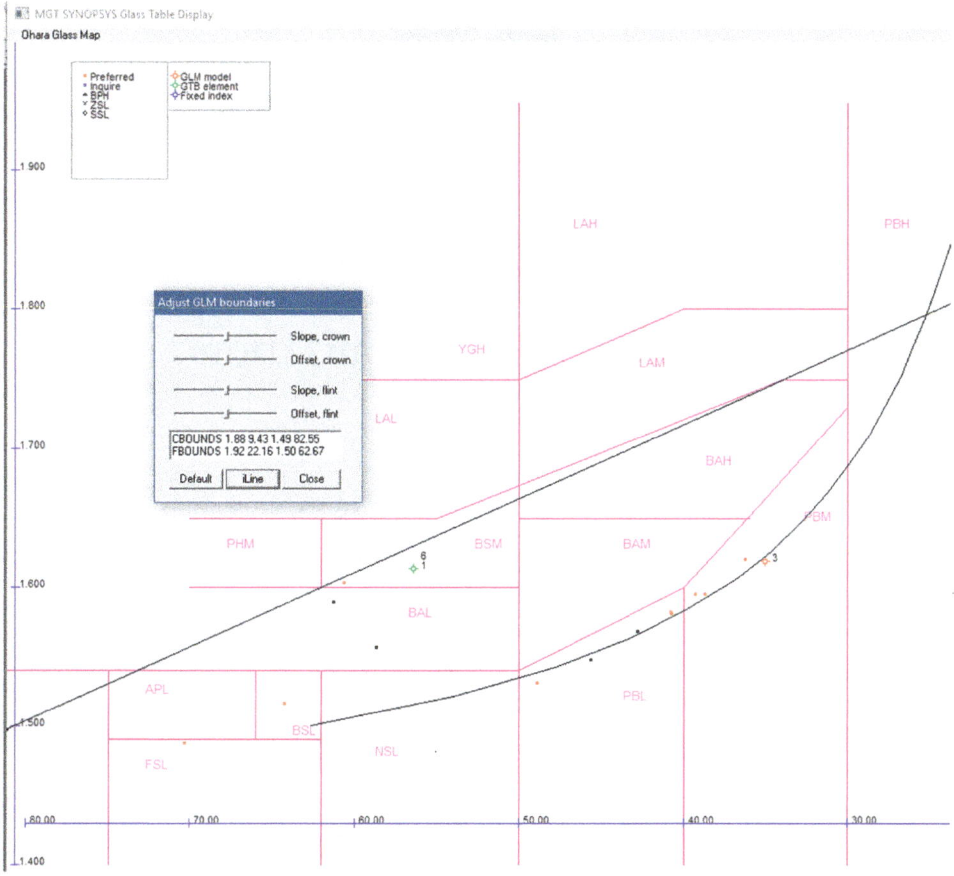

Figure 26.7. Glass map display with boundaries appropriate for iLine glasses.

the *output* from the model, and the two can be slightly different. If the lens is assigned the CDF spectral lines, they will be very close—but if your spectrum is anything else, then you can expect the index listing from SPEC (which gives the model *input*) to be a little different from the output of PRT (which lists the *output* indices).

This glass model has proven to be invaluable for finding where on the glass map the lenses work best. In some cases, the program has even managed to correct secondary color by a good choice of glasses, all by itself. If the **ARG** dialog cannot find glasses that maintain the excellent correction provided by the glass models, the search program **GSEARCH** will usually do the job. An example is given in chapter 47.

IOP Publishing

Lens Design
Automatic and quasi-autonomous computational methods and techniques
Donald Dilworth

Chapter 27

Chaos in lens optimization

Effect of PSD optimization; three-parameter plots; automatic ray-failure correction

In this chapter, we will explore a powerful feature of SYNOPSYS: it can perform parametric studies showing the effects of two variables on a third. In this case, we want to see how the end point of a lens optimization run depends on the starting point. In a perfect world, every starting point would go to the best of all possible optima, but the world is not perfect yet. There are usually many local minima for any given problem, and the best we can expect is that a good optimization algorithm should go reliably to the closest one. (Of course, *global* optimization algorithms such as DSEARCH can find a variety of solutions, but that is a different topic. Here we will analyze the process that simply minimizes the merit function, starting with a single configuration.)

One would therefore expect that two starting points that are almost exactly the same would go to the same local minimum, even if it is not global. How well do current algorithms perform on this score? Some interesting results were discovered by Dr Florian Bociort of TU Delft. He ran a simple case, the doublet shown in figure 27.1.

To make the job very simple, he corrected rays at three field points in the major color only, ignoring edge violations. Then he varied the starting value of radii 2 and 3 in a raster fashion and made a plot where the color of each pixel on the grid encodes the final value of the merit function. He found that there are several local minima, which is not surprising, even for so simple a problem—but what was completely unexpected was how the merit function varied in a chaotic manner in many places. Thus, nearby starting points often go to very different end points. (He did this analysis on a different program, with a different optimization algorithm than the PSD method used in SYNOPSYS.) A figure from his article is shown on the right in figure 27.1.

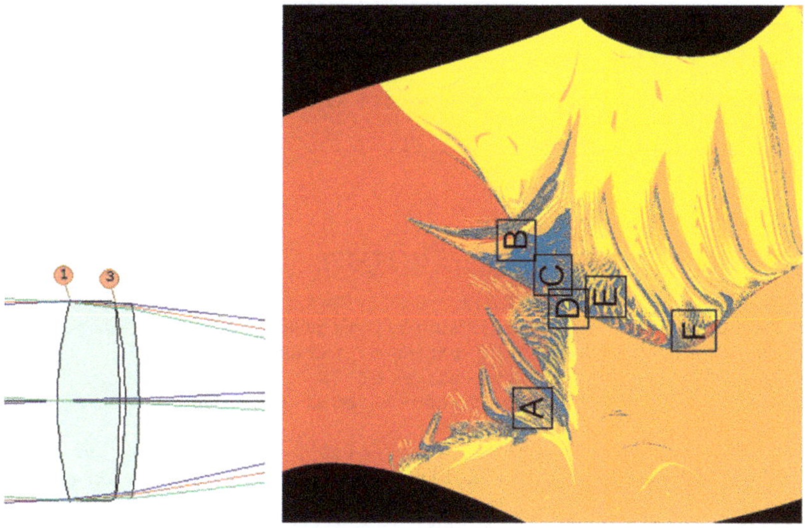

Figure 27.1. Doublet lens and a map of the merit function as a function of starting radii. Right-hand panel reproduced with permission from van Turnhout M and Bociort F 2009 Instabilities and fractal basins of attraction in optical system optimization *Opt. Express* **17** 314–28.

(We have turned this picture on its side so it will line up with the SYNOPSYS analysis below.)

Notice how the results near the boundaries of the zones of attraction are complex and chaotic. The black areas show starting points that gave ray failures and were not analyzed.

It would be interesting to see whether the PSD III algorithm in SYNOPSYS is more reliable and stable than the method used for the above picture. Here is the input to run the three-parameter evaluation feature **PA3**, to find out.

Here is the starting doublet (**C27M1**):

```
RLE
ID FLORIAN STARTING DOUBLET
WA1 .5876000
 WT1 1.00000
 APS        1
 UNITS MM
 OBB 0.000000  3.00000  16.66670  0.00000  0.00000  0.00000  16.66670
   0 AIR
   1 CV    0.0146498673770  TH    10.34600000
   1 N1 1.61800000
   1 GID 'GLASS       '
   2 RAD   -174.6512432672814  TH    1.00000000 AIR
   2 AIR
   3 RAD   -80.2251653581521  TH    2.35100000
   3 N1 1.71700000
   3 GID 'GLASS       '
   4 RAD   -111.8857786363961  TH    92.41206276 AIR
   4 AIR
   4 CV    -0.00893769
```

```
    4 UMC   -0.16667000
    4 TH   92.41206276
    4 YMT   0.00000000
    5 CV   0.0000000000000  TH   0.00000000 AIR
    5 AIR
  END
STORE 5
```

This is the input for the PA3 program (**C27M2**):

```
ON 78                    ! use finer grid (118x118 points)
PA3 LOOP COLOR           ! initialize PA3, request color boxes for output
RZ1 -.025 .04            ! set the range of variable Z1
RZ2 -.045 .075           ! set the range of Z2
RZ3 0 3.7                ! display results over this range of merit function
                         ! values
NOSMOOTH                 ! there will be steps in the output; do not smooth
XLAB "2 CV -.025 .04"          ! define the label for the X-axis, which is
                               ! variable Z1
YLAB "3 CV -.045 .075"         ! label for Y-axis, Z2
ZLAB "MERIT"             ! label for Z-axis, the final merit function
LOOP                     ! tell PA3 to loop over the above raster of data

GET 5                    ! get the starting lens each time
2 CV = Z1                ! set curvature 2 to the value of variable Z1
3 CV = Z2                ! and CV 3 to Z2, using the artificial-intelligence
                         ! parser

PANT                     ! initialize the variable list
VLIST RAD 2 3            ! and vary two radii
END                      ! end of the variable list

AANT                     ! initialize the merit function definition
GSR .5 10 3 P 0          ! correct a sagittal fan, three rays, on axis
GNR .5 1 3 P .75         ! correct a full grid of rays, primary color, 0.75
                         ! field point
GNR .5 1 3 P 1           ! same, at full field.
END                      ! end of merit function definition

DAMP 10000               ! initial damping (see below)
SNAP 50                  ! watch what happens, but not too often, in order to
                         ! keep it fast
SYNOPSYS 100             ! optimize until it converges

Z3 = MERIT               ! assign the current merit function value to Z3
PA3                      ! tell PA3 to cycle to the next case.
```

Why the high damping? (The default is 1.0 or 0.01, depending on mode switches.) The first iteration in SYNOPSYS is a DLS (damped-least-squares) cycle[1], and we want to avoid any chaos that results on the first pass from that algorithm; the high damping will ensure that the lens changes very little on that pass. The more powerful

[1] This is true if switch 67 is off. Somewhat better results are sometimes obtained if that switch is on, causing the first iteration to use calculated second derivatives. The PSD III method is employed from the second pass in either case.

PSD algorithm keeps track of changes in the first derivatives from pass to pass and deduces information about higher-order derivatives. That is the magic behind the PSD method, as described in detail in appendix B—but it only works starting at the second pass.

The results of this study are shown in figure 27.2. The violet areas on the left and near the bottom show that the program went to the same minimum for very different starting points, whereas in Florian's study those areas went to different minima. There is no evident chaos at the boundaries of the zones of attraction, as we expected would be the case with the PSD method, although scattered poles do show up in the central red region. We attribute the latter to the nonzero changes made by the DLS method on the first pass. Indeed, if we run this again with different initial damping, those random spots appear in different places.

The black areas at top and bottom show where the starting points yielded ray failures, in the same place where they did in Florian's study. We are curious what would happen if we activate the automatic ray-failure correction feature. Change the SYNOPSYS command to

```
SYNOPSYS 100 0 FIX
```

and rerun this job. The result is in figure 27.3.

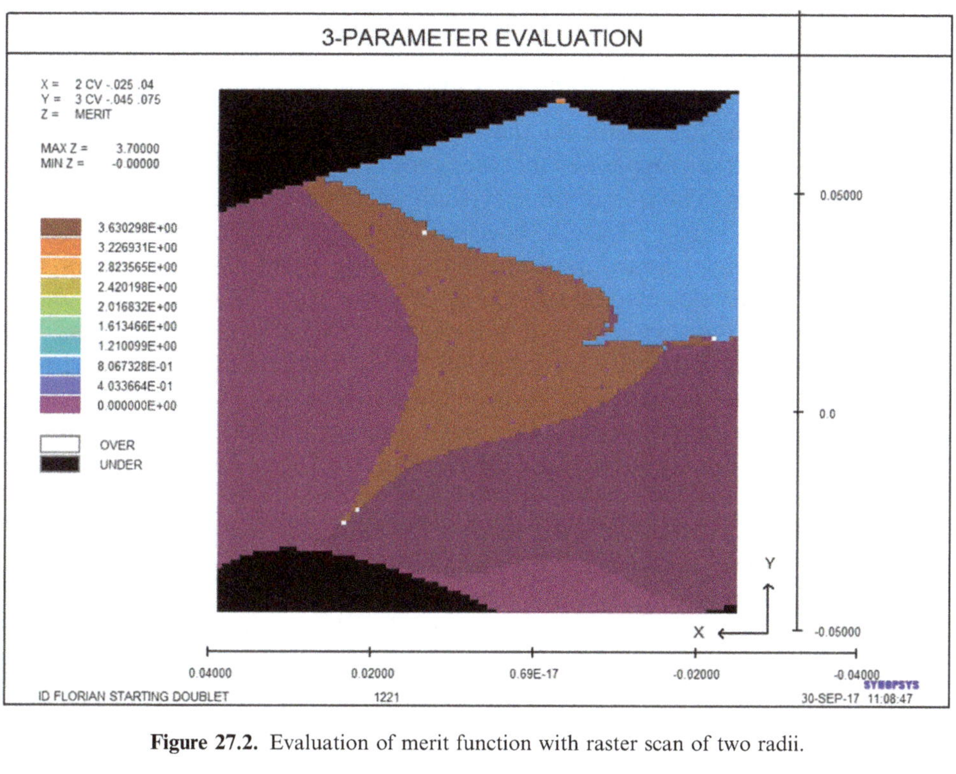

Figure 27.2. Evaluation of merit function with raster scan of two radii.

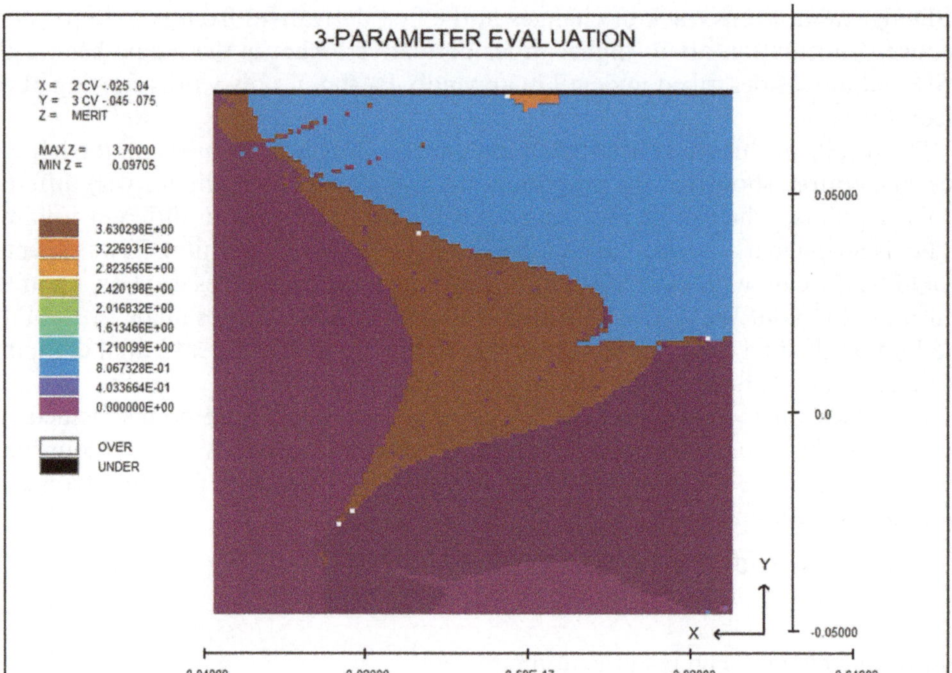

Figure 27.3. Evaluation with automatic ray-failure correction activated.

Now the program has corrected the ray failures over every point where they occurred before. The starting lenses that were not optimized by Florian now all yield respectable solutions. There is some very slight chaos evident at the boundaries in areas that were previously all black, however, and we attribute this to changes the ray-failure correction program made to that starting point. Those changes sometimes moved the lens closer to another zone of attraction.

This very simple study involved optimizing with only two variables. What happens if you add CV 1 to the variable list? Try it and see. (The boundaries shift about somewhat, and the scattered spots no longer show up.) For those interested in looking further into the subject of chaos in lens design, I quote here an analysis by Florian Bociort[2], who has studied the effect.

> I use the word 'Chaos' in the strict sense of Nonlinear Systems Theory. In that theory it is shown that when the basins of attraction are fractal, then the succession of iterations of the optimization algorithm that leads to the color of a point is always a 'temporary' chaotic attractor (the technical term is 'chaotic saddle'). The iteration process is first attracted to something that seems to be a

[2] Personal communication.

chaotic attractor (which is something like what you find in the popular Lorenz Butterfly Effect metaphor). However, it turns out that this chaotic attractor has ‘escape’ holes, so when the iteration finds such a ‘hole’ and ‘escapes’, it finally converges as expected to some minimum. The finely interwoven structure of the basins results when iterations starting at neighboring points find different escape holes and land therefore in different minima.

Chapter 28

Tolerance example with clocking of element wedge errors and AI analysis of an image error

Statistics of tolerance budget with wedge clocking; adding AI commands to a Monte-Carlo simulation

This chapter applies some of the features discussed earlier and adds some interesting and powerful new options. Here, we will use BTOL to calculate a tolerance budget for an eight-element lens, then look at the image quality statistics for the case where the wedge errors are compensated by clocking the elements in the cell. Finally, we will examine the statistics of the lateral color of a set of 100 lenses subject to the budget, after refocusing the lens and clocking the elements.

Here is a MACro that will create the tolerance budget (**C28M1**). The lens (**C28L1**) is shown in figure 28.1:

```
FETCH C28L1            ! Get out the starting lens.
STORE 5
BTOL 90                ! Ask for 90% confidence level.
TPR ALL                ! All surfaces are matched to testplates.
EXACT ALL INDEX        ! Assume melt data are received,
EXACT ALL VNO          ! so the index and dispersion tolerances are zero.
TOL WAF .18 .32 .18    ! Ask for this wavefront variance at three fields.
FOCUS REAL             ! Focus the on-axis image point
ADJUST 14 TH 100       ! with thickness 14 (the last airspace).
PREP MC                ! Prepare the input for Monte-Carlo evaluation.
GO                     ! Start BTOL.
```

This MACro will get the lens out with the **FETCH** command and store a copy in library location 5. Then it creates a BTOL budget, which is listed on the monitor. Now we need to use the Monte-Carlo program MC. An adjustment MACro was

doi:10.1088/978-0-7503-1611-8ch28

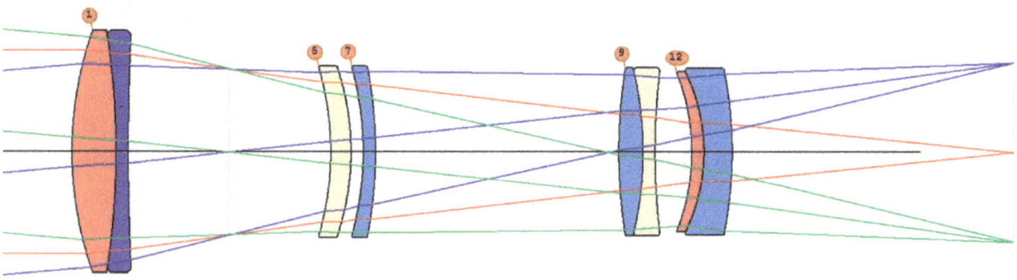

Figure 28.1. Example lens to demonstrate wedge clocking.

prepared by BTOL, called **MCFILE.MAC**, that will be part of the MC analysis. Let us see what it contains. Type **LM MCFILE** to load that MACro:

```
PANT
VY  14 TH
END
AANT
M    0.000000E+00 0.3333    A  2 XC  0.000 0  .1     0.000
M    0.297888E-05 0.3333 SR A  2 YC  0.000 0  .1     0.000
M    0.000000E+00 0.3333    A  2 XC  0.000 0 -.1     0.000
M   -0.297888E-05 0.3333 SR A  2 YC  0.000 0 -.1     0.000
M    0.297888E-05 0.3333    A  2 XC  0.000 .1 0      0.000
M    0.000000E+00 0.3333 SR A  2 YC  0.000 .1 0      0.000
M   -0.297888E-05 0.3333    A  2 XC  0.000 -.1 0     0.000
M    0.000000E+00 0.3333 SR A  2 YC  0.000 -.1 0     0.000
M   -0.177179E-02 0.3333    A  2 XC  0.000 -.64 .64  0.000
M    0.177179E-02 0.3333 SR A  2 YC  0.000 -.64 .64  0.000
M    0.177179E-02 0.3333    A  2 XC  0.000 .64 .64   0.000
M    0.177179E-02 0.3333 SR A  2 YC  0.000 .64 .64   0.000
M    0.177179E-02 0.3333    A  2 XC  0.000 .64 -.64  0.000
M   -0.177179E-02 0.3333 SR A  2 YC  0.000 .64 -.64  0.000
M   -0.177179E-02 0.3333    A  2 XC  0.000 -.64 -.64 0.000
M   -0.177179E-02 0.3333 SR A  2 YC  0.000 -.64 -.64 0.000
M    0.000000E+00 0.6667    A  3 XC  0.000 0  0.     0.000
M    0.000000E+00 0.6667    A  3 YC  0.000 0  0.     0.000
M    0.000000E+00 0.6667    A  3 XC  0.000 0  .1     0.000
M    0.149917E-03 0.6667    A  3 YC  0.000 0  .1     0.000
M    0.000000E+00 0.6667    A  3 XC  0.000 0 -.1     0.000
M   -0.149917E-03 0.6667    A  3 YC  0.000 0 -.1     0.000
M    0.149917E-03 0.6667    A  3 XC  0.000 .1 0.     0.000
M    0.000000E+00 0.6667    A  3 YC  0.000 .1 0.     0.000
M   -0.149917E-03 0.6667    A  3 XC  0.000 -.1 0     0.000
M    0.000000E+00 0.6667    A  3 YC  0.000 -.1 0     0.000
END
SYNOPSYS 10
MC
```

As requested, the last airspace is varied in the PANT file, and the AANT file defines a merit function that will converge to zero if the adjustment is able to restore exactly the same ray pattern as the nominal design. (Note that we are not trying to *improve* the image here, just trying to keep it as it is.) Now we need to prepare an MC MACro. (This is the file where we specify what kind of Monte-Carlo analysis we want; the file MCFILE.MAC, shown above, specifies the *adjustments* we want to run on each case. They are separate files.)

To start with, we will run MC with random wedge orientation. Here is the MACro (**C28M2**):

```
MC ITEMIZE
SAMPLES 1              ! One case, please.
LIBRARY 5              ! We saved the initial lens in library location 5.
WORST ALL 1            ! Later we may want to see a worst case.
THSTAT UNIFORM         ! Uniform thickness statistics.
WEDGES RANDOM          ! Wedges have random orientation.
TEST                   ! Let's just look at a perturbed example.
GO                     ! Run MC.
```

Here, we do not optimize anything, just prepare a single perturbed example so we can examine it. (The elements all have a wedge error now, so the PAD display does not color the elements as before.) This is shown in figure 28.2.

Let us run a set of 100 lenses and look at the statistics. First, we **GET 5** (to restore the nominal lens), and then comment out the TEST directive and change the samples number:

```
MC ITEMIZE
SAMPLES 100 ! Ask for a set of 100 lenses.
LIBRARY 5
QUIET
WORST ALL 1
THSTAT UNIFORM
WEDGES RANDOM
! TEST
GO
```

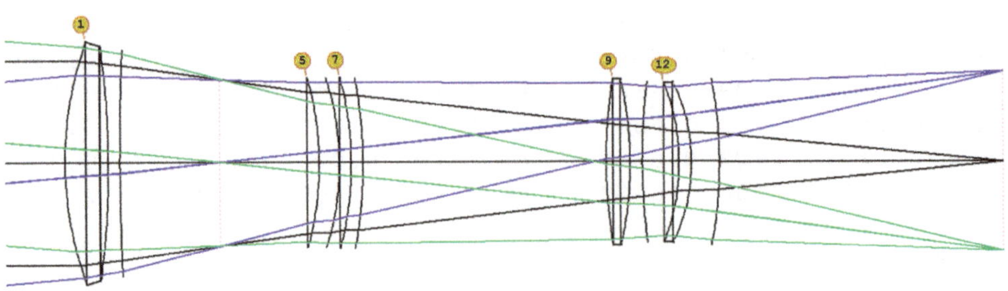

Figure 28.2. Lens perturbed by MC, with wedges on all elements.

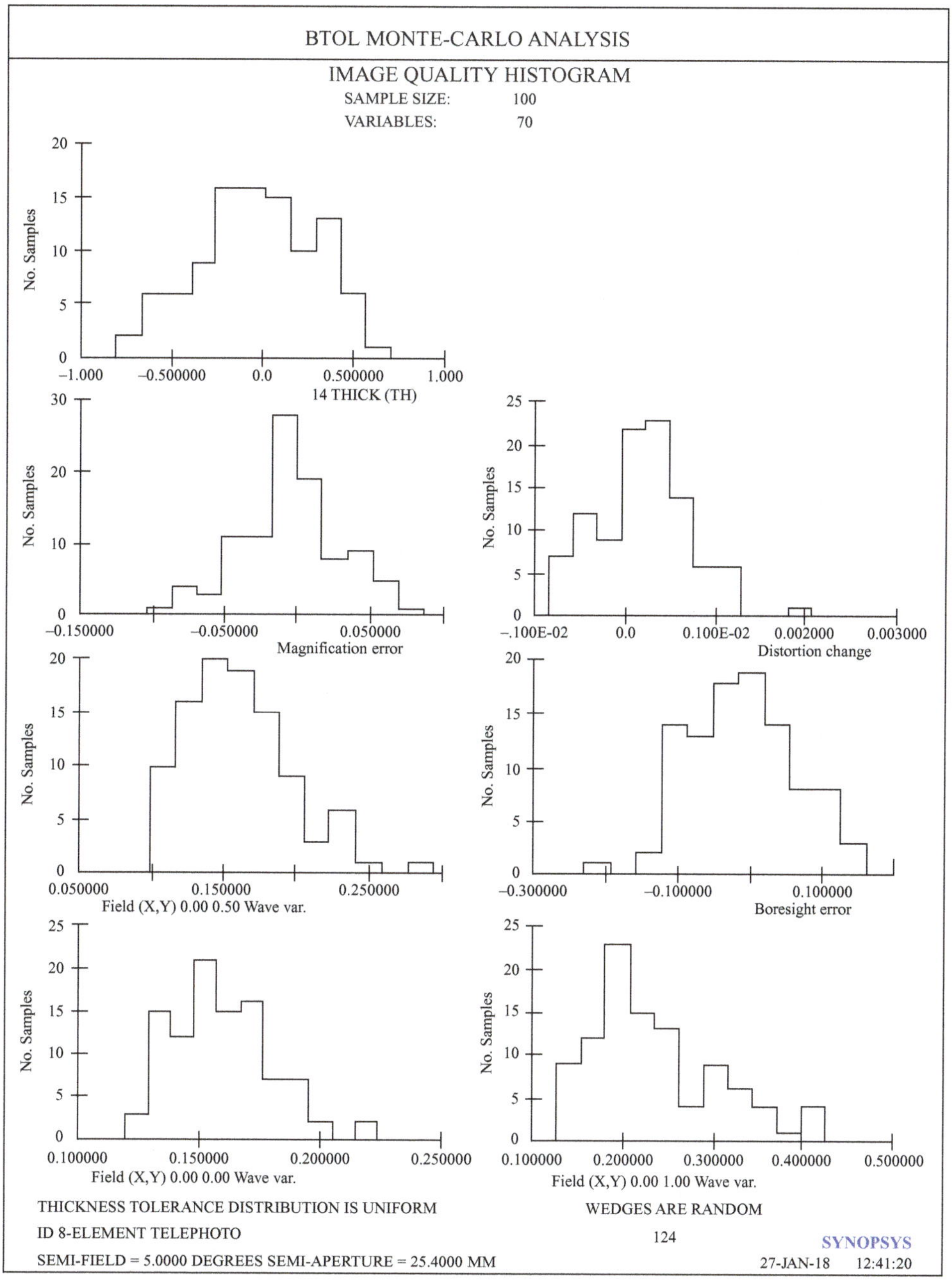

Figure 28.3. Monte-Carlo statistics of lens without wedge clocking.

When MC finishes, make a plot of the statistics with **MC PLOT**, as shown in figure 28.3.

All of this is fairly simple, so now we make things more interesting. Change the MACro as follows:

```
MC ITEMIZE
SAMPLES 100
LIBRARY 5
QUIET
WORST ALL 1
THSTAT UNIFORM
WEDGES CLOCK          ! Clock the wedge errors for each case.
TEST                  ! Again make a single TEST case.
GO
```

Now the program will model the element tilts with a different protocol, using GROUP instead of RELATIVE tilts. This frees up the gamma tilt on each element to be used for clocking the wedge errors. ('Clocking' refers to the practice of rotating each element in the cell, aligning them so that the effects of the wedge error of a given element are compensated by the orientation of the other elements.) Here, we ask for a test case so we can examine how the errors have been defined. When this has executed, look at the **ASY** listing for the perturbed lens:

TILT AND DECENTER DATA
LEFT-HANDED COORDINATES

SURF TYPE	X	Y	Z	ALPHA	BETA	GAMMA
TILT OR DECENTER GROUP SIZE:	3					
1 GROUP	0.01787	-0.00406	0.00000	0.0065	0.0000	0.0000
2 REL	0.00000	0.00000	0.00000	0.0000	0.0072	0.0000
3 REL	0.00000	0.00000	0.00000	0.0114	0.0000	0.0000
TILT OR DECENTER GROUP SIZE:	2					
5 GROUP	-0.00969	0.03471	0.00000	0.0643	0.0000	0.0000
6 REL	0.00000	0.00000	0.00000	0.0000	0.0259	0.0000
TILT OR DECENTER GROUP SIZE:	2					
7 GROUP	0.01027	-0.01232	0.00000	0.0000	0.0735	0.0000
8 REL	0.00000	0.00000	0.00000	0.0000	-0.0029	0.0000
TILT OR DECENTER GROUP SIZE:	3					
9 GROUP	0.01898	0.03135	0.00000	-1.85E-05	0.0000	0.0000
10 REL	0.00000	0.00000	0.00000	0.0852	0.0000	0.0000
11 REL	0.00000	0.00000	0.00000	0.0000	-0.0034	0.0000
TILT OR DECENTER GROUP SIZE:	3					
12 GROUP	0.00808	-0.01227	0.00000	0.0000	-0.0285	0.0000
13 REL	0.00000	0.00000	0.00000	0.0000	0.0712	0.0000
14 REL	0.00000	0.00000	0.00000	0.0000	-0.0550	0.0000
15 REL	0.00000	0.00000	0.00000	0.0400	0.0000	0.0000

From this listing, you see that surfaces 1, 5, 7, 9, and 12 have been assigned a group tilt. You will vary the clocking angle (gamma tilt) all of them except for surface 1, which furnishes a reference direction.

We need to modify the file MCFILE.MAC, adding the gamma tilt variables. We also elect to do a more sophisticated optimization while we are at it. Then save the new MACro so MC will be able to open it and see the changes. (**Do not run this file yourself; it will be used by MC as it optimizes each case.**) It now looks like this:

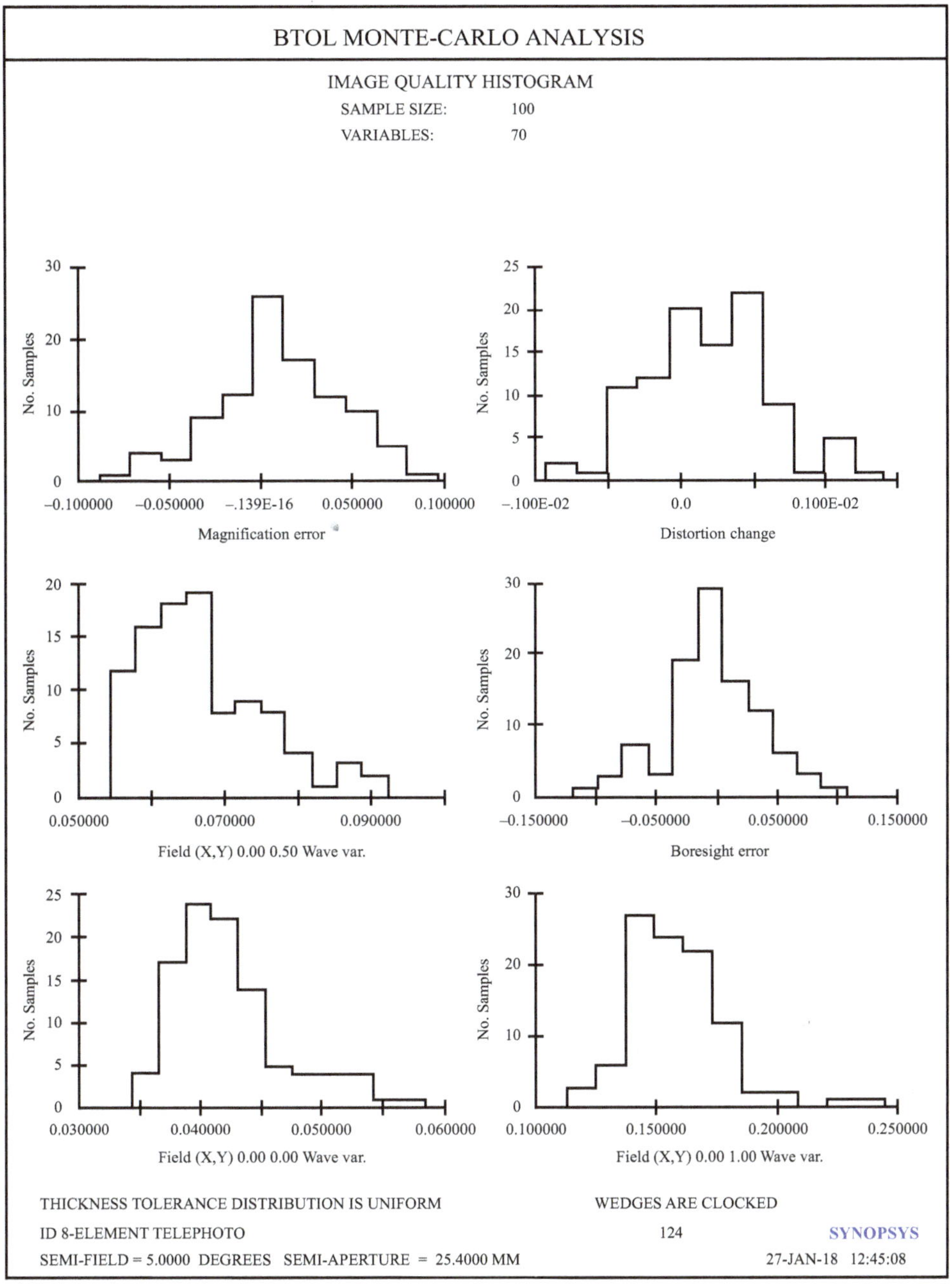

Figure 28.4. Statistics of tolerance budget when wedges are clocked.

```
PANT
VY 14 TH
VY 5 GPG        ! Vary group gamma tilt on surfaces 5, 7, 9, and 12
                !(but not surface 1).
VY 7 GPG
VY 9 GPG
VY 12 GPG
END

AANT
M 0 1 A P YA           ! Control the boresight error this way.
M 0 1 A P XA
GSR .5 10 5 M 0 0 0 F        ! Correct over the full pupil since the
                             ! lens no longer has
GNR .5 2 3 M .7 0 0 F        ! bilateral symmetry.
GNR .5 1 3 M 1 0 0 F

GNR .5 2 3 M -.7 0 0 F       ! For the same reason, we also control
                             ! the negative field.
GNR .5 1 3 M -1 0 0 F
END
SYNOPSYS 10
MC
```

Save this version, type **GET 5** so you start with the original lens, and rerun the MC MACro, requesting 100 cases and deleting the TEST directive. When you run it, you obtain improved statistics, shown in figure 28.4. Indeed, clocking the elements improves performance, as expected.

This exercise is almost finished—but suppose the lens is used in a device where the lateral color must be well controlled. You would like to know the statistics of the resulting aberration as each case is optimized. Add to the file MCFILE.MAC some AI input, and it now reads as follows:

```
PANT
VY 14 TH
VY 5 GPG
VY 7 GPG
VY 9 GPG
VY 12 GPG
```

```
END
AANT
M 0 1 A P YA
M 0 1 A P XA
GSR .5 10 5 M 0 0 0 F
GNR .5 2 3 M .7 0 0 F
GNR .5 1 3 M 1 0 0 F
GNR .5 2 3 M -.7 0 0 F
GNR .5 1 3 M -1 0 0 F
END
SYNOPSYS 10

Z1 = XA IN COLOR 1   ! Get the actual X coordinate of the chief ray in
                     ! color 1.
RMS 1 0 555          ! Run the RMS command, which also finds the centroid.
Z2 = FILE 4          ! This is the X-centroid location, relative to the
                     ! chief ray,
Z3 = FILE 5          ! and this is the Y.
Z4 = YA IN COLOR 1   ! Also get the actual Y coordinate.

Z5 = XA IN COLOR 3   ! Do the same thing in color 3.
RMS 3 0 555
Z6 = FILE 4
Z7 = FILE 5
Z8 = YA IN COLOR 3

= SQRT((Z1 + Z2 - Z5 - Z6)**2 + (Z3 + Z4 - Z7 - Z8)**2)
Z9 = FILE 1                 ! Load it into variable Z9, and tell MC
MC IZ9 "RedCen-BlueCen"     ! to gather the statistics and plot Z9 with
                            ! this label.
MC
```

Save this, as before. Now, when you run your MACro, MC adds the statistics of the lateral color to the second plot page, which also shows the adjustment statistics, as seen in figure 28.5.

If you run these examples with switch 98 turned on, you should obtain similar results. (However, if you turn on the multicore option (with **CORE nb**) the statistics will be different, since each core has to run a different set of cases.)

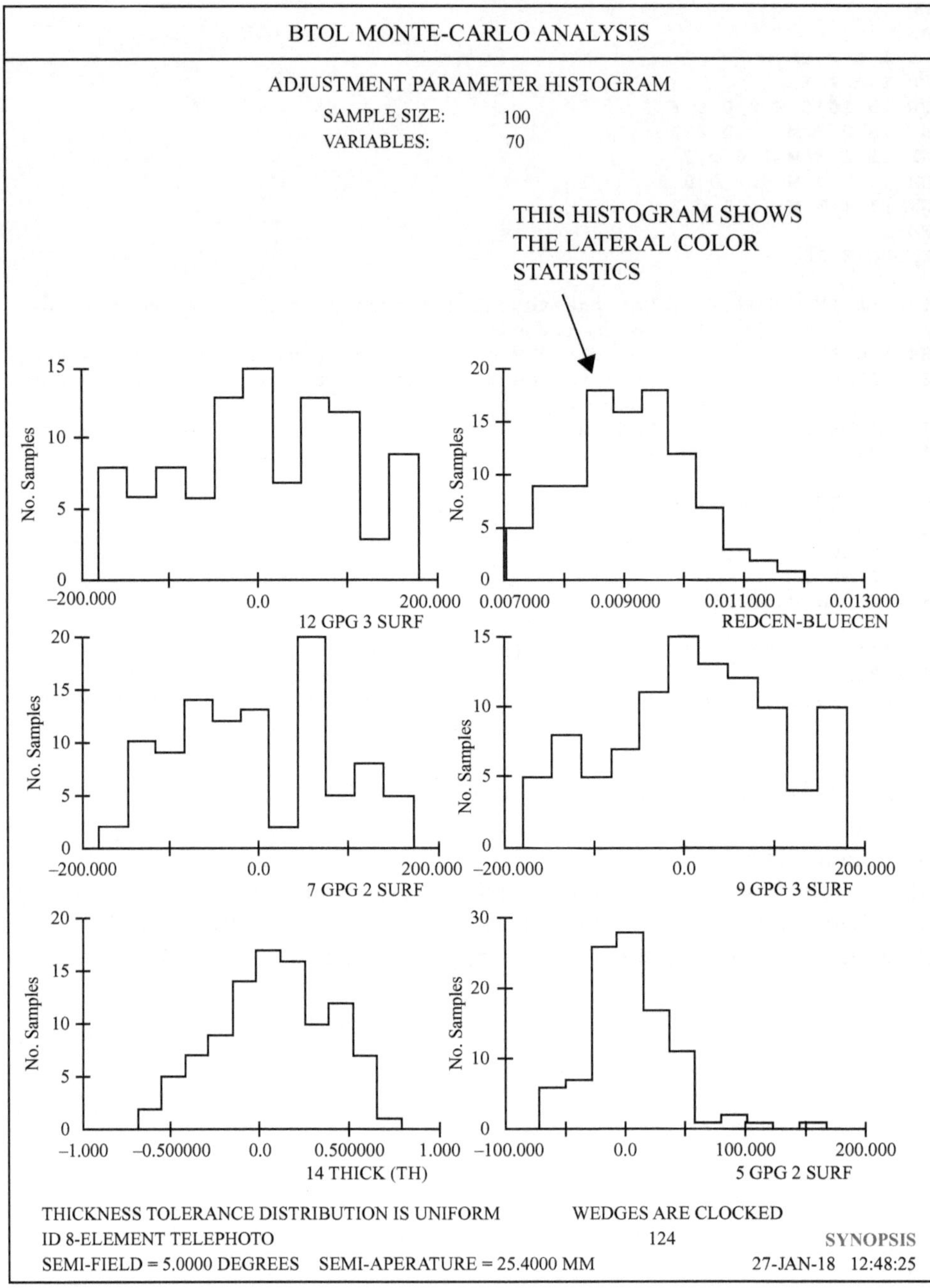

Figure 28.5. MC plot showing statistics of adjustments and lateral color when 100 lenses are manufactured.

IOP Publishing

Lens Design
Automatic and quasi-autonomous computational methods and techniques
Donald Dilworth

Chapter 29

Tips and tricks of a power user

Time-saving strategies for effective use of lens optimization software

If you have gone through the previous chapters in this book, you already know the basic procedures for running SYNOPSYS. This chapter goes a little deeper and shows some ways you can obtain just what you are after, ways that may not be obvious.

Here is an example of a sophisticated optimization MACro. By now, you should be able to read this kind of file fluently:

```
LOG
AWT: 1.0
CHG
NCOP
END

PANT
VY 0 YP1 50 -50
VY 0 BTH
VLIST RD ALL
VLIST TH ALL
END

AANT P
AEC
ACC
M 100 10 A FOCL
LLL 2 1 1 A BACK
AAC 49.5 .5 5
```

doi:10.1088/978-0-7503-1611-8ch29

```
SKIP
GSR    AWT    6.000000    4  1    0.000000
GNR    AWT    3.000000    4  1    0.100000
GNR    AWT    3.000000    4  1    0.300000
GNR    AWT    3.000000    4  1    0.500000
GNR    AWT    3.000000    4  1    0.70000
GNR    AWT    3.000000    4  1    0.80000
GNR    AWT    3.000000    4  1    0.90000
GNR    AWT    3.000000    4  1    1.000000
EOS

!SKIP
GSO   0    0.8     4  1    0.000000
GNO   0    0.27    4  1    0.100000
GNO   0    0.27    4  1    0.300000
GNO   0    0.27    4  1    0.500000
GNO   0    0.27    5  1    0.70000
GNO   0    0.27    5  1    0.80000
GNO   0    0.37    5  1    0.930000
GNO   0    0.27    5  1    0.950000
GNO   0    0.27    5  1    1.000000
EOS

!SKIP
LUL 29 1 1
A BLTH 8

EOS

END
!EVAL
!EDS

SNAP/DAMP 1
SYNOPSYS 40
```

Tip 1. In this example, we vary the paraxial quantity YP1. The lens does not have an explicit stop defined, and this variable will let the chief-ray intercept on surface 1 vary—thus sending it into the lens at the current location and *implying* a stop position wherever it happens to cross the axis. This is a powerful way to find out where the stop should go. If the design looks promising, then it is simple to just assign the stop at that location or close to it and reoptimize.

Tip 2. Note the monitor **`AAC 49.5 .5 5`**. This lens has to fit inside a tube of 100 mm diameter, and this monitor will penalize the lens if any of the clear apertures exceed the value 49.5 mm in radius. The other two arguments give the relative weight and the monitor window. You can adjust them according to the importance of that control (and you can always *look up* that topic in the User's Manual if you want to see how it works. Type **HELP AAC** in the Command Window. There are 12 monitors available, and it pays to know how to use them.)

Tip 3. Notice how the file defined a symbol, **AWT: 1.0**. That symbol shows up in the AANT file as the aperture weighting parameter on some ray sets, a parameter we have used in previous chapters. A value of 0 says to weight all rays in the generated grid by the same amount, the value given by the second argument on the ray-grid request. A weight of 1.0 weights the central rays rather more heavily than the edge, and for this lens that is a good idea. Uniform weights tend to give an image of high contrast, while a higher weight often gives better resolution. Here, too, you can experiment to see what works best with your lens. I often start with a value of 0.5. The point of making this a *symbol* is so you can try a different value just by changing that symbol and reoptimizing—so you do not have to change all the weights on every line of the MACro every time you want to try a new value.

Tip 4. The shop making the lens in figure 29.1 has a curious problem: they already have a lens blank for the element at surface 8, and have measured its thickness at 30 mm. So that lens must be controlled during optimization to be sure it does not require a thicker blank.

That is where the AANT entries below come in:

```
LUL 29 1 1
A BLTH 8
```

Here we specify a maximum value (with LUL) of 29 mm for the thickness of the blank for this element. LUL means Limit, Upper Limit, and you can read about this

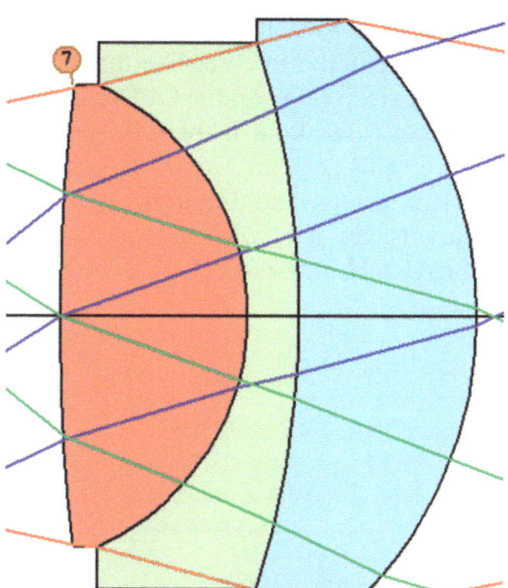

Figure 29.1. Example lens for which the blank thickness must be controlled.

useful feature by typing **HELP LUL**. You can also make your own aberration by combining the TH and the sag at any surface that happens to be concave at the current clear aperture. Section 10.3.3 of the User's Manual describes the target SCAO (the sag at the current aperture). It is useful to know how to make aberrations like this.

Tip 5. Notice the use of the **SKIP** directive in this MACro. Ray-grid definitions and weights are easily generated with the buttons  on the MACro editor toolbar, which generate rays whose transverse intercepts are to be controlled, or rays with OPD targets. But which should you select? The **SKIP** directive in the above example lets you choose one or the other (or both) just by commenting or uncommenting that directive. As shown above (in bold), the first set of ray grids, which target transverse aberrations, will be skipped, because it is within a SKIP block. When the program reaches the **EOS** (End Of Skip) line, it stops skipping—so the ray grids targeting OPDs are then in effect. To see the effect of transverse targets instead of OPD, just comment out the first SKIP and uncomment the other (delete the '!'). You see how simple it is to switch from one to the other. With a few clicks and keystrokes you can study the effect.

Tip 6. More on the subject of TAP targets versus OPD targets: look at the ray-fan plots in figures 29.2 and 29.3.

You might think this is a terrible image, with the rays flying off like that at the ends of the fans. However, look at the OPD fans in figure 29.3.

It cannot get much better than that. This is a good lesson on why you should switch to OPD targets if your lens is close to the diffraction limit. Someone targeting only ray intercepts would likely look at the top curves and then discard the lens and start over—or give a heavy weight to the marginal rays and keep optimizing. However, that would not work very well. Specifying a large value of the aperture weight parameter (which we called **AWT** in the above MACro) might work somewhat better, but still, OPD targets are better in cases like this. Remember, the OPD fan is the *integral* of the TAP fan. A steep *slope* on the OPD curve means a large transverse aberration—which you do not care about if the OPDs are small enough.

Tip 7. So why use transverse targets, anyway? Well, for the lenses I have studied, those targets can change a lens faster—and thus arrive at a good configuration—more quickly than can OPD targets. So start with transverse and switch to OPD when you are almost there. DSEARCH and ZSEARCH both have an option to include

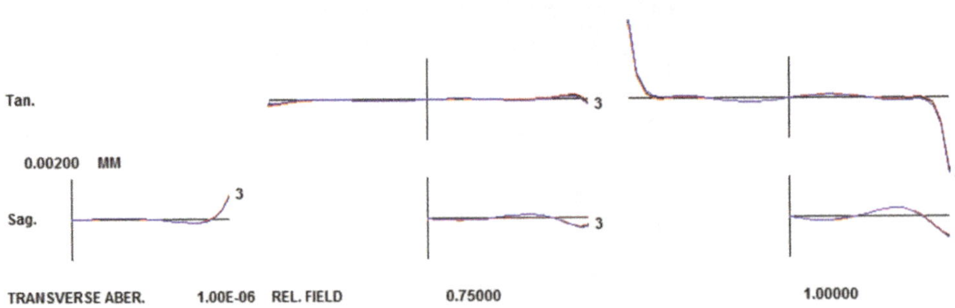

Figure 29.2. Transverse aberration plot example with large aberrations at the ends of the TFAN.

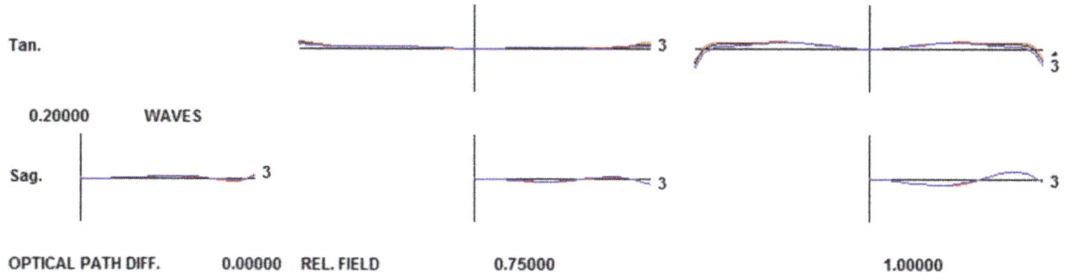

Figure 29.3. OPD plots for the same lens. The large transverse aberrations turn into a large *slope* of the curve, but the OPD errors are small.

both kinds of targets (using the **TOPD**, **OPD**, **OPSHEAR**, and **TOSHEAR** directives), and if you want a diffraction-limited image, those options sometimes prove helpful. Try it both ways; you will probably find some interesting configurations.

Tip 8. Suppose you only want to evaluate the current lens, to spot the largest aberrations—but do not want to change anything just yet. It is simple—just uncomment the lines

```
!EVAL
!EDS
```

and run the MACro. The program will evaluate the merit function and then end the data set (the MACro) at the **EDS** line. Later, if you want to optimize the lens, just comment those lines out again and rerun it.

Tip 9. When you optimize your lens, assuming you have switch 1 turned on (the default), it is a good idea to run the program twice. Thus,

```
SYNOPSYS 50
SYNOPSYS 50
```

is sometimes better than

```
SYNOPSYS 100
```

This is because if any of the variables in the first run encounters a boundary, that variable is dropped for the remaining iterations (according to switch 1). The rule saves time, since usually that variable will continue to try to violate the same boundary, and if not dropped it will just slow things down. However, if the lens changes its form significantly during the run, the same variable may now want to move in the other direction. The second run frees all variables so it can do so. This is especially true for glass model variables (GLM), which usually hit a boundary early on.

By the way, we often see new users asking for, say 500 iterations of optimization. While this may be necessary when running a less powerful program, it is almost never necessary with SYNOPSYS. Ask for a smaller number and use the simulated annealing button thereafter.

Tip 10. The MACro given at the start of this lesson includes both transverse aberrations and OPDs. Notice the difference in weights. This reflects the fact that an

OPD error of one unit (one wave) usually gives a much better image than a transverse error of one inch (or one millimeter). Sometimes one can obtain excellent results from giving targets to *both* kinds of errors—but now the relative weights are important. If you have, say, a target for a mechanical property, such as an aperture or spacing somewhere, you want to be sure that your very nice balance is not upset when the program thinks the one-wave OPD is just horrible and tries to bring it down at the expense of the other errors. To make finding a suitable weight easier, the program provides two tools. If you click the '**Ready-Made Raysets**' button in the MACro editor and choose option 8, the program will create ray grids for both kinds of targets, assigning weights to the OPD errors that reflect the current wavelength and F/number. These weights ensure that the difference in units is accounted for in a reasonable way, and of course you can adjust things from there after you see the effects. The button also lets you select OPD targets, and in this case, you assign relative weights as usual and then click the box 'Calculate special OPD weights'. The OPD weights will be modified by the same rule when the ray-grid requests are added to the MACro. (These calculations are a function of the F/number of the current lens, which must be well defined already.)

Tip 11. You have seen several pupil definitions in the earlier chapters, from simple paraxial to sophisticated wide-angle options. Some users try to start a new design with the most sophisticated option, such as WAP 3—and then wonder why the runs take so long. We advise them to start instead with the simplest setup that makes sense, a paraxial pupil if possible, and add complications only when clearly necessary. I almost never use the complicated options, and if vignetting is an issue and the paraxial pupil is clearly not adequate, you should switch to a real pupil (with a negative value on the APS input). If the stop is not filled properly, then switch temporarily to WAP 2 with a CAO on the stop (or CSTOP in effect), and then use the **FVF** utility to find a set of vignetted apertures (the VFIELD) that duplicate the WAP 2 results. This removes the WAP option, and everything runs much faster thereafter since all the pupil searching has already been done. If the lens changes form and the current VFIELD is no longer appropriate, just run **FVF** again, and the apertures are updated. If you have not done so, it would be wise to type **HELP VFIELD** and follow the links to the explanation of that feature at this time.

Tip 12. We mention here an obscure topic that occasionally has proved very mysterious. If you optimize your lens with OPD errors only, in FOCAL mode there are actually two solutions. One is the one you want, and the other is a lens with collimated output, which you probably do not want. The OPD errors are calculated by comparing the path length of a given ray with that of the chief ray—and if the beam enters and leaves the lens collimated and hits a flat image plane, those paths may again be equal. The program will faithfully find this spurious solution if it should head that way. It is just doing its job. The moral is, it is a good idea to add at least one transverse intercept error to the MF. It can have a low weight, but it makes the collimated solution look unattractive. This situation rarely happens, but it is a mystery when it does—now you know why.

Tip 13. If your PC has more than a single CPU, you can save much time by authorizing more than a single core when running the search programs or some of the image analysis features—but the gain in speed is not a simple function of that number. It turns out that, if you have *N* cores, for large values of *N* the incremental gain in speed by adding yet one more core goes as $1/N^2$, while the overhead required to start, stop, and manage the data from the extra core is a linear function of *N*. Mathematics tells us those two functions have to cross somewhere, and after that point, adding yet more cores actually makes the run take *longer*.

So the biggest saving of time does not always occur when you activate the maximum number of cores in your PC, if that number is large. Increasing from one to two cores cuts the time by 0.5, while Increasing from 10 to 11 cuts it by only 0.009 09, and so on. Experiment with your PC to find the core number that gives you the fastest performance. For our eight-core hyperthreaded PC, which can in principle run 16 processes simultaneously, we get the fastest processing with about 14 authorized. Keep in mind that, if a program requires a single process for half of its calculations, the maximum improvement will be 50%, even with an infinite number of cores.

Tip 14. Much effort has gone into figuring out how to maximize the MTF of a lens, and some designers routinely target the wavefront variance in the MF for this purpose. While that works, and can be accomplished with the GNV ray set option in the AANT file, it is not necessarily the best approach since it often converges very slowly. Instead, read about the **GSHEAR** ray sets in the User's Manual. That feature usually works much better and faster. The DSEARCH examples in chapter 20 show how that program can utilize that technique.

Tip 15. Changing a curvature usually has a much larger effect on the MF than changing a thickness or airspace. In most of these examples, we have not put any monitors on lens thicknesses at all, and the results often came back with elements that were too thick or too thin to be practical.

If an element is too thin, it is hard to keep it from flexing under the pressure of polishing, which ruins the figure. If too thick, it is more expensive and can absorb light. The program has default limits and monitors, but those are often too forgiving, and you should reduce the target values in those cases. A lens maker looking at the preliminary results would likely exclaim at how impractical they are—but he would not understand that one approaches the goals in steps, and that issue is corrected later. Also, be aware that, when the vendor first makes an element, it always has a larger diameter than the print calls for. That excess is ground away later, after the lens is carefully centered on a precision lathe and both sides running true—to remove the residual wedge—and at that time the diameter is cut to the desired dimension. However, that does not work unless there is some glass outside of the desired clear aperture to cut away, so be sure the edge thickness on the print is not too thin, or there might not be. Some lens makers greatly prefer elements with a rather thick edge, to facilitate mounting on the polishing machine. Of course, that is not often a good idea for other reasons, so here you must engage the shop in constructive dialog.

When the design is in good shape, it is then time to gently corral those properties. The ADT monitor is very helpful, but should be used gingerly: assign a low weight and large window at first, such as

```
ADT 7 .01 10
```

and increase the weight gradually if you need to. That targets the ratio of the lens diameter to thickness to the entered value. Adjust the target value and weight to see which combination works best. Also useful are the ACM and ACC monitors, which simply penalize any element thickness that is outside the limits on those entries; a minimum thickness for ACM and maximum for ACC. You should apply these monitors gently, as you need them. The goal is not to perturb the design so strongly that it leaves the very nice solution region you are in at the moment.

In rare instances, a lens returned by DSEARCH simply does not work as well when edges are modified with these tools. That is a good time to explore some of the other results from the search routines; or you can put an ADT monitor in the SPECIAL AANT section, with a low weight, and expect to obtain a different set of candidates. Experiment with these settings; there are a great many solutions out there.

IOP Publishing

Lens Design
Automatic and quasi-autonomous computational methods and techniques
Donald Dilworth

Chapter 30

FLIR design, the narcissus effect

FLIRS; Retroreflection from a cooled detector produces a dark area on a FLIR display; how to correct it

Night-vision systems can see in total darkness. That works because all matter in the Universe radiates energy in the form of photons, following the Planck function in the case of a perfect blackbody radiator or approximating that function to some degree otherwise. Since your skin is close to room temperature at 20 °C or 293 K, you emit radiation according to the curve in figure 30.1, as calculated by the SpectrumWizard[1]. Note the peak at about 10 μm. (Type **MSW** to open the Wizard.)

Although the atmosphere absorbs much infrared radiation, it has a window of transparency centered at a wavelength of just over 10 μm, which nicely fits the spectrum shown below.

Night-vision systems sense this radiation by means of a detector that converts IR photons into an electrical current. A common material for this purpose is HgCdTe, which has the spectral sensitivity shown in figure 30.2. The exact sensitivity range depends on the relative proportions of the ingredients. We are fortunate that the source, atmosphere, and detector all work well over the required spectral window.

In order to obtain a high signal-to-noise ratio, one must ensure that the optics, and even the detector itself, do not radiate unwanted flux at the same wavelengths. This is accomplished by cooling the detector, often with liquid nitrogen, and using high-quality anti-reflection coatings on the lens surfaces. If those steps are not taken, the situation is like what you would see looking through a telescope if the lenses and housing were all glowing white: it would be hard to distinguish what you are looking at. In this chapter we will develop a new FLIR design and then analyze the narcissus properties.

Designing forward-looking infrared radiometers (FLIRS) is not difficult. We will ask DSEARCH to design a five-element lens, using Germanium for the positive elements and ZnSe for the negative. Here is the input file **(C30M1)**:

[1] SpectrumWizard is a trademark of Optical Systems Design, Inc., a Maine, USA corporation.

doi:10.1088/978-0-7503-1611-8ch30

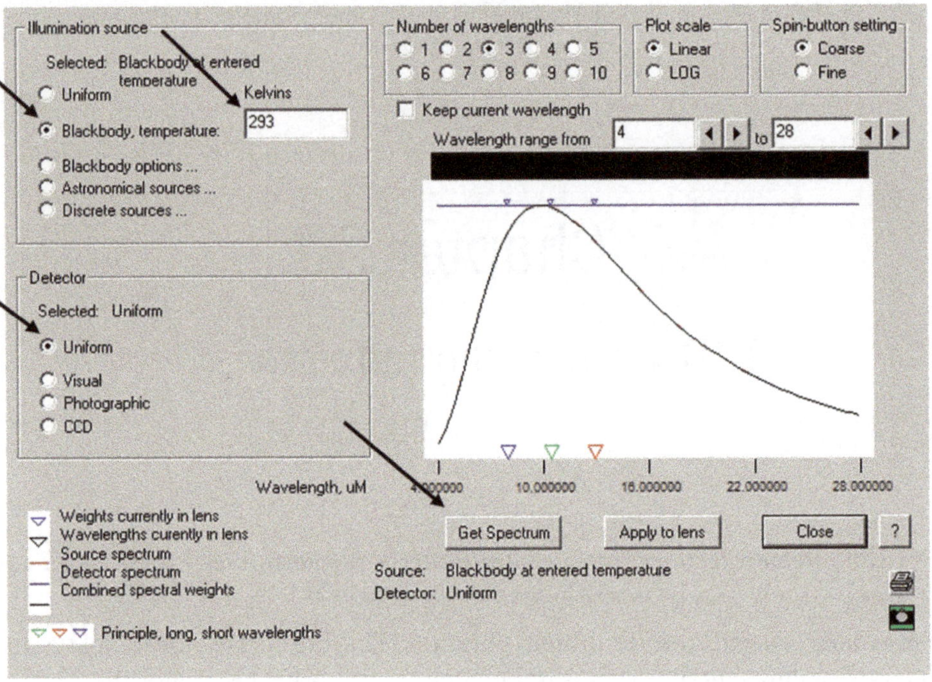

Figure 30.1. Spectrum Wizard showing the blackbody curve at 293 K.

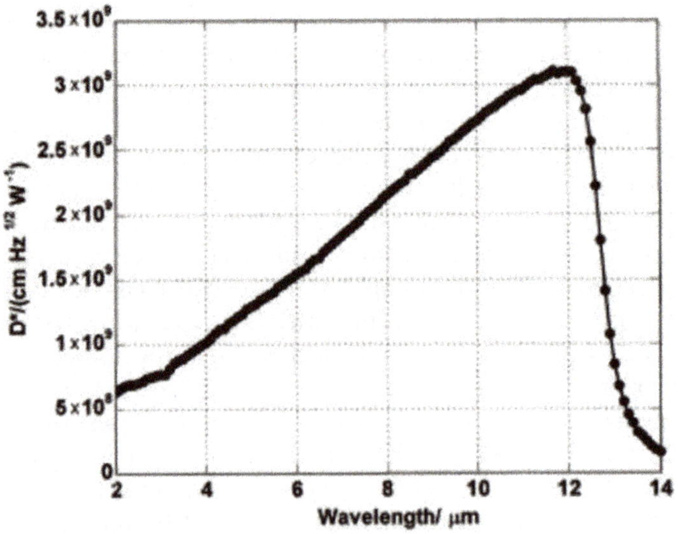

Figure 30.2. Sensitivity of HgCdTe detectors. Reproduced from Theocharous *et al* 2005 A comparison of the performance of a photovoltaic HgCdTe detector with that of large area single pixel QWIPs for infrared radiometric applications *Infrared Phys. Technol.* **46** 309–22. With permission from Elsevier.

```
CORE 14
 DSEARCH 5  QUIET
 SYSTEM
 ID FLIR DSEARCH
 AFOCAL
 OBB  0.000000     17.50000      6.35000
 WAVL 12 10 8

 UNITS MM
 END
 GOALS
 ELEMENTS 5

 TOTL 350 .1

 STOP FIRST
 STOP FREE
 RT 0.5
RSTART 300
TSTART 5
 FOV 0.0 0.75 1.0 0.0 0.0
 FWT 5.0 3.0 1.0 1.0 1.0

 NPASS 100
 ANNEAL 200 20 Q
 TSTART 5 ! This thickness on each element to start with
 QUICK 60 90

GLASS POS ! positive elements will use this glass type
U GE
GLASS NEG ! and negative this type.
U ZNSE
END

SPECIAL AANT
ACC 10 .1 1
AAC 26 .1 1
M 13.37 1 A P YA 1 0 0 0 2
M 0 1 A P YA 1 0 0 0 10
M -.076 10 A P HH 1
ADT 7 .1 10
END
GO
```

The SPECIAL AANT section is where we request the chief ray at full field to hit surface 2 a height of 13.37 mm (to allow clearance for a scanning prism at surface 1) and hit surface 10 at the axis (since that is the objective lens). This system is set up backwards, and light actually comes through the other way. It can be designed either way. The ACC monitor keeps thicknesses under control, and the AAC keeps the apertures not more than 26 mm in radius. The ray targets we have added are crucial for this kind of design: those control the first-order properties. The afocal magnification, for example is controlled by the target on HH, which is the tangent of the full-field ray angle coming out.

After this file runs, the program returns the lens in figure 30.3 (**C30L1**). It is already in the diffraction limit and one might think it requires no more optimization. However, we have not yet controlled the *narcissus*.

30.1 Narcissus correction

The narcissus effect is too often overlooked, showing up in scanning IR systems as a dark smudge in the center of the displayed image, as in the simulated image in figure 30.4.

This effect occurs because at the center of the field the detector can see a ghost image of itself, reflected from a lens surface somewhere. This ghost is very cold—because the detector is very cold—and the total background signal seen by the detector is therefore lower at the center than it is at other parts of the field, where the ghost image is vignetted by other lens apertures, scanned out of the field, or so aberrated that it cannot form a sharp image. Only at the center are all the ghosts lined up.

Let us start with the lens designed above and get the narcissus under control. It is designed for the 8–12 μm band and uses AFOCAL mode, which means that ray output is given in *angles* instead of transverse coordinates. (No 'perfect lens' is required in SYNOPSYS.)

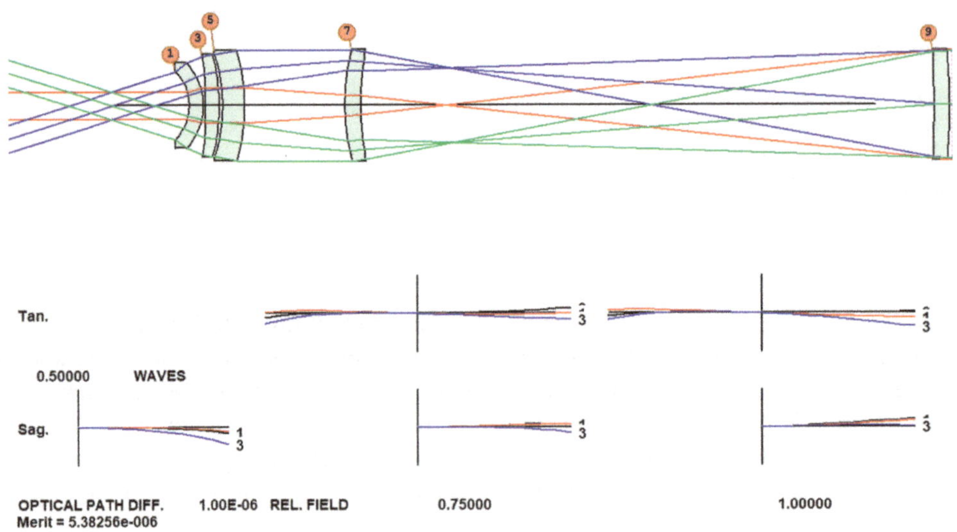

Figure 30.3. FLIR design returned by DSEARCH.

Figure 30.4. Example of the narcissus effect.

To analyze the narcissus properties of this lens, use the command **NAR**:

```
SYNOPSYS AI>NAR

ID FLIR DSEARCH                                2158          13-FEB-18   13:55:44

NARCISSUS ANALYSIS

SURF          YNI          Imarg/Ichief
----------------------------------------
  1         1.8300           1.4401
  2         1.7778           1.3718
  3       6.1560E-04      3.0729E-04
  4         1.5312           1.5831
  5         1.6345           1.3040
  6         1.3634           1.5349
  7         0.1504          -0.1138
  8         0.3572          -1.1011
  9         1.7827           0.9957
 10         1.9784          -0.8887
 11       7.0565E-16      3.5245E-16
SYNOPSYS AI>
```

The column YNI shows the value of the quantity $\phi A_0 N$, where the arguments are defined in figure 30.5. From this we can calculate the approximate size of the retroreflected blur on the detector: $Y1 = 2\phi A_0 N/\alpha$, where α is the half-angle of the converging beam at the detector, and Y1 is the ray height of the blur at the image after reflection at the given surface.

To control the narcissus, one must ensure that the value of YNI never becomes lower than a limiting value, which is a function of the sensitivity of the scanner and user acceptance. A larger value means the ghost image is more out of focus and therefore less intense. According to the above table, the worst narcissus is from surface 3, where the value is 6.1E−4. That is very small. We can use the GHPLOT program to visualize what that means. Type **MGH** in the CW or navigate to that dialog in **MLI**. Enter the data shown in figure 30.6, and then click the '**GHPLOT**'

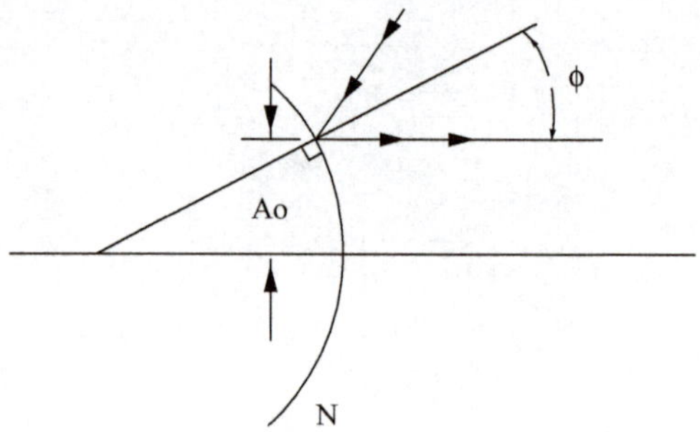

Figure 30.5. Geometry of the narcissus equation.

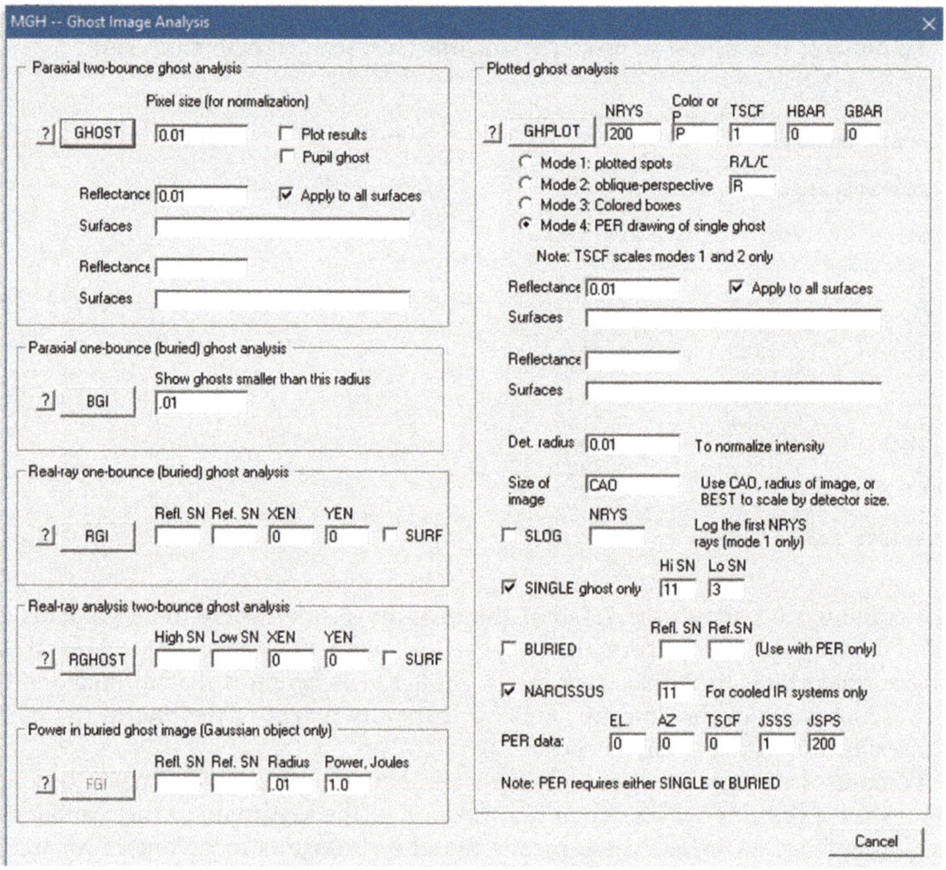

Figure 30.6. Data entered into the MGH dialog.

button. You obtain the picture in figure 30.7. Here, we want to look at the beam that reflects from surface 11, then from 3, and then goes to the output.

Select the area shown on the left and click inside the selection rectangle to zoom in to that portion, as shown in figure 30.8.

Light comes in from the left, shown in red, until it reaches surface 11, the flat surface at the end of the lens, which is off the page to the right, and which we declared a NAR surface on the dialog. (That assigns a reflection coefficient of 1.0.) We requested a single ghost, from 11 back to 3, and from there returning to the last surface again. The signal flux there is collimated, and we assume that any light that returns there and is again collimated will show up on the detector sharply focused. After reflecting from 11, the rays are shown in blue, and following the second reflection, at surface 3, they are drawn in green. The retroreflected beam falls almost exactly over the incoming rays. This will be a very bad narcissus unless we do something about it. How to control it?

To understand that question, you need to know the angular size of the Airy diffraction disk. Ask for a **PSPRD** analysis, on-axis. Go to the **MDI** dialog (`MDI`) and click the '**PSPRD**' button. You obtain the plot in figure 30.9.

Note the value of the Airy disk radius: 0.000 24. That is an angular value, since the lens is declared AFOCAL. We know that if the narcissus beam returns with an angle equal to or less than that value the beam will be in sharp focus and cause a very objectionable narcissus.

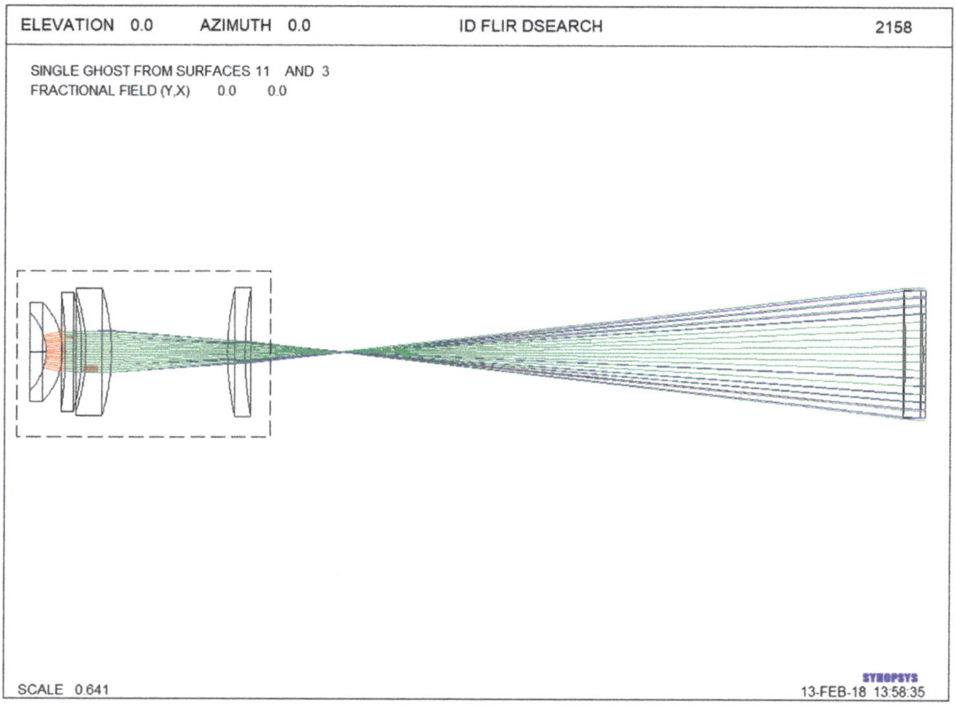

Figure 30.7. GHPLOT showing retroreflected beam at surface 8.

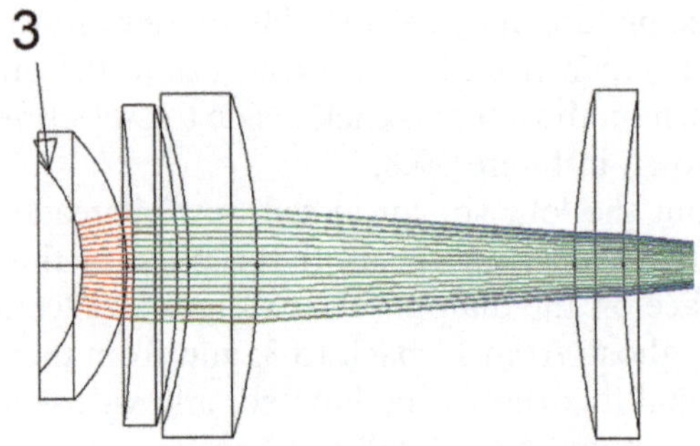

Figure 30.8. Portion of the lens responsible for a bad narcissus.

So what is the current value of the returning beam angle? Go back to the MGH dialog, and this time click the '**RGHOST**' button, with data shown in figure 30.10.

```
--- RGHOST 11 3 0 0 SURF
                    RAY VECTORS                      (X DIR TAN)(Y DIR TAN)(INC. ANG.)
SURF            X               Y            Z            ZZ          HH          UNI
-------------------------------------------------------------------------------------
 OBJ     0.000000     0.000000      0.000000    0.000000    0.000000
   1     0.000000     0.000000      0.000000    0.000000    0.000000    0.000000
   2     0.000000     0.000000      0.000000    0.000000    0.000000    0.000000
   3     0.000000     0.000000 -4.440892E-16    0.000000    0.000000    0.000000
   4     0.000000     0.000000  7.105427E-15    0.000000    0.000000    0.000000
   5     0.000000     0.000000 -3.108624E-15    0.000000    0.000000 1.707547E-06
   6     0.000000     0.000000      0.000000    0.000000    0.000000    0.000000
   7     0.000000     0.000000  1.421085E-14    0.000000    0.000000 8.537736E-07
   8     0.000000     0.000000  2.220446E-14    0.000000    0.000000 1.207418E-06
   9     0.000000     0.000000  5.684342E-14    0.000000    0.000000 3.721513E-06
  10     0.000000     0.000000  1.243450E-14    0.000000    0.000000    0.000000
 --- RAY REVERSES AFTER NEXT SURFACE ---
  11        0.000000        0.000000        0.000000        0.000000        0.000000
  10        0.000000        0.000000        0.000000        0.000000        0.000000
   9        0.000000        0.000000    1.243450E-14        0.000000        0.000000
   8        0.000000        0.000000        0.000000        0.000000        0.000000
   7        0.000000        0.000000   -6.217249E-15        0.000000        0.000000
   6        0.000000        0.000000   -2.131628E-14        0.000000        0.000000
   5        0.000000        0.000000   -3.552714E-15        0.000000        0.000000
   4        0.000000        0.000000   -6.661338E-16        0.000000        0.000000
   3        0.000000        0.000000        0.000000        0.000000        0.000000
   4        0.000000        0.000000    7.105427E-15        0.000000        0.000000
   5        0.000000        0.000000   -3.108624E-15        0.000000        0.000000
   6        0.000000        0.000000        0.000000        0.000000        0.000000
   7        0.000000        0.000000    1.421085E-14        0.000000        0.000000
   8        0.000000        0.000000    2.220446E-14        0.000000        0.000000
   9        0.000000        0.000000    5.684342E-14        0.000000        0.000000
  10        0.000000        0.000000    1.243450E-14        0.000000        0.000000
  11        0.000000        0.000000        0.000000        0.000000        0.000000
GHOST REFLECTED FROM SURFACES     3    11 AT SURFACE   12
     X                   Y                   ZZ                  HH
-------------------------------------------------------------------------
   0.00000             0.00000             0.00000             0.00000
 Type <ENTER> to return to dialog.
 SYNOPSYS AI>
```

The program creates and runs the RGHOST command for you, and you see the tangent of the ray (HH) when it returns to surface 11 equals zero, to five places; but

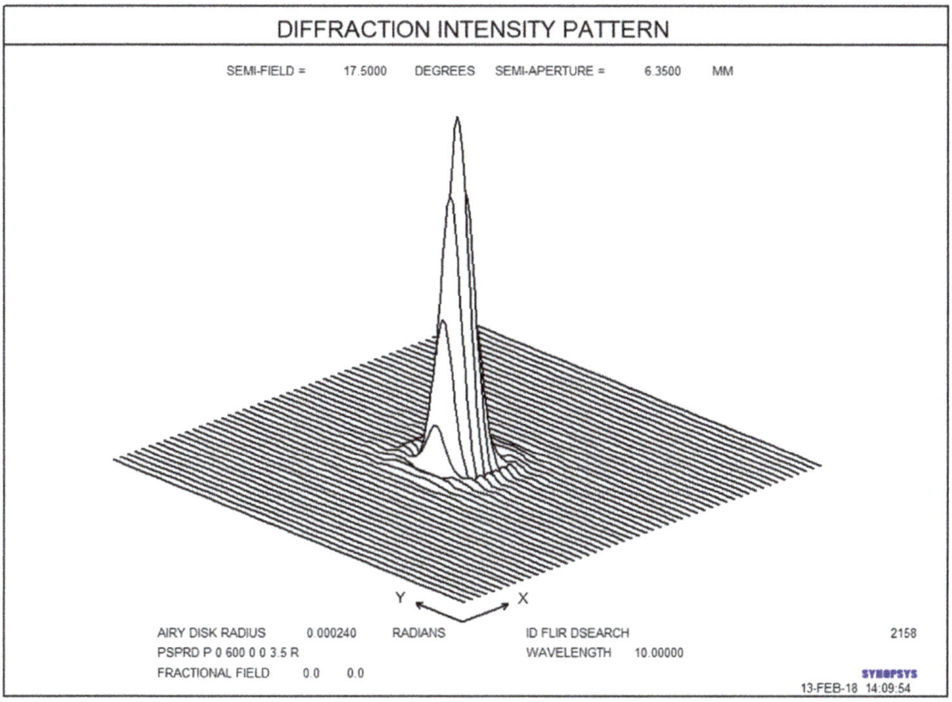

Figure 30.9. Diffraction point-spread image from the IR telescope.

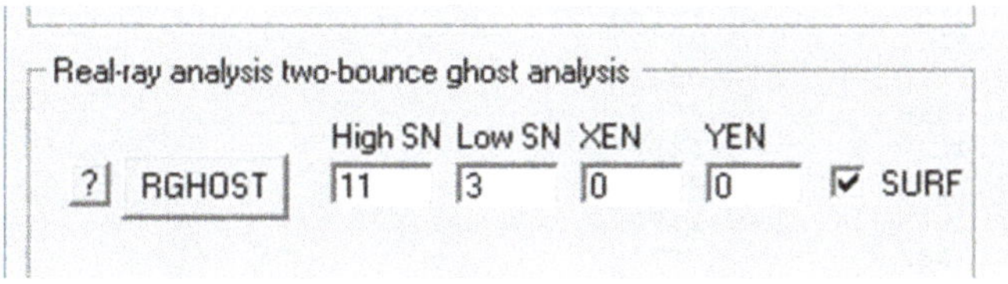

Figure 30.10. MGH selection to show a single ghost path.

the Airy disk radius is 0.00024, and we want the narcissus blur to be much larger than that. Experience says that this lens will display a very severe narcissus on the display. We have to do better.

Again, from experience, we learn that the minimum value of the YNI should be about 0.009 if the lens is in units of inches, and 0.229 for a lens in millimeters. (Even though the lens is AFOCAL and ray output is in angles—which are independent of lens units, the quantity YNI has units of length, and therefore scales with those units.)

Let us correct this lens, hoping to obtain a better narcissus value. Here are the PANT and AANT files (**C30M2**):

```
PANT
VY 0 YP1
VLIST RD ALL
VLIST TH ALL
END
AANT P
AEC
ACC
GSR      0.500000      5.000000       4  M      0.000000
GNR      0.500000      3.000000       4  M      0.750000
GNR      0.500000      1.000000       4  M      1.000000
M    0.350000E+03  0.100000E+00 A TOTL
ACC 10 .1 1
AAC 26 .1 1
M 13.37 1 A P YA 1 0 0 0 2
M 0 1 A P YA 1 0 0 0 10
M -.076 10 A P HH 1
ADT 7 .1 10

LLL .23 1 .1 A NAR 3
END
SNAP   0/DAMP 1
SYNOPSYS  100
```

Run this job, and the lens changes very little. What happened to the narcissus?

```
NAR

 ID FLIR DSEARCH                                   2158            13-FEB-18   14:21:30

 NARCISSUS ANALYSIS

 SURF          YNI          Imarg/Ichief
 ----------------------------------------
   1          1.8412          1.6395
   2          1.9141          1.3703
   3          0.2292          0.0949
   4          1.3468          2.6760
   5          1.4198          1.1497
   6          1.1825          1.3306
   7          0.1574         -0.1371
   8          0.2883         -0.7876
   9          2.1443          1.2589
  10          1.6034         -0.7234
  11      7.0376E-16      3.5150E-16
 SYNOPSYS AI>
```

The lens is much improved, and the narcussus on surface 3 has reached our target. Narcissus is usually quite easy to control.

However, now the narcissus from surface 7 is below our limit—that happens. So add a target for that surface as well and reoptimize and—great! Now all surfaces are close to or above the limit. That is how it is done: indentify the problem and fix it. The final lens is shown in figure 30.11 (**C30L2**).

That is what the narcissus is all about. It is usually not difficult to control, but if you forget to look at the NAR listing and do not control the values, you might end up with a very poor display without expecting it.

A final word about this system: germanium is heavy and expensive, so you want the elements to be as thin as practical. Also there is some scatter, again calling for

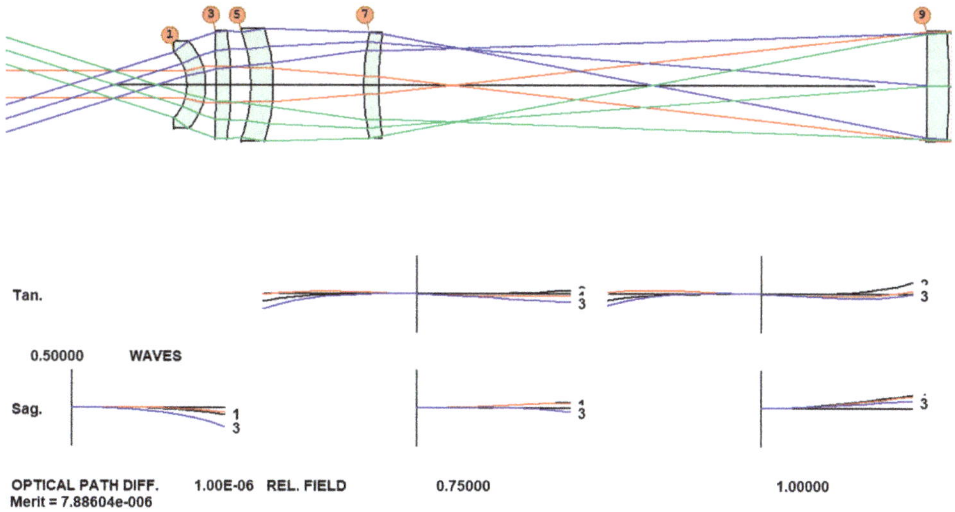

Figure 30.11. The final FLIR design.

thin elements. Also, be careful the system does not get too hot; germanium starts to absorb IR light when it becomes hot, causing the heat to increase ever more, and so on. This is called 'thermal runaway', and you should be aware of it, especially if you design optics for a CO_2 laser, which can have high power. In this lens, element 3 is made of ZnSe, which is also expensive. So we would probably ask for a reduced thickness on that element.

IOP Publishing

Lens Design

Automatic and quasi-autonomous computational methods and techniques

Donald Dilworth

Chapter 31

Understanding artificial intelligence

Natural language input; changing lens parameters with AI; evaluation loops

In earlier chapters you have seen a few of the AI features available in SYNOPSYS. This chapter will present a more complete picture of what this tool can do and how you use it. AI mode is turned on by typing the command **AI**, or by clicking on the 'AI' button. It can be turned off by the command **INTERACTIVE**, or clicking the 'AI Off' button. Turn on AI, if it is not already on, and then **FETCH C31L1** and make a checkpoint. The lens is shown in figure 31.1.

What is the airspace distance after surface 3? Ask AI:

```
SYNOPSYS AI>3 TH?

The thickness or spacing of surface number  3  is      26.3666993
```

Change it to 27.0:

```
SYNOPSYS AI>TH 3 = 27

The thickness or spacing of surface number  3  is      27.00000000
```

What is the third-order spherical aberration? Well, you actually can ask just that question:

```
SYNOPSYS AI>What is the third-order spherical aberration?

The third-order spherical aberration sum (SA3) is      -0.02340108
```

Or you can type a much shorter sentence:

```
SYNOPSYS AI>SA3?

The third-order spherical aberration sum (SA3) is      -0.02340108
```


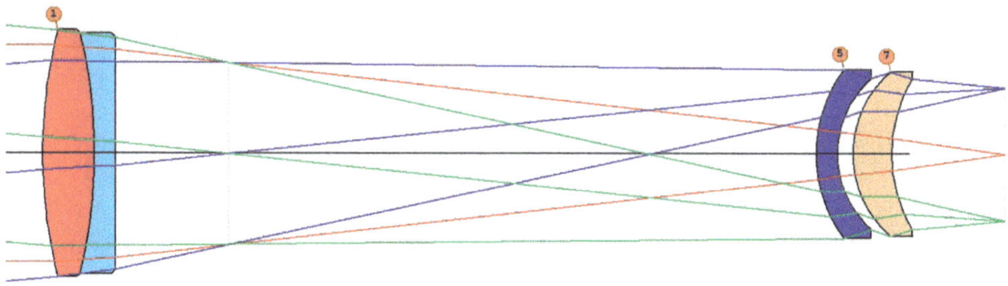

Figure 31.1. Example lens for AI exercises.

The last question, **SA3?**, is grammatically the same as the previous sentence, and of course you would prefer to do as little typing as possible, so that is what you would enter. However, you see that the input is extremely flexible, and the way you type it is often not critical. The program parses the sentence, finds the subject and verb, satisfies any conditions, and then tries to answer the question.

Since many tasks in SYNOPSYS can be done in several ways, you naturally want to find the easiest. Suppose you want to know the global z-coordinate of surface 7. You can type the **ASY GLOBAL** command and pick out the answer from the listing:

```
SYNOPSYS AI>ASY GLOB

THIS LENS HAS NO SPECIAL SURFACE TYPES
THIS LENS HAS NO TILTS OR DECENTERS
Global mode has been turned on.

GLOBAL COORDINATE DATA

GLOBAL COORDINATE SURFACE LOCATION IN COORDINATE SYSTEM OF SURFACE  1
```

SURF	X	Y	Z	NOTES	ALPHA	BETA	GAMMA
1	0.000000	0.000000	0.000000		0.00000	0.00000	0.00000
2	0.000000	0.000000	12.000000		0.00000	0.00000	0.00000
3	0.000000	0.000000	17.000000		0.00000	0.00000	0.00000
4	0.000000	0.000000	43.366699		0.00000	0.00000	0.00000
5	0.000000	0.000000	179.512319		0.00000	0.00000	0.00000
6	0.000000	0.000000	184.512319		0.00000	0.00000	0.00000
7	0.000000	0.000000	188.168005		0.00000	0.00000	0.00000
8	0.000000	0.000000	197.168005		0.00000	0.00000	0.00000
9	0.000000	0.000000	223.717528		0.00000	0.00000	0.00000

However, it is easier to just ask AI:

```
SYNOPSYS AI>7 ZG?

Surface number 7 is not controlled by any tilt or decenter.
Surface number 7 has a global Z-coordinate of   188.16800509
```

Suppose you want to change that value. Surface 7 is not assigned global coordinates at present, so you could go to the SpreadSheet and enter data into sub-menus, or use the WorkSheet or a CHG file and enter the data in the correct format. In this case, however, AI is better:

```
7 ZG = 200
```

That simple sentence assigns the global coordinate.

Probably the most useful of the AI features is for making a plot of something against something else. Restore the lens to the checkpoint you made earlier, and then remove the paraxial solves:

```
CHG
NOP
END
```

Then type the sentence below to look at the color correction, shown in figure 31.2:

```
PLOT DELF FOR WAVL = .4 TO .8
```

Since the lens has no solves now, the paraxial defocus (DELF) varies with wavelength. If the lens were assigned a YMT solve, then DELF would have been zero at all wavelengths, and you would instead have plotted the back focus (BACK). (Note that we removed the *curvature* solve as well, by using the **NOP** entry, since we do not want the last radius to change with wavelength too!)

Suppose you are working on a lens where secondary color is an issue. You want to make this kind of plot more than once, but you do not want to type that long sentence every time. What can you do? Make a symbol, of course. Here is how to define a new command, SC:

```
SC: PLOT BACK FOR WAVL = .4 TO .8
```

Now just type **SC**, and the program gives you the plot again. Put this definition in your CUSTOM.MAC MACro, and it will come back every time you start the program.

The principle is simple: ask AI for what you want with an English sentence. If that sounds too good to be true, you are right—probably all the sentences in Shakespeare would baffle the program. However, applied to a limited range of issues, it can do a great deal. There are five categories of sentences that the program recognizes:

1. Questions about something.
2. Changes to something.
3. Looping, changing something, and evaluating something else each time through, usually, making a plot of the results.
4. Assigning symbols to equal a character string.
5. Evaluating an equation.

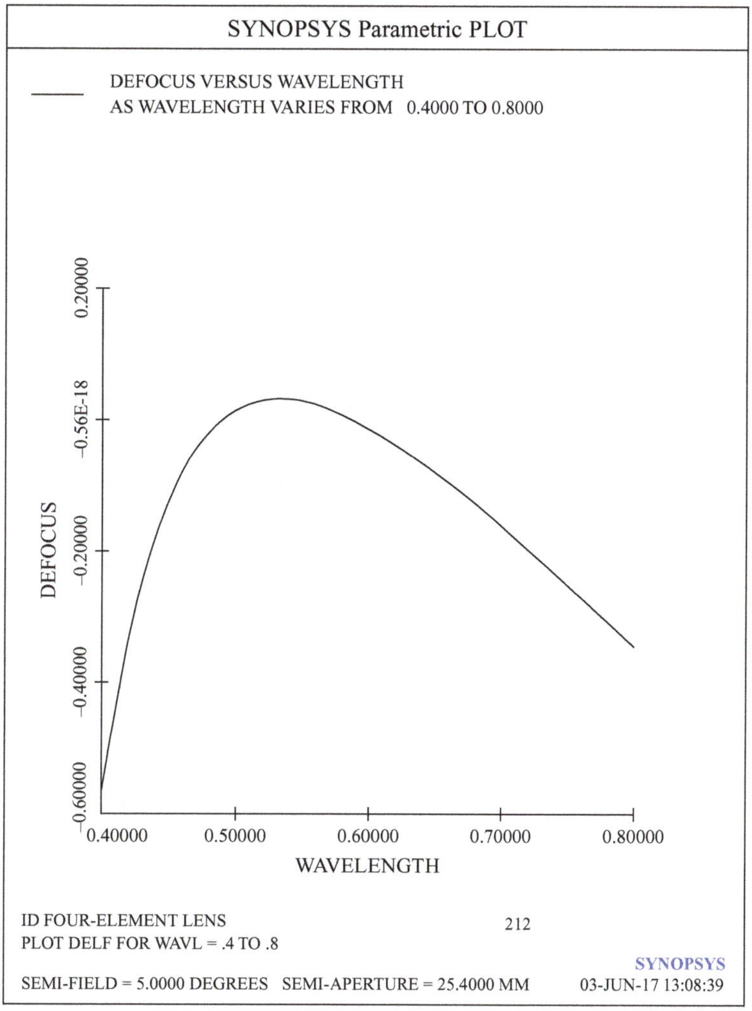

Figure 31.2. Color-correction curve plotted by the AI program.

All of this is explained in chapter 15 of the User's Manual, and when you finish this lesson you should read the primer that starts in section 15.2.

The AI vocabulary is a few hundred words, and you can see a list on the monitor if you ask AI to **SHOW SUBJECTS**, **SHOW VERBS**, or **SHOW CONDITIONS**.

Changes are requested in an intuitive manner:

```
4 RAD = 123.456
Change radius 4 to 123.456
Increase 4 RD by 12.66
Increase 4 RAD to 33.5.
```

Note the difference between the last two examples. The last sentence actually will not work unless radius number 4 is currently *less* than 33.5. AI watches what you are doing and makes helpful suggestions if it sees what looks like an error.

31.1 Error correction

Speaking of errors, as a new user you will make lots of them. That is one reason the program has an extensive set of menus and dialogs. Those dialogs submit commands for you when you click a button, and the format is of course correct in that case. However, some features can be run with a very simple command, and we usually get those things done faster by typing those commands manually; and, yes, we make mistakes sometimes, but do not worry—simple errors can be corrected quickly and you usually do not have to type the whole sentence again. Suppose by mistake you typed

```
4 RRD = 123.456.
```

The characters **RRD** are not in the vocabulary, and the program will immediately ask you to re-enter four characters starting at **RRD**. So you type **RAD** (notice the space after **RAD**: the program replaces *four* characters with whatever you type). It repairs the sentence and proceeds correctly. This error correction works both for AI sentences and for ordinary SYNOPSYS commands. Thus, typing

```
DDW 0 1 123 HBAR 0 1 -1
```

also produces the same error message, and if you then type **DWG**, the drawing command is properly executed.

Lastly, if your input is so garbled up that you just want to start over, simply hit the <Esc> key, and AI will discard the sentence.

31.2 MACro loops

The AI looping feature is powerful and general. Suppose you want a plot of the wavefront variance over the field. There are many kinds of diffraction image analysis, which you will see in the **MDI** dialog (`MDI`), but if the one you want is not already there, you can make it yourself. Here we will make a plot of the variance over field.

There actually *is* a command to do this analysis—but the lesson below is a good example of how to make your own feature using the tools of AI, for cases where there is no command.

Select the 'Multicolor' option on the VAR entry, and click the 'VAR' button, as shown in figure 31.3.

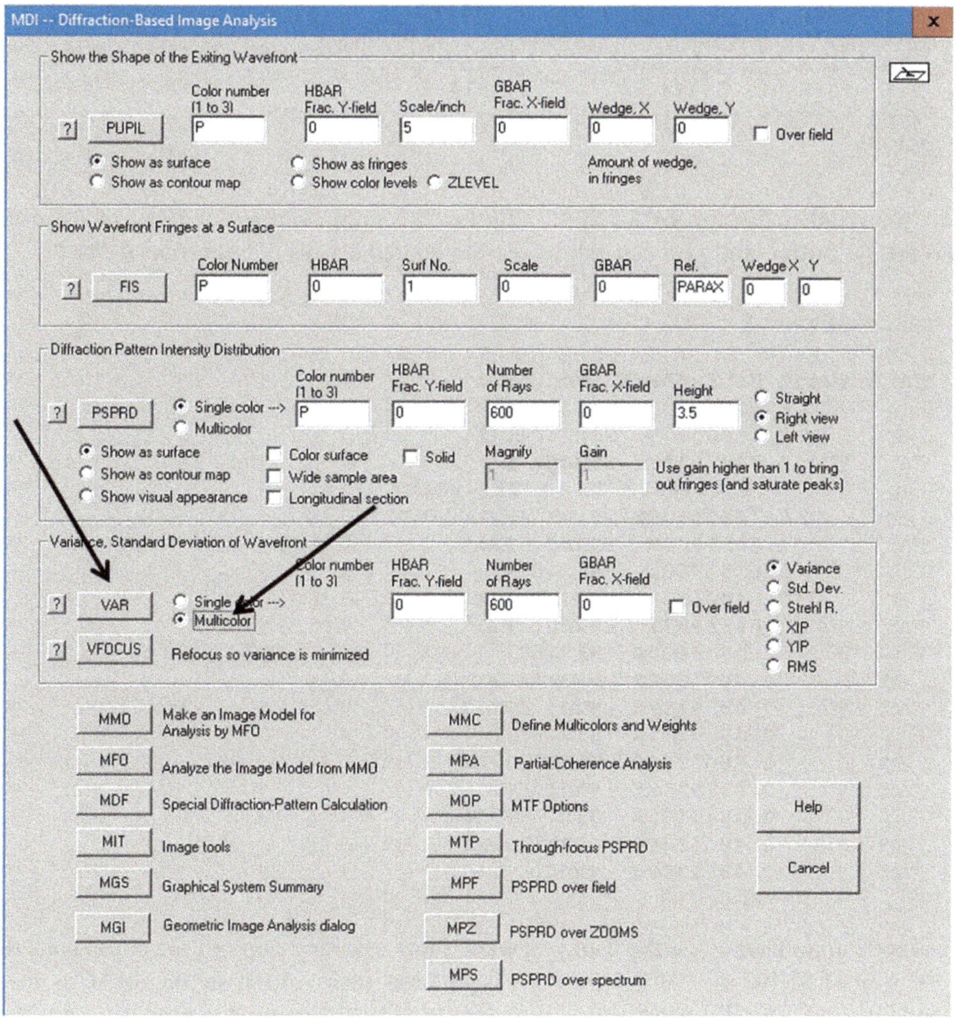

Figure 31.3. The MDI dialog, with multicolor variance selected.

The program prints the VAR value:

```
  VARIANCE      STD. DEV.        STREHL R.      XIP            YIP
 0.287577E-01  0.157605       0.428754      -0.492627E-20  0.874175E-21

VARIANCE IN EACH COLOR AT ABOVE IMAGE POINT:

WAVELENGTH, WEIGHT      0.587560     1.000000
  VARIANCE      STD. DEV.        STREHL R.
 0.591760E-01  0.243261       0.966967E-01
WAVELENGTH, WEIGHT      0.656270     1.000000
  VARIANCE      STD. DEV.        STREHL R.
```

```
 0.910363E-02  0.954130E-01  0.698097
WAVELENGTH, WEIGHT      0.486130     1.000000
  VARIANCE      STD. DEV.        STREHL R.
 0.179936E-01  0.134140      0.491468
 Type <ENTER> to return to dialog.
IMAGE>
```

Like many other features of SYNOPSYS, the VAR command puts a copy of its results into the AI buffer, and you can ask to see the contents with the question **BUFF?**:

```
IMAGE>BUFF?

The current FILE BUFFER contains
  1         0.02875775  VARIANCE
  2         0.15760478  STD. DEVIA.
  3         0.42875398  STREHL R.
  4         -4.92627452E-21  X IM. POINT
  5         8.74175408E-22  Y IM. POINT
  6         1.00000000  TRANS. FRAC.
  7         0.05917602  VARIANCE
  8         0.24326121  STD. DEVIA.
  9         0.09669675  STREHL R.
  10         0.58756000  WAVEL.
  11         0.00910363  VARIANCE
  12         0.09541296  STD. DEVIA.
  13         0.69809714  STREHL R.
  14         0.65627000  WAVEL.
  15         0.01799359  VARIANCE
  16         0.13414018  STD. DEVIA.
  17         0.49146805  STREHL R.
  18         0.48613000  WAVEL.
```

File location number 1 has the data you want. Now ask for a copy of the commands that were submitted by the 'VAR' button. Type **LMM** (also found in the MACro menu dropdown list). The EE editor opens, with the **VARIANCE** command properly formatted, as shown in figure 31.4.

You need to tell AI to vary the relative field for every point on the plot. Which argument is that? *Select* the characters 'VAR', and then look down at the tray, shown in figure 31.5.

The program displays the format of the command and you see that the relative field ('hbar' on the tray) is in word 3. Edit the command in the editor, replacing that word with the characters 'AIP', which stands for 'AI Parameter', as in figure 31.6. Then tell AI that the ordinate on the plot is to be taken from file location 1 in the AI output buffer.

Figure 31.4. The `VARIANCE` command formatted automatically by LMM after running it from the **MDI** dialog.

Figure 31.5. Selecting the characters 'VAR' displays the command format in the tray.

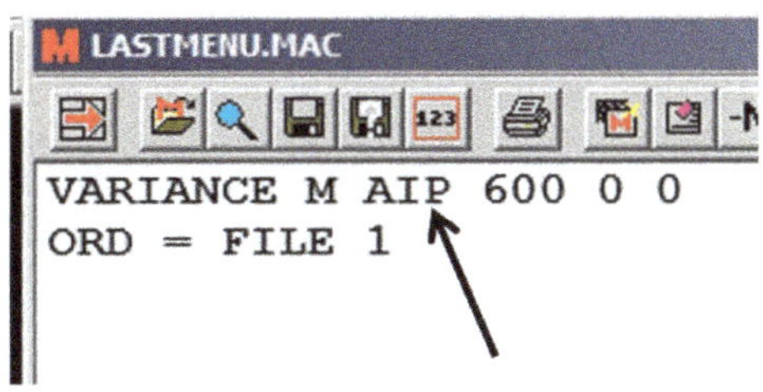

Figure 31.6. The relative field request is replaced by the characters 'AIP', and the result will come from the first file location in the AI buffer.

Load this MACro into memory by clicking the '**Load This**' button (or you can click the '**Run MACro**' button, which also loads it, running the current value of AIP). We are almost there. Now type the AI sentence

```
IMAGE>DO MACRO FOR AIP = 0 TO 1
```

The program loops for the default 100 cases and then displays the desired plot, shown in figure 31.7.

It is also easy to change the labels on the axes:

```
ALAB = "REL. FIELD"
AGAIN
```

Using the built-in command to do this task, one would simply go to the MDI dialog (`MDI`), select the 'Over field' checkbox, enter the appropriate data, and click the 'VAR' button.

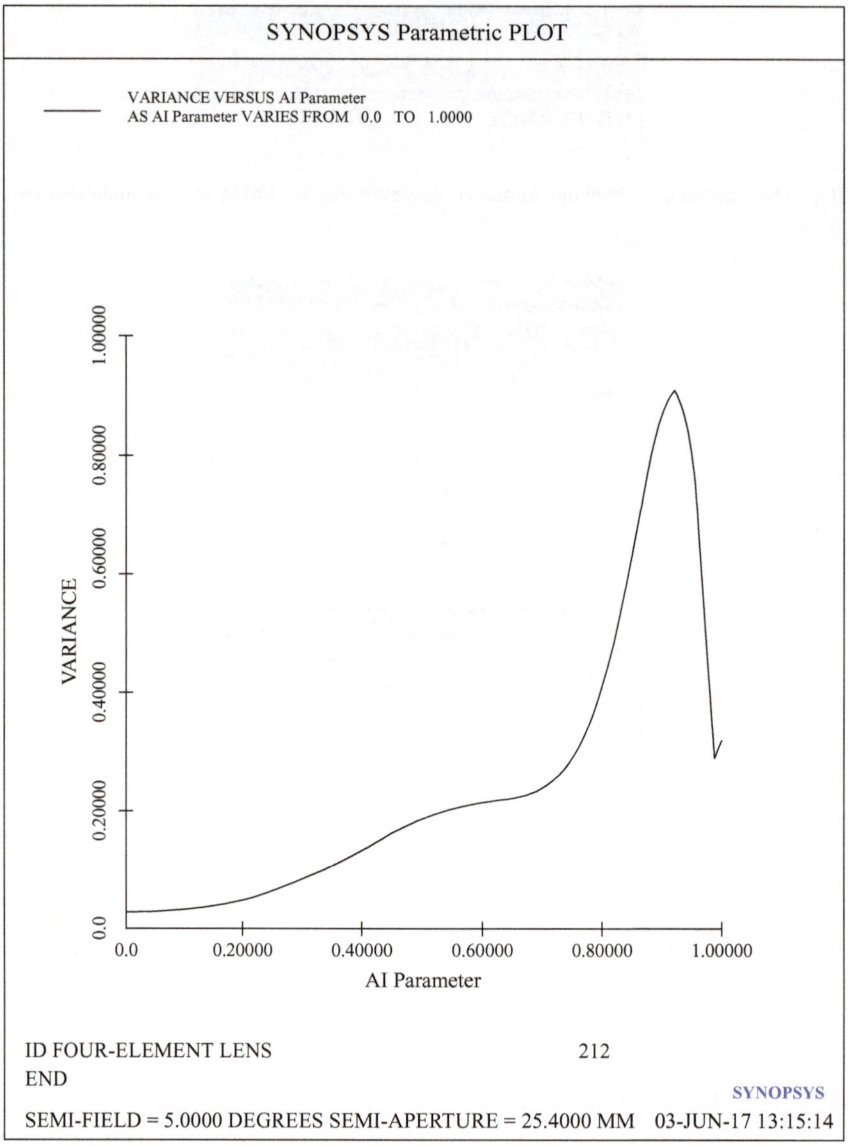

Figure 31.7. Plot of wavefront variance over field, produced by AI.

You can loop over many kinds of things. For example, if you have designed a zoom lens, you could type

```
PLOT DISTORTION FOR ZOOM = 1 TO 9.
```

One more very useful feature of AI lets you perform simple calculations involving output from other features. Fetch the lens saved as 4.RLE, in figure 31.8 (**FETCH 4**).

Look at the current clear apertures with the **CAP** command:

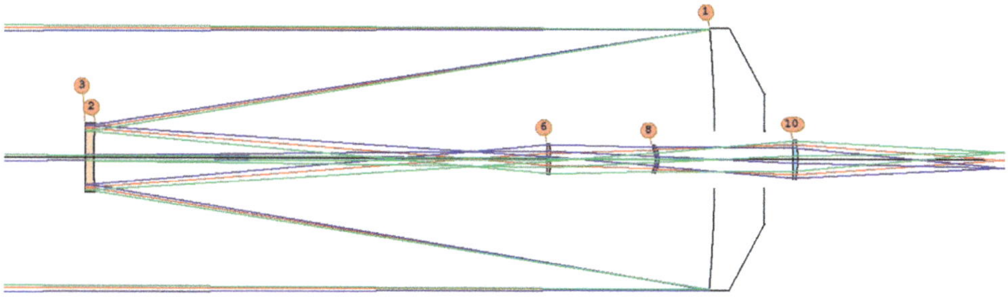

Figure 31.8. A telescope with an obscuration.

```
SYNOPSYS AI>CAP

ID RELAY FLAT                                    141

CLEAR APERTURE DATA

SURF     X OR R-APER.     Y-APER.     REMARK        X-OFFSET      Y-OFFSET   EFILE?
--------------------------------------------------------------------------------------
   1           8.0014              Soft CAO                                     *
   1           1.7500             *User CAI                                     *
   2           2.0456              Soft CAO                                     *
   3           2.0070              Soft CAO                                     *
   4           1.9644              Soft CAO                                     *
   5           0.4628              Soft CAO
   6           0.8797              Soft CAO                                     *
   7           0.8849              Soft CAO                                     *
   8           0.7373              Soft CAO                                     *
   9           0.7781              Soft CAO                                     *
  10           1.1225              Soft CAO                                     *
  11           1.1340              Soft CAO                                     *
  12           0.4898              Soft CAO
```

Right now, the mirror has an assigned clear aperture inside (CAI) of 1.75. Suppose you want that to equal the outside aperture of surface 2. Here is how AI can do it:

```
SYNOPSYS AI>Z1 = CAO OF 2

The semi-aperture on surface number  2  is       2.04561850

SYNOPSYS AI>CAI OF 1 = Z1

Surface number  1  has an inside semi-clear aperture       2.04561850
```

Here we use one of the 20 *Z*-parameters to transfer the value from one place to another. Now the CAI on 1 equals the CAO of 2, as desired.

Lastly, AI can perform simple calculations. Just enter a sentence that starts with an equals sign, =, and involves only constants, the *z*-parameters, and any currently defined symbols that equate to a number. For example, if the variable Z1 currently equals 2.055619 and AIP has the value 3.66, one could calculate as follows:

```
SYNOPSYS AI>AA: 4.147

 SYMBOL  43 DEFINED: AA*
 4.147
 SYNOPSYS AI>= Z1 + AIP + AA

 = Z1 +     3.66000000000     + AA
 = Z1 +     3.66000000000     + 4.147

 The composite value is       10.76261900
```

You are encouraged to read chapter 15 of the User's Manual, where you will find other examples of how to use AI.

IOP Publishing

Lens Design

Automatic and quasi-autonomous computational methods and techniques

Donald Dilworth

Chapter 32

The Annotation Editor

Annotating graphics drawings; adding tolerances to element drawings

In this chapter, you will learn how to use the Annotation Editor of SYNOPSYS, a tool that can add many kinds of symbols and text to a graphics drawing. Get out the lens saved as 1.RLE, and make a drawing, as in figure 32.1:

```
FETCH 1
DWG
```

Let us add a warning message on the drawing. Click the 'Annotate' button Ab on the Graphics Window toolbar. The Annotation Editor toolbar opens up. Click the leftmost button, as in figure 32.2.

Then click in the drawing above element 3. Enter the text shown in figure 32.3, select size 14, and click 'OK'.

Now the text shows up on your drawing, as in figure 32.4.

We are not quite done. Click on the 'Arrow' button, as in figure 32.5.

Then click below the text line and drag down to element 3. You have added an arrow, shown in figure 32.6.

Now click the red box on the toolbar and draw another arrow pointing at the last element, and add some more text there, as seen in figure 32.7.

Now click the red box again (to turn it off) and click the leftmost hashmark button, shown in figure 32.8.

Click several times in the area showing element 2. The lens element now shows hashmarks, as in figure 32.9.

Try the right hashmark button and add marks to element 3. If you hold down the <Ctrl> key when you click the element, the hash marks are smaller, which would be appropriate for smaller elements, as shown in figure 32.10.

You can probably figure out how to use the line, circle, and rectangle buttons. Give them a try. Those all let you drag in the drawing to define the size and position

doi:10.1088/978-0-7503-1611-8ch32

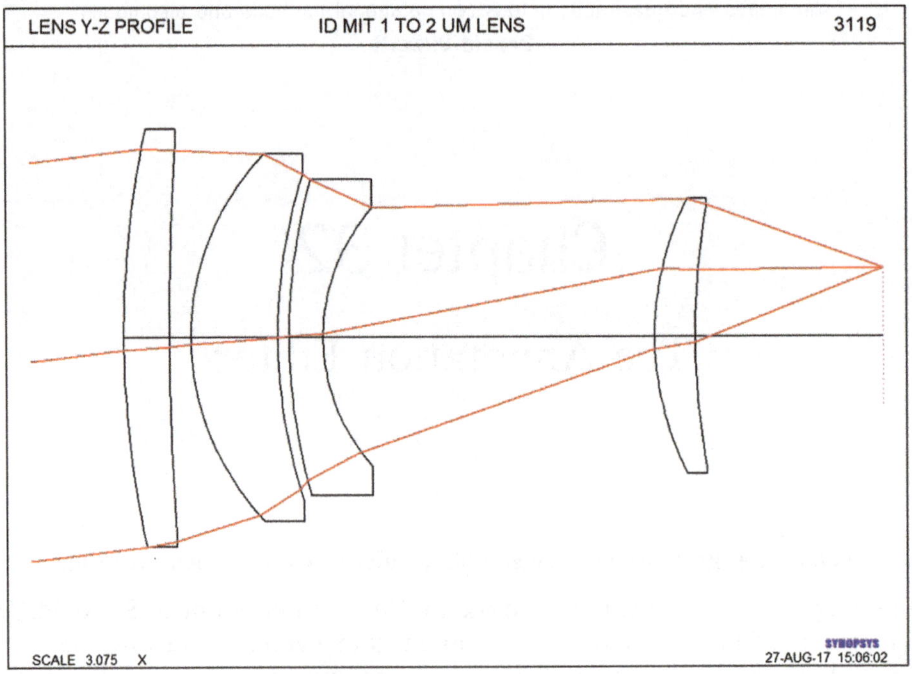

Figure 32.1. Sample DWG drawing.

Figure 32.2. Opening the Annotation Editor.

of the annotation. What if you change your mind? Simple: click on the 'Edit' button, shown in figure 32.11.

All of the annotation you have added sprouts an edit handle. Click on one and it turns black. You can then press the <Delete> key to remove it, drag it to a different location, or double-click to edit it. All of this is fairly simple. Now we will get more advanced. Make a MACro as follows, run it, and then open the dialog (**MPL**) and enter the data as shown in figure 32.12:

```
CCW
FETCH 1
BTOL 2
TPR ALL
DEGRADE WAVE 0.2
GO
```

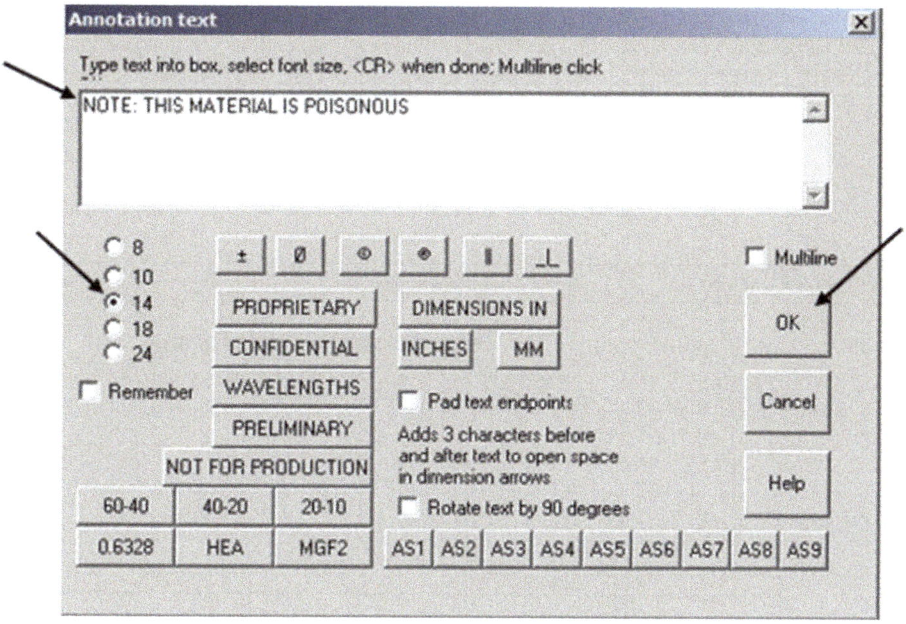

Figure 32.3. Entering text in the Annotation Editor.

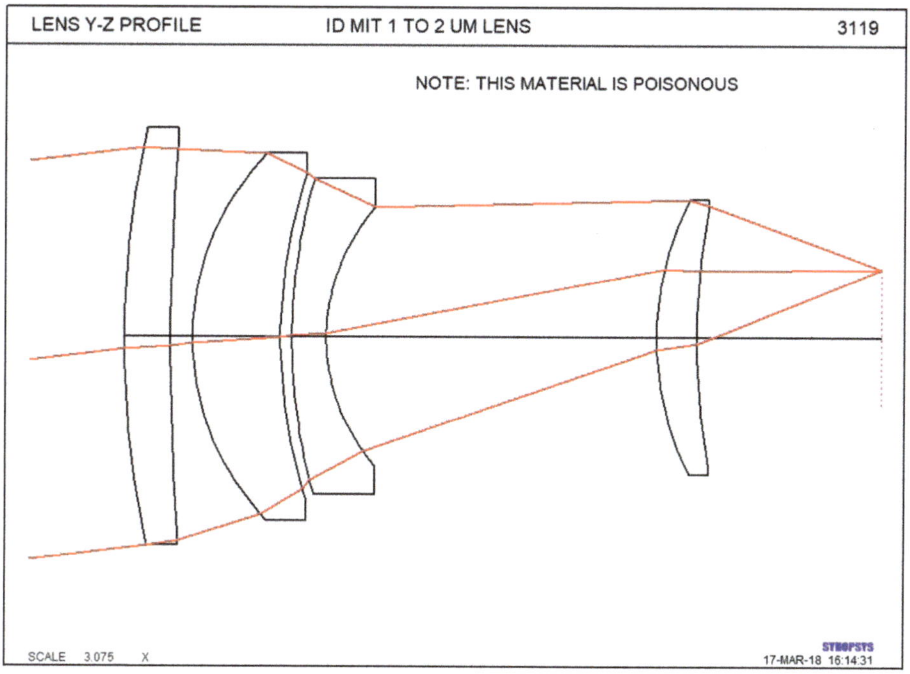

Figure 32.4. DWG drawing with added annotation.

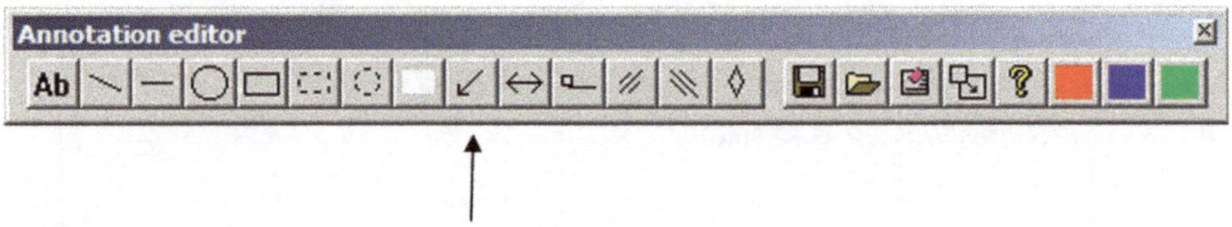

Figure 32.5. Selecting the 'Arrow' option.

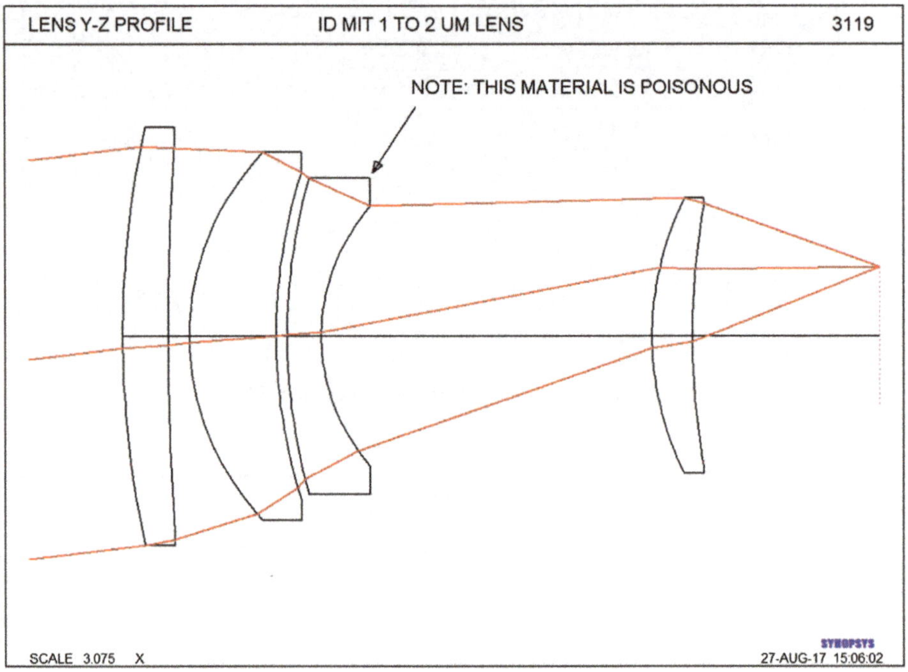

Figure 32.6. DWG drawing with added arrow and line of text.

The program has made a tolerance budget for the lens, and the ELD command makes a drawing for element 2.

Click the 'ELD' button. The drawing shows up, complete with the tolerances for that element produced by BTOL, as shown in figure 32.13. That is the function of the **USE BTOL** command.

All of the tolerances are added by the program as *annotation* rather than graphics text, so if you want to change or customize anything, you can do it with the Edit button. The drawing at the moment does not specify the surface finish or coating, but you can add those data with the annotation editor as well. Finish is usually specified by a scratch–dig specification, such as 60–40, which is a standard quality good for most lenses, or 20–10, which is a very high quality used mainly for reticles, where any defect would be sharply in focus. An uncoated glass surface reflects about 4% of the incident light, so all surfaces except those to be cemented to another element are routinely coated to minimize unwanted reflections. For a multi-

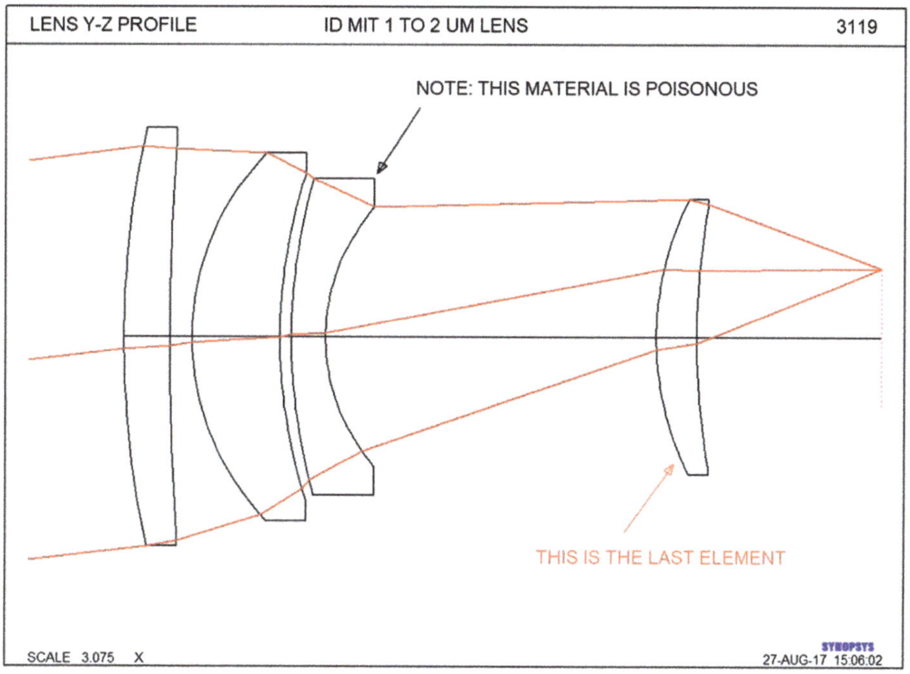

Figure 32.7. DWG drawing with added text in red.

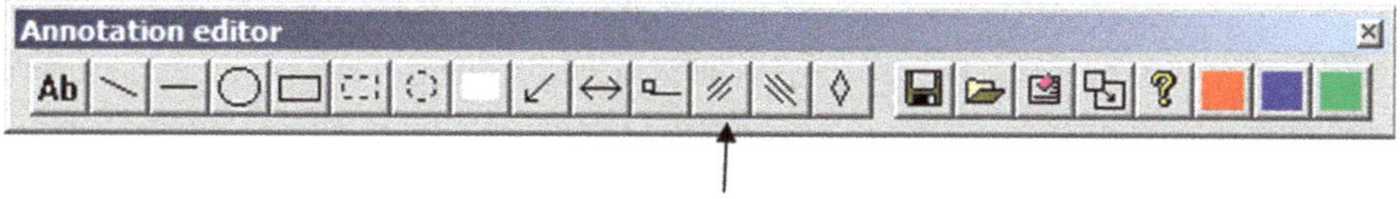

Figure 32.8. Selecting the hashmark option.

element lens, the light loss would quickly become unacceptable otherwise, and the reflected light has to go somewhere, often ending up as veiling glare at the final image. The cheapest coating is a 1/4-wave layer of MgF_2, but today one usually specifies a high-efficiency antireflection coating (HEA) and gives the wavelength range over which it should be designed. Such coatings can achieve less than 0.1% loss per surface.

Here is a nice trick: we often want to add a comment or note to every element drawing. Here is how you can type it only once. There are nine annotation strings that you can define with a command. We will define the first one (note the quotation marks; the program needs to know that those characters are not commands):

```
AS1 "GET MELT DATA FOR ALL ELEMENTS"
```

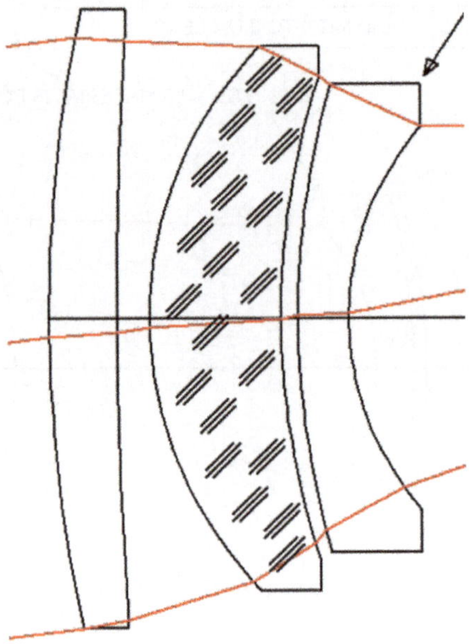

Figure 32.9. Lens element with added hashmarks.

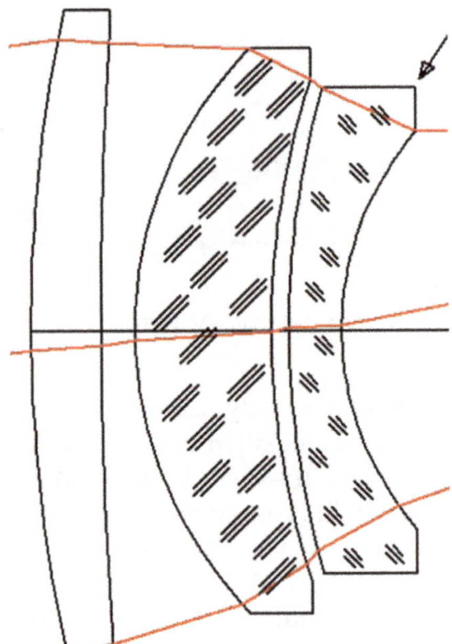

Figure 32.10. Lens elements with two kinds of hashmarks.

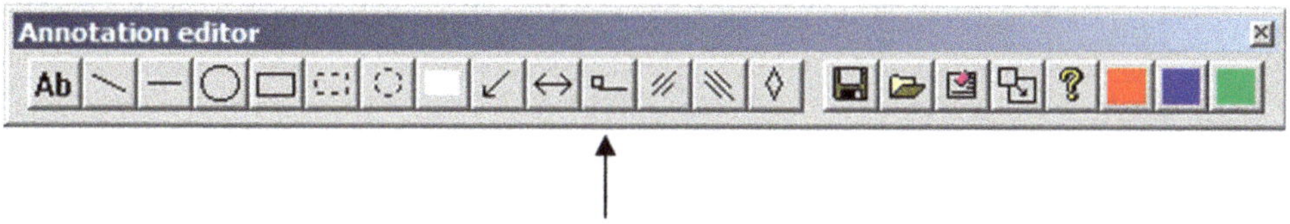

Figure 32.11. Selecting the edit handle.

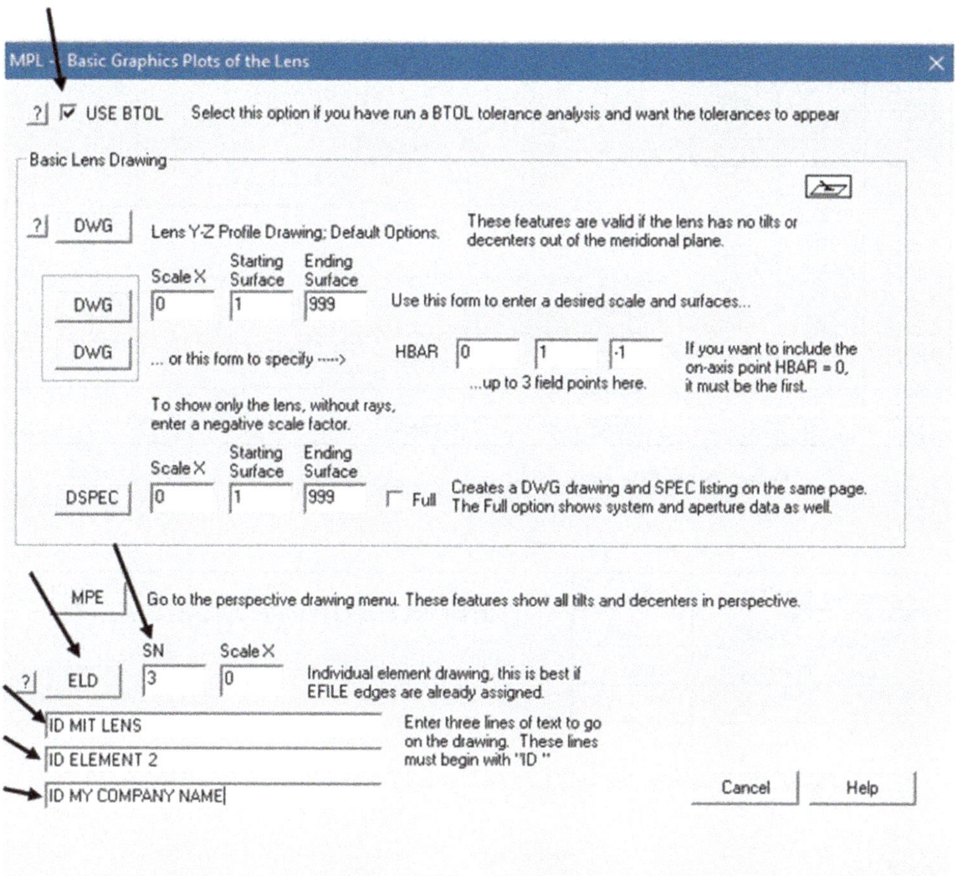

Figure 32.12. Selections in the MPL dialog to make an element drawing with tolerances.

Now, open the Annotation text editor, click in the drawing, and then click the 'AS1' button . Your string pops up in the text window. Click 'OK' and it is on the drawing, as shown in figure 32.14.

Here is another nice trick. Sometimes you would like to list the MACro that produced a particular drawing. Make a MACro as follows:

PARAMETERS	SIDE 1	SIDE 2
RADIUS OF CURVATURE	R1 23.3200	R2 46.0900
RADIUS TOLERANCE	TESTPLATE	TESTPLATE
FRINGE TOLERANCE	1.82	1.41
CYLINDER FRINGES	0.39	0.50
EDGE ROLL FRINGES	0.18	0.19
FINISH		
COATING		
CLEAR AP. DIAMETER	32.2420	32.2420
SAGITTA		S2 ± 0.0130 2.33547
DIA. TO FACE		Y2 28.9710
DIA. TO BEVEL	B1 34.2740	B2 34.2740
FACE WIDTH TO BEVEL		D2 2.6515
BEVEL WIDTH	C1 0.2540	C2 0.2540
FACE ANGLE		
THICKNESS	TH 7.6487	
TH. TOL.	0.0130	
WEDGE TOL.	1.63 MIN.	
FLAT TIR	0.0136	
DIAMETER	DIA 34.7820	
DIA. TOL.	0.0095	
MATERIAL	ZNS	
GRADE		
ANNEAL		
SLOPE	0.069 FR/MM	

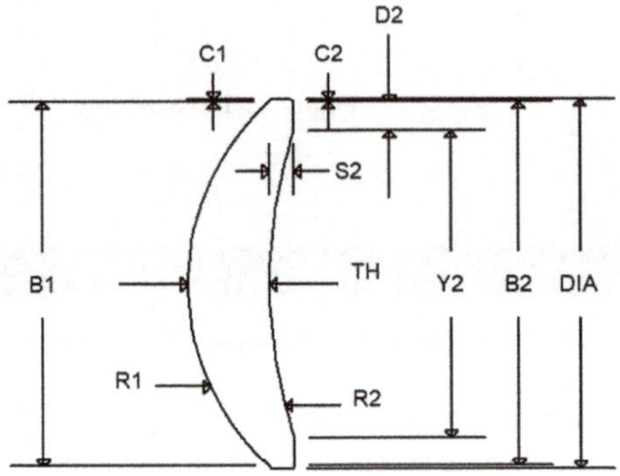

SCALE 2.000 X	NUMBER	
DATE 09-FEB-18	REV.	
DESIGNER	APPROVED	
CHECKER		
TEST WAVL		
DIMENSIONS MM		SYNOPSYS

MIT LENS
ELEMENT 2

MY COMPANY NAME

Figure 32.13. Element drawing with tolerances added automatically.

PARAMETERS	SIDE 1	SIDE 2
RADIUS OF CURVATURE	R1 23.3200	R2 46.0900
RADIUS TOLERANCE	TESTPLATE	TESTPLATE
FRINGE TOLERANCE	1.84	1.42
CYLINDER [illegible]	[illegible]	[illegible]

GET MELT DATA FOR ALL ELEMENTS

Figure 32.14. Annotation string added automatically by the annotation editor.

```
OFF 88
PER 20 30 2 1 99
PLOT
RED
RAY P
BLUE
PUP 2 1 20
TRA P 1 0 20
END
```

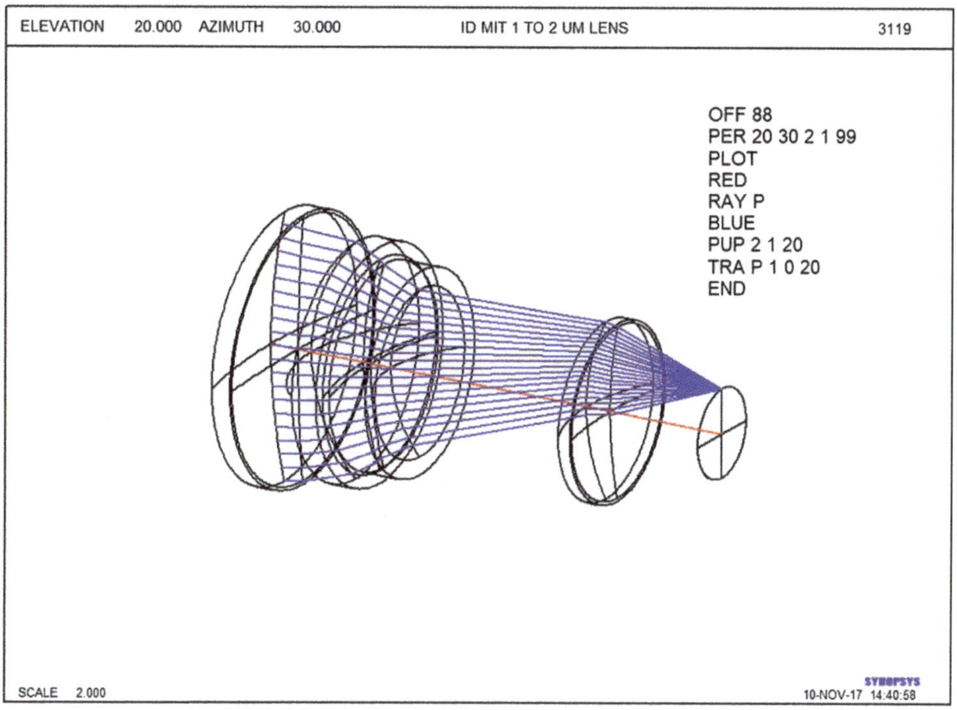

Figure 32.15. Drawing with command strings annotated.

Select the text and press <Ctrl>+C to put a copy on the clipboard. Now run the MACro, and when the picture shows up, open the Annotation Editor, select 'Text', and click in an empty space. Now paste the clipboard into the edit pane, with <Ctrl>+V, select size 14 and click 'OK'. The annotation shows up, as seen in figure 32.15.

If you save your picture (with the 'Save' button), the annotation is saved too. Once you have saved a picture, you can easily copy the annotation and put it on a new picture. Just make the new picture, and then, in the Annotation Editor, click the 'Fetch' button and select the file you wish to copy. All the annotation comes back. That button erases any annotation that was previously there, but if you want to *add* to what was there before, use the 'Copy' button instead.

IOP Publishing

Lens Design
Automatic and quasi-autonomous computational methods and techniques
Donald Dilworth

Chapter 33

Understanding Gaussian beams

Propagation of Gaussian beams; modeling with real rays

Lasers, which are frequently used as light sources for all kinds of optical systems, produce a beam that is usually very small in diameter. The intensity of such beams is nonuniform, following a Gaussian profile in the ideal case, hence the name Gaussian beam, and deviating from that profile in a characteristic manner in many real cases. There are two issues that must be accounted for when designing and analyzing systems with this kind of illumination: the shape of the profile and the fact that beams of very small diameter exhibit strong diffraction effects as they propagate.

33.1 Gaussian beams in SYNOPSYS

As with most sophisticated features, the program aims to obtain accurate results with as few complications as possible. For this reason, it addresses the peculiar properties of such beams in a novel fashion.

The main issue is the fact that, if the beam diameter is small, diffraction plays a role all along the beam. Rays going through normal lenses, on the other hand—where the beam diameter is much larger than the wavelength of the light—follow straight lines to a very good approximation, and we can then deal with 'rays' of light. That is not the case with Gaussian beams wherever the beam is small. Then the path of a light ray is *curved* and requires special attention in ray tracing.

Consider the following system (**C33L1**), shown in figure 33.1:

```
RLE
ID OBG DEMO
OBG .15 2
UNI MM
WA1 .6328
1 TH 50
```

doi:10.1088/978-0-7503-1611-8ch33

```
2 RD -2.55 TH 2 GTB S
BK7
2 CAO 2
3 CAO 2
3 RD -55 TH 100
4 RD 100 TH 2 PIN 2
5 TH 50 UMC
4 CAO 10
5 CAO 10
7
AFOC
END
```

Following the rules for this kind of beam, the object is declared type **OBG**, with a waist on surface 1 and a radius of 0.15 mm. We are concerned with rays out to a point that is twice the $1/e^2$ point, according to the third word on the **OBG** line. The marginal rays shown in the picture above originate at that point in the beam. The example also includes two simple lenses, to expand and then recollimate the beam.

If we took the beam at surface 1 to be exactly collimated, the ray intercept on surface 2 would be the same as on surface 1. But that would be incorrect, since diffraction enlarges the beam by the time it gets there. To account for this effect, the program considers the beam at the waist to be slightly curved, just enough so that a *real ray*, traced from surface 1, hits surface 2 at about the same place as the diffracted Gaussian beam, and with the same divergence angle. From that point, we can treat a diffracting beam with the usual ray tracing methods, provided diffraction thereafter is minimal. How accurate is this trick?

Ask for a beam trace, which evaluates the beam everywhere according to paraxial Gaussian beam theory:

```
SYNOPSYS AI>BEAM

ID OBG DEMO                                33262          13-MAY-13   14:16:08

GAUSSIAN BEAM ANALYSIS
```

SURF	BEAM RADIUS	WAIST LOCATION	WAIST RADIUS	DIVERGENCE
1	0.150000	-7.5157030E-15	0.150000	0.001343
2	0.164341	-7.368983	0.005965	0.022287
3	0.208892	-6.563589	0.006332	0.031811
4	3.389933	-357.899054	0.014036	0.009472
5	3.408876	-2087.561971	3.406641	5.9127598E-05
6	3.408985	-2137.561971	3.406641	5.9127598E-05
7	3.408985	-2137.561971	3.406641	5.9127598E-05

Figure 33.1. Laser system using a Gaussian beam.

Notice that the beam radius on surface 2 is larger than on surface 1 because of diffraction. Now trace a real ray at pupil point (0, .5), which is at the $1/e^2$ point:

```
SYNOPSYS AI>RAY P 0 0 .5 SURF

INDIVIDUAL RAYTRACE ANALYSIS

FRACT. OBJECT HEIGHT                HBAR       0.000000   GBAR       0.000000
FRACT. ENTRANCE PUPIL COORD.         YEN       0.500000    XEN       0.000000
COLOR NUMBER                           1

                            RAY VECTORS          (X DIR TAN)  (Y DIR TAN)   (INC. ANG.)
SURF          X              Y            Z           ZZ           HH            UNI
-------------------------------------------------------------------------------------
 OBJ      0.000000       0.000000     0.000000     0.000000      0.000549
  1       0.000000       0.136910     0.000000     0.000000      0.000549      0.031434
  2       0.000000       0.164338    -0.005301     0.000000      0.022307      3.663636
  3       0.000000       0.209062    -0.000397     0.000000      0.031846      1.060103
  4       0.000000       3.395560     0.057666     0.000000      0.009449      3.769940
  5       0.000000       3.413463    -0.047616     0.000000 -5.576629E-05      1.057009
  6       0.000000       3.410672     0.000000     0.000000 -5.576629E-05      0.003195

          REDUCED RAY ANGLES IN RADIANS AT IMAGE SURFACE
           PSI (X)          PHI (Y)              Z
          0.000000 -5.576629E-05         0.000000
```

The path of this real ray closely approximates the **BEAM** trace. We now have a tool that lets you analyze and optimize such a system using real rays. As long as the beam is expanded early on in the system (so diffraction has little effect thereafter) this real-ray approximation is useful and simple to set up.

33.2 Complications

But there are sometimes complications. Suppose, for example that there is an element right at the waist. If thickness number 1 is zero, or if that surface is not a dummy, the program cannot make the adjustments described above. Instead, it then adjusts the geometry so that it can trace an OBA object (finite object distance) where

```
THO  = 1.0E14
YPO  = THO * DIV
YMP1  = WAIST * RBS
YP1  = 0.0
```

Thus, the object is then at infinity and the entrance pupil radius is a function of the input OBG waist. In this situation, the program can still run the BEAM analysis, but diffraction is not taken into account with real rays, as it was before. If the first element expands the beam, however, then diffraction plays little role anyway and this is still a useful approach.

But what if there are one or more surfaces or elements in the beam where it is still very small? Suppose an expander is located, say one meter from the waist, and there are several fold mirrors along the way. The trick first described only operates between surfaces 1 and 2, and diffraction between other surfaces would be neglected

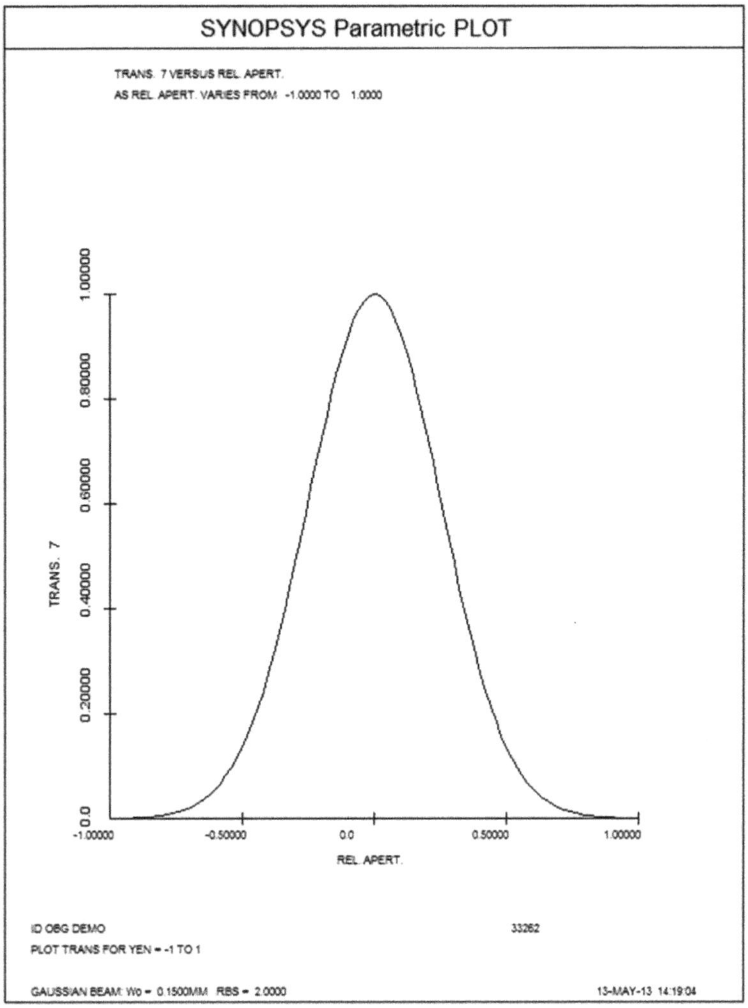

Figure 33.2. Intensity profile of a Gaussian beam.

in that case but should not in this one. Fortunately, there is another trick, and it is very simple.

What you do is assign a thickness of one meter to surface 1 (or whatever the distance is to the expander), put a dummy surface 2 at that distance, and then assign a thickness of *minus* a meter (or whatever is needed to get back to the first element or mirror) to surface 2. Now the program can adjust the beam properties at the waist so that diffraction is taken into account at the dummy surface 2. If you trace a real ray, it will hit surface 2 at the same place as the Gaussian beam does, and the path, once the beam actually reaches the expander, will be correct thereafter.

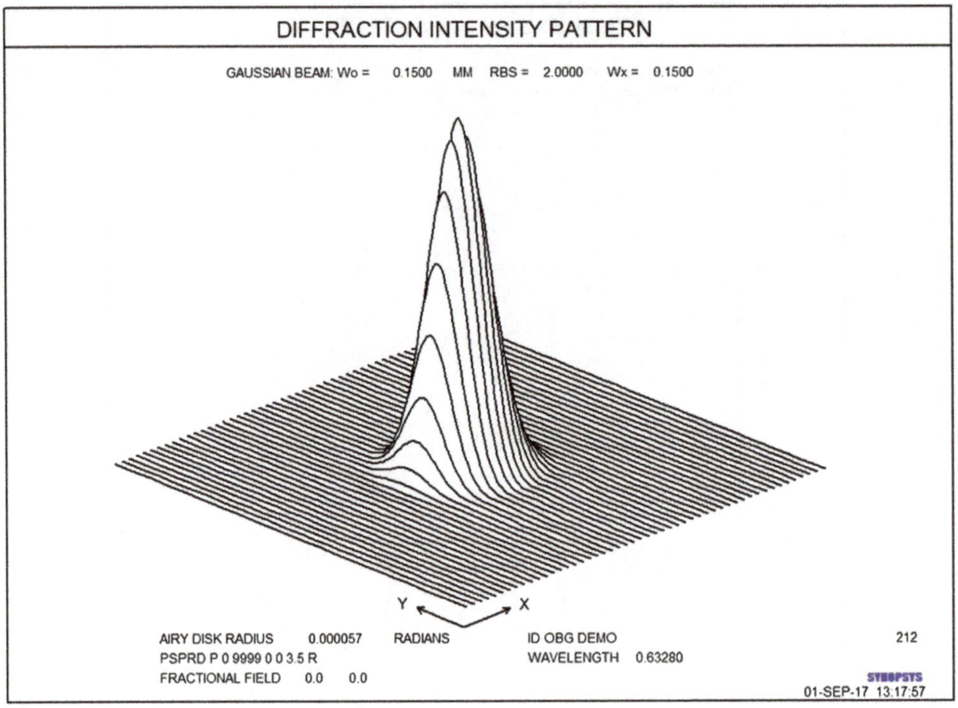

Figure 33.3. Diffraction pattern of Gaussian beam.

33.3 Beam profile

Let us look at that Gaussian profile. Type the following AI sentence to obtain the Gaussian profile shown in figure 33.2:

```
STEPS = 100
PLOT TRANS FOR YEN = -1 TO 1
```

This shows a beautiful Gaussian shape. There are other ways to see the shape too. Chapter 15 showed how to make a MACro using the COMPOSITE aberration format to plot the profile, and how to design a simple system to expand the beam and yield uniform intensity at the same time. It also showed how the diffraction-propagation program DPROP can analyze the improved energy distribution, giving yet another way to analyze this kind of beam.

33.4 Effect on image

To complete this lesson, let us make a diffraction pattern of the output. Since the beam is Gaussian, the far-field image is also Gaussian in shape. Go to the MDI dialog (**MDI**), ask for a PSPRD plot, shown in figure 33.3, and specify 9999 rays. (The energy is all concentrated near the center of the beam and analyzing the image with fewer rays will be less accurate.)

Indeed, we see there are no diffraction rings at all. That is a property of Gaussian beams. Diffraction takes place mainly around the edge of a beam, and if that edge is very fuzzy, having fallen off to a much lower value than at the center, then diffraction at the edge plays no role.

To read about other subtleties regarding Gaussian beams, including noncircular beams and the effects of beam quality, type **HELP OBG** in the Command Window.

Chapter 34

The superachromat

Correcting chromatic aberration with Herzberger's theory

This chapter will explore a unique feature of SYNOPSYS that can be helpful when you need exceptional color correction, better even than an apochromat. Chapter 12 showed how to select three glass types that make it possible to correct axial color at three wavelengths. For many tasks, that is as good as you will need.

However, this is not always the case. Suppose you are designing a lens to be used over the range 0.4–1.0 μm. Can you do it with an apochromat? Let us find out. Here is the RLE file for a starting system (**C34L1**), where all surfaces are flat except for the last, which will give you an F/8 telescope objective of 6 inch aperture:

```
RLE
ID WIDE SPECTRAL RANGE EXAMPLE
OBB 0 .25 3
UNITS INCH
1 GLM 1.6 50
3 GLM 1.6 50
5 GLM 1.6 50
6 UMC -0.0625 YMT
7
1 TH .6
2 TH .1
3 TH .6
4 TH .1
5 TH .6
END
```

This file did not specify the wavelengths yet, so you obtain the default CdF lines. We have to change this. Open the Spectrum Wizard (`MSW`), and change the data indicated in figure 34.1.

doi:10.1088/978-0-7503-1611-8ch34

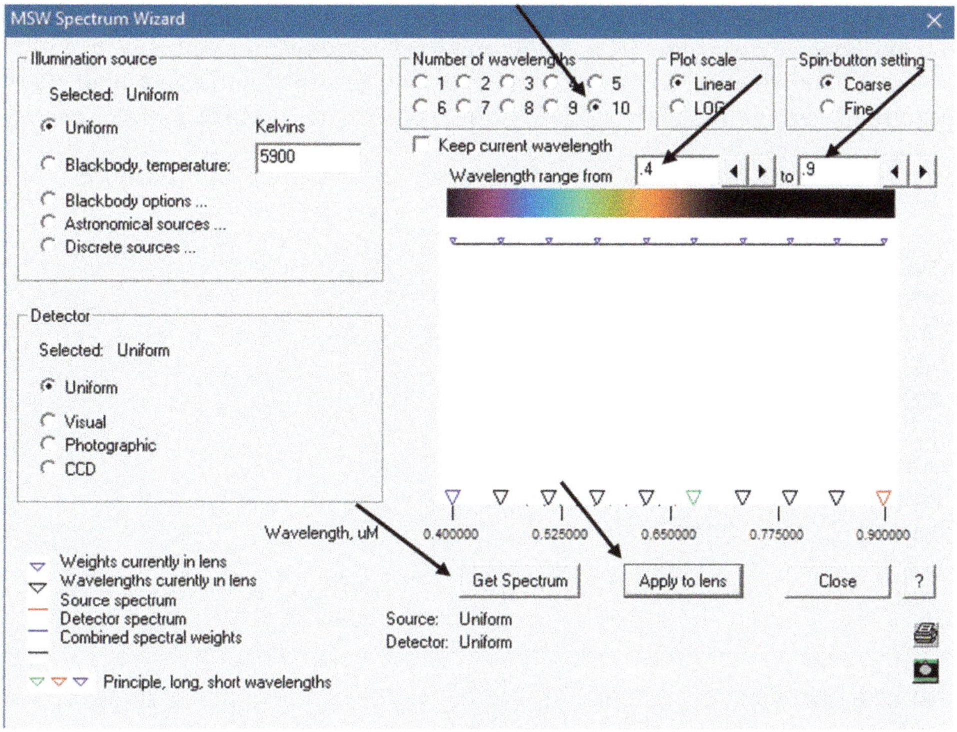

Figure 34.1. Spectrum Wizard with ten wavelengths selected over the range 0.4–0.9 µm.

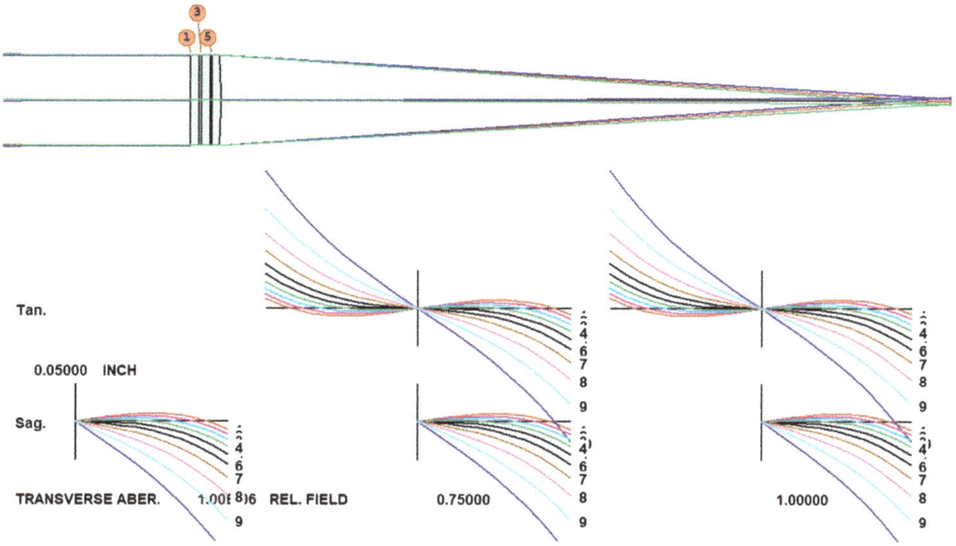

Figure 34.2. Starting lens with ten wavelengths defined.

After clicking the '**Get Spectrum**' button, click the '**Apply to lens**' button. The lens now has a wider spectrum. Here is the starting lens, shown in figure 34.2.

Yes, it is terrible (of course: all curves are flat except for the last one). Let us optimize it, varying the glass models, radii, and airspaces. Make a MACro:

```
LOG
STO 9

PANT
VLIST RAD 1 2 3 4 5
VLIST TH ALL
VLIST GLM ALL
END

AANT
ACM .5 1 1
LUL 5 1 1 A TOTL

END

SNAP
SYNOPSYS 50
```

Now put the cursor on the blank line in the AANT section and click the  button. Merit function number 6 is selected by default, so just click the 'Back to MACro editor' button. This gives you a simple merit function, and the requirement on TOTL is there because the airspaces will vary and you want the lens to remain a compact group. Modify the AEC monitor so lens edges stay above 0.1 inches in thickness:

```
...
AANT
ACM .5 1 1
LUL 5 1 1 A TOTL
AEC .1 .1 1
ACC
GSR .5 10 5 M 0
GNR .5 2 3 M .7
GNR .5 1 3 M 1

END
...
```

Here, you correct all ten colors via the **M** in the AANT file. Now it is time to optimize. Run the MACro and then open the annealing dialog. In this case, select the '**Free GLM**' option in the dialog, because the glass models are likely to become pinned to either the crown or flint boundary of the glass map almost immediately, and we want them to be free to leave the boundary later on as the design form changes. This option is generally not advisable if the lens already has a reasonable construction at the start, but that is not the case here. Select a temperature of 50, cooling 2, 50 passes. The lens is vastly better, as shown in figure 34.3.

How good is this state of correction? We can ask AI to show us the defocus over wavelength—but that would be unwise at the moment. This lens has a curvature solve and at each wavelength the program would recalculate it. So instead, we make a second MACro, as follows:

```
STORE 9
STEPS = 50
CHG
NOP
END
PLOT DELF FOR WAVL = .365 TO 0.9
GET 9
```

This file removes all of the solves (and pickups, if there are any) via the **NOP** entry, and then plots the defocus. Afterwards, it gets back the lens the way it was. Here is the color correction curve, in figure 34.4.

This lens already looks like a 'superachromat', a term coined by Herzberger and McClure in 1963. Their theory[1] says that, if you make a graph of the glass catalog where the axes are the values of P^* and P^{**}, and then select three glasses that lie on a straight line, it is possible to correct at *four* wavelengths at the same time. The term P^* refers to the partial dispersion $(NF - N^*)/(NF - NC)$, where F and C are the

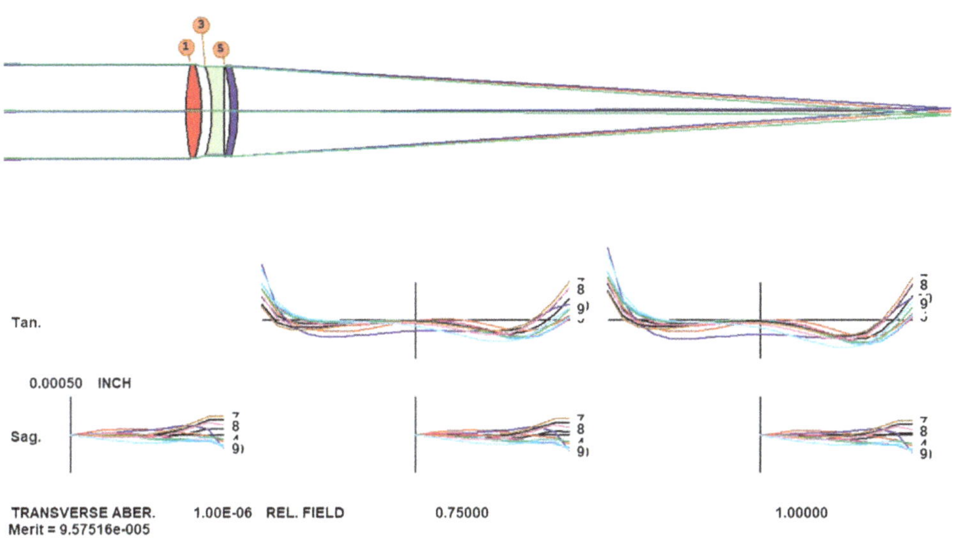

Figure 34.3. Lens optimized at ten wavelengths with glass model variables.

[1] Herzberger M and McClure N 1963 The design of superachromatic lenses *Appl. Opt.* **2** 553–60.

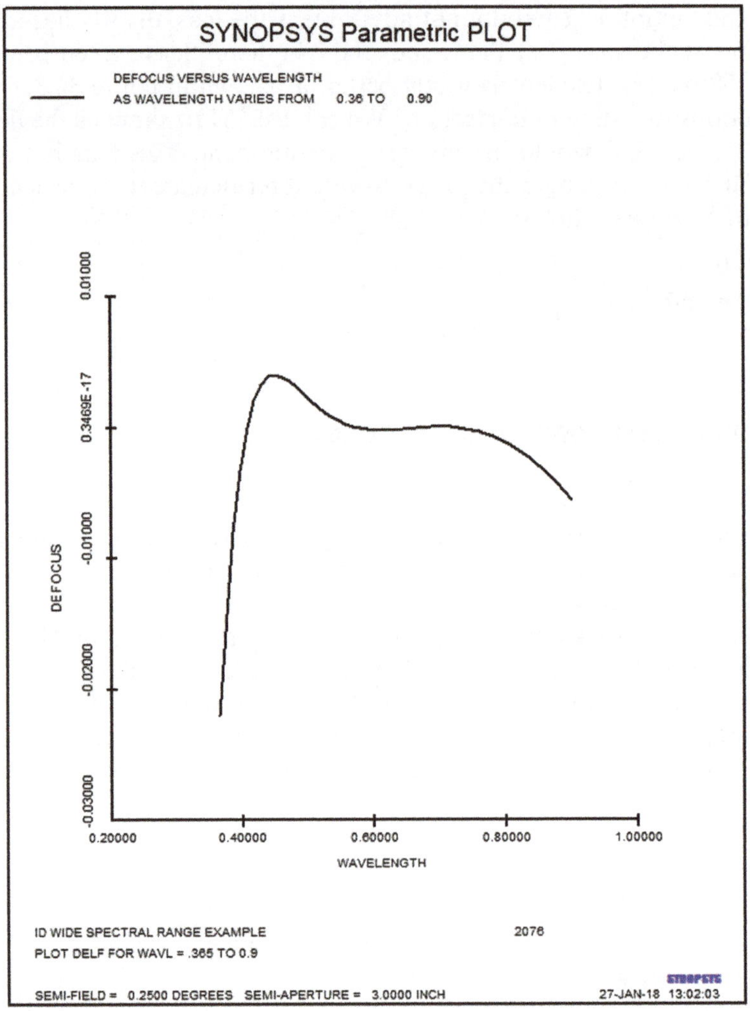

Figure 34.4. Color correction curve for the reoptimized lens.

Fraunhofer lines at 0.4861 and 0.6563 μm, and N^* is the IR line at 1.014 μm. N^{**} is the UV line at 0.365 μm, giving you a similar equation for P^{**}. In this case, the program found a great combination of glass models automatically.

Now it is time to learn about making a superachromat yourself, for cases when the program does not find one automatically.

First, we show how to find suitable glass combinations by hand, using the glassmap feature of SYNOPSYS. Then we will show another way the program can perform the task automatically, which is a real time-saver.

The on-screen glassmap of SYNOPSYS can show just the kind of plot you need. Type **MGT** to open the Glass Table Selection dialog and select the O (Ohara) catalog, and when the map is displayed, click the 'Graph' button and select the bottom option (as shown in figure 34.5) to see the display shown in figure 34.6.

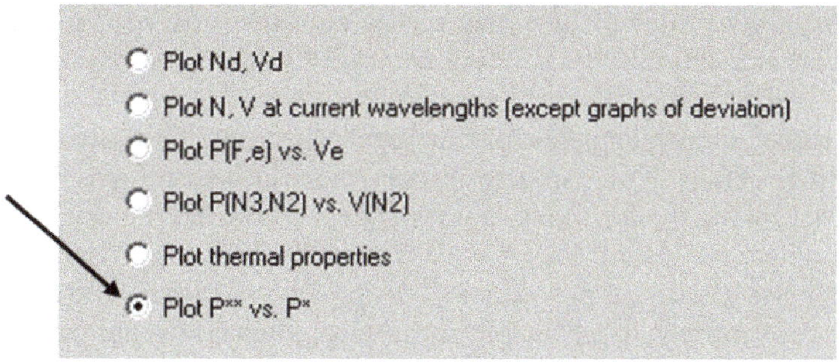

Figure 34.5. Selecting the graph of P^{**} versus P^{*} on the glass map.

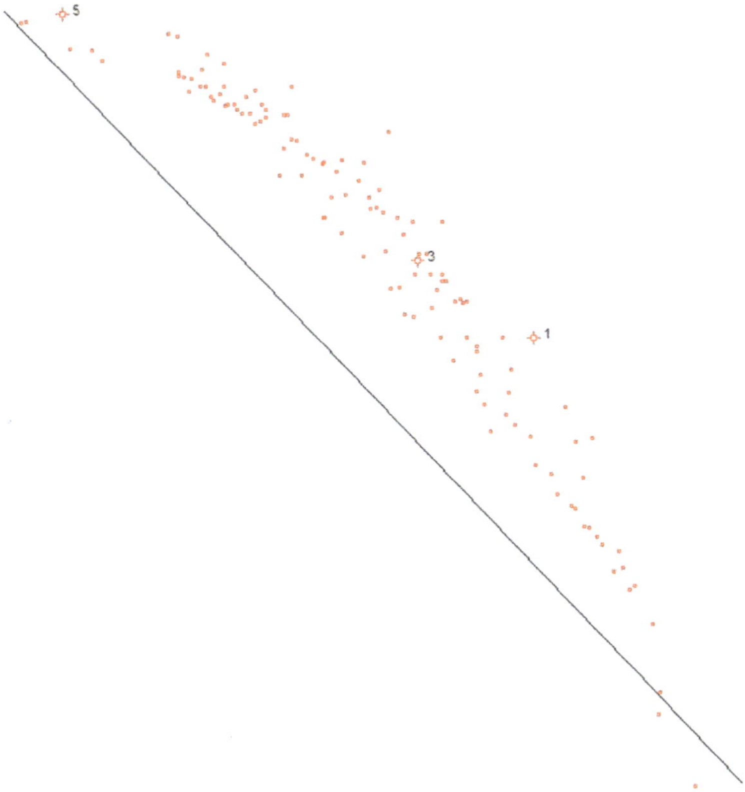

Figure 34.6. Glass map display showing plot of P^{**} versus P^{*} for the Ohara glass catalog.

On this graph, you see the current location of the model for each element (the red circles). They are lined up quite well, but the line between them is quite short. What you have to do is to adjust the line so it connects three glass types, preferably with a line as long as practical. Select a glass near the bottom, where the flints tend to be, and <Ctrl>+click one of them. That puts the bottom of the black line on that glass and displays the glass name in the 'Glass' box. Then select a glass near the top of the distribution and <Shift>+click that one, to put the top of the line there. Now select a third glass near the center of that line and as close to it as you can find. Click the symbol so you can see that glass name too. Write down the names of those three glasses. Our selection is shown in figure 34.7.

We have three potential glasses for our superachromat. They are types S-PHM52, S-NPH5, and S-TIL27. You can also display the relative cost and other properties, to help you select three acceptable glasses. Then you insert those three glasses into the lens and optimize. If that does not yield a satisfactory lens, you select a different three according to the same procedure. This process is rather tedious but works.

The other procedure is to let the program select glass combinations for you. Type, in the CW,

```
FST
PREF
CAT O
CAT S
GO
```

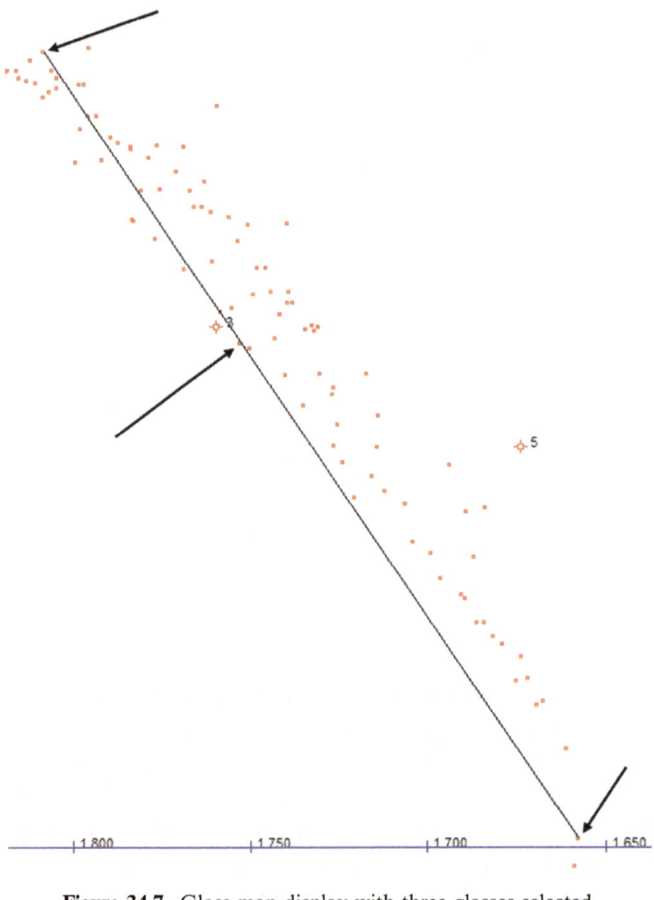

Figure 34.7. Glass map display with three glasses selected.

FST means Find Superachromat Triplets. This input will examine all combinations of glass types from the Ohara and Schott catalogs and rate the ten most suitable for a superachromat. The program finds the following:

```
SYNOPSYS AI>FST
FST>PREF
FST>CAT O
FST>CAT S
FST>GO
SUPERACHROMAT GLASS SEARCH RESULTS (LOWER SCORES ARE BETTER)
     SCORE       UPPER        MIDDLE      LOWER        OFFSET
1    0.02026308  S N-FK58     S N-SSK8    S SF4        0.00000296
2    0.02008505  O S-FPL53    O S-LAL8    O S-NPH1     0.00000923
3    0.01810522  O S-FPL53    O S-BAL42   O S-NBH53    0.00000154
4    0.02120605  O S-FPL53    O S-LAL13   O S-TIM28    0.00000424
5    0.02082027  O S-FPL55    S N-KF9     S SF10       0.00000567
6    0.02008505  O S-FPL53    O S-LAL8    O S-NPH1W    0.00000923
7    0.01881642  O S-FPL55    O S-TIL27   O S-TIH23    0.00000071
8    0.02171385  O S-FPL55    O S-LAM54   S SF57       0.00000914
9    0.02139100  O S-FPL53    S N-SK4     S SF56A      0.00000909
10   0.02147608  O S-FPL55    S N-SSK8    S SF1        0.00000460
```

This method is superior to doing it by hand since it can combine glasses from different manufacturers. Combination number 10, for example, is made from one Ohara glass and two from Schott. Let us try that combination. Go back to the original lens, with ten wavelengths selected. Edit the optimization MACro as shown below (**C34M1**; here, we used ready-made merit function number 8, which corrects a combination of transverse and OPD aberrations, and then adjusted the weights):

```
LOG
STO 9

CHG
1 GTB O 'S-FPL55'
3 GTB S 'N-SSK8'
5 GTB S 'SF1'
END

PANT
VLIST RAD 1 2 3 4 5
VLIST TH ALL
END

AANT
ACM .5 1 .1
LUL 5 1 1 A TOTL
```

```
AEC .1 1 1
ACC
GSR .5 10 5 M 0
GNR .5 5 3 M .7
GNR .5 4 3 M 1
GSO 0 0.003916 5 M 0
GNO 0 0.003 3 M .7
GNO 0 0.002 3 M 1

END

SNAP
SYNOPSYS 90
```

After running this and then annealing (**50, 2, 50**), you obtain a lens corrected to about 1/10 wave on-axis and 1/2 wave at full field, although color 10 (at 0.4 μm) is not as well corrected as the others, as shown in figure 34.8.

However, we only guessed the order of our three glasses. There are six possible combinations, and next we will try the order 5, 1, 3. The result is shown in figure 34.9, after annealing. This is not good at all. What happened? It looks like the lens is stuck in a poor solution region; the first two elements are bent to the left, and we suspect that they really should be bent the other way. The program sometimes cannot get out of this kind of local minimum without some guidance—but there is a tool for this too. Type, in the CW,

```
BFO 3
```

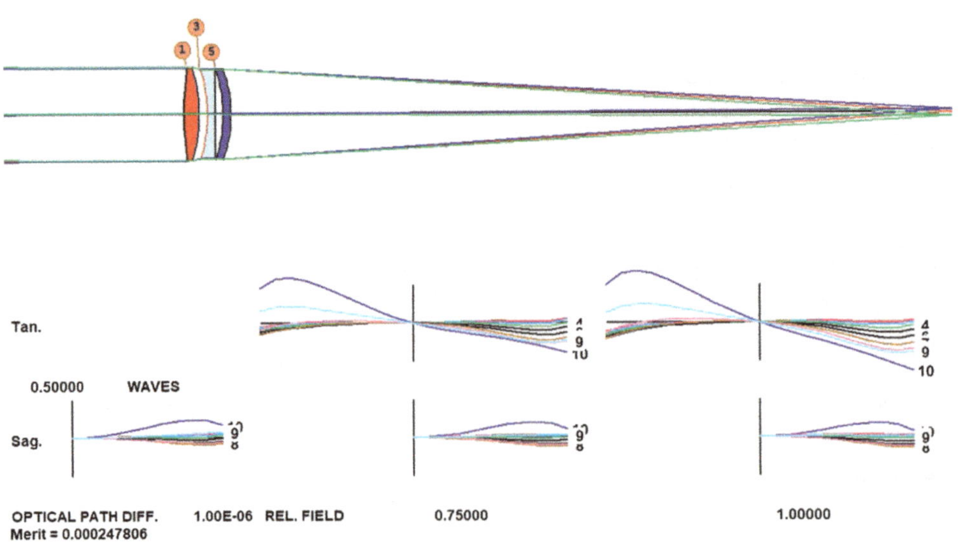

Figure 34.8. Lens optimized with three glasses found by FST.

This runs the **B**end-**F**lip **O**ptimization program, which forces the bending of the element at surface 3 to flip the other way.

The lens is improved, so anneal again, and now it is superb, as shown in figure 34.10.

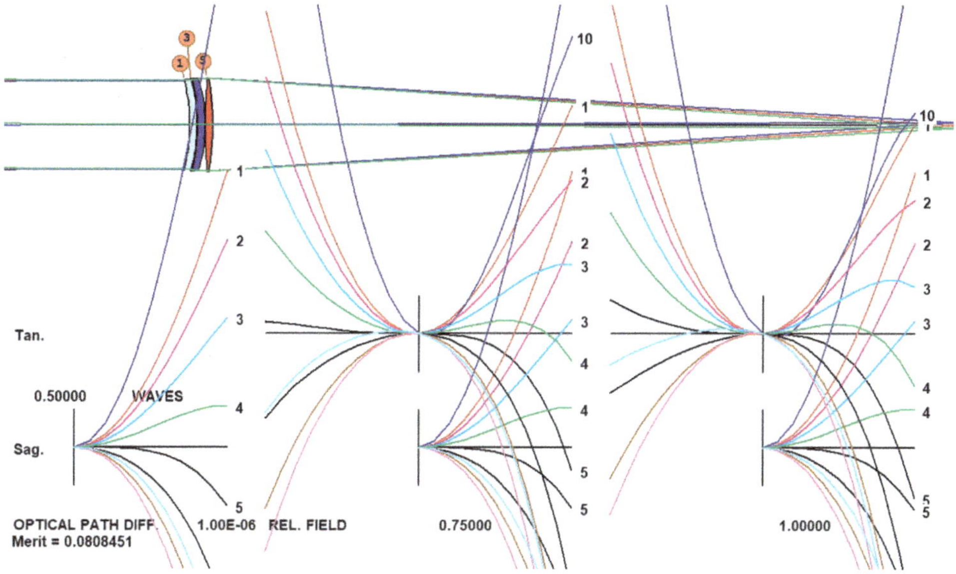

Figure 34.9. Lens optimized with a different order of glasses from FST. The lens appears to be stuck in a local minimum.

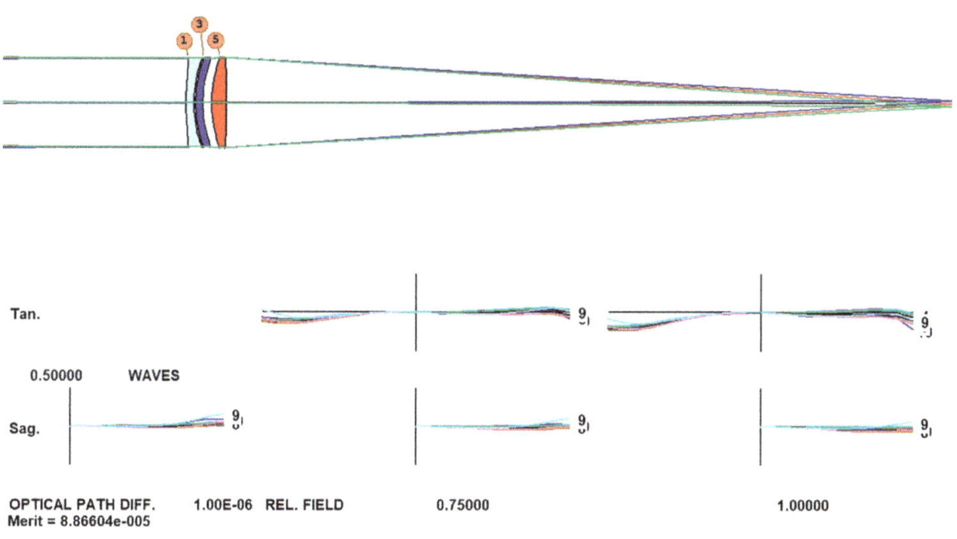

Figure 34.10. Lens improved with BFO, then annealed.

Now the lens **(C34L2)** is corrected to about a quarter wave over the entire (very wide) spectral region. What does the second MACro show now? The curve is in figure 34.11.

Well, it is corrected at three wavelengths for sure—but we were aiming for four. How come the curve does not go up again at the right end—as a true superachromat would? It is simple: as usual, the program is balancing everything in the merit function, not just axial color, and the other aberrations make it depart slightly. Still, this is a great lens and, if you look at the OPD fans of the first lens, where the program found a superachromat by itself (with model glasses), you see that the OPD errors in that design are slightly *larger*. The beautiful plot for that lens showed the *paraxial* focus, and when you consider real rays over the whole pupil it is not

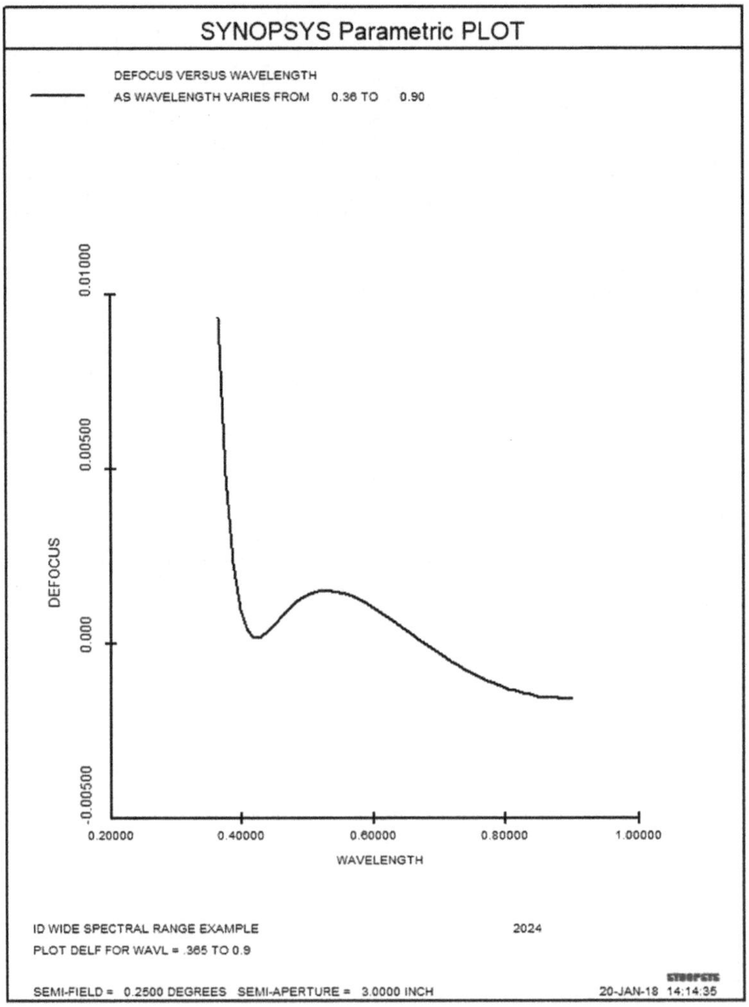

Figure 34.11. Color correction of the final lens.

necessarily the best solution. Here is another example of when classic theory may not tell you all you need to know.

The next chapter will design an even more demanding superachromat, and chapter 47 shows how DSEARCH and GSEARCH together can come up with great glass combinations that correct secondary color.

Chapter 35

Wide-band superachromat microscope objective

Challenges correcting a very wide spectral range; polarization; vector diffraction

In this chapter, we will undertake a rather difficult lens design task that will utilize some of the many powerful tools that you have learned about in previous chapters. (You will need a license to run this example, since it requires more than 12 surfaces and requires saving lens files.) As you read each of the paragraphs below, be sure to look up any topics you are not yet familiar with in the help file so you understand what the arguments mean and what other possibilities you have available.

This lens must work over the wavelength range of 0.38–0.9 μm—which is a challenge right off the bat. In addition, we want the lens to work at a speed of F/0.714. That is also not too easy. Here are the requirements:

1. Object at infinity, 0.8 degree semi-field, 1.26 mm semi-aperture.
2. Spectral range 0.38–0.9 μm.
3. F/number 0.714.
4. Total track length not more than 25 mm.
5. Good distortion correction.
6. Telecentric at image.
7. No feathered edges, center thicknesses not over 8 mm.

doi:10.1088/978-0-7503-1611-8ch35

We guess that this job will require perhaps ten elements, but want to get there gradually. Set up the input for DSEARCH, asking for eight elements, as shown below. That will give you some potential configurations, and you can increase the complexity as needed once you see how things are going. Since the spectral range is so wide, specify five wavelengths instead of the usual three in order to avoid large focus errors at in-between wavelengths. Here is the MACro (**C35M1**):

```
CORE 14
TIME
DSEARCH 3 QUIET

SYSTEM
ID EXAMPLE WIDE-SPECTRUM FAST LENS
UNI MM
OBB 0 0.8 1.26
WA1 0.9 0.77 0.64 0.51 0.38
CORDER 3 1 5
END

GOALS
ELEMENTS 8
FNUM 0.7143 100
BACK 0 0
TOTL 0 0
STOP FREE
COLORS M
RSTART 400
THSTART 1
ASTART 1
RT 0.0
OPD
PASSES 50
QUICK 70 70
ANNEAL 200 20 Q 70
END

SPECIAL PANT
SLIMIT 100 0.1          ! SMALL ELEMENTS; CAN BE CLOSE TOGETHER
END

SPECIAL AANT
AEC .1 1 .05            ! edge monitor
ACM .1 1 .05            ! minimum element TH
ACC 8 1 0.5                     ! maximum TH
ACA 70 1 1                      ! avoid critical-angle refraction

LUL 25 1 1 A TOTL               ! limit track length
A BACK
M 0.5 1 A BACK          ! want image clearance of 0.5 mm
M 0 1 A P YA 1          ! control distortion
```

```
S GIHT
M 0 1 A P HH 1          ! and make telecentric
END

GO
TIME
```

Notice the use of a weighting factor on the FNUM request in word 3. This has subtle consequences: if it were left out, the program would satisfy the request exactly, with a UMC solve on the last radius. However, with a lens with a very low F/number like this one, that is likely to yield a very short radius of curvature on that surface and produce ray failures when real rays are traced. So in cases like this, it is better to enter a weighting factor. Then the radius becomes an ordinary variable and the F/number is controlled via an entry in the MF.

Run this file, and in two minutes DSEARCH returns a set of potential starting points. It also creates an optimization MACro, and after running it and then annealing (**50, 2, 50**), you obtain the design shown in figure 35.1. (If you turned on the **Free GLM** option in the previous chapter, be sure to turn it off for this one; the example below was prepared with that option off.)

Since color correction is going to be a challenge, the next step is to find some glasses that have the potential to make a superachromat. We will do this in two ways: first using the superachromat theory, and then by letting GSEARCH find glass combinations automatically. Save this version so you can go back to it later:

```
STORE 1
```

Chapter 34 explained the theory of superachromats; for now, open the glass map with the command **MGT**, select the Schott catalog, click the 'Graph' button, and select the bottom option, '**plot P* vs. P****'. We need three glasses that lie on a long line. <Ctrl>+click the glass P-SF68, which defines the bottom of the line, and then <Shift>+click the glass N-PK52A, defining the top, as shown in figure 35.2.

See the glass N-F2? It is near the center of the line. That gives us three types, but we do not know which glass to assign to which element yet. Never mind; GSEARCH can tell us.

Next, create two files. The first is a normal optimization file. Using the MACro that DSEARCH has created, just edit it a little: request that the optimization program run the automatic ray-failure fixing routine (**C35M2**) if any of the combinations will not trace initially (and well they might; large changes to the index of refraction send rays in a different direction, which can cause failures):

```
PANT
SLIMIT 100 0.1   ! SMALL ELEMENTS; CAN BE CLOSE TOGETHER
VY 0 YP1
VLIST RD ALL
```

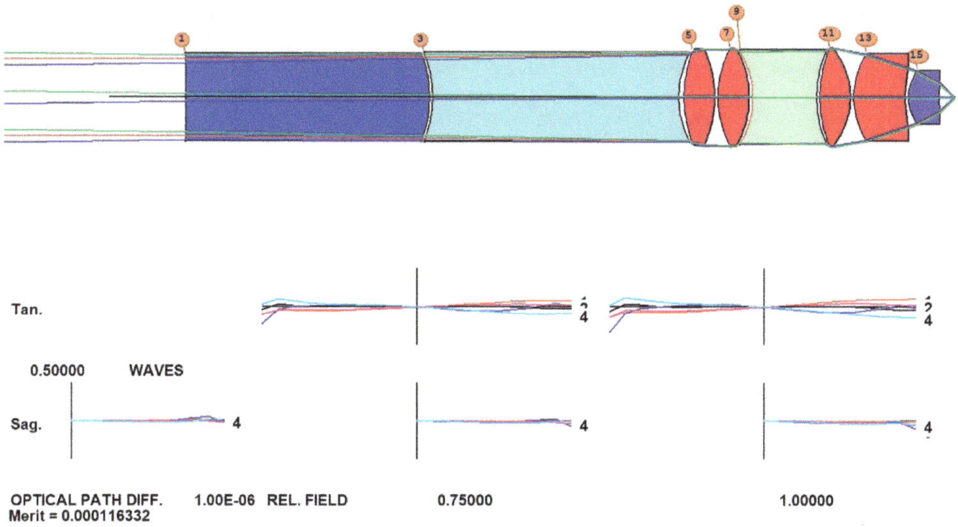

Figure 35.1. Lens returned by DSEARCH, optimized and annealed, with model glasses.

```
VLIST TH ALL
VLIST GLM ALL
END
AANT P
AEC
ACC
M    0.139997E+01  0.100000E+03 A CONST 1.0 / DIV FNUM
GSO      0.000000      0.039182       4   M      0.000000
GNO      0.000000      0.023509       4   M      0.750000
GNO      0.000000      0.007836       4   M      1.000000
AEC .1 1 .05   ! EDGE MONITOR
ACM .1 1 .05   ! MINIMUM ELEMENT TH
ACC 8 1 0.5    ! MAXIMUM TH
ACA 70 1 1    ! AVOID CRITICAL-ANGLE REFRACTION
LUL 25 1 1 A TOTL  ! LIMIT TRACK LENGTH
A BACK
M 0.5 1 A BACK  ! WANT IMAGE CLEARANCE OF 0.5 MM
M 0 1 A P YA 1  ! CONTROL DISTORTION
S GIHT
M 0 1 A P HH 1  ! AND MAKE TELECENTRIC
END
SNAP/DAMP 1
SYNOPSYS  40 FIX 30
```

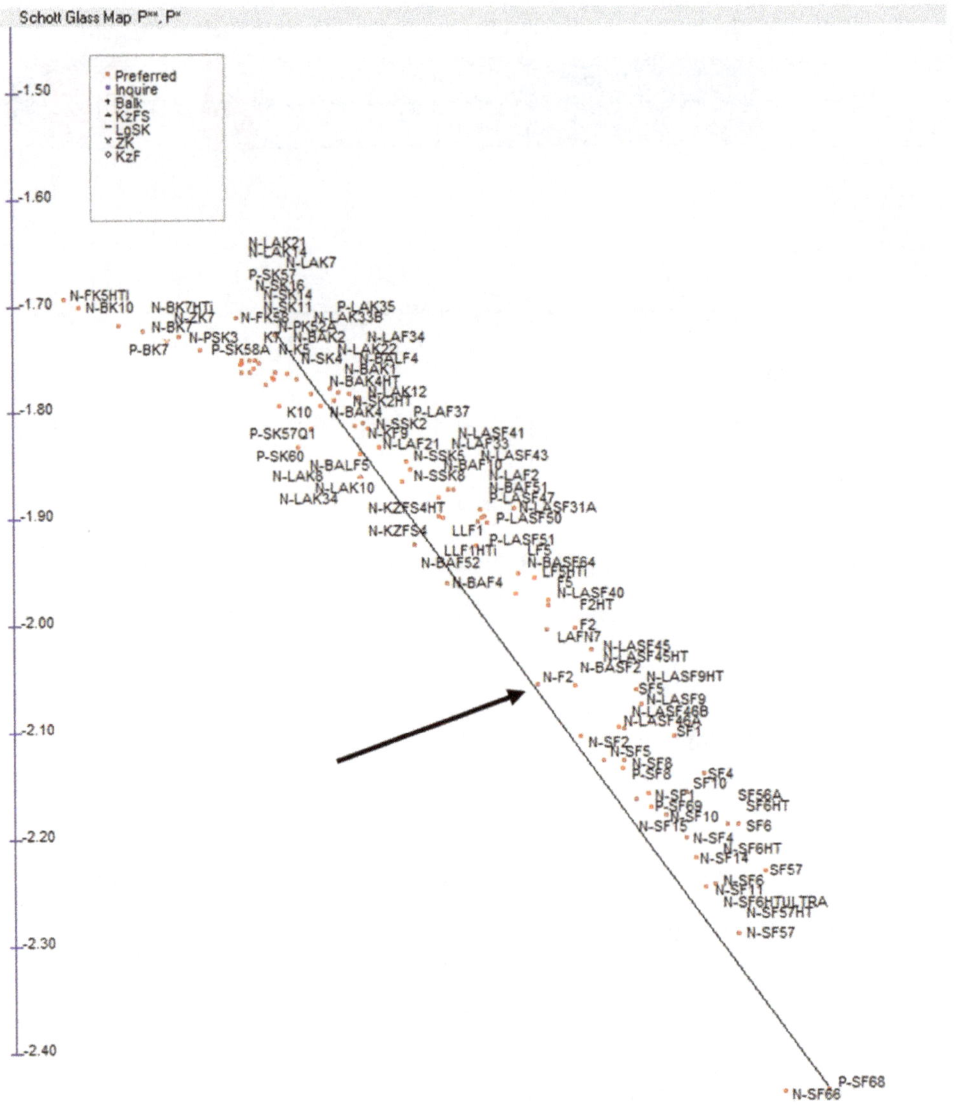

Figure 35.2. Glass map display showing *P** versus *P***.

Save this file with the name GSOPT.MAC, and then create a second MACro (**C35M3**) to tell GSEARCH what you want it to do:

```
GSEARCH 3 QUIET LOG
SURF
1 3 5 7 9 11 13 15
END
NAMES
```

```
S N-PK52A
S N-F2
S P-SF68
END
USE 3          ! only allow cases that use all three glass types
GO
```

Then run this file.

With 14 cores activated, this runs for about 15 minutes, producing the design in figure 35.3, after optimization and annealing (**C35L1**).

This is a rather good design, which is somewhat surprising since the superachromat theory applies only to thin lenses, and these elements are clearly not thin. Let us see what happens if we let GSEARCH find its own glasses. Go back to the version you saved, and then edit your MACro so GSEARCH searches combinations of the three nearest glasses from the Guangming catalog instead of the three that we picked out above (note the **SKIP** directive, which ignores input up to the **EOS** line; the USE directive will not apply when the **NEAREST** option is used):

```
GSEARCH 3 QUIET LOG
SURF
1 3 5 7 9 11 13 15
END
SKIP
NAMES
S N-PK52A
```

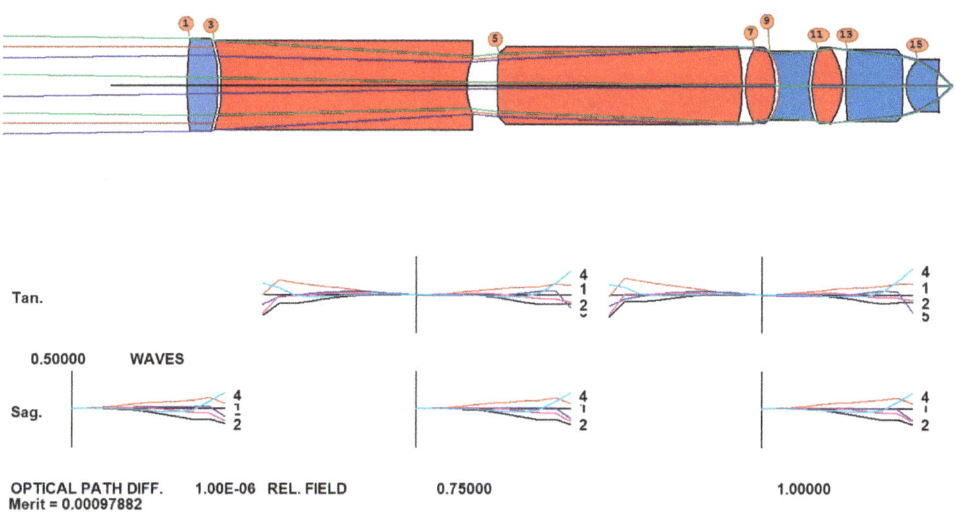

Figure 35.3. Lens returned by GSEARCH when three glass types were specified according to superachromat theory.

```
S N-F2
S P-SF68
END
EOS
NEAREST 3 P
G
END
USE 3          ! only allow cases that use all three glass types
GO
```

This version produces a better lens. Optimize and anneal again, and you obtain the result shown in figure 35.4. GSEARCH does not make any thin-lens assumptions, using numerical methods that do not depend on superachromat theory. This is another example of when autonomous methods can be as good as or better than classic techniques.

This lens is essentially perfect. However, we instinctively ask: can we do it with fewer elements? It is easy to find out with the Automatic Element Deletion feature. Go back to the version you saved, with model glasses, and add a new line at the top of the optimization MACro:

```
AED 3 QUIET 1 16.
```

Then run it again. The program detects that you can remove element 7. Accept the suggestion (which deletes that element), remove the AED line from the MACro, and reoptimize and anneal again. Then match this version to the G glass catalog (delete surface number 15 from the list, since it no longer exists), optimize and anneal, and you obtain the lens in figure 35.5 (**C35L2**).

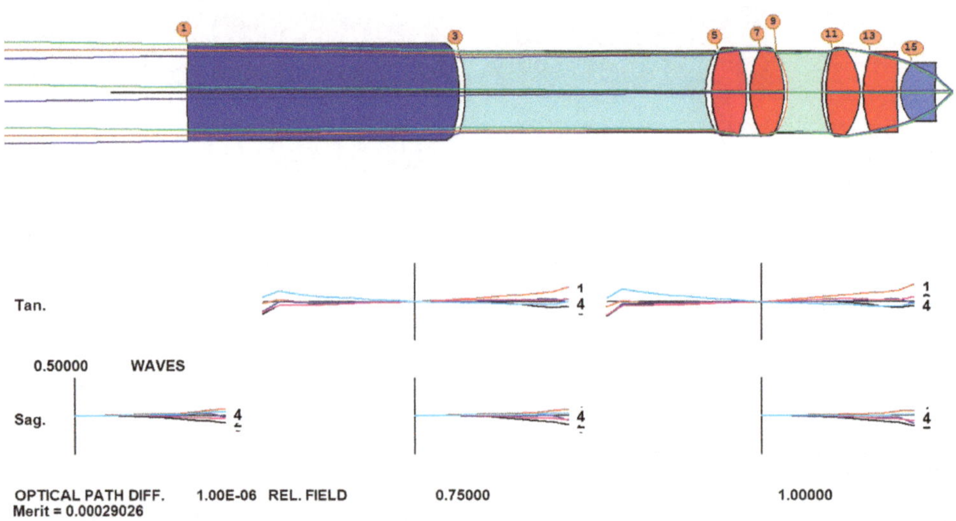

Figure 35.4. Results when GSEARCH matched the three nearest glasses.

Again, nearly perfect—and requiring only seven elements! Let us see what the MTF looks like over the field. Type **MMF**, select 'Multicolor', and click 'Execute'.

The result is shown in figure 35.6.

This is as close to perfect as most people will ever need.

Are we done? Let us see how stable the back focus position is as a function of wavelength. Enter the AI sentences

```
STEPS = 100
PLOT BACK FOR WAVL = .38 TO .9
```

Since the lens has a YMT solve on the final airspace, this will show us the color-correction curve. The plot is shown in figure 35.7.

The paraxial focus position indeed varies by just over 1 μm over this wide range—this is an excellent lens. Before you actually make the lens, it would be a good idea to move the stop to surface 8—but that is enough for now. Even a difficult challenge like this responds well to these new tools.

We also tried asking DSEARCH for a seven-element lens, rather than finding an eight-element lens and then deleting an element with AED as we did above. It is hard to predict which path through the chaotic design tree will turn out best, and all one can do is to try. The results in this case were almost as good as the previous, as shown in figure 35.8 (**C35L3**), when we tried several of the lenses and matched them to the Ohara glass table.

There are other things we could have tried. If the results were not good enough with seven elements, we could try the Automatic Element Insertion feature, adding the line

```
AEI 3 1 14 CONLY 100 1 10 50
```

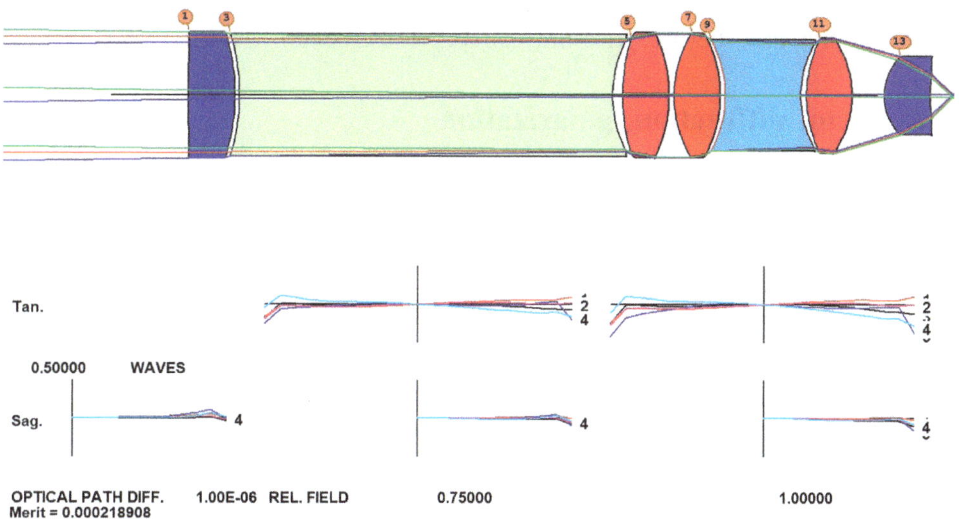

Figure 35.5. Lens reoptimized after an element is removed by AED and matched to real glass.

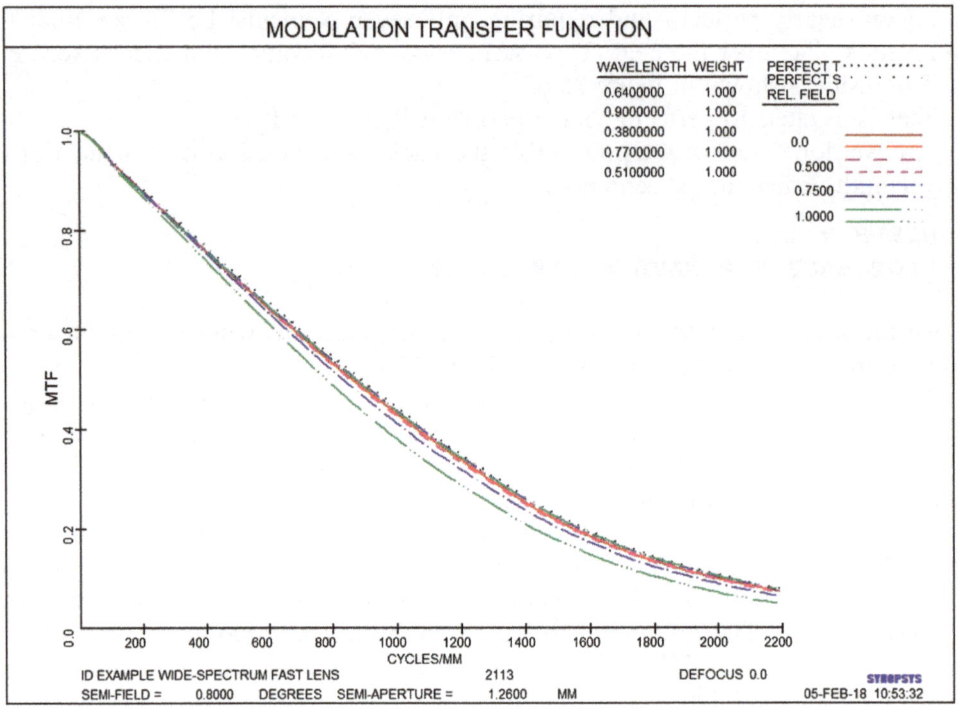

Figure 35.6. MTF of the lens in figure 35.5.

to the top of the MACro. That will add a cemented element on each side of all the current lenses in sequence and then come back with the combination that worked best. With these tools, you can go either way. If you also want to try airspaced elements, change **CONLY** to **CEMENT**. Then they will be tested as well (the entry '0', which we have used before, tries only airspaced elements).

35.1 Vector diffraction, polarization

This is a good place to discuss some of the finer points of image analysis. Look closely at the rays converging to the image in the above designs, shown in figure 35.5. This lens works at a very 'fast' F/number, the term originating in the early days of photography, when it was discovered that a lens with a low F/number did not require the long exposure needed by other lenses and was therefore 'fast'.

Ordinary diffraction-based image analysis employs what is called 'scalar diffraction' theory, which says that if two wavefronts are in phase they add, and if out of phase they cancel. However, the electric vectors $E1$ and $E2$ in figure 35.9 are almost at right angles to each other, and they cannot add or cancel completely. To accurately analyze the image in such a system one must use *vector diffraction* theory, which involves resolving the electric vector into three (x,y,z) components and performing the diffraction calculation three times, adding the results scalarly. However, now we have to take into account the polarization of the light. If this

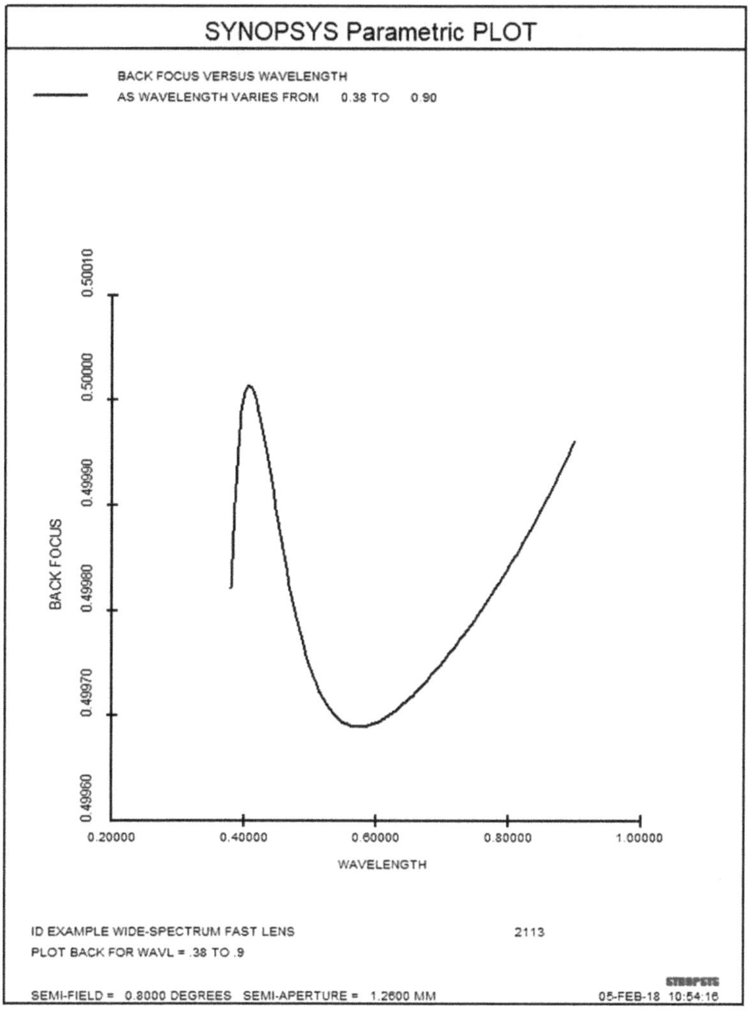

Figure 35.7. Color-correction curve for the lens in figure 35.5.

beam is formed with light polarized in the *y*-direction (in the plane of the picture), the *y*-components of the *E* vectors add if they are in phase, while the *z*-components, which are almost in opposite directions, will cancel. Light polarized in the *x*-direction, on the other hand, follows scalar rules in this case since the vectors on those two rays point in and out of the page and are parallel to each other.

To model this case, first put the lens in polarization mode, with the input

```
CHG
POL LIN Y
END
```

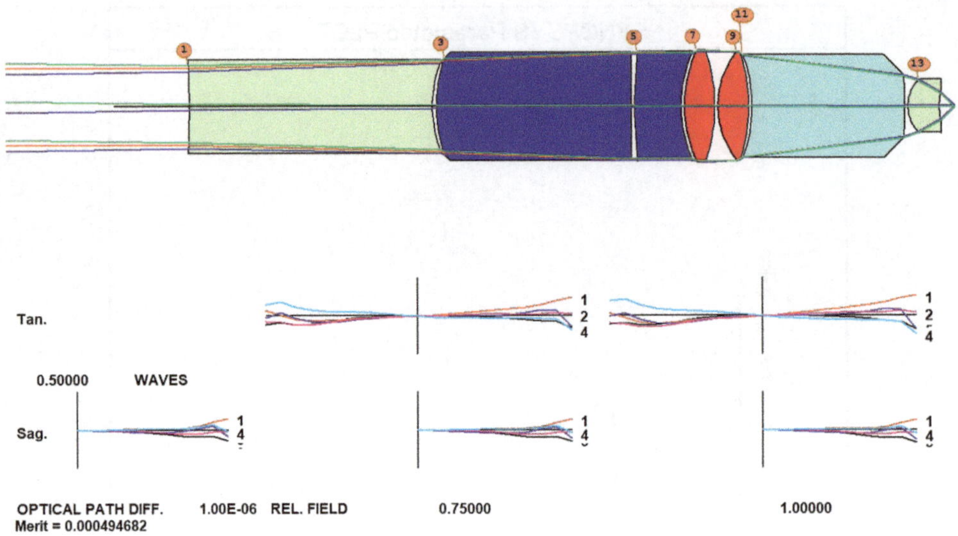

Figure 35.8. Lens returned by DSEARCH with seven elements, optimized, annealed, and matched to Ohara glasses.

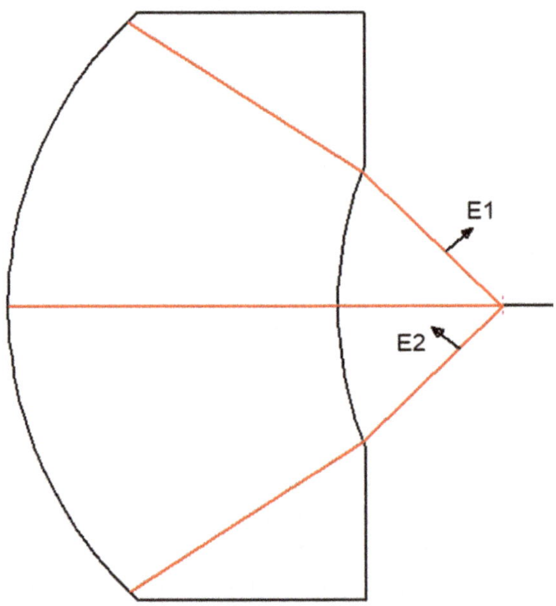

Figure 35.9. Rays converging at the image of a very fast lens.

Now one can perform a vector-diffraction analysis, using the Fourier-transform version of the MTF calculation:

```
DMTF M 0 6000 1 0 P
```

The result, shown in figure 35.10, tells us that the MTF in the *y*-direction is lower than in *x*. (The MTF rises past the diffraction cutoff frequency because the short wavelengths have their own cutoff at a higher frequency.)

If the light is *unpolarized*, the results are a little different, as shown in figure 35.11. The calculation is again based on vector-diffraction theory, but is performed in both *x*- and *y*-polarizations and the results added scalarly:

```
CHG
POL UNPOLAR
END
DMTF M 0 6000 1 0 P
```

While we are on the subject of polarization, there is another subtle effect worth knowing about. Fetch the file **AMICI.RLE**, shown in figure 35.12, where an SFAN of rays is traced. This is one of several kinds of prisms containing a roof surface, where rays are reflected from one side and then immediately from the other.

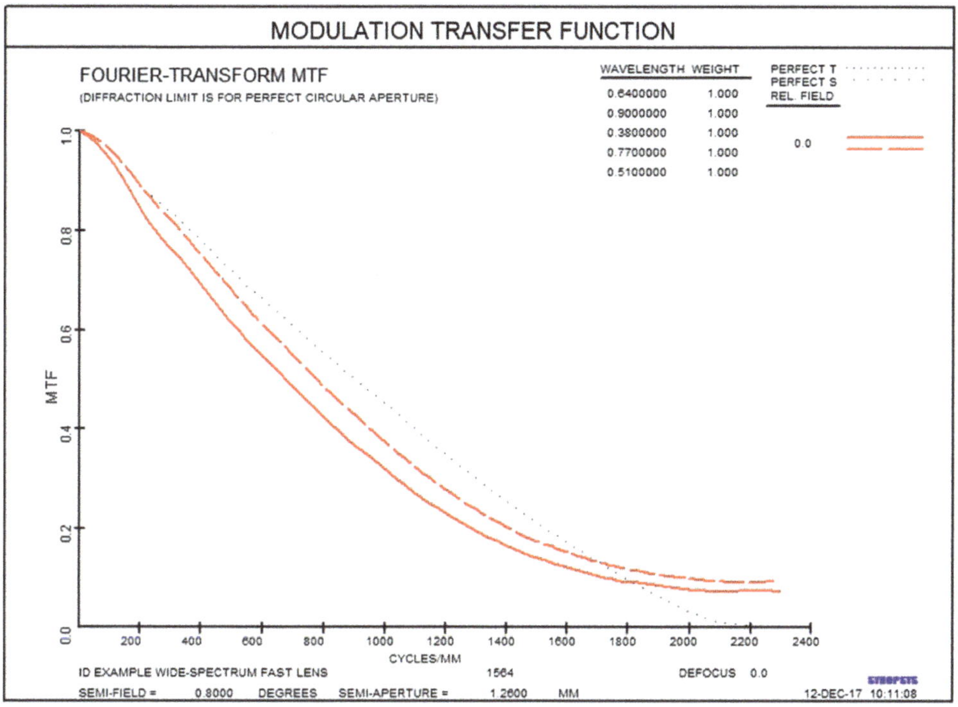

Figure 35.10. Vector-diffraction calculation of the on-axis MTF of the fast lens, in *y*-polarization.

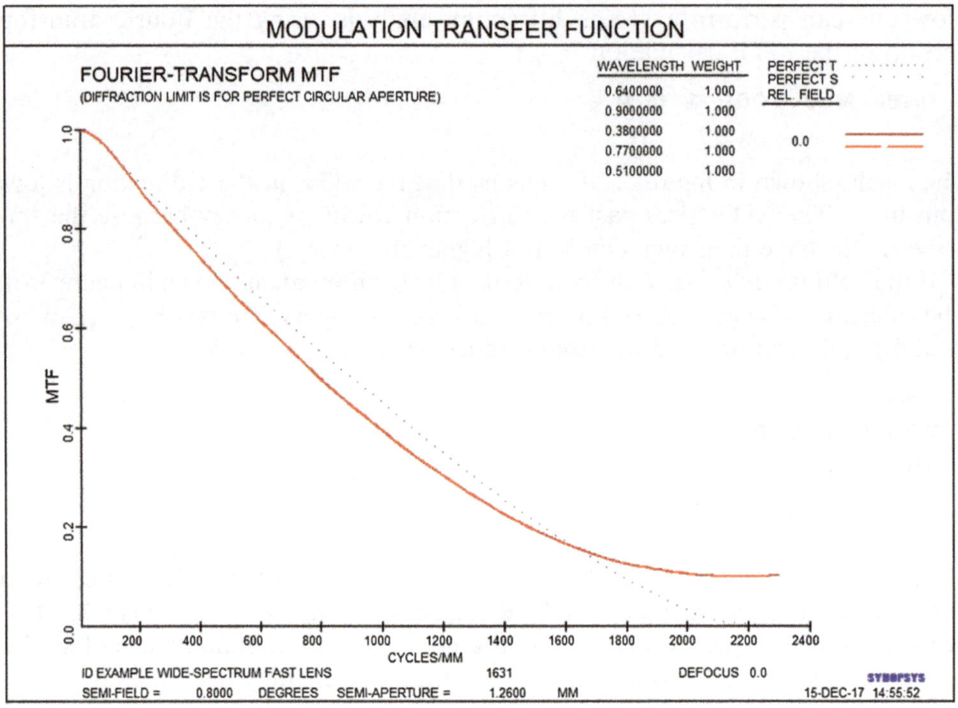

Figure 35.11. Vector-diffraction MTF calculation with unpolarized light.

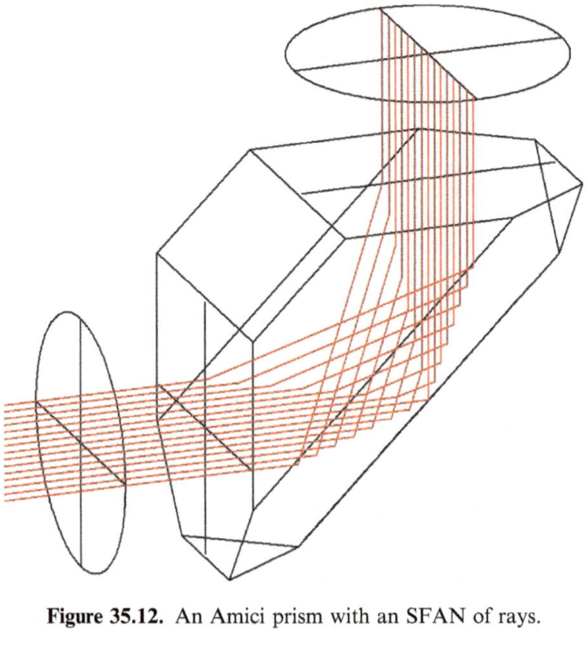

Figure 35.12. An Amici prism with an SFAN of rays.

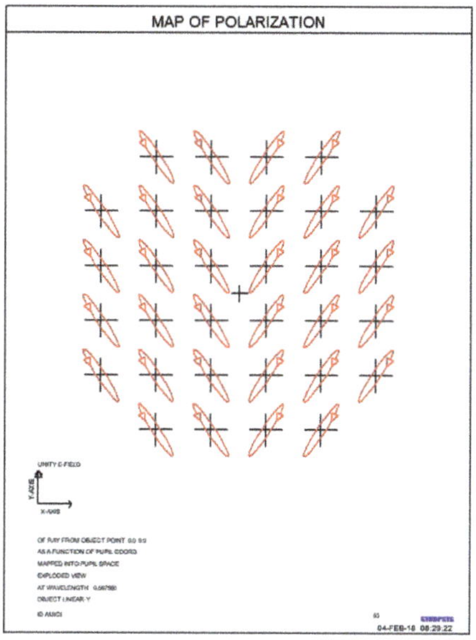

Figure 35.13. MAP of polarization emerging from the Amici prism when the roof surfaces are uncoated.

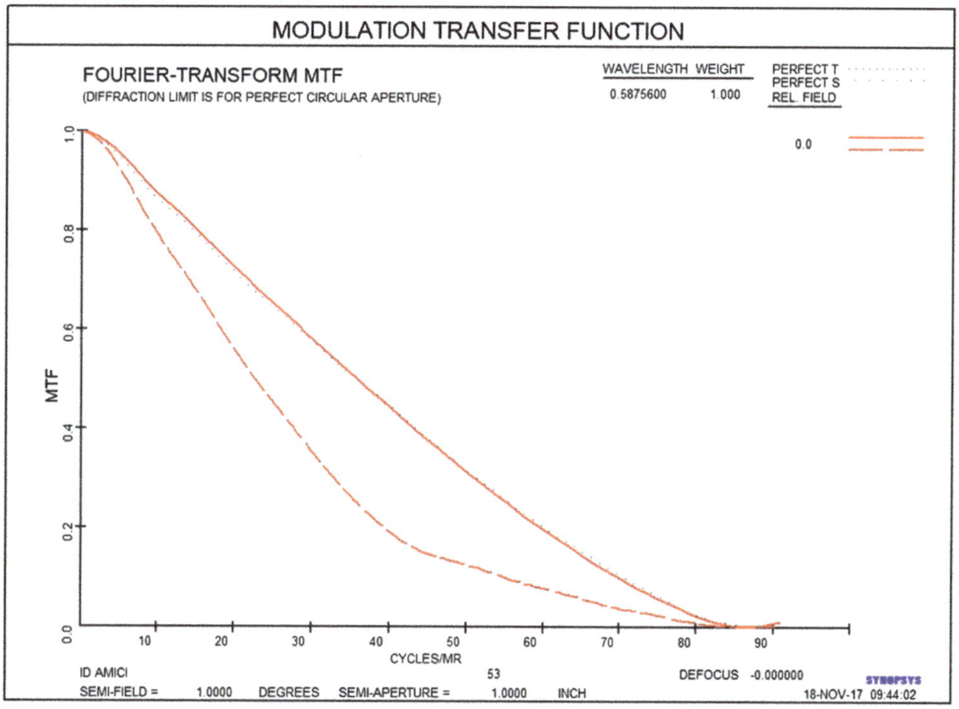

Figure 35.14. Effect of polarization rotation on the MTF of the Amici prism when the surfaces are uncoated.

Special tools are required to analyze this kind of system. First, the system is in NONSEQUENTIAL mode, because the order in which the surfaces are encountered differs depending on which side of the roof is hit first. That issue can be ignored in optimization, since the MF is the same either way if the roof angle is perfect, but not when the final image is analyzed. Then one must account for the effect of the roof on the polarization of the light. If you send in light linearly polarized in the y-direction and make a map of the polarization coming out, you obtain the picture in figure 35.13. Here again, the polarization vectors on each side of the pupil are at a steep angle with respect to those on the other side, and again the MTF suffers, as shown in figure 35.14. Someone unaware of this effect might wonder why the design did not yield the MTF that was expected (and might blame the prism maker!).

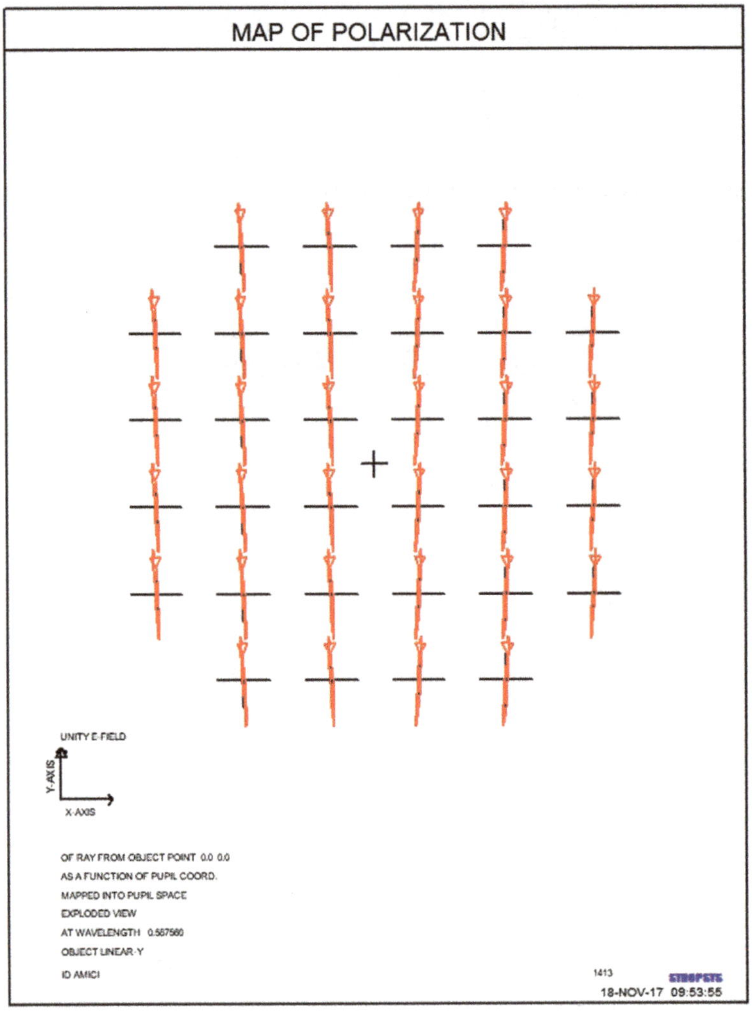

Figure 35.15. Polarization from the Amici prism when the roof surfaces are coated with aluminum. This yields a nearly perfect MTF curve.

The command **PCOAT** shows that the roof surfaces are uncoated at the moment:

```
SYNOPSYS AI>PCOAT
SURF. NO.  COATING
________________________
1 Dummy surface
2 UNCOATED
3 UNCOATED REFLECTOR
4 UNCOATED REFLECTOR
5 UNCOATED
6 Dummy surface
```

However, if you assign a reflective coating to the roof, the polarization is restored almost perfectly, as seen in figure 35.15, and the MTF is nearly perfect as well. This input assigns an aluminum coating to the roof surfaces:

```
SYNOPSYS AI>CHG
RLE>3 COAT AL
RLE>4 COAT AL
RLE>END
```

Chapter 36

Ghost hunting

Ghost images; analyzing; correcting in the merit function

Your lens looks great on paper and the shop did a fine job of building it. However, when you test it, you see a horrible ghost image whenever a bright source enters the field. Not a good scenario, but one that happens all too often. The customer will not be pleased.

To prevent this sort of surprise, use the tools in the **MGH** dialog (MGH; Menu, GHost image), with which you can find problems such as this early in the design process and correct them as you go.

Put briefly, a ghost image is a concentration of light at the image arising from two unwanted reflections within the lens system. If you have three lenses, there are 15 possible ghosts. With six elements you have 66, and so on. That is a lot to keep track of. To see what some of these tools can do, **FETCH** the lens 1.RLE. Then look at the PAD display, shown in figure 36.1.

It is not obvious that you will be bothered with ghost images, but then it seldom is. Let us see about that. Open the dialog **MGH**, shown in figure 36.2.

At the upper left is the '**GHOST**' button. That feature is used to find ghosts using only paraxial ray tracing, and the ghosts it finds of course differ somewhat from those formed by real rays. Nonetheless, the results are usually close enough for you to see where problems will show up. You can assign reflectance coefficients to any or all surfaces in the lens, and the program will take the values into account when it estimates the intensity of each of the ghosts it finds. The dialog opens with a default reflectance of 1%, applied to all lens surfaces. This is about what you can expect with some antireflection coatings.

doi:10.1088/978-0-7503-1611-8ch36

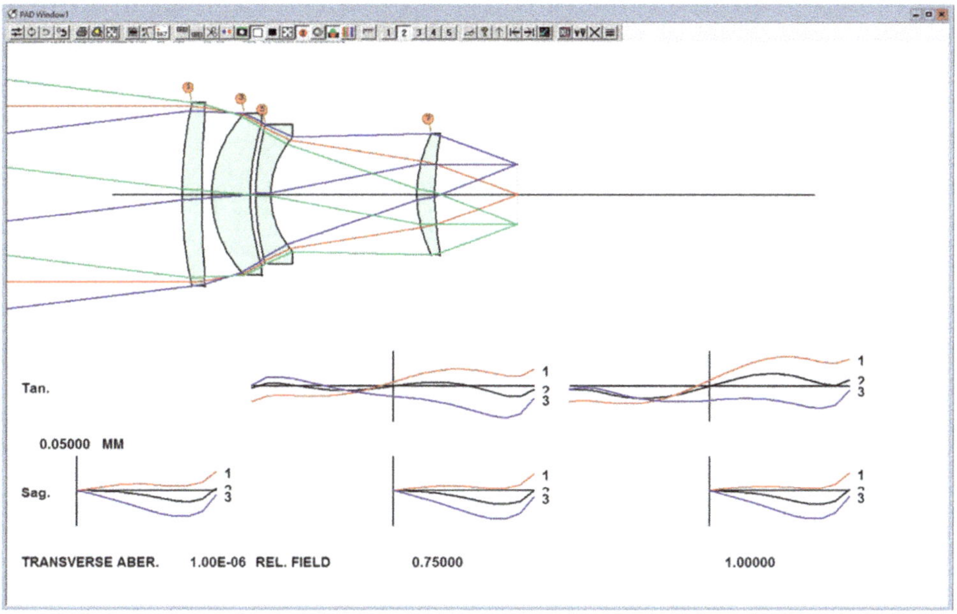

Figure 36.1. PAD display of a lens with a bad ghost image.

Click the '**GHOST**' button. You obtain two tables of numbers. The first analyzes all combinations of surfaces; a portion of the output is shown below:

```
--- GHOST R 0.01
 ID MIT 1 TO 2 UM LENS

 GHOST IMAGE ANALYSIS

 --- R 0.01 ALL
 --- END
```

NO.	GHOST SURF	Ymarg	U'marg	Ychief	INTENSITY
1	2 - 1	-30.0244	-0.3778	7.3095	1.10930E-11
2	3 - 1	27.2543	-0.3984	4.6907	1.34627E-11
3	3 - 2	63.5945	-0.2721	4.1819	2.47264E-12
4	4 - 1	-46.5167	-0.2920	4.9249	4.62149E-12
5	4 - 2	-25.1674	-0.3088	6.6272	1.57879E-11
6	4 - 3	-75.6813	-0.4937	4.4156	1.74591E-12
7	5 - 1	-45.8712	-0.2619	4.0455	4.75247E-12
8	5 - 2	-27.0481	-0.2839	6.4039	1.36686E-11
9	5 - 3	-72.6068	-0.4577	2.7558	1.89691E-12
10	5 - 4	-3.4550	-0.3402	6.5304	8.37735E-10
11	6 - 1	0.5515	-0.5074	4.2824	3.28758E-08
12	6 - 2	44.1259	-0.4028	8.4078	5.13585E-12
13	6 - 3	-38.8865	-0.5677	1.4386	6.61305E-12
14	6 - 4	67.9018	-0.0930	9.4364	2.16889E-12
15	6 - 5	66.9185	-0.0241	8.9874	2.23310E-12
16	7 - 1	-26.2150	-0.5336	-37.0598	1.45513E-11
17	7 - 2	17.4775	-0.4677	31.7392	3.27372E-11
18	7 - 3	-71.3906	-0.6925	-107.3067	1.96208E-12

. . .

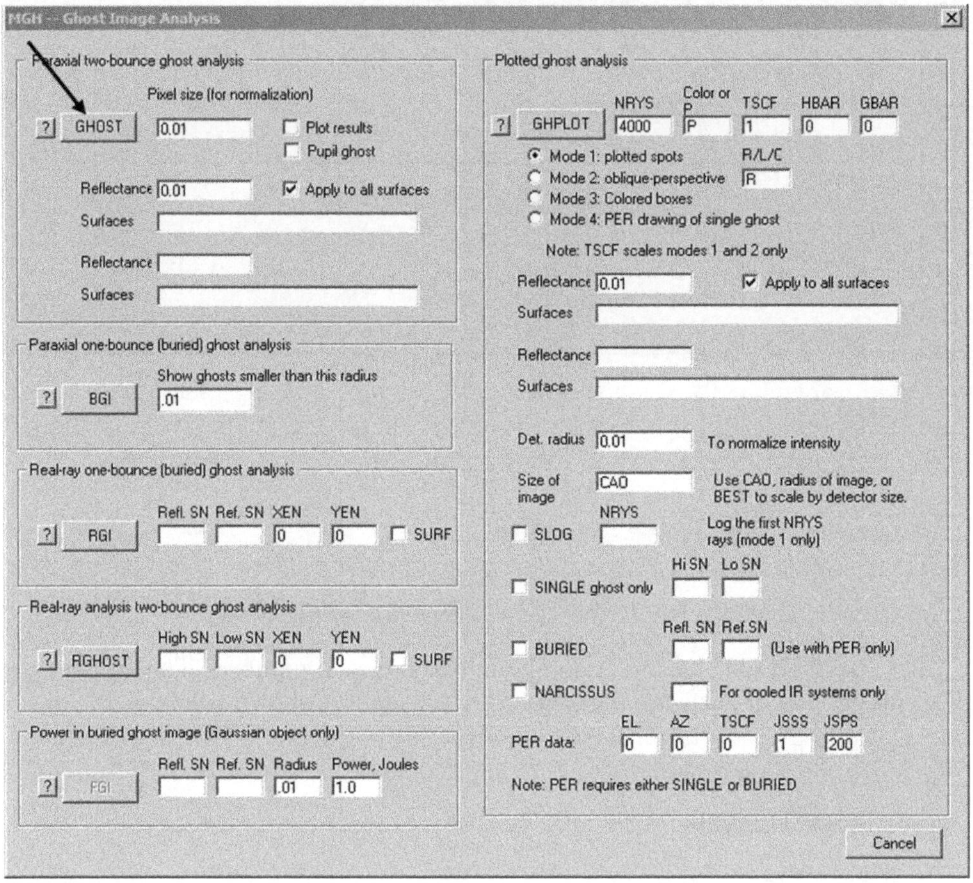

Figure 36.2. MGH dialog for analyzing ghost images.

Under the heading Ymarg, notice the smallest value, 0.5515, for the combination of surfaces 6 and 1. This tells you that light reflected from surface 6, and then from surface 1, will arrive at the image plane forming a (paraxial) blur of about 0.5 mm in radius. That may be a problem.

If your lens is lengthy, it is easier to pick out the problem ghosts by examining the second listing:

```
CUMULATIVE GHOST DISTRIBUTION
NORMALIZED FOR DETECTOR SEMI-APERTURE          0.0100

 NO.   GHOST INTENS.   ACCUM. INTENS.   SURFACES
_________________________________________________
   6   1.74591E-12    1.74591E-12      4      3
   9   1.89691E-12    3.64282E-12      5      3
  18   1.96208E-12    5.60491E-12      7      3
  14   2.16889E-12    7.77380E-12      6      4
  15   2.23310E-12    1.00069E-11      6      5
   3   2.47264E-12    1.24795E-11      3      2
  20   3.94955E-12    1.64291E-11      7      5
  19   4.15611E-12    2.05852E-11      7      4
   4   4.62149E-12    2.52067E-11      4      1
   7   4.75247E-12    2.99592E-11      5      1
  12   5.13585E-12    3.50950E-11      6      2
  13   6.61305E-12    4.17081E-11      6      3
  21   6.62606E-12    4.83341E-11      7      6
   1   1.10930E-11    5.94271E-11      2      1
   2   1.34627E-11    7.28898E-11      3      1
   8   1.36686E-11    8.65584E-11      5      2
  16   1.45513E-11    1.01110E-10      7      1
   5   1.57879E-11    1.16898E-10      4      2
  25   2.20749E-11    1.38973E-10      8      4
  26   2.25477E-11    1.61520E-10      8      5
  17   3.27372E-11    1.94257E-10      7      2
  24   5.08281E-11    2.45085E-10      8      3
  23   5.72162E-11    3.02302E-10      8      2
  28   8.28536E-11    3.85155E-10      8      7
  10   8.37735E-10    1.22289E-09      5      4
  27   3.67946E-09    4.90235E-09      8      6
  22   1.09304E-08    1.58327E-08      8      1
  11   3.28758E-08    4.87085E-08      6      1
```

Here the ghosts are sorted, with the most severe ones at the bottom, and the accumulated intensity is calculated, printed, and summed. Indeed, the cumulative ghost intensity, 4.87E-08 is mostly due to that single ghost, which has the intensity 3.29E-8. Now you know where the ghost is coming from. Let us see the effect.

To do this, we will use a MACro that exercises several ghost analysis features. Here is that MACro (**C36M1**):

```
; GHPLOT.MAC
; THIS EXAMPLE EXAMINES THE GHOST IMAGE IN A LENS
; IT RUNS GHPLOT IN ALL FOUR MODES.

CCW            ! CLEAN UP FIRST; CLEAR COMMAND WINDOW
KAG            ! AND CLOSE GRAPHICS WINDOWS
```

```
FET 1
CHG
CFIX              ; FIX CLEAR APERTURES TO DELETE VIGNETTED GHOSTS
VIG               ; AND TURN ON VIG MODE IF OFF
END

OFF 27            ! SPOTS SHOWN AS SYMBOL
SSS .01           ; SMALL SPOT SIZE HERE

GAW               ; NEED NEW WINDOW FOR EACH PICTURE (GRAPHICS ADD
                  ! WINDOWS)
GHPLOT 4000 P 10 .5 0 1       ; SELECT MODE 1, INDIVIDUAL RAYS
R .01 ALL                ; THIS IS ALSO THE DEFAULT REFLECTANCE
PLOT

GHPLOT 20000 P 1 .5 0 2 L ! NOW GET AN OBLIQUE !PERSPECTIVE VIEW
DRAD .0004
PLOT

GHPLOT 20000 P 1 .5 0 3 L     ; THIS MAKES COLORED BINS
DRAD .0004
PLOT

GHPLOT 400 P 1 .5 0 4 L                ! AND THIS DRAWS A SINGLE GHOST
                                       ! WITH PERSPECTIVE
SINGLE 6 1
PER 0 0 0 1 99
PLOT

GRW    ; RESTORE GRAPHICS OPTION (GRAPHICS REUSE WINDOW)
```

GHPLOT has four modes, and you are encouraged to read about them before we go any further. Load this MACro in the editor, *select* the characters ‘GHPLOT’, and then look at the TrayPrompt, shown in figure 36.3.

Since this is a multiline command, the prompt cannot show the entire format, but if you press the <**F2**> key when the prompt is displayed, the help file opens to that section in the index.

You will use all four modes in this exercise. The first call to GHPLOT uses mode 1, producing a picture of all the ghost images, superimposed for an object point at HBAR = 0.5, at the image plane, as shown in figure 36.4. There is indeed a dark blob about halfway up in the field. That is certainly the ghost we flagged earlier. The mode-2 analysis shows the same energy distribution as an oblique–perspective plot, as seen in figure 36.5.

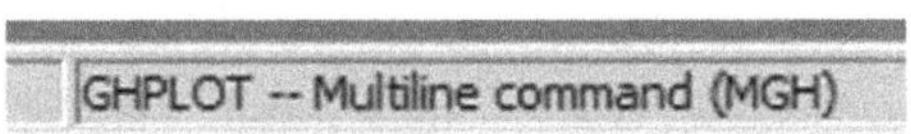

Figure 36.3. The TrayPrompt display when the characters ‘GHPLOT’ are selected.

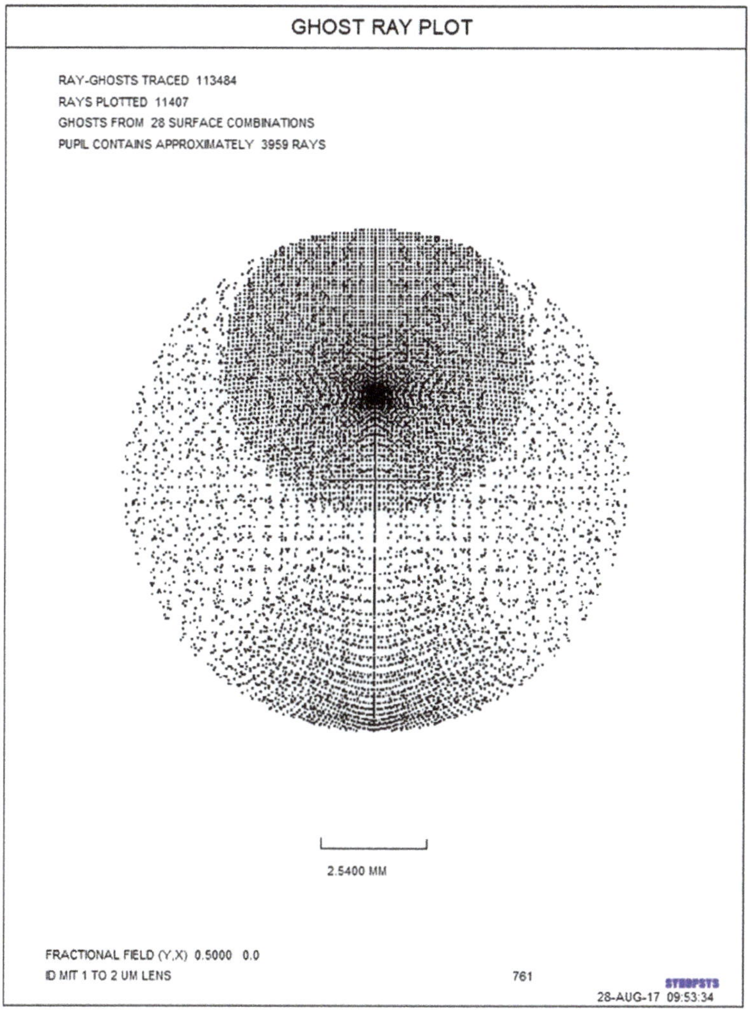

Figure 36.4. Mode 1 GHPLOT plot, showing all ghosts superimposed.

That sharp peak is indeed the ghost. Yet another way to view it is the mode-3 plot, shown in figure 36.6.

Lastly, the mode-4 plot singles out that a particular set of reflections (which we asked for) and draws a tangential fan of rays. This is shown in figure 36.7.

Here, light comes in from the left drawn in red, turns blue after reflecting from surface 6, and then turns green after the second reflection, at surface 1. It is indeed almost in focus at the image, but there is a lot of spherical aberration and coma so the ghost is not very sharp.

Looking again at the **MGH** dialog, you see four more features we have not used yet. Let us trace the path of a real ghost ray at zone 0.5. Fill in the boxes as shown in figure 36.8 and click the '**RGHOST**' button.

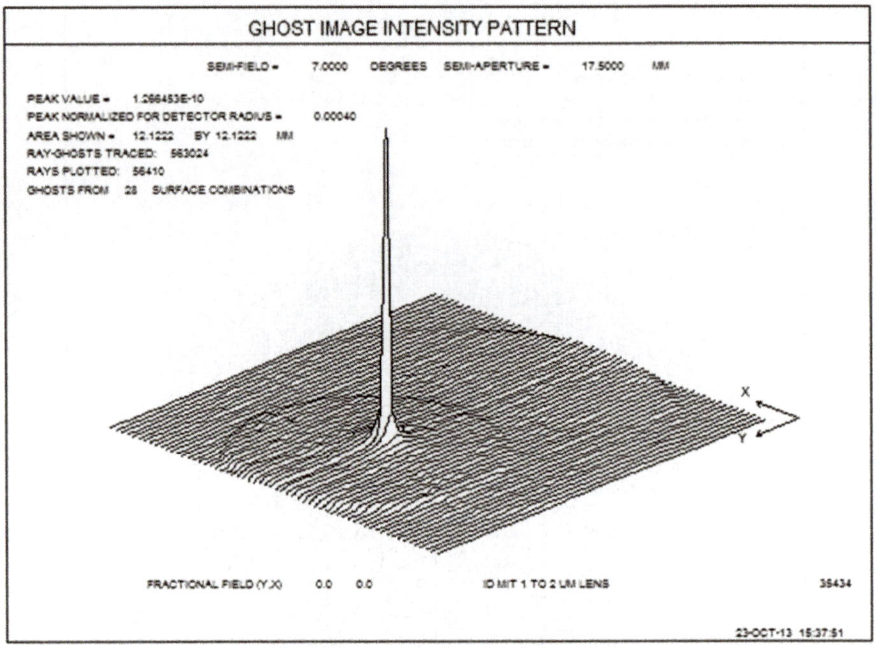

Figure 36.5. Superimposed ghosts plotted in oblique perspective with Mode 2.

This produces the output shown below:

```
--- RGHOST 6 1 0 .5 SURF
                      RAY VECTORS       (X DIR TAN)  (Y DIR TAN) (INC. ANG.)
 SURF            X            Y            Z            ZZ           HH           UNI
-------------------------------------------------------------------------------------
  OBJ      0.000000     0.000000     0.000000     0.000000     8.750000E-12
    1      0.000000     8.750000     0.442564     0.000000      -0.056702     5.790967
    2      0.000000     8.540210     0.142402     0.000000      -0.086503     1.334779
    3      0.000000     8.256701     1.510615     0.000000      -0.246653    15.791928
    4      0.000000     6.624679     0.478579     0.000000      -0.384916     5.591649
    5      0.000000     6.274262     0.388953     0.000000      -0.226690    13.957837
 --- RAY REVERSES AFTER NEXT SURFACE ---
    6      0.000000     5.481374    -0.886627     0.000000       0.444814
    5      0.000000     6.994957    -0.483893     0.000000       1.206445
    4      0.000000     7.951450    -0.691073     0.000000       0.498903
    3      0.000000    10.791512    -2.647176     0.000000       0.470102
    2      0.000000    12.783436    -0.319172     0.000000       0.221387
    1      0.000000    13.505397     1.05809      0.000000      -0.095157
    2      0.000000    13.193103     0.339970     0.000000      -0.151262
    3      0.000000    12.414364     3.579028     0.000000      -0.404621
    4      0.000000    10.296378     1.164810     0.000000      -0.674892
    5      0.000000     9.767857     0.947929     0.000000      -0.369535
    6      0.000000     8.241826     2.077523     0.000000      -0.148082
    7      0.000000     4.247415     0.334421     0.000000      -0.153884
    8      0.000000     3.743737     0.107527     0.000000      -0.282387
GHOST REFLECTED FROM SURFACES     1     6 AT SURFACE     9
     X               Y                ZZ               HH
-----------------------------------------------------------------
  0.00000       -0.828738         0.00000         -0.282387
```

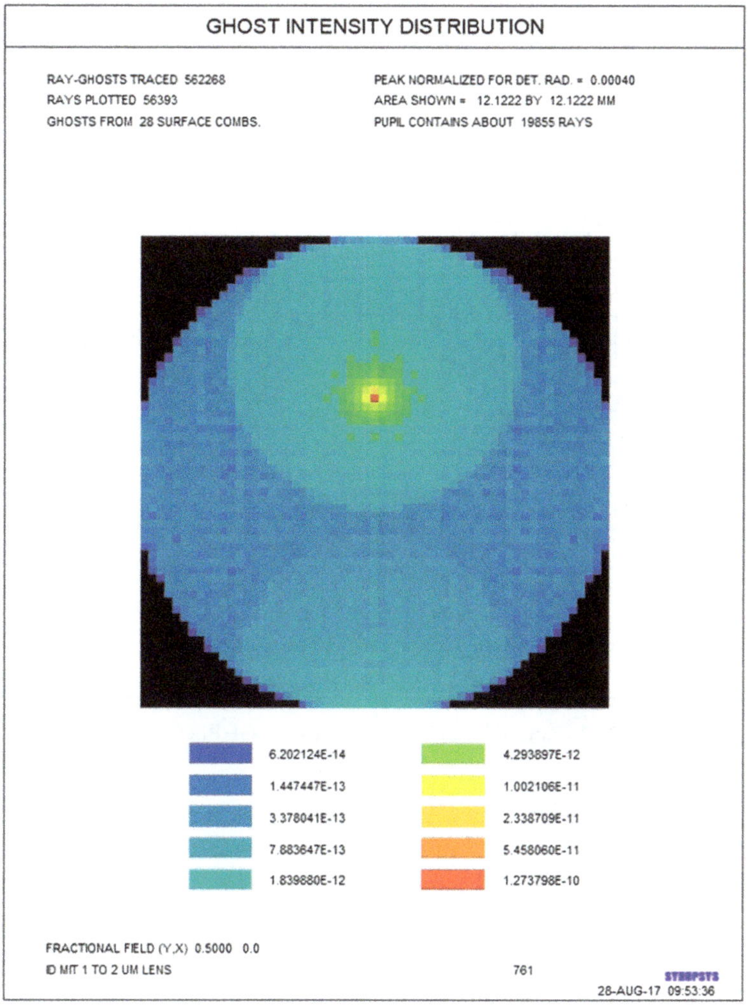

Figure 36.6. Superimposed ghosts plotted with color scale, with mode 3.

The ray reflects from surface 6, then again at 1, and proceeds to the image, where its y-coordinate is −0.829 mm. That may indeed be a serious ghost.

If you catch this problem early in the design process, it is usually easy to control. Type **HELP GHOST**, and select the link that describes controlling the ghost, as shown in figure 36.9.

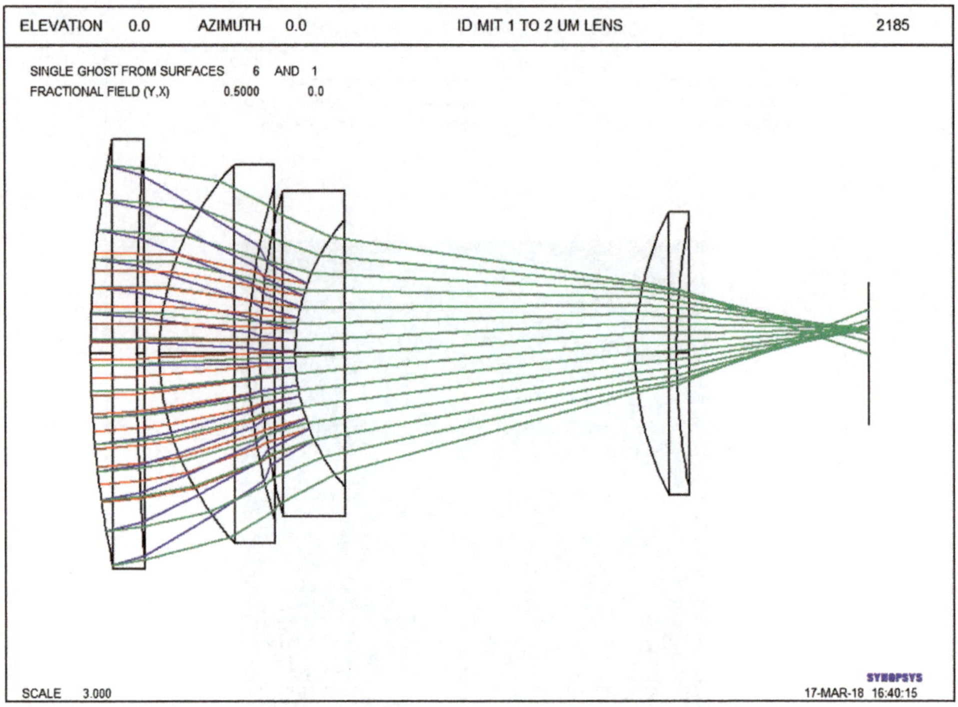

Figure 36.7. Path of a single ghost plotted by GHPLOT in mode 4.

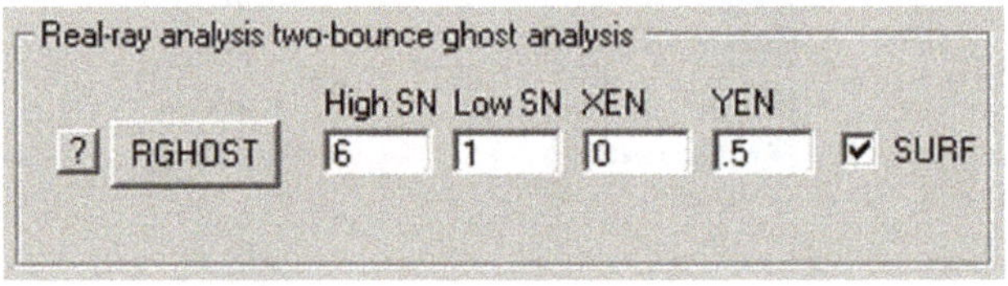

Figure 36.8. Data for the 'RGHOST' command.

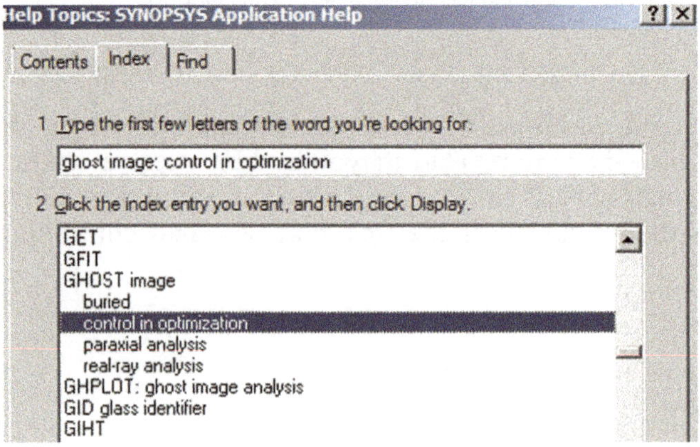

Figure 36.9. Selecting a ghost option in the help file.

This opens a page that describes how to control ghosts.

10.3.1.5 Ghost-image control Next Previous TOC

A ghost image is caused by a reflection from one or more refracting surfaces. SYNOPSYS can evaluate and control two types: The GHOST program can show which combinations of surfaces are responsible for ghost images at the image surface, and BGI can evaluate the properties of a ghost image that is formed at another place within a lens system.

To control the size of the blur at the image from a selected paraxial ghost, the input is

M TAR WT A PGHOST JREFH JREFL

...

Here you see the simple input required to control the ghost. A suitable request in your AANT file might be

```
M 5 0.1 A PGHOST 6 1
```

with the weighting adjusted to balance nicely with your other aberrations. If you achieve this goal, the ghost will be about 1% as intense as before. The target of 5 is somewhat arbitrary; a larger ghost is a weaker ghost, and this is a good guess to start with.

This procedure usually yields a great improvement in the specified ghost. Often, however, another combination of reflections then produces its own ghost, requiring another evaluation with GHOST and another PGHOST aberration in the merit function. Add them as they show up until you reach a point where many ghosts are roughly the same intensity. We have never encountered a situation where this intensity was high enough to be a problem. If it were, it would be time to invest in better coatings on the problem surfaces.

Chapter 37

Importing a Zemax[1] file into SYNOPSYS

Converting a file from one program to another requires user input

SYNOPSYS can open most lens files created by the programs Zemax and Code-V[2]. As is true with most conversions from one protocol to another, however, the results are often incomplete, and the user must then edit the lens file to restructure certain parameters according to the rules of the target program.

However, some things cannot be converted. The two programs utilize very different descriptions of the entrance pupil—although both in the end achieve much the same result. Also, not all surface shapes that can be defined in Zemax can be defined in SYNOPSYS (and *vice versa*). Nonetheless, all of the most popular shapes work well in either program, so most users will not encounter difficulties for that reason. While a Zemax file contains a whole lot more information than does a SYNOPSYS file—such as definitions of variables, the merit function, and tolerances—the conversion will capture only the basic lens data, since the RLE file in SYNOPSYS is just a lens description and the other data are stored as separate files. Anyone converting from one program to the other will naturally want to exploit the advantages of SYNOPSYS and will create their own data files, so it makes no sense to attempt to import those other items.

A more common question is properly identifying the names of commercial glass types. The two programs have extensive glass tables, but the names often differ. So the most common user task, after importing a .zmx file, is to edit the RLE file and insert the correct glass names. An example will illustrate some of these issues.

(You are encouraged to read section 5.42 of the User's Manual before importing a file, where you will find even more information.)

[1] Zemax™ is a trademark of Focus Software.
[2] Code-V is a trademark of Synopsys, Inc. (not affiliated with Optical Systems Design, Inc.).

doi:10.1088/978-0-7503-1611-8ch37

To illustrate this feature, we will convert a file describing a diffractive optical element that is stored in the Dbook directory with the name doe.zmx. This file contains the following lines:

```
VERS 91012 185 25430
MODE SEQ
NAME Achromatic singlet
NOTE 0 Notes...
NOTE 4
NOTE 0
NOTE 4
NOTE 0
UNIT MM X W X CM MR CPMM
ENPD 5.0E+1
ENVD 2.0E+1 1 0
GFAC 0 0
GCAT SCHOTT
RAIM 0 0 1 1 0 0 0 0 0
PUSH 0 0 0 0 0 0
SDMA 0 1 0
FTYP 1 0 3 3 0 0 0
ROPD 2
PICB 1
XFLD 0 0 0
XFLN 0 0 0 0 0 0 0 0 0 0 0 0
YFLD 0 3.5 5.0
YFLN 0 3.5 5.0 0 0 0 0 0 0 0 0 0
FWGT 1 1 1
FWGN 1 1 1 1 1 1 1 1 1 1 1 1
ZVDX 0 0 0
VDXN 0 0 0 0 0 0 0 0 0 0 0 0
ZVDY 0 0 0
VDYN 0 0 0 0 0 0 0 0 0 0 0 0
ZVCX 0 0 0
VCXN 0 0 0 0 0 0 0 0 0 0 0 0
ZVCY 0 0 0
VCYN 0 0 0 0 0 0 0 0 0 0 0 0
ZVAN 0 0 0
VANN 0 0 0 0 0 0 0 0 0 0 0 0
WAVL 4.861E-1 5.876E-1 6.563E-1
WAVN 4.861E-1 5.876E-1 6.563E-1 5.5E-1 5.5E-1 5.5E-1 5.5E-1 5.5E-1 5.5E-1
5.5E-1 5.5E-1 5.5E-1
WWGT 1 1 1
WWGN 1 1 1 1 1 1 1 1 1 1 1 1
WAVM 1 4.861E-1 1
WAVM 2 5.876E-1 1
WAVM 3 6.563E-1 1
WAVM 4 5.5E-1 1
WAVM 5 5.5E-1 1
WAVM 6 5.5E-1 1
WAVM 7 5.5E-1 1
WAVM 8 5.5E-1 1
WAVM 9 5.5E-1 1
WAVM 10 5.5E-1 1
WAVM 11 5.5E-1 1
WAVM 12 5.5E-1 1
```

```
WAVM 13 5.5E-1 1
WAVM 14 5.5E-1 1
WAVM 15 5.5E-1 1
WAVM 16 5.5E-1 1
WAVM 17 5.5E-1 1
WAVM 18 5.5E-1 1
WAVM 19 5.5E-1 1
WAVM 20 5.5E-1 1
WAVM 21 5.5E-1 1
WAVM 22 5.5E-1 1
WAVM 23 5.5E-1 1
WAVM 24 5.5E-1 1
PWAV 2
POLS 1 0 1 0 0 1 0
GLRS 1 0
GSTD 0 100.000 100.000 100.000 100.000 100.000 100.000 0 1 1 0 0 1 1 1 1 1 1
NSCD 100 500 0 1.0E-6 5 1.0E-6 0 0 0 0 0 1 1000000 0
COFN COATING.DAT SCATTER_PROFILE.DAT ABG_DATA.DAT PROFILE.GRD
SURF 0
  TYPE STANDARD
  CURV 0.0 0 0 0 0 ""
  HIDE 0 0 0 0 0 0 0 0 0 0
  MIRR 2 0
  SLAB 1
  DISZ 2.5E+2
  DIAM 5.0 0 0 0 1 ""
  POPS 0 0 0 0 0 0 0 0 1 1 1 1 0 0 0
SURF 1
  STOP
  TYPE STANDARD
  CURV 7.5762934618539999OOE-003 0 0 0 0 ""
  HIDE 0 0 0 0 0 0 0 0 0 0
  MIRR 2 0
  SLAB 2
  DISZ 2.5E+1
  GLAS BK7 0 0 1.69673 5.6419998E+1 -7.4E-3 1 1 1 0 0
  DIAM 3.0E+1 1 0 0 1 ""
  POPS 0 0 0 0 0 0 0 0 1 1 1 1 0 0 0
  FLAP 0 3.0E+1 0
SURF 2
  TYPE BINARY_2
  CURV -6.6766952605729997OOE-003 0 0 0 0 ""
  HIDE 0 0 0 0 0 0 0 0 0 0
  MIRR 2 0
SLAB 3
  PARM 0 1
  PARM 1 0
  PARM 2 0
  PARM 3 0
  PARM 4 0
  PARM 5 0
  PARM 6 0
  PARM 7 0
  PARM 8 0
  XDAT 1 3.000000000000E+000 0 0 0.000000000000E+000 0.000000000000E+000 0 ""
  XDAT 2 3.000000000000E+001 0 0 0.000000000000E+000 0.000000000000E+000 0 ""
  XDAT 3 -2.993832387049E+003 0 0 0.000000000000E+000 0.000000000000E+000 0 ""
```

```
   XDAT 4 1.135544608547E+003 0 0 0.000000000000E+000 0.000000000000E+000 0 ""
   XDAT 5 -5.932105454300E+001 0 0 0.000000000000E+000 0.000000000000E+000 0 ""
   DISZ 2.5073834507E+2
   DIAM 3.0E+1 1 0 0 1 ""
   POPS 0 0 0 0 0 0 0 0 1 1 1 1 0 0 0
   FLAP 0 3.0E+1 0
 SURF 3
   TYPE STANDARD
   CURV 0.0 0 0 0 0 ""
   HIDE 0 0 0 0 0 0 0 0 0 0
   MIRR 2 0
   SLAB 4
   DISZ 0
   DIAM 5.175465768436 0 0 0 1 ""
   POPS 0 0 0 0 0 0 0 0 1 1 1 1 0 0 0
 BLNK
 TOL TOFF  0  0       0       0  0 0 0
 MNUM 1 1
 MOFF  0  1 "" 0 0 0 1 1 0 0.0 ""
```

Begin by typing the command **ZMC** (ZeMax Convert). A warning message is displayed, shown in figure 37.1.

(This is to prevent naïve users from blindly selecting a file and expecting the lens to open every time exactly as it does in Zemax. Yes, some users are that naïve, but life is not simple.) If you click the 'No' button, you go immediately to the help file describing ZMC.

When you click the 'Yes' button, you are shown a list of the .zmx files in the current directory, and you can select the one mentioned above. (The file to be imported must reside in the current directory.) The above lines scroll past on the Command Window, and you see a warning message:

```
**************************************************
******************  WARNING  *****************
***** THE DOE EMULSION INDEX IS NOT GIVEN ****
*** IN THE ZEMAX FILE AND MUST BE ENTERED BY **
*** HAND IN THE RLE FILE AFTER CONVERSION ****
****** IN WORD THREE OF THE DOE ENTRY ********
**************************************************
```

Figure 37.1. Warning message displayed by ZMCONVERT.

This is an example of the difference in protocols. In a SYNOPSYS RLE file, the exact index values of a material are given along with the glass-catalog name, if any, and then the properties of the DOE are listed. Because of this protocol, anyone reading an RLE file created by SYNOPSYS knows the index of the material, even if, years from now, that glass type has become obsolete and is no longer in the catalog. Zemax lists the glass name but not the index values. So when the DOE input is read (and converted by ZMC) the index data are not yet known. The glass name shows up later, but the conversion is then way past that stage already. Since the DOE specification in SYNOPSYS requires the index of the emulsion when the input is first read in, the program inserts a dummy index of 1.517 just to avoid input errors. It turns out that this DOE is in fact made from BK7, so the index, just by chance, is correct. Otherwise you would have to edit the file and change that number to the correct index for the correct glass. (In SYNOPSYS, the index values are retrieved from the glass tables *after* the RLE file is completely processed and are not available while ZMC is running.)

At the end of the conversion, the program displays an informative message:

```
NOTE: OBJECT AND PUPIL DEFINITIONS MAY DIFFER. THE PROGRAM PUTS THE
WAP 3 PUPIL IN EFFECT TO BE SAFE. BUT THIS LENS MAY OR MAY NOT REQUIRE
THAT OPTION. YOU SHOULD DELETE IT IF IT IS NOT NECESSARY.
IF ANY GLASS-TABLE GLASSES WERE NOT FOUND, IT MAY BE DUE TO DIFFERENT
SPELLINGS. CHECK THE LISTING ABOVE TO SEE WHAT THE NAME WAS, AND CHANGE
TO THE APPROPRIATE SPELLING IF THAT GLASS IS IN ONE OF THE GLASS TABLES.
```

Here you learn that the program has (by default) implemented the WAP 3 option, which is sometimes a safe bet but is almost never actually required. By all means try to understand the geometry that underlies the pupil definition used in Zemax, and if it does not really require WAP 3, try the simpler WAP 0 instead.

The next job is to look at the RLE file it has created and loaded into a MACro editor:

```
RLE
ID ACHROMATIC SINGLET
ID1 NOTES...
ID2
ID3
 UNITS MM
TEMPERATURE 20.000
PRESSURE  100.000
 GTZ
WT1  1.00000     1.00000     1.00000
WA1 0.486100    0.587600    0.656300
CORDER  2  3  1
 POLAR OFF
   0 CV 0.0
OBA   250.000     5.00000     25.0000     0.00000     0.00000     0.00000     0.00000
APS   -1
 WAP 3
```

```
  1 RD   131.991
  1 TH   25.0000
  1 GTB S
BK7
  1 CAO   30.0000
  3 RD  -149.775
  3 TH   250.738
  3 CAO   30.0000
  2 PIN    1
  2 DOE  0.587600       1.51700
 RNORM   30.0000
 A11   476.483     -180.728      9.44124      0.00000       0.00000
  3 DC1  0.0000000E+00 0.0000000E+00 0.0000000E+00 0.0000000E+00 0.0000000E+00
  4 CV 0.0
  4 TH   0.00000
 END
```

To see this lens, change WAP 3 to WAP 0, run the RLE file, and then open **PAD**. The lens in figure 37.2 appears.

Converting this file was rather simple (as it should be). If you now ask SYNOPSYS to create a proper RLE file (with the command **LEO**), you obtain:

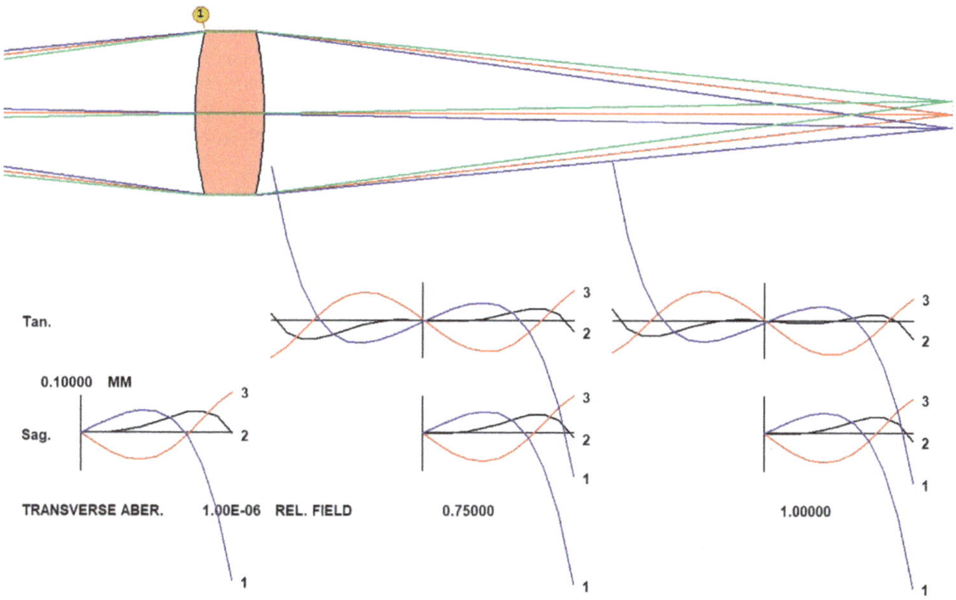

Figure 37.2. Lens converted from a Zemax file to SYNOPSYS format.

```
RLE
ID ACHROMATIC SINGLET          378
ID1 NOTES...
ID2
ID3
 LOG   378
 WAVL .4861000 .5876000 .6563000
 CORDER  2  3  1
 APS       -1
 GTZ
 UNITS MM
 OBA 250. 5. 25. 0 0 0 25.

  0 AIR
  1 CAO    30.00000000     0.00000000     0.00000000
  1 RAD  131.9906634000000   TH    25.00000000
  1 N1 1.52237223 N2 1.51679274 N3 1.51431609
  1 CTE   0.710000E-05
  1 GTB S  'BK7        '
  2 N1 1.52237223 N2 1.51679274 N3 1.51431609
  2 CTE   0.710000E-05
  2 GID 'BK7        '
  2 DOE    0.587600    1.517000    55.000000
  RNORM  30.0000
  A11 4.7648E+02 -1.8073E+02 9.4412E+00 0.0000E+00 0.0000E+00
  A12 0.0000E+00 0.0000E+00 0.0000E+00 0.0000E+00 0.0000E+00 0.0000E+00
  A13 0.0000E+00 0.0000E+00 0.0000E+00 0.0000E+00 0.0000E+00 0.0000E+00
  2 PIN  1
  3 CAO    30.00000000     0.00000000     0.00000000
  3 RAD  -149.7746955999999   TH   250.73834510 AIR
 3 DC1 0.00000000E+00 0.0000000E+00 0.0000000E+00 0.0000000E+00 0.00000000E+00
 3 DC2 0.00000E+00 0.00000E+00 0.00000E+00 0.00000E+00 0.00000E+00 0.00000E+00
 3 DC3 0.00000E+00 0.00000E+00 0.00000E+00 0.00000E+00 0.00000E+00 0.00000E+00
 3 DC4 0.00000E+00 0.00000E+00 0.00000E+00 0.00000E+00 0.00000E+00
  4 CV    0.0000000000000   TH    0.00000000 AIR
 END
```

Notice that the OPD coefficients have been altered. Zemax expresses the coefficients in units of *radians*, while all OPD expressions in SYNOPSYS are in units of *cycles*, or waves. So the values must be different, as you can see.

This example gave us little difficulty, so let us show a more difficult one. We open a file describing an IR lens (**IR_EXAMPLE.ZMX**) and see an error message when we run the conversion, shown in figure 37.3.

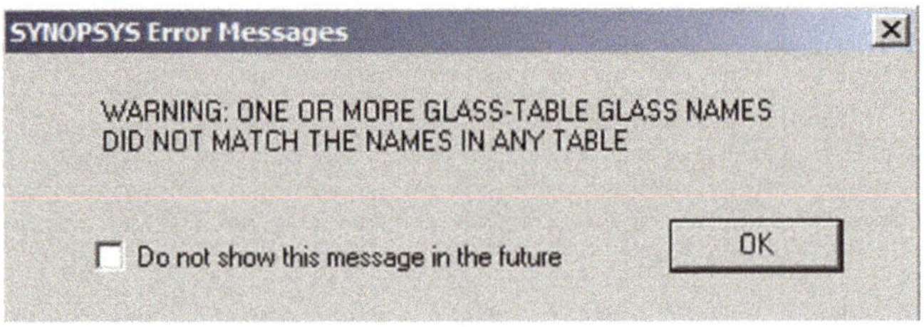

Figure 37.3. Warning message displayed by ZMCONVERT when a glass name does not match.

Reading the listing that scrolled by, you see some details:

```
SURF 12
 COMM OBJ EL1
 TYPE STANDARD
 CURV -4.12711514651300020OE-001 0 0.000000000000E+000 0.000000000000E+000 0
 HIDE 0 0 0 0 0 0 0 0
 MIRR 2 0.000000000E+000
 SLAB 4
 DISZ -2.362204724409E-001
 GLAS CLEARTRAN_WANDA 0 0 3.46217496 0.000000 0.000000 0 0 0 0.000000 0.000000
 **************************************************
 ***********  GLASS TYPE NOT FOUND *************
 CLEARTRAN_WANDA

 ***** A GLASS MODEL (GLM) IS USED INSTEAD ****
 *** SOME GLASS TABLES USE DIFFERENT SPELLING **
 *** CHECK THE NAME CAREFULLY. GLM DATA MAY **
 ************* NOT BE APPROPRIATE *************
 **************************************************
```

This surface needs a material whose name is not found in the SYNOPSYS glass tables. The program assigns a glass model since it has no other information at this point, but when you edit the resulting RLE file, you have to change to the right material. Here is what is assigned at this point:

```
 12 SID 'OBJ EL1              '
 12 RD    -2.42300
 12 TH   -0.236220
 12 GLM     1.50000        55.0000
 12 CAO    0.745000
 13 RD     7.82870
```

If you do not know the name of the material to use, you might look at the Unusual glass catalog. Type **HELP UNUSUAL** and follow the link. In the list that opens you find a likely candidate:

```
...
NACL        Sodium chloride                      0.2         22.3
NAFL        Sodium fluoride                      0.186       17.3
PBFL        Lead fluoride                        0.2909      11.9
SAPPHIRE    Aluminum oxide                       0.193       5.263
SILICON     Silicon; see SILICON-NIR, below      1.4         16.0
ZNS         Zinc sulfide                         0.42        18.2
CLEARTRAN   Zinc sulfide, higher grade           0.4047      13.0
ZNSE        Zinc selenide                        0.54        18.2
CRQUARTZ    Crystal quartz, ordinary ray   0.198       2.053
```

Now you can edit the RLE file:

```
12 SID 'OBJ EL1                    '
12 RD    -2.42300
12 TH   -0.236220
12 GTB U
CLEARTRAN
12 CAO    0.745000
13 RD     7.82870
13 TH   -0.100000E-01
13 CAO    0.745000
```

The same error showed up on several other surfaces, and you can correct them as well, perhaps with a `PIN 12` directive. Yet another surface wanted a material with the name `SILICON_FIT`. Of course you change this to `SILICON`. Proceeding in this manner, you identify all materials whose names you must update, and then run the MACro with the corrected RLE file.

Be particularly wary of glasses in Zemax from the Chinese company Guangming (CDGM). That company uses many of the same glass names as the Schott company, even though the index and dispersion are very different. Clearly, this is a disaster waiting to happen, and you must verify carefully which glass from which catalog you want to use. (SYNOPSYS identifies all of the Chinese glasses that have conflicting names with a prefix. Thus F2 becomes G-F2 and so on.)

IOP Publishing

Lens Design

Automatic and quasi-autonomous computational methods and techniques

Donald Dilworth

Chapter 38

Improving a Petzval lens

Designing a diffraction-limited spy camera

Joseph Max Petzval (1807–91) was a German mathematician whose name is attached to a type of camera lens consisting of two separated positive groups. (Legend says he used artillery gunners to perform *six months* of ray tracing for the first examples.) This lens form was widely used during the Cold War for spy cameras that flew over the Soviet Union. A typical example is shown in figure 38.1.

This example has two lenses added near the focal plane to correct field aberrations. Can DSEARCH design this kind of lens? The original required eight elements, but since we like a challenge, we will attempt it with only seven.

Here are the requirements:

- 13 inch focal length.
- F/3.5.
- Field ±6 degrees.
- Spectral range 0.7–0.52 μm.
- Total length 17.06 inches.
- Back focus 0.7 inches.

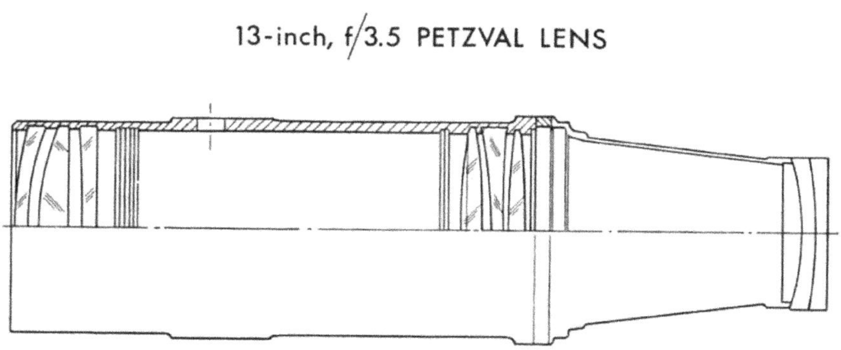

Figure 38.1. A Petzval spy camera design. Reproduced with permission from Berge Tatian.

Here is a suitable input MACro (**C38M1**):

```
CORE 14
TIME
DSEARCH 3 QUIET
 SYSTEM
 ID DSEARCH SAMPLE
 OBB 0 6 1.857
 WAVL 0.7 .6 .52
 UNITS INCH
 END
 GOALS
 ELEMENTS 7
 FNUM 3.5
 BACK 0.7 1
 TOTL 17 .1
 STOP FIRST
 STOP FREE
TSTART .5
ASTART 1.0
 RT 0.5
 FOV 0.0 0.75 1.0 0.0 0.0
 FWT 5.0 3.0 3.0 1.0 1.0
 NPASS 55
 ANNEAL 200 20 Q 50
 TOPD
 SNAPSHOT 10
 QUICK 40 40
 END
 SPECIAL PANT
 END
SPECIAL AANT
ADT 7 .01 10
M 0 10 A GIHT
S P YA 1
END
GO
TIME
```

Run this file, and in about two minutes the program returns a lens corrected to about 0.2 waves, which, after optimization and annealing (**50, 2, 50**) is even better, as shown in figure 38.2.

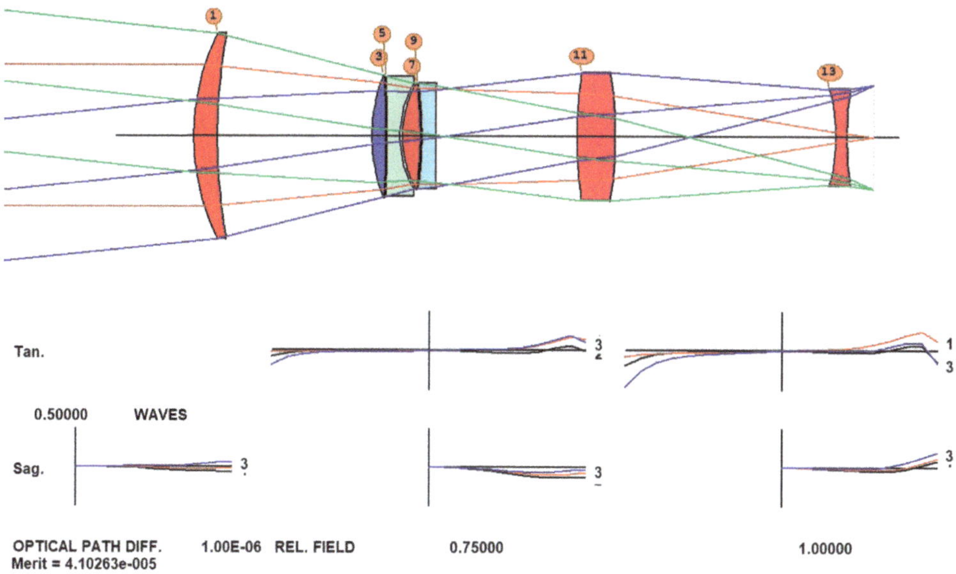

Figure 38.2. Lens returned by DSEARCH, optimized and annealed.

This lens is already at the diffraction limit, and it does not resemble the Petzval configuration at all. It is always pleasant to discover a lens form the old masters missed.

The lens is superb, with almost no secondary color. Note the positive flint element at surface 3. The program often finds this a good way to reduce secondary color. Next, we will insert real glasses with ARGLASS—but first, save this version so you can go back if necessary.

Open the **MRG** dialog, select 'Schott' and 'Sort' and the lens comes back with excessive secondary color. That program assigns glasses by finding the closest real glass to the model and optimizing the result. It often works very well, but in this case we need more horsepower.

Let us see what GSEARCH can do. Get back the saved version, save the optimization MACro with the name GSOPT.MAC, and prepare another MACro:

```
GSEARCH 5 QUIET LOG
SURF
1 3 5 7 9 11 13
END
NEAREST 3 P
S
END
GO
```

Run this, and GSEARCH finds an excellent selection of glass types from the Schott catalog. The program will examine, in this case, 3^7 combinations, or 2187 altogether. The result (**C38L1**), after more optimization and annealing, is shown in figure 38.3.

We also tried matching this design to the Guangming and Ohara catalogs, but those, in this case, were not quite as good. This is no criticism of those vendors, by the way; sometimes the reverse is true. When running GSEARCH, it pays to try several catalogs, since the results depend on which glasses are close to which GLMs in the lens, and they all differ.

Open the MMF dialog (**MMF**), select 'Multicolor', and click 'Execute'. The MTF curves for this lens are shown in figure 38.4. The performance is considerably better than that of the original Petzval design, and with only seven elements—enough said.

We repeated this exercise, changing the DSEARCH input from

```
FNUM 3.5
BACK 0.7 1
```

to

```
FNUM 3.5 10
BACK 0.7 SET
```

which opens up yet more possiblities. Now the F/number will be a target in the MF instead of the result of a UMC solve on the last element, and the back focus distance will be set at 0.7 inches and will not be controlled by a YMT solve. With this change,

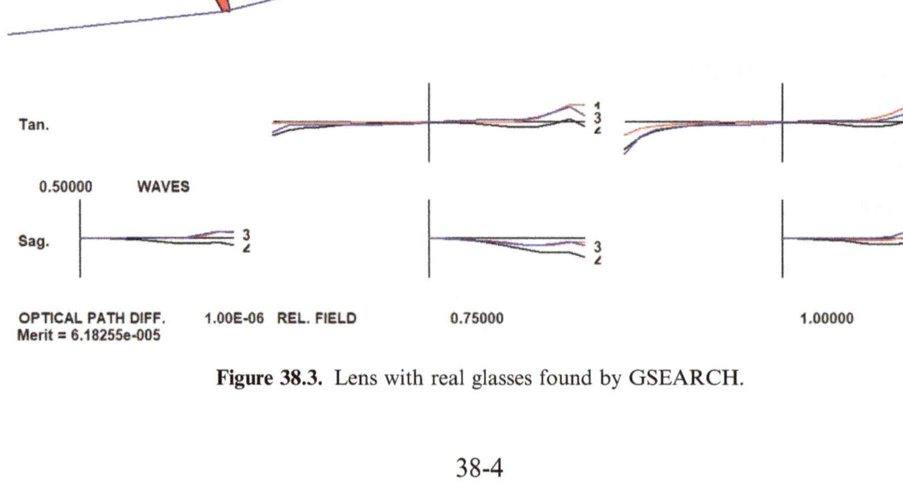

Figure 38.3. Lens with real glasses found by GSEARCH.

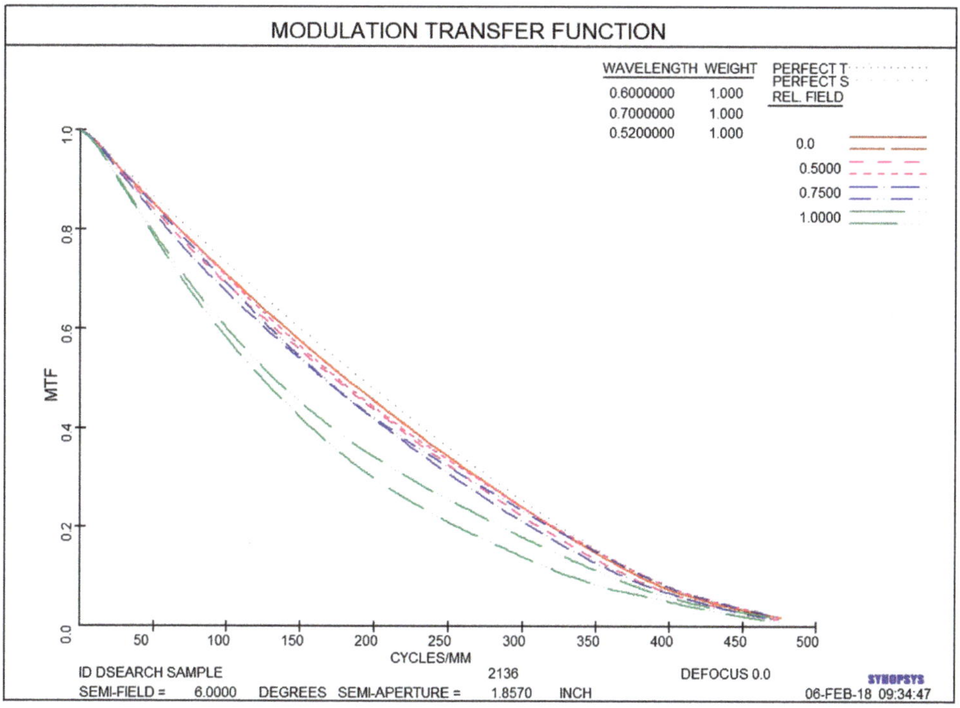

Figure 38.4. MTF of the lens with real glass.

DSEARCH returned a different design—but when real glasses were inserted with GSEARCH, the quality was not quite as good as before. That happens. We are counting on at least some of the glass models being near a real glass with just the right properties, but this is not guaranteed to happen. Still, it is an alternative way to describe the task to DSEARCH and is worth trying. Another thing to try is giving a value for RSTART different from the default 100 inches. That will also explore different branches of the design tree.

In preparing this chapter, we tried several combinations of the paramters RT, ASTART, and RSTART. Each produced different results, and while most were in the 1/4 wave region, some were not. This is the nature of the task we are addressing. As Kingslake observed, 'I tried everything …'. It would be great if our algorithm always found the best of all possible results, but we are not quite there yet. Still, the tools we have are already impressive, even considering their limitations. We are searching a very large and very bushy tree, and those tools are able to find excellent designs most of the time.

We are curious what would happen if we ask DSEARCH for a six-element lens. Can it do better? Here is the input file (**C38M2**); we want all the power we can obtain, so the **QUICK** directive is commented out:

```
CORE 14
TIME
DSEARCH 3  QUIET
 SYSTEM
 ID DSEARCH SAMPLE
 OBB 0 6 1.857
 WAVL 0.7 .6 .52
 UNITS INCH
 END
 GOALS
 ELEMENTS 6
 FNUM 3.5
 BACK 0.7 1
 TOTL 17 .1
TSTART 1.0
ASTART 1.0
 STOP FIRST
 STOP FREE
 RT 0.5
 FOV 0.0 0.75 1.0 0.0 0.0
 FWT 5.0 3.0 3.0 1.0 1.0
 NPASS 90
 ANNEAL 200 20 Q 90
 TOPD
 SNAPSHOT 10
 !QUICK 40 40
 END
 SPECIAL PANT
 END
SPECIAL AANT
ADT 7 .01 10
M 0 10 A GIHT
S P YA 1
END
GO
TIME
```

When DSEARCH finishes, save the file DSEARCH_OPT as GSOPT as you did before. Then make a new MACro to run GSEARCH and find the best glass combination from the Schott catalog:

```
GSEARCH 5 QUIET LOG
SURF
1 3 5 7 9 11
END
NEAREST 3 P
S
END
GO
```

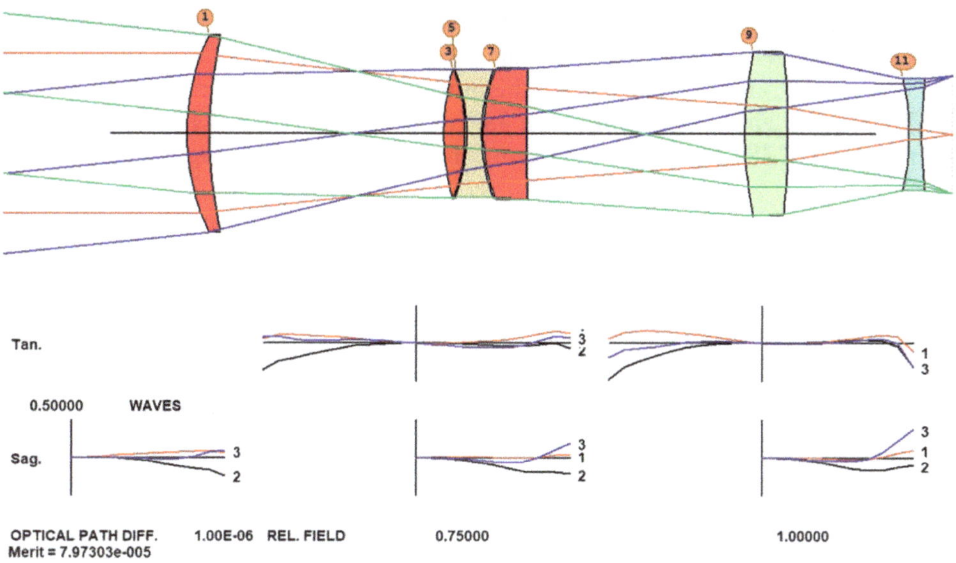

Figure 38.5. Six-element lens found by DSEARCH and then GSEARCH.

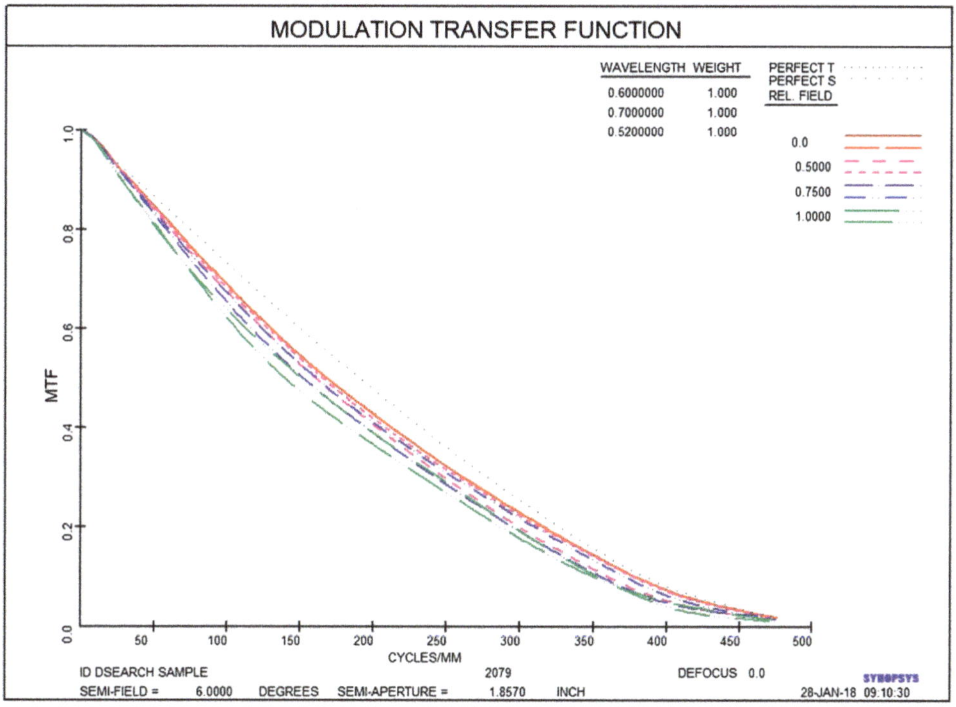

Figure 38.6. MTF of the final six-element lens, with real glass.

Now open each of the lenses found by DSEARCH and run this GSEARCH MACro. Do the same with the Ohara and Guangming catalogs. The best one came back with a MF of 0.000 08, matched to the Guangming catalog, then optimized and annealed (**50, 2, 50**). The lens is shown in figure 38.5, and the MTF is in figure 38.6. This is a superb design (**C38L2**). Only six elements, and better than the original eight-element Petzval.

Different glasses have different partial dispersions, so it makes sense to try more than a single catalog. If you obtain a great design with one catalog, you can sometimes then change to equivalent but less expensive glasses from a different vendor with little loss of quality.

IOP Publishing

Lens Design
Automatic and quasi-autonomous computational methods and techniques
Donald Dilworth

Chapter 39

Athermalizing an infrared lens

Calculating spacings and materials so thermal effects are minimized

In this chapter, we will examine what happens to the image of a mid-infrared lens as the temperature changes. Start with the lens in figure 39.1 (**C39L1**).

Here is the RLE file for this example:

```
RLE
ID FOUR ELEMENT INFRARED OBJECTIVE
 WAVL 4.000000 3.250000 2.500000
 APS        1
 UNITS MM
 OBB 0.000000  3.0000 30.0000  0.0000  0.0000  0.0000  30.0000
 MARGIN    1.270000
 BEVEL    0.254001
  0 AIR
  1 RAD  163.0500000000000  TH   4.50000000
  1 N1 3.42403414 N2 3.42836910 N3 3.43782376
  1 DNDT 1.336E-04 1.336E-04 1.336E-04 1.4000E+00 7.5000E+00 1.6000E+01
  1 CTE  0.255000E-05
  1 GTB U  'SILICON      '
  1 EFILE EX1  31.417334  31.417334  31.671335   0.000000
  1 EFILE EX2  31.014427  31.417334   0.000000
  2 RAD  255.4500000000000  TH   5.55000000 AIR
  2 AIR
  2 EFILE EX1  31.014427  31.417334  31.671335
  3 RAD  -721.5000000000000  TH   3.60000000
  3 N1 4.02415626 N2 4.03741119 N3 4.06419029
  3 DNDT 4.100E-04 4.100E-04 4.100E-04 2.0500E+00 1.1000E+01 2.200E+01
  3 CTE  0.550000E-05
  3 GTB U  'GE       '
  3 EFILE EX1  30.633643  30.633643  30.887644   0.000000
  3 EFILE EX2  30.633643  30.633643   0.000000
  4 RAD -1590.0000000000000  TH   65.70000000 AIR
  4 AIR
  4 EFILE EX1  30.633643  30.633643  30.887644
```



```
 5 RAD  145.5000000000000  TH   3.15000000
 5 N1 4.02415626 N2 4.03741119 N3 4.06419029
 5 DNDT 4.100E-04 4.100E-04 4.100E-04 2.0500E+00 1.1000E+01 2.200E+01
 5 CTE  0.550000E-05
 5 GTB U  'GE          '
 5 EFILE EX1  27.236976  27.236976  27.490977   0.000000
 5 EFILE EX2  26.712556  27.236976   0.000000
 6 RAD  120.4500000000000  TH   13.20000000 AIR
 6 AIR
 6 EFILE EX1  26.712556  27.236976  27.490977
 7 RAD  255.0000000000000  TH   4.50000000
 7 N1 3.42403414 N2 3.42836910 N3 3.43782376
 7 DNDT 1.336E-04 1.336E-04 1.336E-04 1.4000E+00 7.5000E+00 1.600E+01
 7 CTE  0.255000E-05
 7 GTB U  'SILICON     '
 7 EFILE EX1  27.355510  27.355510  27.609511   0.000000
 7 EFILE EX2  27.165926  27.355510   0.000000
 8 RAD  2025.0000000000000  TH  107.272545 AIR
 8 AIR
 8 EFILE EX1  27.165926  27.355510  27.609511
 9 RAD  -405.0000000000000  TH   0.00000000 AIR
 9 AIR
END
```

Let us assume this lens must stay in focus over the temperature range of 20–100 °C. What does it do now? To find out, run the THERM program, first testing to see if all required coefficients are present:

```
SYNOPSYS AI>THERM TEST

 WARNING - NO DEFAULT CTE HAS BEEN ASSIGNED TO AIRSPACES
 ALL GLASSES IN THIS LENS HAVE BEEN ASSIGNED THERMAL-INDEX COEFFICIENTS
```

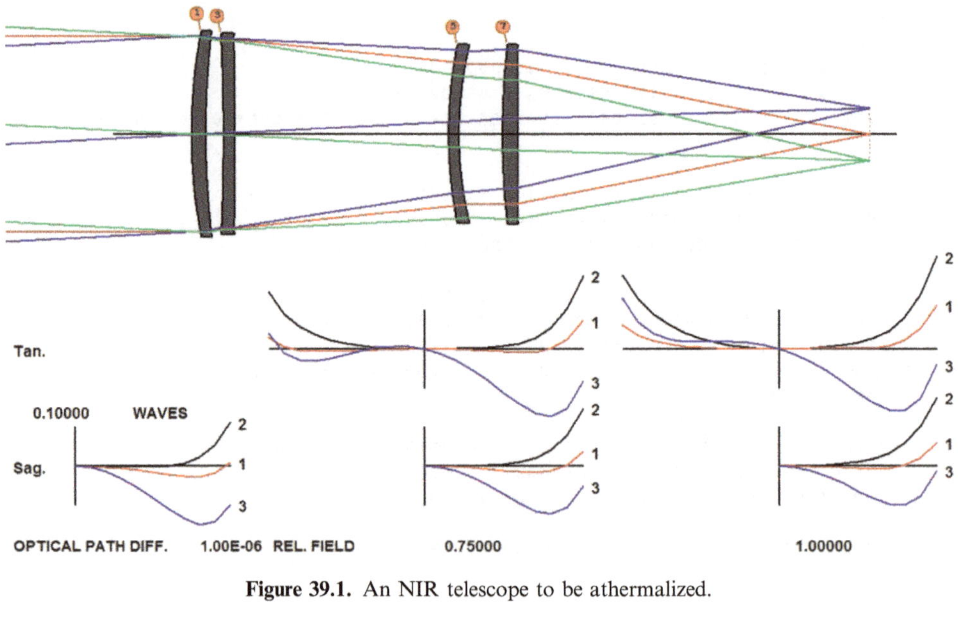

Figure 39.1. An NIR telescope to be athermalized.

Indeed, this lens was never assigned a coefficient for the airspaces. Fix that with a CHG file, assigning the coefficient of aluminum type 6061:

```
CHG
ALPHA A6061
END
```

Now you can activate thermal shadowing. Create and run a new MACro:

```
THERM
ATS 100 2
END
```

This puts a copy of the lens in configuration 2, with all parameters altered as required by the change in temperature from the default 20–100 °C. Figure 39.2 shows what ACON 2 looks like at that temperature.

The lens is out of focus—we have to correct that.

Here is an easy way to tell where an axial shift of an element might do some good. First, make a checkpoint in ACON 2, by clicking the button. Now open the WorkSheet (click on the button), and then click on surface 4 in the PAD display. We suspect that a change of that airspace might alter the focus position. To be

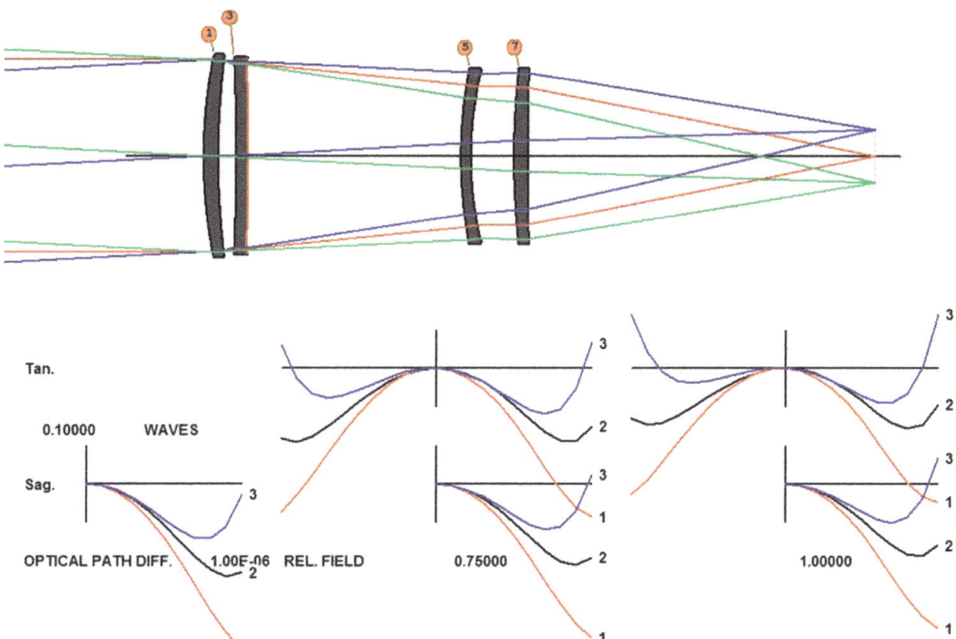

Figure 39.2. NIR telescope at 100 °C, before correcting.

practical, the required motion must be quite small, so slide the speed slider closer to the bottom, and then slide the 'Spacing' slider to the right, as shown in figure 39.3.

Indeed, the image comes back nearly in focus, and the motion was quite small, from 65.7 to 65.569—we are getting close.

Now we have to figure out a way to make element 3 move in that way with temperature. One trick that sometimes works is to design the cell with an outer sleeve that extends from surface 4 to the right, past the next elements, and then with an inner sleeve that comes back partway and holds those elements. If the outer sleeve is made of aluminum and the inner one of plastic, the net motion of element 3 will be less than it would be with an all-aluminum cell.

Go back to ACON 1 again, with the WorkSheet still open, make a checkpoint, and click the '**Add Surface**' button, . Now click on the axis in the lens drawing in between surfaces 4 and 5. A dummy surface is inserted, shown in figure 39.4.

Now you must tell the program that the expansion coefficient from 5 to 6 is different from the default aluminum, which we assigned above to all airspaces. Close WS and make a new THERM file:

```
THERM
COE 1 STYRENE
TCHANGE 1
5
ATS 100 2
END
```

This file says to define coefficient number 1 as that of styrene, and then assign thickness coefficient number 1 to surface 5. Then alter the lens to 100 °C and put the result in ACON 2.

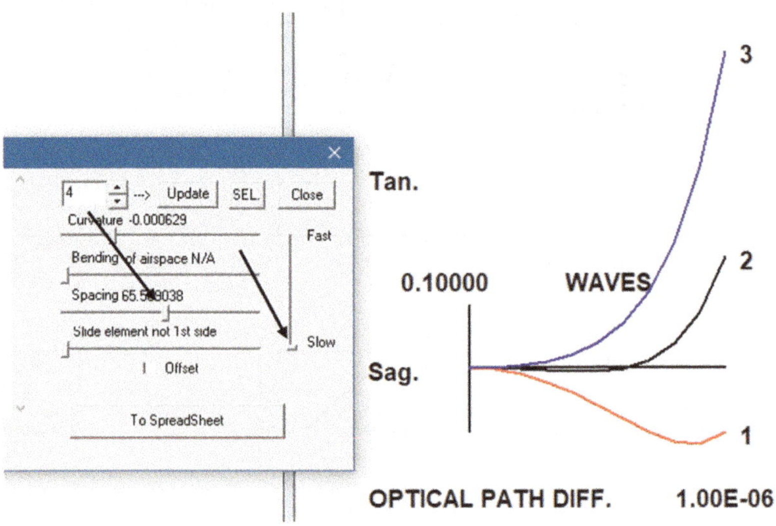

Figure 39.3. WorkSheet sliders for adjusting an airspace in the lens.

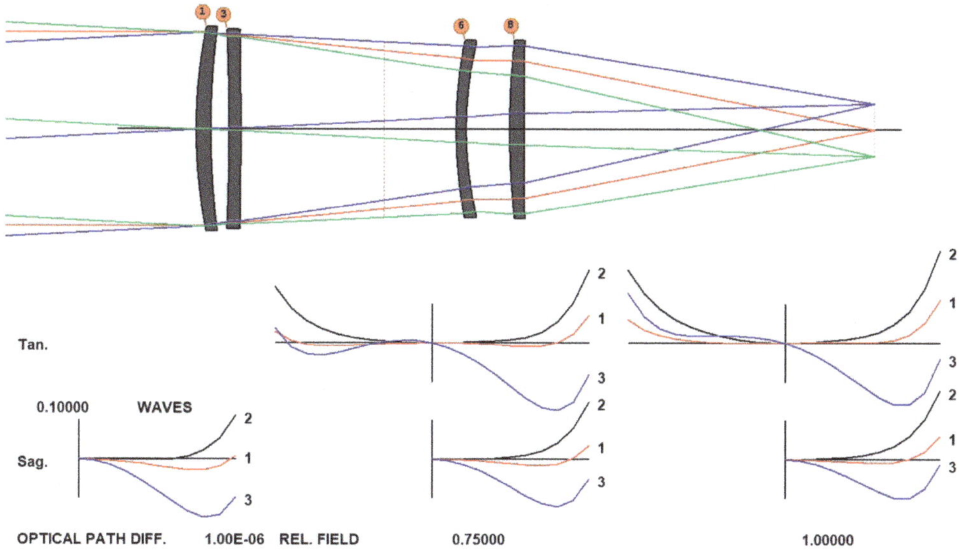

Figure 39.4. NIR lens with dummy surface inserted in preparation for athermalization.

Run this, and ACON 2 has indeed changed. The trick now is to find the length of the outer and inner sleeves that will best compensate for this thermal change. For this task, we use the optimization program. Here is the MACro **(C39M1)**:

```
ACON 1
PANT
VY 4 TH 1000 -1000
VY 5 TH 1000 -1000
END

AANT
ACON 1
M 0 1 A DELF
M 8.103249 1 A P YA 1

GSO 0.5 5.332000 3 M 0
GNO 0.5 1 3 M 0.5
GNO 0.5 1 3 M 1.0

ACON 2
M 0 1 A DELF

GSO 0.5 5.332000 3 M 0
GNO 0.5 1 3 M 0.5
GNO 0.5 1 3 M 1.0
END
```

```
SNAP
SYNO 20 MULTI
```

This will attempt to keep the system in focus at both temperatures and try to hold image quality at the same time, while varying thicknesses 4 and 5. Note the MULTI declaration on the optimization command. That allows the program to optimize more than a single configuration. Run this, and now the lens in ACON 2 is better than before, as shown in figure 39.5.

There is some image degradation, but within reason, and the focus remains where it should even with this change in temperature. Note the position of surface 5 now. That tells you where the two sleeves must extend to and where they should connect. Athermalization does not have to be difficult.

Some comments are in order. We entered explicit limits for the TH variables since the program will not let a positive TH become negative otherwise. To keep the magnification constant, we added a target for the YA of the chief ray. We did not implement the options to account for whether the cells hold the lenses on the right or left side of an element, because for this example the expansion is applied in the right place by default. Yes, sometimes athermalization is more complicated, and you are referred to the User's Manual for a complete description of the options you have available for more demanding tasks.

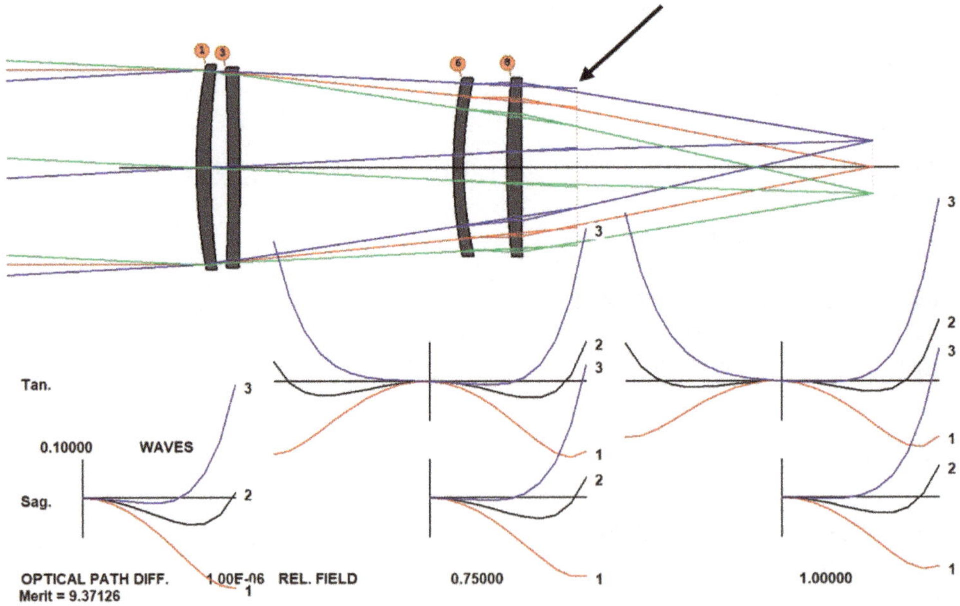

Figure 39.5. Lens at 100 °C with athermalization.

IOP Publishing

Lens Design

Automatic and quasi-autonomous computational methods and techniques

Donald Dilworth

Chapter 40

Edges

Defining lens edges and bevels

Before you send element drawings to the shop, you have to carefully define the shape and dimensions of the edges. This is achieved with the Edge Wizard[1] (**MEW**).

Get out lens shown in figure 40.1 (**C40L1**).

This lens has already been assigned reasonable edges with the Edge Wizard. To show how it works, we will first delete all the edge definitions—and then show how to put them back. Type in the Command Window:

```
EFILE
ERASE
END
```

Now you see the default edges in figure 40.2, assigned by the program so they clear the upper and lower marginal rays at full field. Those are reasonable edges to use during lens optimization, so you can see what is going on, but when you manufacture the elements they have to be larger and defined more carefully to allow for mounting in a cell. Open the Edge Wizard, either by typing **MEW** or by clicking the button  on the PAD toolbar. At the moment, the lens has no edge definitions anymore. Click the '**Create all**' button, and you obtain a set of reasonable edges. Click '**Yes**' to the prompt, and the picture has changed. Enter the number 1 in the 'From surface' box and you can see the dimensions that have been applied to the first element, shown in figure 40.3.

The program creates five reference points at the edge of each element, labeled A through E in the diagram on the dialog. You usually have to edit those default

[1] Edge Wizard™ is a trademark of Optical Systems Design, Inc., a Maine, USA corporation.

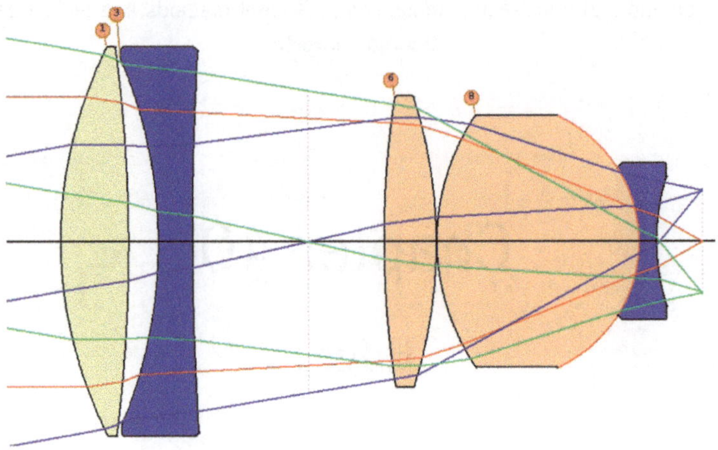

Figure 40.1. Lens with EFILE edges assigned.

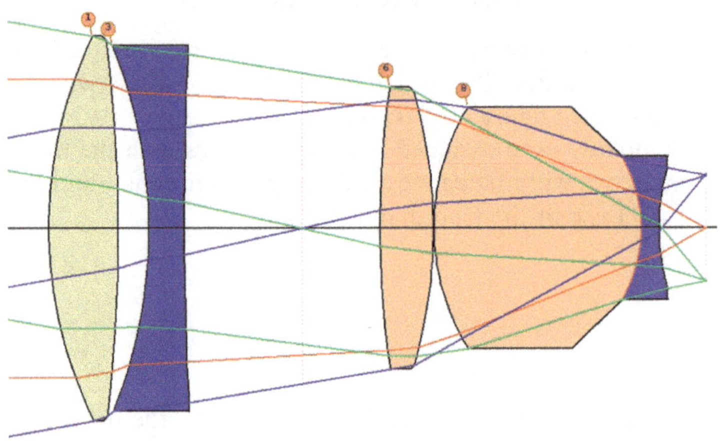

Figure 40.2. Lens with default edges.

dimensions, and the data for element 1 show one reason why. The first surface is convex, and you may not want a bevel on that surface. The program defined the default edges and put into effect the 'Explicit' rules, which work for most lenses, with data you can edit with the edit boxes and spin buttons on the dialog if you want.

The default point C is currently at 34.2198 mm from the axis, while the clear aperture on surface 1 is 31.9355. This element has a rather thin edge, so let us reduce the diameter slightly. Enter the number 34 in the box for dimension C and click 'Update'. We will also remove the bevel on that surface. Click the box '-B' to the left of the C dimension. '-B' means remove the bevel on that side. Then click the '-F' button too. When you remove the bevel, you leave point A where it was, which may be appropriate for some plastic elements that have a mounting flange molded in place, but is not welcome here. That button removes the flat portion from A to B.

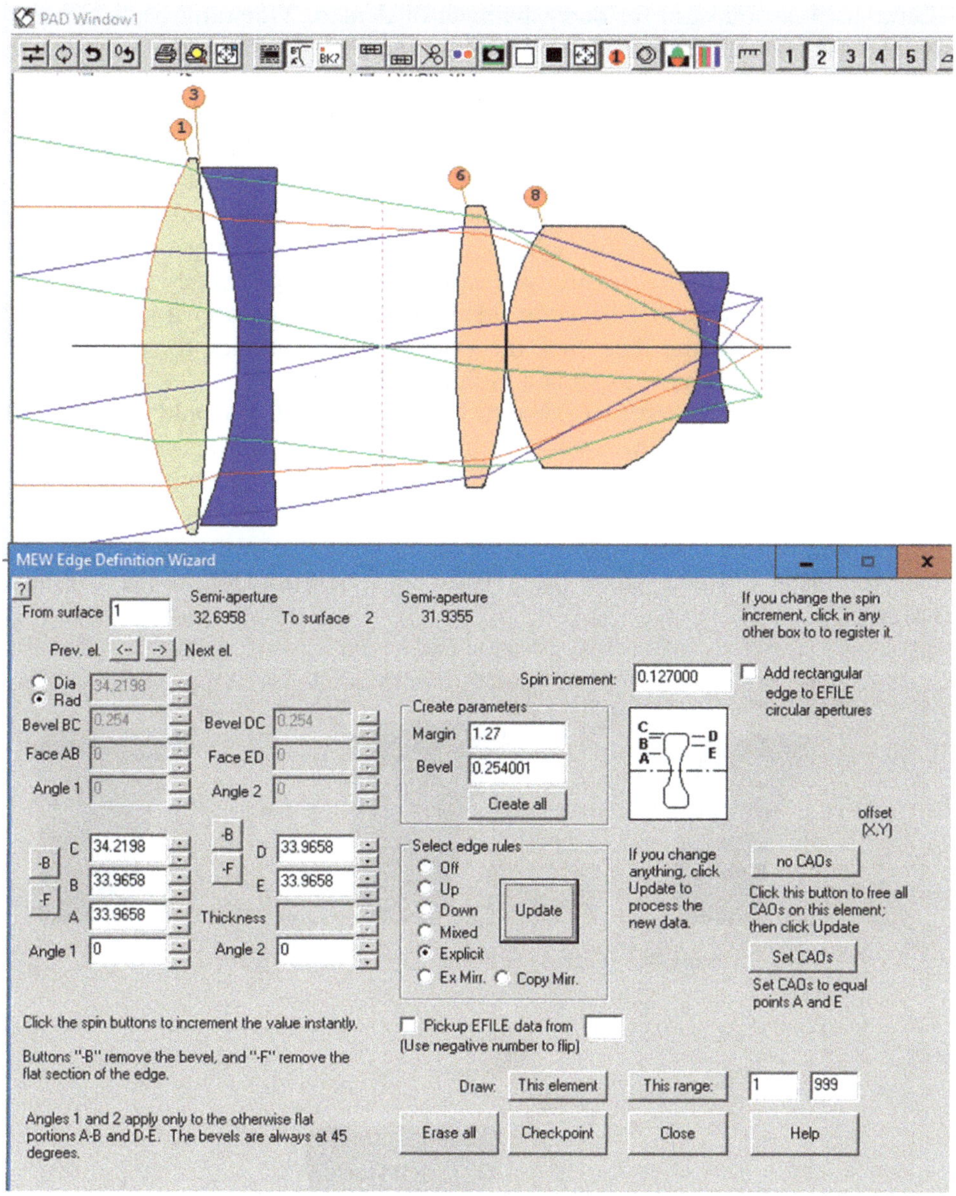

Figure 40.3. Edge Wizard, showing data for element 1.

Side 1 is now reasonable, and you may want to remove the bevel and flat on side 2. (For thicker positive elements with shallow curves, we usually leave the bevels in place.)

Element 2 is negative, and here we want a flat portion on side 1 and a bevel—but no flat—on side 2. Click the '**Next el.**' button to see the data on that element, shown in figure 40.4 on the left.

Let us assume you want the outer diameter of element 2 the same as on element 1. Just enter the same dimension, 34 in the C box and click '**Update**'. The edge changes, as shown in the center.

This increases the element diameter, but again has left dimensions A and B where they were. Let us reduce the size of the bevel on surface 3. To the right of the edit box for dimension B there are two spin buttons (see figure 40.3). Click the upper of the two a number of times, watching the bevel becoming smaller and the flat portion becoming larger, shown on the right in figure 40.4. By clicking on the two spin buttons while watching the picture, you can define the edge exactly as you want.

If you also want a smaller bevel on side 2 of that element, adjust it with the spin buttons for points D, and then click the **–F** button for that side to remove the flat portion. The edges are now as shown in figure 40.5.

Now the edges of the first two elements look about right. It would be a good idea at this point to click the '**Checkpoint**' button on the MEW dialog. As you work on the other elements, you might make a mistake and want to go back to a previous version.

Proceeding in this fashion you can define all the edges just as you want them. When you have finished, close the Wizard and type **`ELIST`** in the Command Window:

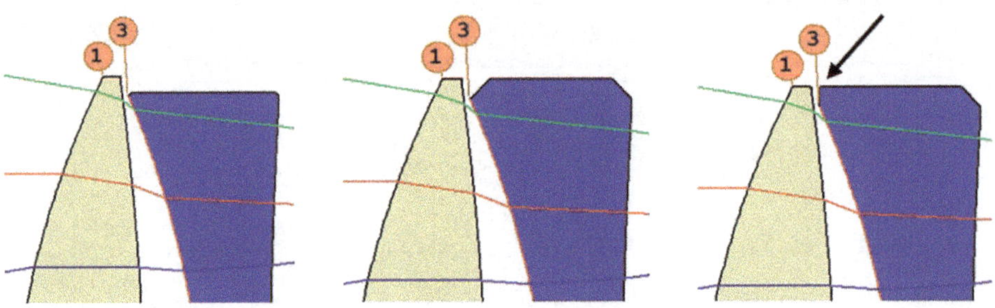

Figure 40.4. Edges to be modified by the Edge Wizard.

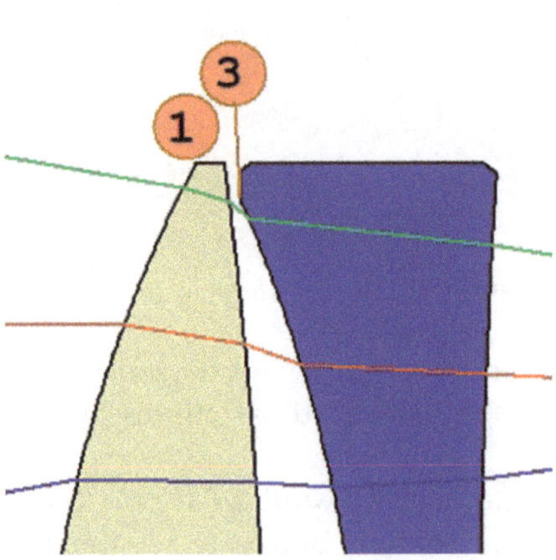

Figure 40.5. Finished edges of first two elements.

```
SYNOPSYS AI>ELIST

CURRENT EFILE DATA:

Surf.  A            AB            BC            C          ANG          CAO      TYPE
       E            ED            DC            C          ANG2         CAO

----------------------------------------------------------------------------------
  1   34.000       0.0000       0.0000       34.000      0.0000       32.696    EXPL
      33.966       0.0000      0.34200E-01   34.000      0.0000       31.936

  3   32.345       1.5240       0.13071      34.000      0.0000       31.075    EXPL
      33.615     0.12192E-05    0.38471      34.000      0.0000       29.602

  6   25.242       0.0000       0.25400      25.496      0.0000       23.972    EXPL
      25.242       0.0000       0.25400      25.496      0.0000       23.349

  8   21.809       0.0000       0.25400      22.063      0.0000       20.539    EXPL
      21.809       0.0000       0.25400      22.063      0.0000       12.174

  9   13.444       0.0000       0.25400      13.698      0.0000       12.174    EXPL
      11.590       1.8532       0.25400      13.698      0.0000       10.320

CURRENT BEVEL IS    0.2540010
CURRENT MARGIN IS     1.270000
SYNOPSYS AI>
```

These edges become part of the lens file and show up in the RLE data as EFILE parameters. For element 2, those data look like this:

```
...
   3 RAD     -81.3505230000000    TH       6.00000000
   3 N1 1.83648474 N2 1.84664080 N3 1.87201161
   3 CTE    0.830000E-05
   3 GTB S     'SF57                    '
   3 EFILE EX1      32.345300     33.869287     34.000000       0.000000
   3 EFILE EX2      33.615288     33.615289      0.000000
   4 RAD     553.8617899999995    TH      19.92504900 AIR
   4 AIR
   4 EFILE EX1      33.615288     33.615289     34.000000
...
```

While it is *possible* to edit the edge dimensions in the WorkSheet, it is not recommended. Some of them are coupled to others and the result is not always intuitive. Use the Wizard to edit the data if you need to. Everything is shown on the dialog and it is very simple to use.

40.1 A mirror example

This has been a useful example, but now we will look at a system with fold mirrors. Those can be assigned edges—and thicknesses—too. The lens is in **C40L2**.

Fetch this lens and type **CAP** to look at the current apertures:

```
SYNOPSYS AI>CAP

 ID EXAMPLE FOLDED SYSTEM                 28301            01-SEP-17    14:25:22

 CLEAR APERTURE DATA
 (Y-coordinate only)
```

SURF	X OR R-APER.	Y-APER.	REMARK	X-OFFSET	Y-OFFSET	EFILE?
1	0.2621		*User CAO			
2	0.6611		Soft CAO			*
3	0.6870		Soft CAO			*
4	0.7014		Soft CAO			*
5	0.7064		Soft CAO			*
6	0.7071		Soft CAO			*
7	0.6302		Soft CAO			*
8	0.5566		Soft CAO			
9	1.2000	1.6000	*User RAO			*
10	0.6779		Soft CAO			
11	0.9358		Soft CAO			*
12	0.9595		Soft CAO			*
13	1.5000	2.2000	*User RAO			*
14	0.9705		Soft CAO			
15	2.0000	2.4000	*User RAO			*
16	0.9793		Soft CAO			
17	1.0306		Soft CAO			*
18	1.0406		Soft CAO			*
19	1.0424		Soft CAO			
20	1.0424		Soft CAO			

```
 NOTE: CAO, CAI, EAO, and EAI input is semi-aperture.
       RAO and RAI input is full aperture.
 SYNOPSYS AI>
```

This system currently has EFILE edges, as indicated by the '*' on the listing above. Now open the Edge Wizard again and click the '**Erase all**' button on the Wizard to revert to default edges. A portion of the system is shown in figure 40.6.

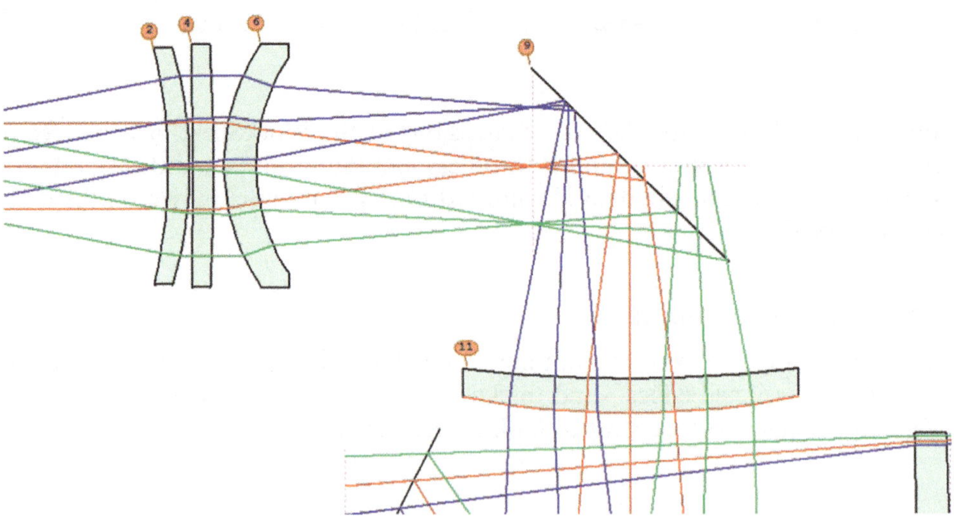

Figure 40.6. System with fold mirrors and default edge definitions.

Surface 9 is a fold mirror that has been assigned a rectangular outside aperture of dimensions 1.2 inches × 1.6 inches. (Those are the full dimensions of the rectangle: circular apertures are given by the radius, rectangular by the side lengths.) However, without assigned EFILE data, it shows up as just a straight line on the PAD display. The Edge Wizard can create suitable dimensions on this mirror, with the aspect ratio taken from the RAO data. In the Wizard, navigate to surface 9, and you see that nothing is assigned yet. Select the '**Ex Mirr**' option and click '**Update**'. A default edge is created, and now the mirror has a thickness, as shown in figure 40.7.

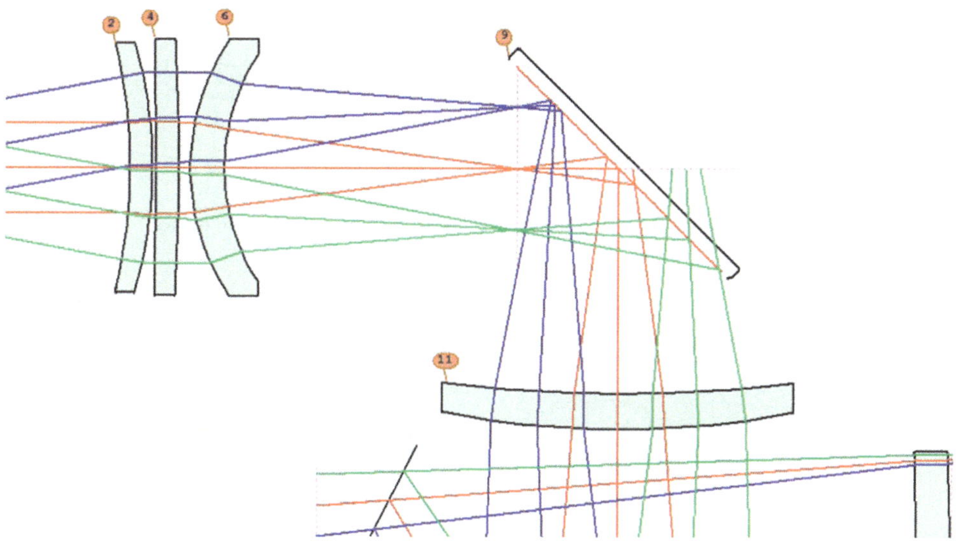

Figure 40.7. A thickness has been assigned to the fold mirror at surface 9.

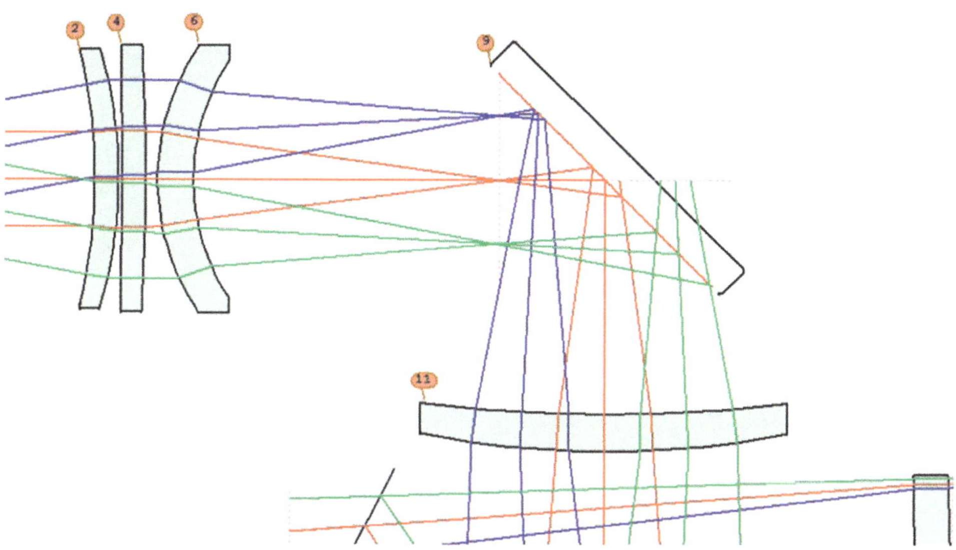

Figure 40.8. Mirror thickness increased with the Edge Wizard.

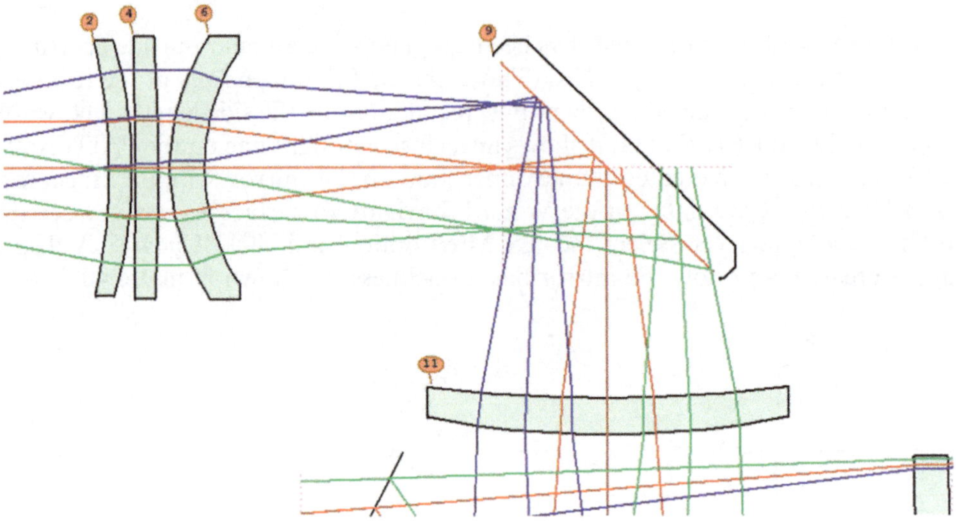

Figure 40.9. A bevel has been added to the back side of the fold mirror.

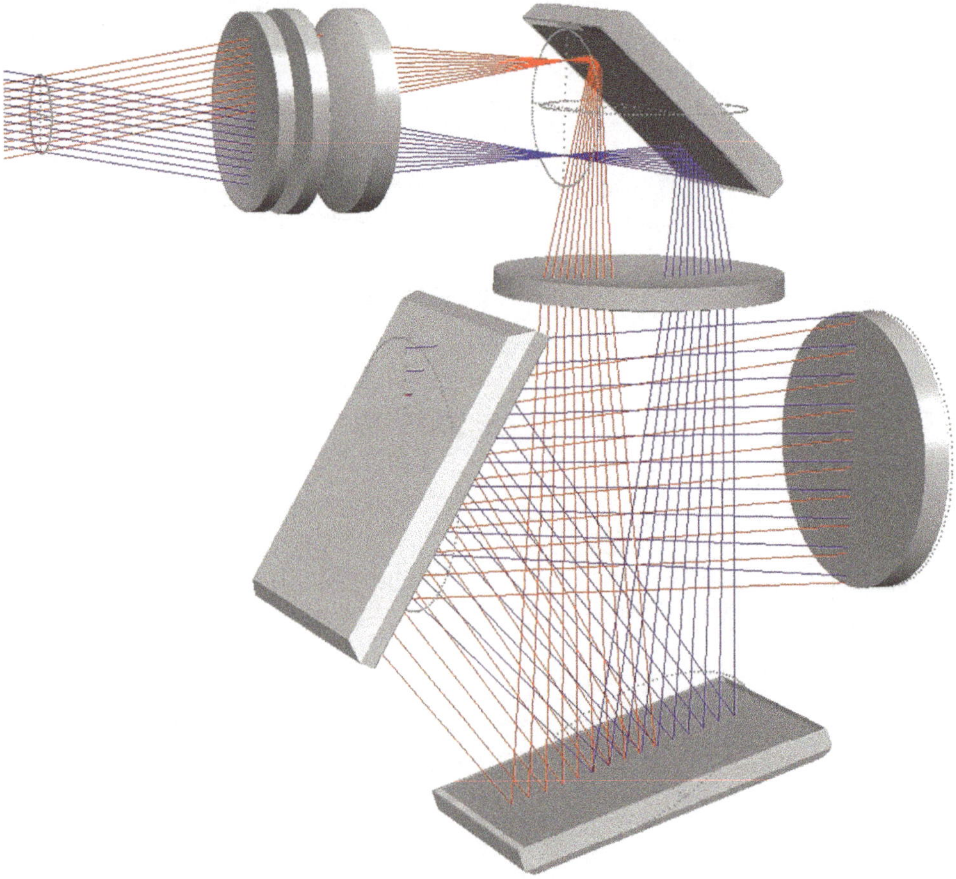

Figure 40.10. RSOLID drawing of the system.

Let us assume you want it to be thicker. You can enter a larger number in the 'Thickness' edit box or click the upper spin button on that box. The thickness increases, as shown in figure 40.8. The amount by which the spin buttons change the dimension is given in the 'Spin increment' box.

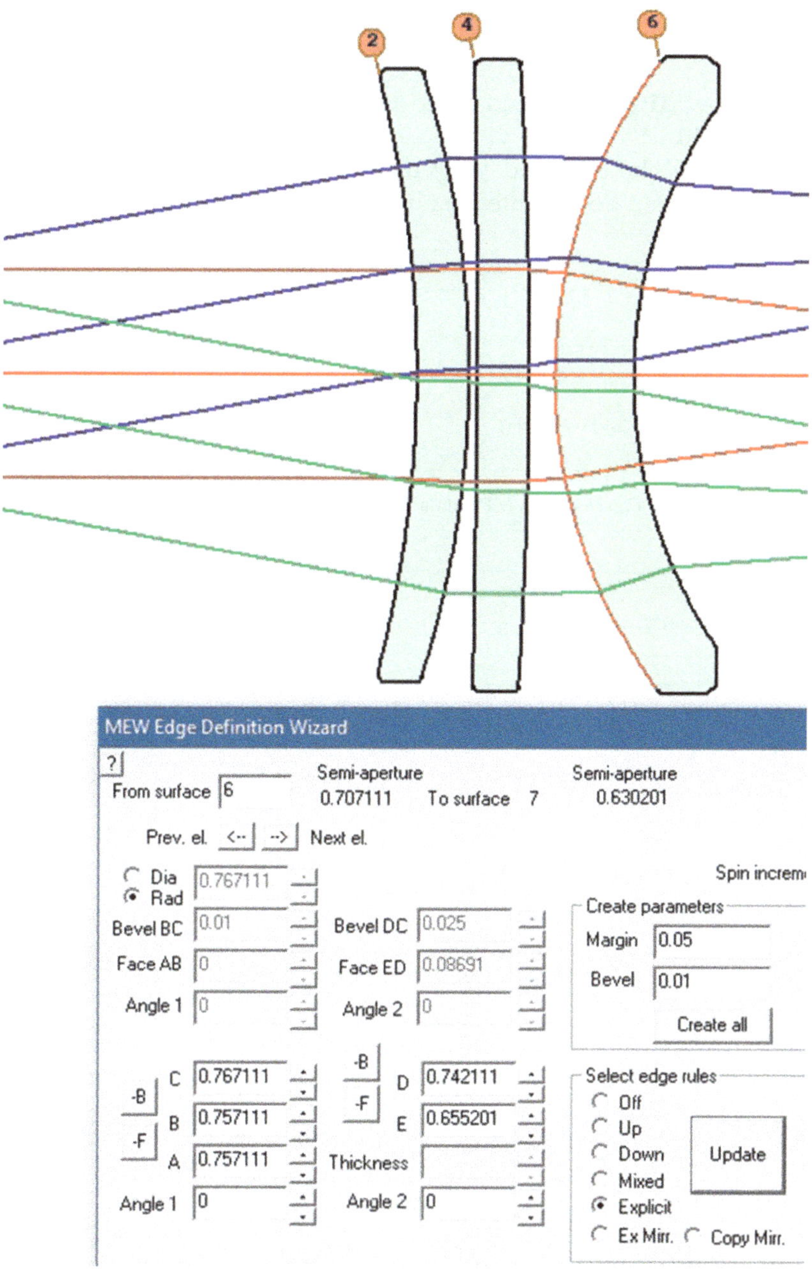

Figure 40.11. Edge Wizard with data for adjusting edges on surfaces 6 and 7.

Change the increment to 0.02, click ‘Update’, and use the lower spin button in box D to add a bevel to the back side of the mirror. The bevel is shown in figure 40.9.

Add edges to the other fold mirrors, at surfaces 13 and 15, in the same way, and close the Wizard. Now make an RSOLID picture with the dialog **MPE**, shown in figure 40.10 (**C40L3**).

Your mirrors are shown, beveled exactly as you wanted.

Go back to the Wizard and define the edge for element 3, at surface 6, as shown in figure 40.11.

Now open the **MPL** dialog and enter data for an ELD drawing at surface 6, as shown in figure 40.12.

Click the ‘**ELD**’ button, and your drawing shows up—with all of the edge dimensions shown and documented, as in figure 40.13.

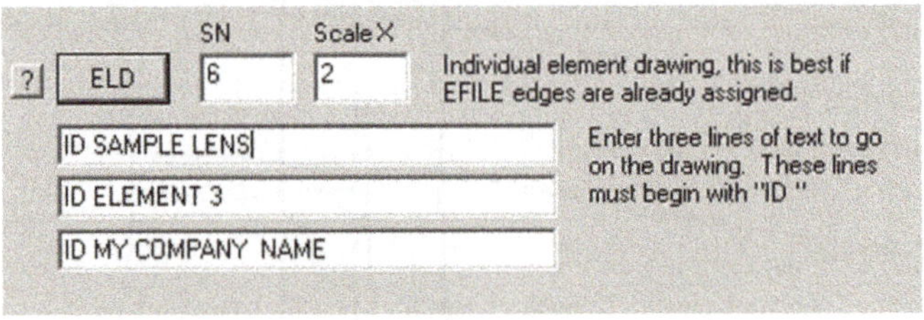

Figure 40.12. MPL data for drawing a lens element.

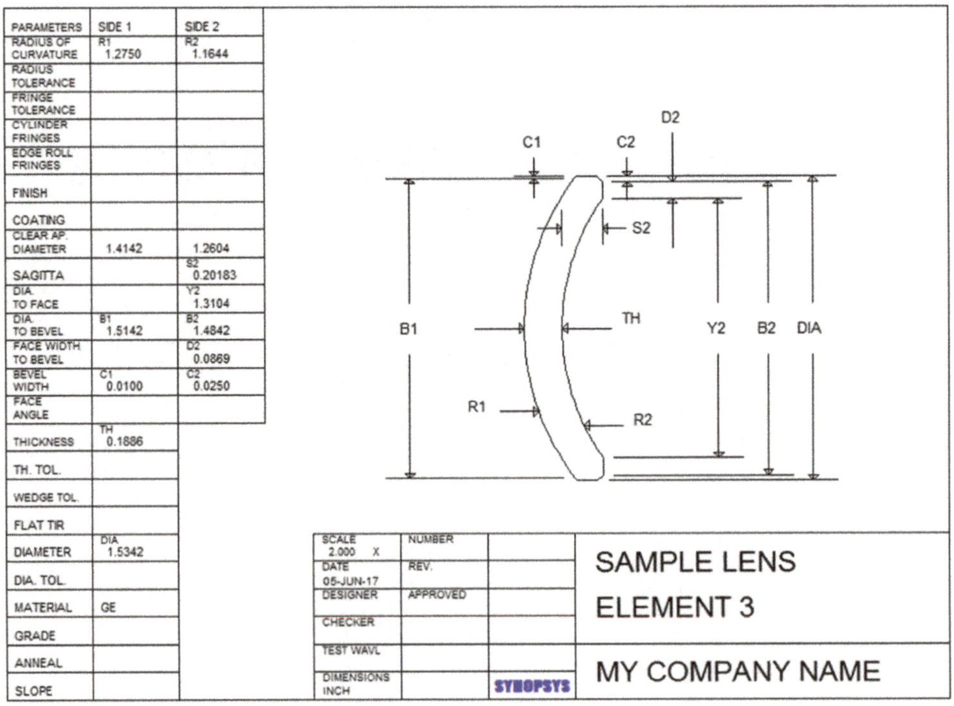

Figure 40.13. Sample element drawing with user-defined edge geometry.

This has been a brief introduction to the Edge Wizard. There are many more options, including some that will automatically change dimension C if the lens CAO changes. You may never need that much horsepower, but by all means read section 7.8 of the User's Manual. There you will learn how to create edges like this (**C40L4**):

```
RSOLID 22 33 0 0 0
PLOT
PUPIL 1
RED
TRACE P 0 0 200
END
```

The RSOLID drawing is shown in figure 40.14.

Figure 40.15 shows a system with an off-axis aspheric mirror (**C40L5**). Study the RLE file to see how the edges were defined. This system has a real stop on surface 5, the first refractive element, and a DCCR directive on mirrors at 2 and 3. Surface 4 is

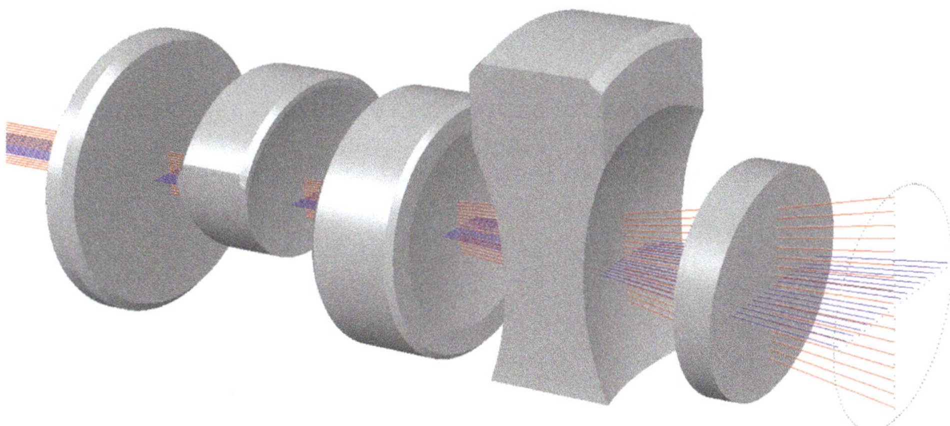

Figure 40.14. RSOLID drawing showing a variety of edge shapes.

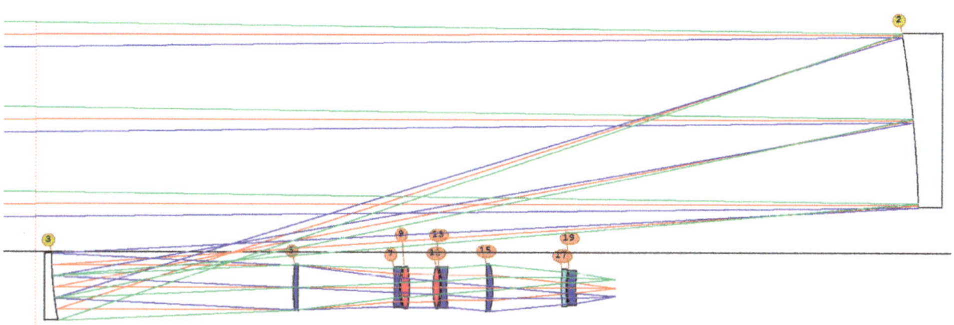

Figure 40.15. A system with a decentered aperture on an off-axis mirror.

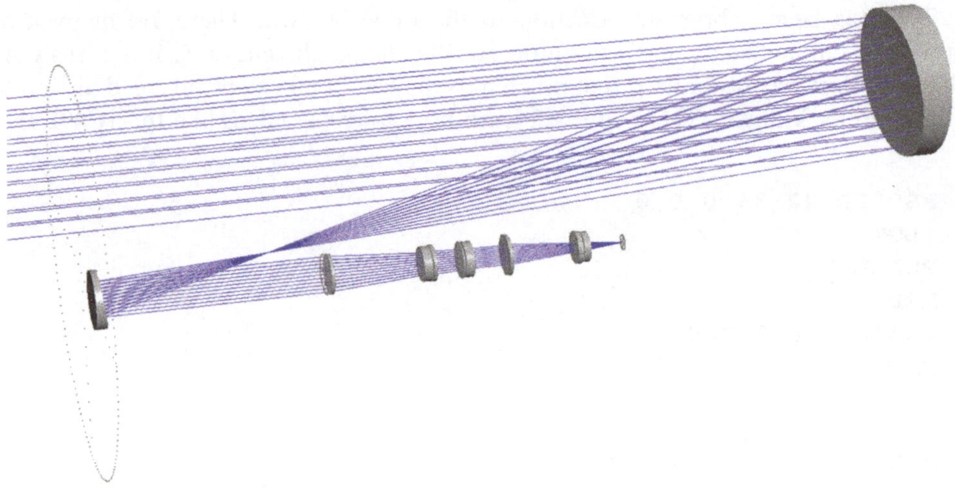

Figure 40.16. RSOLID drawing of the system.

decentered downward, and the two mirrors share a common axis. The chief ray has to come into surface 2 above the axis in order to hit the center of the stop when it reaches 5. The real pupil takes care of that. The clear aperture on 2 is decentered so as to fit the marginal rays at the upper and lower field points, and the same applies to the second mirror. This common geometry is set up with rather simple input, and the DCCR directives take care of the decentered apertures on the mirrors.

The RSOLID view in figure 40.16 was created with the input

```
RSOLID 22 -15 .1 0 0
PLOT
PUPIL 1
BLUE
TRACE P 0 0 20
END
```

IOP Publishing

Lens Design

Automatic and quasi-autonomous computational methods and techniques
Donald Dilworth

Chapter 41

A 90 degree eyepiece with field stop correction

Correcting an intermediate image; controlling exit pupil aberrations

If you have read the previous chapters, you are already familiar with the tools we will use in this exercise—but you may not know how excellent the results can be when you combine them. Here we present a challenging problem and then show how those tools can find an excellent design in a fraction of the time needed by those using classical design methods. Time is money, after all, and this is worth knowing about. In this lesson, you will use DSEARCH to derive a starting configuration, and then use other features to modify the lens construction, always improving its performance. We want the eyepiece to be diffraction-limited, and also have to make sure that the image of the field stop is sharp to the eye. That is more complicated and makes a good exercise.

The problem is to design a wide-angle eyepiece according to the goals summarized below:

- **Field of view: 90 degrees total at the eye.**
- **Eye relief: 15 mm or greater.**
- **F/number of beam from telescope objective: F/7.**
- **Visible spectrum: C, d, and F Fraunhofer lines.**
- **Correction to 1/4 wave or better in d light at 0.587 56 μm.**
- **Correction to 1/2 wave or better in C (0.6563 μm) and F (0.4876 μm) light.**
- **Pupil aberration at eye point no greater than 0.5 mm.**
- **An internal field stop, where tangential image errors must be no greater than twice the Airy disk at the local F/number of the beam.**
- **The telescope objective to be 2000 mm away.**
- **The eyepiece must have no more than 10 elements.**
- **Total length of the eyepiece no more than 200 mm.**

This is not an easy problem—nor should it be. If the problem could be solved entirely by the computer, the results would appear in a manner of minutes and the

doi:10.1088/978-0-7503-1611-8ch41

lesson would not be very interesting. However, these requirements cannot all be input to any lens design code with a single set of commands, so here we will use the computer to do the job that it does best, and the skill of the designer to fill in where needed and to direct the program as necessary.

We will start with absolutely nothing and require the computer to come up with a starting design all by itself, with **DSEARCH**, which you have seen in earlier chapters. The input is listed below. Most of this input can be created in the dialog **MDS**, which will create a MACro **(C41M1)** for you. Then you can edit it as you wish; here is a useful example:

```
LOG
TIME
CORE 14
DSEARCH 5 QUIET
SYSTEM
ID EYEPIECE EXAMPLE
OBD 1.0E9 45 1.27
UNI MM
WAVL CDF
WAP 1
END

GOALS
ELEMENTS 9
TOTL 200 .01
BACK 0 0
FNUM 7.0 10
ASTART 10
THSTART 10
RSTART 1000
RT 0.5
NPASS 80
ANNEAL 100 25 Q
SNAP 10
TOPD
STOP FIRST                  ! keep the stop at the eye point
STOP FREE
QUICK 50 100
FOV 0 .3 .6 .75 .9 1.       ! correct over five field points
FWT 3 1 1 1 1 1
END

SPECIAL AANT
ACA 50 1 1
ADT 10 .1 10
M 15 1 A P YA 1 0 0 0 1     ! control eye relief
M -.008 10 A P HH 1 ! aim light at objective to right
M -.004 10 A P HH .5! and control pupil aberrations
M -.0064 10 A P HH .8
M 0 1 A P YA 1              ! and distortion too
S GIHT
END
GO
TIME
```

Run this, and in about 5 minutes the program displays a picture of the ten best configurations it found. Running the optimization MACro prepared by DSEARCH on the top lens and annealing (**50, 2, 50**) produces a rather good lens, as shown in figure 41.1.

The OPD errors are all less than 1/4 wave, straight out of DSEARCH. So far, so good—but one must also watch and correct pupil aberrations in these wide-angle eyepieces. If those aberrations are too large, the eyepiece will suffer from the notorious 'kidney bean' effect, which causes portions of the field to go black as the user moves his eye about. We have to check.

Prepare a new MACro as follows:

```
STO 9
CHG
NOP
18 TH 2000
19 YMT
20
END
STEPS = 100
PLOT YA ON 19 FOR HBAR = 0 TO 1
GET 9
```

and run it. This will do the following:

1. Remove the YMT solve on surface 18 (via `NOP`, which removes all solves).
2. Put surface 19 at a distance of 2000 mm. This will model the telescope objective, assumed at that distance.
3. Assign a YMT solve to surface 19, which then focuses on 20.
4. Declare surface 20, so it exists.

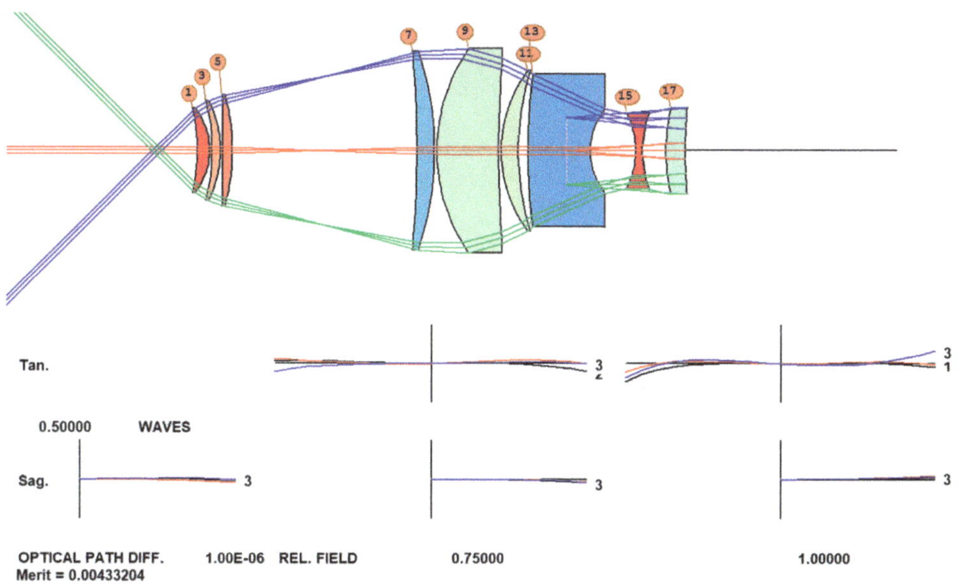

Figure 41.1. Eyepiece design returned by DSEARCH, optimized and annealed.

5. Make a plot of the chief-ray intercept on 19 over the field. If the rays all hit near the center of surface 19, the aberrations will be under control.

Run this MACro, and you see the pupil aberration at the objective, shown in figure 41.2. At F/7, at a distance of 2000 mm, the diameter of the objective will be 285.7 mm. A chief-ray error of 4 mm is therefore only about 2% of the size of the objective, and we are allowed 0.5 mm on an entrance pupil of 2.54 mm, or about 20%, so we judge this degree of correction satisfactory. That did not come for free, of course; the HH targets in the SPECIAL AANT section penalized any solution that

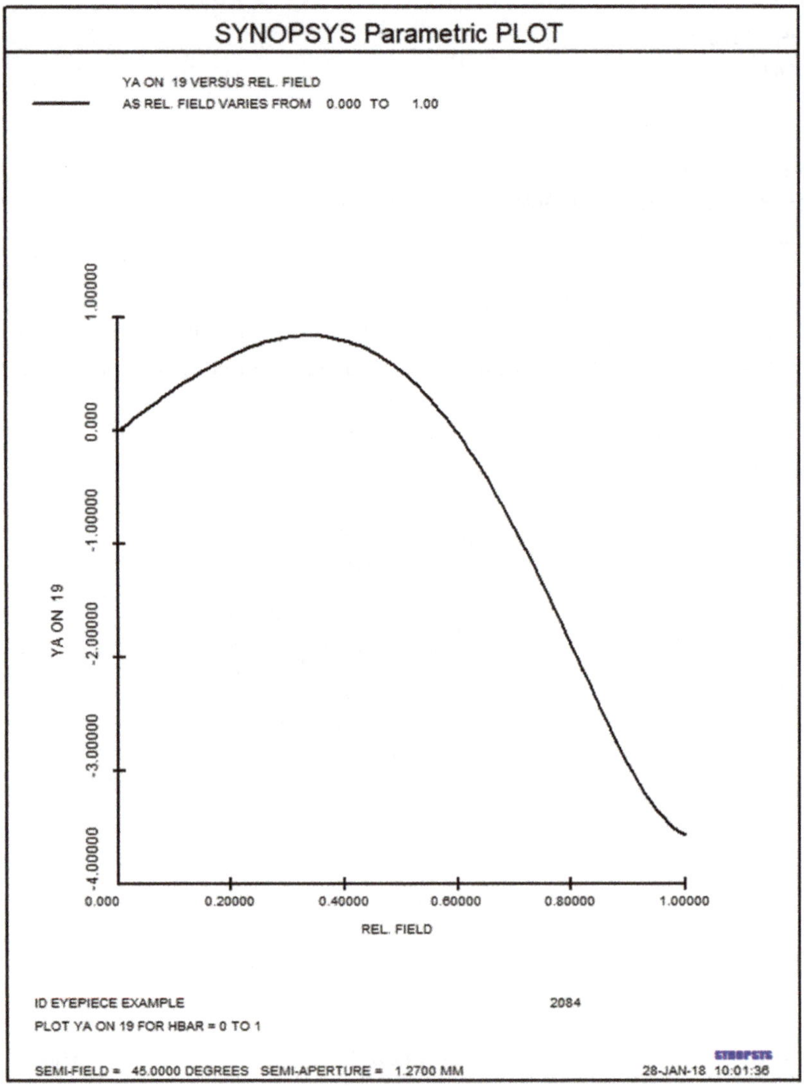

Figure 41.2. Pupil aberration at the objective lens, calculated for the eyepiece.

showed large pupil aberrations. You should feel free to adjust the weights on those targets to balance the errors as you prefer.

The eyepiece is already at the diffraction limit but is not yet finished since we have not controlled image quality at the field stop.

We have to do that—but the lens does not even have a field stop yet. In the WorkSheet, click the 'Add Surface' button, shown in figure 41.3, and then click on the axis between surfaces 6 and 7 (or wherever the intermediate image is in your lens). A surface is added, as shown in figure 41.4.

Now type, in the WS edit pane,

```
7 FLAG
```

and click 'Update'. Now you can refer to that surface in the AANT file with that name.

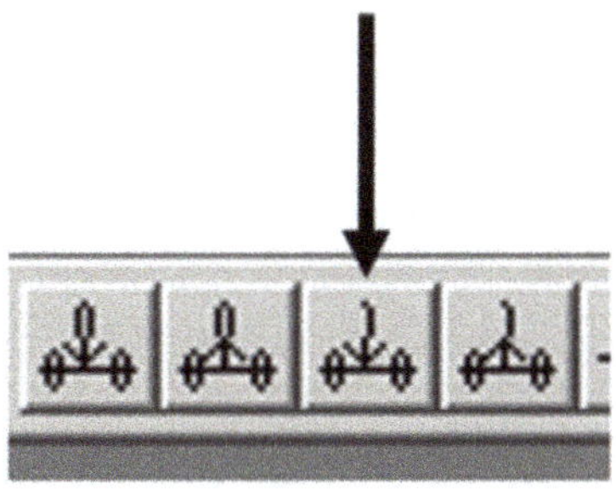

Figure 41.3. Selecting the 'Add Surface' button

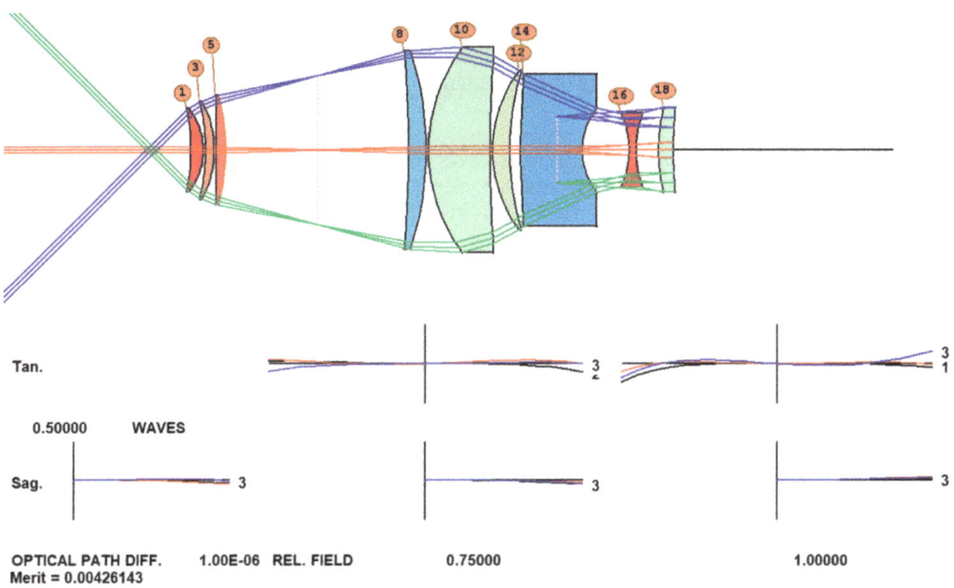

Figure 41.4. Lens with added surface for the field stop.

Edit the MACro that DSEARCH prepared for you. Here, you add some GTR ray sets to control the tangential blur at the flag surface. We do not care about errors in the *x*-direction since they do not affect the sharpness of the field stop as seen by the eye. Also, correct the difference between the full-field chief rays in colors 1 and 3 at the stop, so the image of the stop will not show noticeable color errors:

```
PANT
VY 0 YP1
VLIST RD ALL
VLIST TH ALL
VLIST GLM ALL
END
AANT P
AEC 3 1 1
ACM 3 1 1
ACC
GTR 0 2 4 P 1 0 FLAG
GTR 0 2 4 1 1 0 FLAG
GTR 0 2 4 3 1 0 FLAG
M 0 10 A 1 YA 1 0 0 0 FLAG
S 3 YA 1 0 0 0 FLAG

M    0.125000E+00   0.100000E+02 A CONST 1.0 / DIV FNUM
GSR       0.000000       3.000000        4  M      0.000000
GNR       0.000000       1.000000        4  M      0.300000
GNR       0.000000       1.000000        4  M      0.600000
GNR       0.000000       1.000000        4  M      0.750000
GNR       0.000000       1.000000        4  M      0.900000
GNR       0.000000       1.000000        4  M      1.000000
GSO       0.000000       0.281753        4  M      0.000000
GNO       0.000000       0.093918        4  M      0.300000
GNO       0.000000       0.093918        4  M      0.600000
GNO       0.000000       0.093918        4  M      0.750000
GNO       0.000000       0.093918        4  M      0.900000
GNO       0.000000       0.093918        4  M      1.000000
M    0.200000E+03   0.100000E-01 A TOTL
ACA 50 1 1
ADT 10 .1 10
M 15 1 A P YA 1 0 0 0 1 ! CONTROL EYE RELIEF
M -.01 10 A P HH 1  ! AIM LIGHT AT OBJECTIVE TO RIGHT
M -.005 10 A P HH .5 ! AND CONTROL PUPIL ABERRATIONS
M -.008 10 A P HH .8
M 0 1 A P YA 1   ! AND DISTORTION TOO
S GIHT
END
SNAP/DAMP 1
SYNOPSYS   80
```

Run this MACro, and the image becomes worse. The program has made some tradeoffs. DSEARCH does not know about field stops, so the lenses it returned were not selected for that property. The lens is shown in figure 41.5. Image quality is not good.

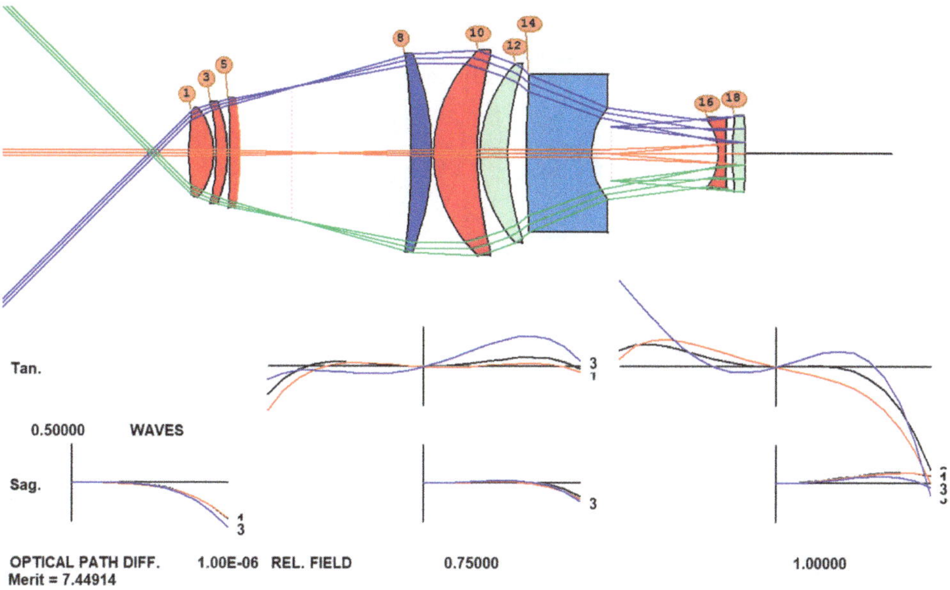

Figure 41.5. Lens optimized with field stop, before adding a new element.

Can we improve this lens?

It is time to run the **Automatic Element Insertion** feature. An expert would observe that you cannot correct lateral color at the field stop without a flint element to the left. Let us see if AEI can figure that out. Add the line

```
AEI 6 1 123 0 0 0 10 2
```

before the PANT command and run the MACro again. The program changes element 3 from a crown to a flint and adds a new element at surface 10. The image is much better. Comment out the AEI line, optimize again, and then anneal. The merit function comes down, as shown in figure 41.6.

One also has to monitor intermediate field points in wide-angle designs like this. Run the PAD scan [↑], and you see the correction remains below 1/4 wave everywhere.

Make a checkpoint and type **MRG** to open the Real Glass Menu. Select the Ohara catalog, 'Library 6', 'QUIET', 'SORT', and then 'OK'. The lens is assigned real glass everywhere:

```
– ARGLASS 6 QUIET
Lens number    6 ID EYEPIECE EXAMPLE
 GLASS S-FPL51      HAS BEEN ASSIGNED TO SURFACE 18; MERIT =  0.103691E-01
 GLASS S-FPL51      HAS BEEN ASSIGNED TO SURFACE  3; MERIT =  0.306426E-01
 GLASS S-FPM3       HAS BEEN ASSIGNED TO SURFACE 10; MERIT =  0.887701E-02
 GLASS S-FPM3       HAS BEEN ASSIGNED TO SURFACE  1; MERIT =  0.798868E-02
 GLASS S-LAH64      HAS BEEN ASSIGNED TO SURFACE 12; MERIT =  0.800182E-02
 GLASS S-LAH65V      HAS BEEN ASSIGNED TO SURFACE 14; MERIT =  0.801335E-02
 GLASS S-NPH4       HAS BEEN ASSIGNED TO SURFACE 16; MERIT =  0.164488E-01
```

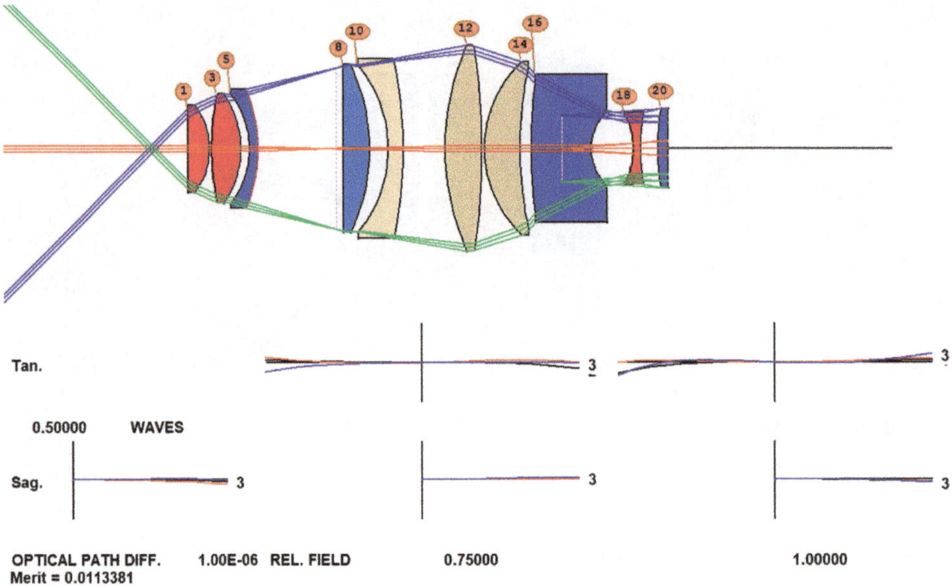

Figure 41.6. Lens reoptimized with added element.

```
GLASS S-TIH13      HAS BEEN ASSIGNED TO SURFACE 20; MERIT =  0.981422E-02
GLASS S-NPH4       HAS BEEN ASSIGNED TO SURFACE  5; MERIT =  0.702720E-02
GLASS S-NPH4       HAS BEEN ASSIGNED TO SURFACE  8; MERIT =  0.671866E-02
Type <ENTER> to return to dialog.
```

The lens is now almost perfect. Let us look at the distortion. Type **GDIS 21 G**. The eye will not notice any distortion at all, as shown by figure 41.7.

Now we have to check how well the image is corrected at the field stop. Make a checkpoint and type

```
CHG
7 MXSF
END
```

This truncates the lens at surface 9 (temporarily, so we can evaluate the image at the field stop). Only the TFAN affects the sharpness of the field stop as seen by the eye. We corrected those rays above, but now want to see how well it worked.

Use the Spectrum Wizard to model 10 wavelengths, visible spectrum, bright light, ‘Get Spectrum’, and ‘Apply to lens’.

Then open the Image Tools Menu (**MIT**), select a reference dimension of 0.2 mm, ‘Coherent’ under ‘Effect’, point source at HBAR = 1, ‘Multicolor’, and click ‘Process’, as shown in figure 41.8.

Indeed, the blur at the field stop is close to the diffraction limit in the *y*-direction. Restore your checkpoint so you can evaluate the final image.

This lens seems to meet every one of our goals. To verify, run the **Spectrum Wizard** (**MSW**) again to define ten wavelengths spaced across the visible spectrum, and then run the **OFPSPRD** feature to show the diffraction pattern over the field. (This is best done

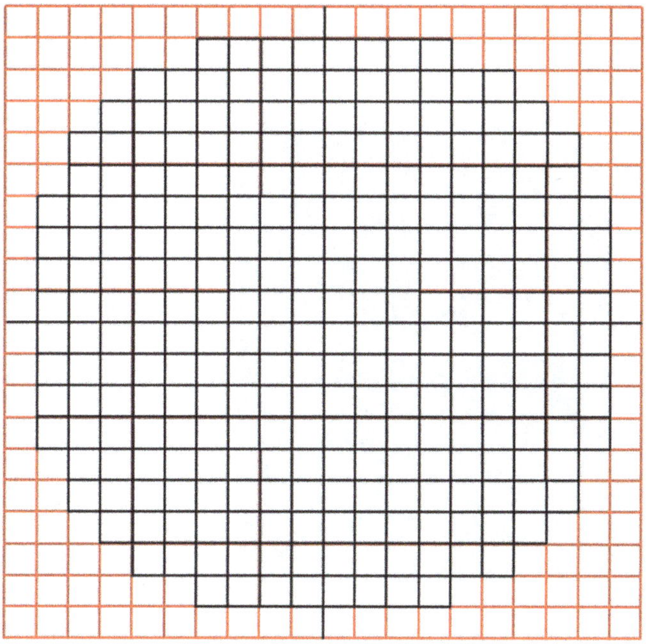

Figure 41.7. Distortion plot for the final design.

with the **MPF** dialog; select 'Show visual appearance', 'Magnify 4'.) The result is in figure 41.9. This is as close to a perfect eyepiece as you will likely ever see.

To the eye, this eyepiece will produce an image that is essentially perfect and undistorted. A check of the pupil aberrations shows that pupil wander is less than the allowed 1/2 mm over the field. It is unlikely you will ever see a telescope as well corrected as this, and if you do, this will be a superb combination.

Since this is a paper exercise, I will stop here. A design for a real market would need more attention, perhaps adjusting the thickness of some of the elements, but the point has been made. The final lens is in **C41L1**.

This result comes back when switch 98 is turned on, but as noted before, for real design work you want that switch turned off. Then you may obtain a different set of design forms every time.

We ran it several times with the switch off. Sometimes the results were not quite as good as this, and in one case we obtained a lens of only *nine* elements that was very nearly as good as this ten-element lens. DSEARCH can explore hundreds of the branches of the design tree in a matter of seconds, and with slightly different input will explore yet others. For investigating the design space, this is the tool to use.

Why did we use the *second* lens from the top in this example? The answer is simple: DSEARCH found great lenses, considering only the requirements we gave it, but when you add the new requirement for correction at a field stop, the top one came out not as good as the next one down—that happens.

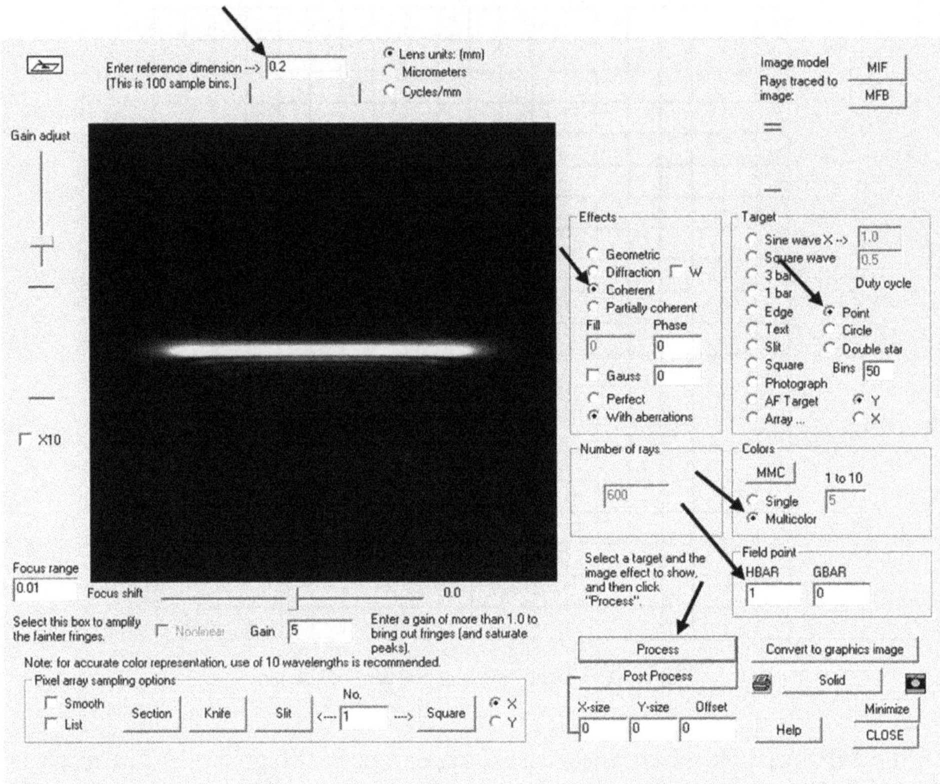

Figure 41.8. MIT dialog with image of a point on the edge of the field stop. This will look sharp to the eye in the *y*-direction.

What happens if you run AED on this lens? We obtained a nine-element lens with almost the same performance. Considering that we are adding a requirement (correction at the field stop) that is not expressed in the DSEARCH input, it would make sense to go back and try some of the other configurations the program returned. It is possible that a design other than the one we chose will turn out to be superior when you are done.

New users may wonder why this lesson called for object type OBD and activated the WAP 1 option. There is some optics here and it is a good idea to understand it. When designing an eyepiece like this, you are after what is called an 'F-theta' lens. In a normal camera lens, one wants the image height to be proportional to the object height; then there is no distortion. But that would not work in an eyepiece, for which one wants the *angles* of the object and image to be proportional, not the heights. Object OBD specifies an object angle (here 45 degrees, tracing from the eye point) and then the field parameter HBAR refers to a fractional *angle*, instead of height, as well. When distortion is corrected, the angles are proportional, and the apparent angular separation between a pair of a double stars is constant, no matter where they appear in the field of view, as one would expect. Since the angular magnification is constant

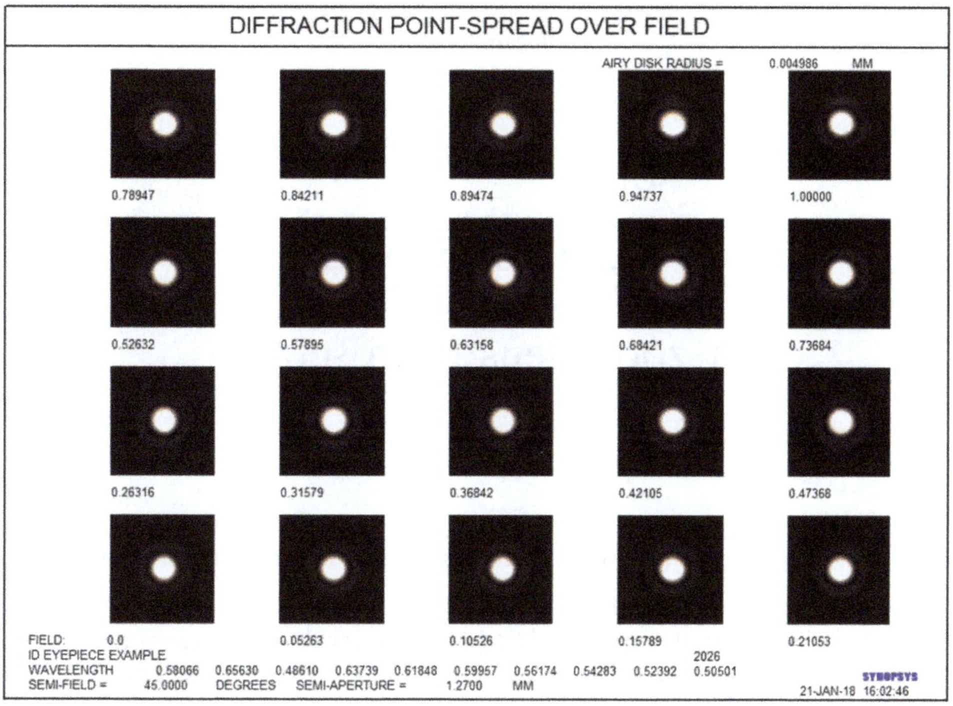

Figure 41.9. Diffraction pattern across the field of the eyepiece.

over the field, by Lagrange's law the diameter of the entering beam (at the eye) should be constant as well. The WAP 1 option takes care of that.

What did we learn from this exercise? It is clear that the numerical approach works. Designers of the classic school will work for many days on a design like this and will rightly be proud of the result if they succeed. They will have developed some insight into which elements are correcting which aberrations and the like. The numerical tools used in this lesson, on the other hand, will produce an excellent design in a fraction of the time. If your goal is to get a product out the door at minimal cost, never mind how it works, then the numerical approach is clearly superior. However, if you do want to know how it works, check out the **`THIRD CPLOT`** feature. It is all there, in living color.

Chapter 42

A zoom lens from scratch

Designing a zoom lens with no starting design; changing zoom numbers

It is noon on Friday. Your boss runs in: 'The customer wants a proposal for an 8× zoom lens by 8:00 Monday morning'. You have never designed a zoom lens in your life. Your job hangs in the balance. He gives you a list of requirements and heads out. What do you do now?

You might go to a patent database and try to find a lens that resembles what you need. That would probably take all weekend—but you have a better plan:

1. Start SYNOPSYS.
2. Type **`HELP ZSEARCH`** in the Command Window. Open that chapter, which is 10.7.3.
3. Read the whole chapter. Yes, read it. To be a professional lens designer you have to know what you are doing, but if you already know how to do other tasks on SYNOPSYS you are almost there, so it is an easy read. Now you know something about zoom lenses—so it is time well spent.
4. Set up your input to ZSEARCH. The lens is to be F/3.5 with a semi-field angle of 14 degrees in the wide setting and a GIHT of 5 mm. This implies a focal length of 20.05 mm, and the semi-aperture is therefore 2.85 mm. The lens must be able to focus over a range of object distances from four meters to infinity.

(This is a good place to underscore what you learned in the manual: SYNOPSYS does not use multiconfigurations to do zoom lenses. A single configuration can model up to 20 zoom settings.)

doi:10.1088/978-0-7503-1611-8ch42

Here is what will be in your MACro (**C42M1**):

```
LOG                     ! to keep track of things later
TIME                    ! to see how long this run took
CORE 14                 ! to run faster

ZSEARCH 3 QUIET         ! save results in library location 3

SYSTEM
ID ZSEARCH TEST
OBB 0 14 2.85           ! infinite object, 14 degrees semi field, 2.85 mm
                        ! semi aperture.
                        ! This defines the wide-field object
UNI MM
WAVL CDF
NOVIG
END

GOALS
ZOOMS 5
GROUPS 2 3 3 3          ! lens has four groups with 11 elements altogether
ZGROUP 0 Z Z 0          ! groups 2 and 3 will zoom

FINAL          ! declare the desired object at the last zoom position,
                        ! which is the narrow field zoom
OBB 0 1.7545 22.8       ! object is 1.7545 degrees semi field and 22.8 mm
                        ! semi aperture.
                        ! This implies an 8X zoom.

ZSPACE APERT            ! other zoom apertures will be evenly spaced between
                        ! the first and last
ZFOCUS 4000 4 15        ! also correct for object at 3 meters
APS 17                  ! put the stop on the first side of the last group
RT 0.25
GIHT 5 5 10             ! the image height is 5 mm for all zooms, with a
                        ! weight of 10.
BACK 5 .01     ! the back focus is 5 mm and will vary.  A target
               ! will be added to the merit function with a low weight.
FOV 0 .4 .6 .85 1       ! correct five field points
FWT 5 4 3 3 3
COLOR M                 ! correct all defined colors
ANNEAL 50 10  Q 40      ! anneal the lens as it is optimized in quick and
                        ! real-ray modes
QUICK 50 100            ! 50 passes in quick mode, 100 in real-ray mode
END

SPECIAL AANT
AAC 30 1 5     ! request a maximum semi aperture on all elements of 30 mm
ACA 50 1 1     ! monitor rays to keep away from the critical angle.
M 212 .01 A SECTION FOCL 1 4
END
GO             ! start ZSEARCH
TIME
```

This zoom lens will consist of four groups, the first with two elements and the others with three. The first group will be used for range focus, and the last is fixed to provide a constant F/number over the zoom range. It is likely you will need more than 11 elements—and you can ask ZSEARCH for more if you want to—but it will run faster if you start here and add elements later where needed.

Run this MACro and watch the progress on a set of windows that monitor each of the cores that have been authorized, a portion of which is shown in figure 42.1.

When quick mode is finished, the program optimizes each of the best 10. In about 21 minutes you see the results shown in figure 42.2.

A glance at the merit function values shows that most of these are promising candidates. The program displays the best one in PAD.

Run the MACro **ZSS**, which ZSEARCH has created, to look at all ten matches. We like the top one. Not perfect yet, to be sure—but not bad considering we gave the program nothing but a list of goals and constraints. The best two want a back focus distance less than our target of 5 mm. Let us assume we can tolerate a shorter distance, since this seems to be useful. This will be our starting lens, shown in figure 42.3.

The program has created an optimization MACro already loaded with a starting merit function definition and a set of variables, but you have to adjust some things. Many of the elements are too thin, so add two monitors:

```
AEC
ACC
AZA
ACA
ACM 3 1 1
ADT 7 .1 10
```

The **ADT** monitor will penalize any variable thickness that is less than 1/7 of the aperture diameter. The paraxial stop on 17 also needs to be changed to a real stop, with **APS -17** in **WS** or a **CHG** file.

Then run the MACro and anneal (**20, 2, 50**). The MF is now at 1.1, as shown in figure 42.4.

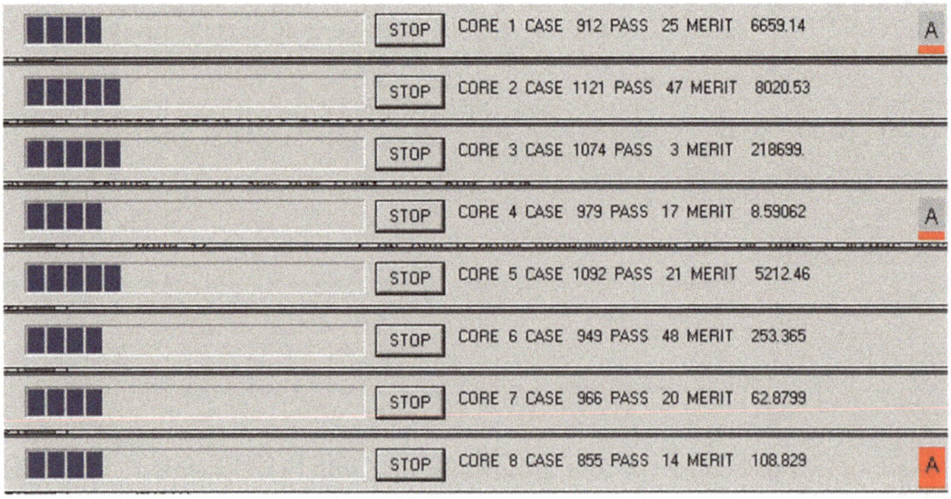

Figure 42.1. Progress bars showing multicore operations. The red blocks show the progress of the annealing stage.

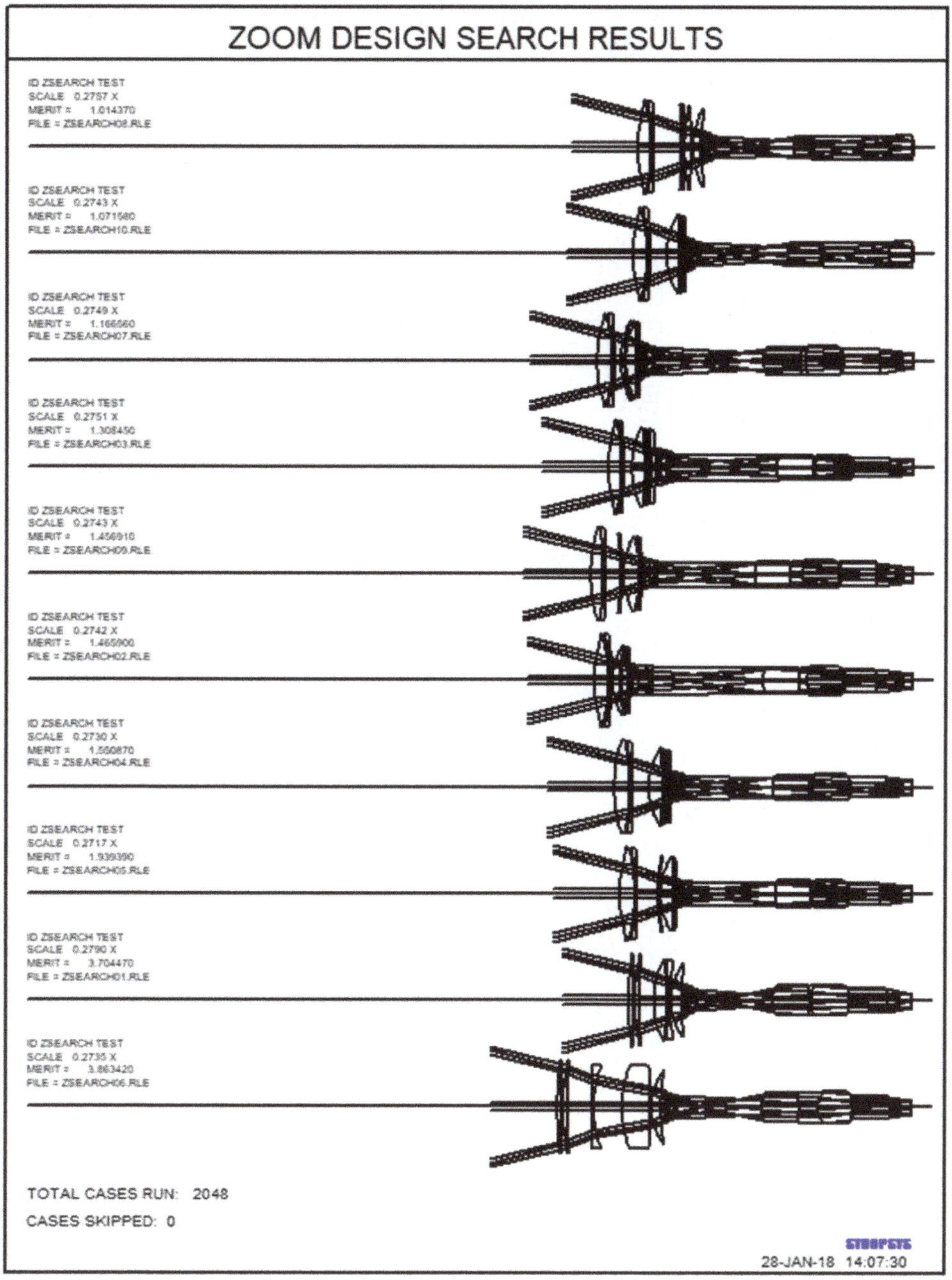

Figure 42.2. The ten best zoom lenses found by ZSEARCH.

Often one wants to know why the MF will not go any lower. Type **FINAL 5,** (or the AI symbol **FF**, which will substitute that string of characters), to show the five largest items in the MF:

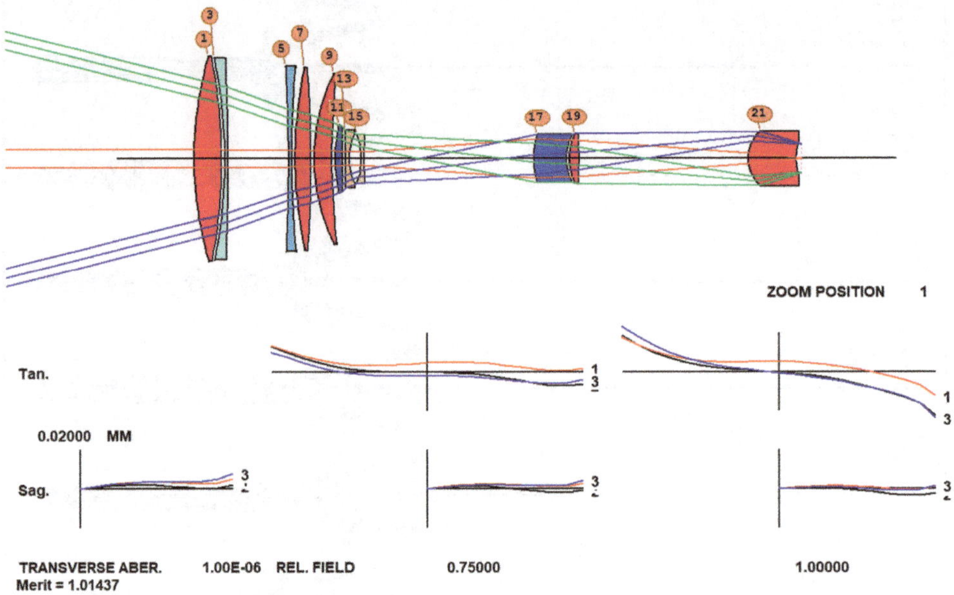

Figure 42.3. A starting zoom lens found by ZSEARCH.

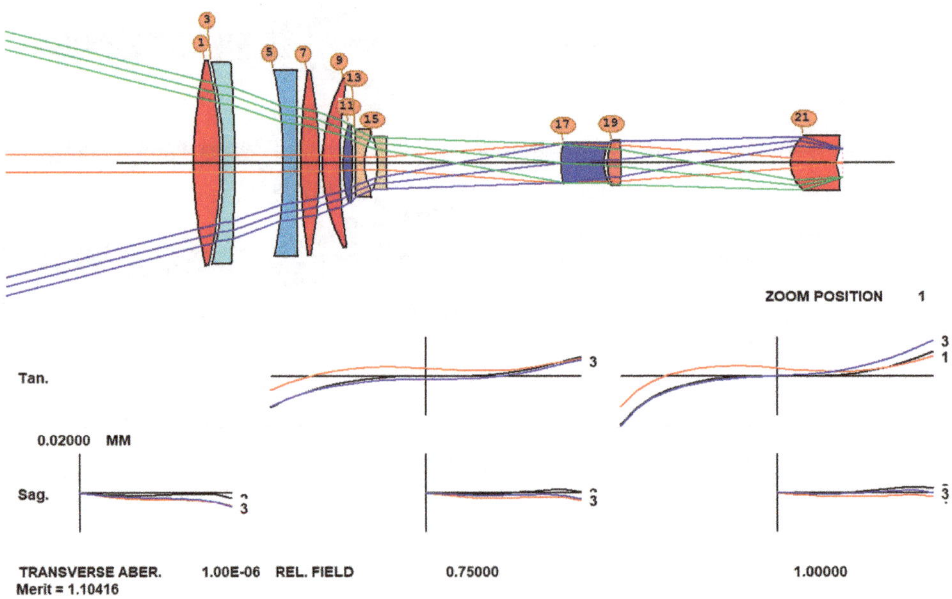

Figure 42.4. Zoom lens reoptimized and annealed.

```
SYNOPSYS AI>FF

FINAL 5
ABERRATION LIST
    NAME            TARGET          WEIGHT              RAW VAL. FINAL ERROR  R. EFFECT

  1 AEC            1.0000000     1.0000000          ------   0.149735       0.020306

  3 AZA            1.0000000     1.0000000          ------   0.107225       0.010413

  5 ACM            3.0000000     1.0000000          ------   0.107870       0.010538

  6 ADT            7.0000000     0.1000000          ------   0.111034       0.011166

  8 AAC           30.0000000     1.0000000          ------   0.375117       0.127439
*** ZFOCUS CHANGED TO      4000.00       4    15.0000          5.00000       ***
SYNOPSYS AI>
```

Here you notice that the **AAC** monitor wants an aperture larger than 30 mm. That aberration is larger than any of the others. Let us assume we can tolerate a larger size.

In the AANT file, change the line

```
AAC 30 1 1
```

to

```
AAC 35 1 1
```

Here is a useful trick: when you reach a stage that you might want to come back to later (if some ideas do not work out as well as you hope) click the '**ACON copy**' button  in the top toolbar. If the current lens is in configuration 1, as it is by default, this puts a copy in alternate configuration 2, or ACON 2. (This is called *bumping* the lens.) Then make a checkpoint in that ACON and develop the design further—and you can instantly go back to ACON 1 with the '1' button 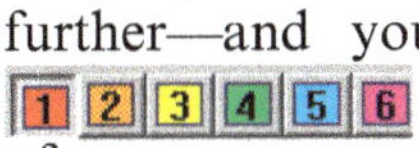 should you want to. I use all six configurations this way quite often.

Make a new checkpoint, then run the MACro and anneal again. The MF comes down to 0.33, as shown in figure 42.5.

It is time to take stock. Click each of the zoom buttons, shown in figure 42.6, to see how the correction holds up over the zoom range.

The image is not bad. We also should check how well the range focus is going to work.

Open a second EE editor (type **AEE**) and enter the following:

```
CHG
VIG
END
ACON BUMP
ZFOCUS 4000 4 15 5
PAD/U
```

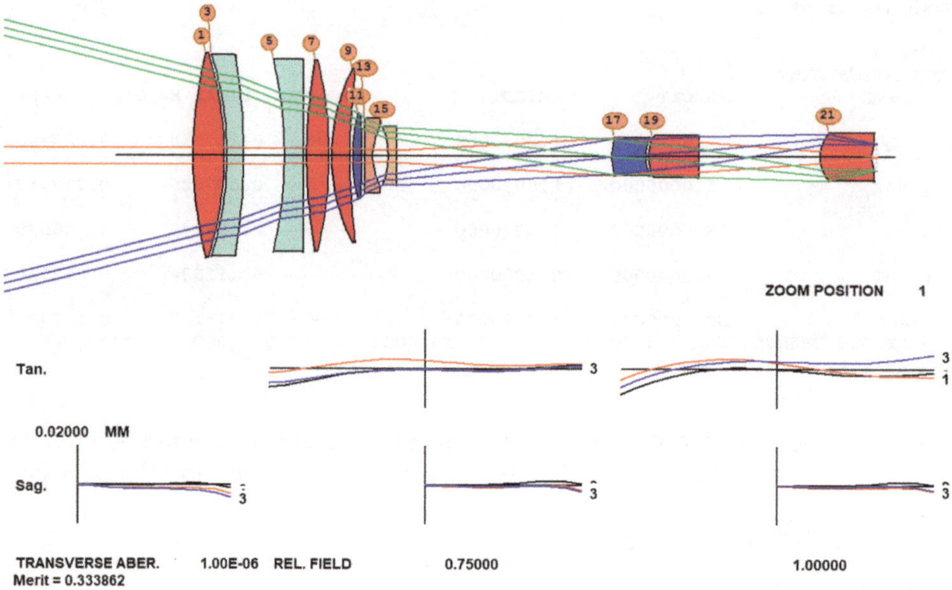

Figure 42.5. Zoom lens reoptimized with larger target on the aperture monitor.

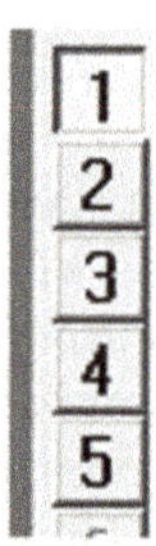

Figure 42.6. The ZoomBar[1] buttons.

Save this as a MACro named BUMP and run it. This will turn on VIG mode (so rays outside feathered edges become vignetted), put a copy of the lens in the next-higher ACON, move the object point from infinity to 4 m, and then move the first two elements to the left by 15 mm, which we specified as the focus adjustment at that range.

Now look at all five zooms in this configuration. All are quite well corrected, with the largest residual aberrations showing up at the ends of the zoom range. Figure 42.7 shows the lens in zoom 5 at the near conjugate.

[1] ZoomBar™ is a trademark of Optical Systems Design, Inc., a Maine, USA corporation.

Let us see if adding an element improves things. Go back to the previous ACON (at infinity conjugate), make a checkpoint, add the line

```
AEI 9 1 123 0 0 0 20 2
```

before the PANT file and repeat the optimization. The program adds an element to group 4 and the MF comes down to 0.23. Aberrations are somewhat reduced. Figure 42.8 shows the lens at near conjugate in zoom 5. We could probably obtain a small improvement with yet another element, but this is already a pretty good lens.

The lens works at five zoom positions, but what happens in between those zooms? We have to find out. Maybe the lens will be poorly corrected somewhere and we will have to add more zooms. The program uses a cam curve to interpolate the defined zooms at in-between points, and there are two types: power-series and piecewise cubic. The latter is often better, so open WS and click on the 'ZFILE' button **Z**. This loads the zoom definitions into the edit pane. Type **CUBIC** where shown in figure 42.9, add the word **DFOCUS** after the group designation 5–10, and click the 'Update' button and 'Close'.

This says that the first zoom group is to be used for focusing at all zoom positions. (This is what zoom lens makers do in practice: move one group and refocus with another, then mark the locations and cut the cam curves accordingly.) We have requested that group 1 (from 5 to 10) adjust itself at each cam position so that the image remains at the same focus relative to the paraxial focus as in the designed lens.

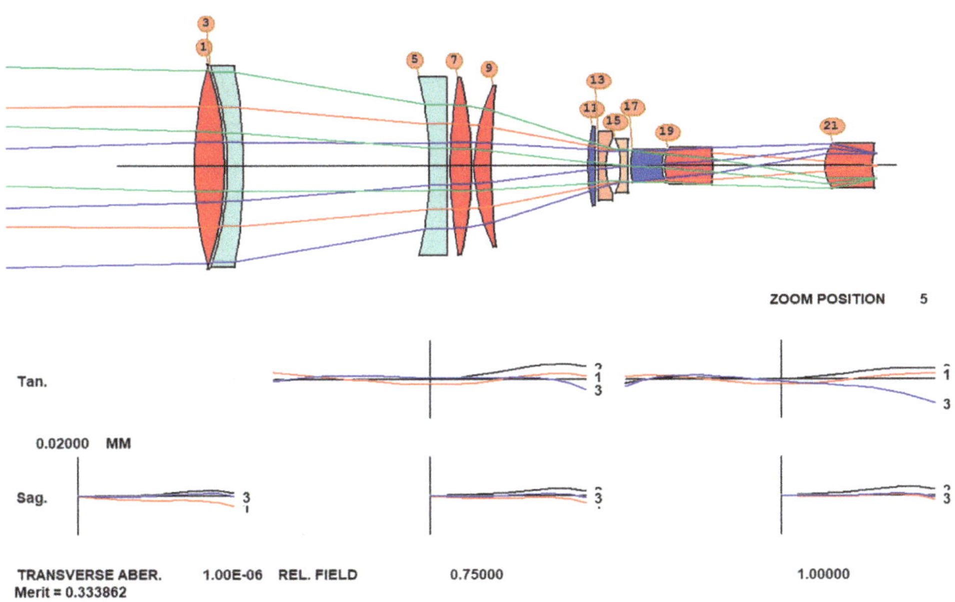

Figure 42.7. Lens in zoom 5 at 4000 mm conjugate.

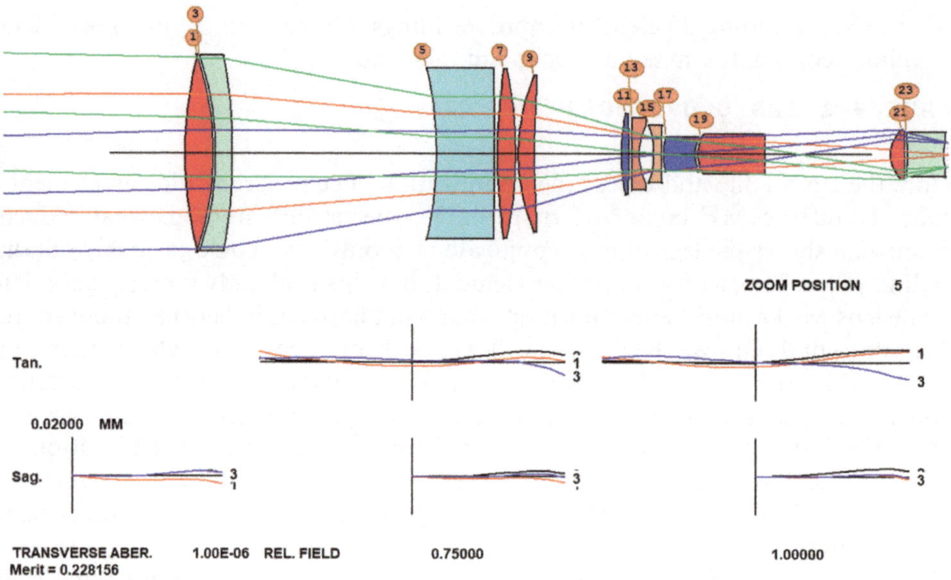

Figure 42.8. Zoom lens with element added to group 4, by AEI, at near conjugate, in zoom 5.

```
WS -- WorkSheet Lens-Edit Window
 OBB  0.000000      14.11659       2.85000     -23.07995      0.00000      0.00000      2.850
 ZFILE   2
 CAM RANK   4
 CAM EXPONENT    0.800000
  5  10 DFOCUS
 11  16
 RSOLVES
CUBIC
 ZOOM    2
 OBB  0.350825E-09     5.18054      7.83750     0.00000      0.00000      0.00000      7.83750
 ZDATA  -9.6514508E-02  2.7313123E+01
 ZOOM    3
```

Figure 42.9. The WorkSheet edit pane for changing zoom settings. Here, we have added a `CUBIC` declaration and specified that group 1 is to be adjusted to maintain the nominal paraxial defocus.

Run the ZoomSlider and examine the image in both conjugates. The image is great everywhere—however, near the left end of the zoom, elements 2 and 3 collide. Figure 42.10 shows group 2 has zoomed to the left and overlaps with group 1.

What happened? Does ZSEARCH not know how to watch out for this kind of error? ZSEARCH only avoids interference at the vertex at the zooms it knows about, honoring the AZA monitor—and those zooms clear as they should. However, in this case, settings *in between* zooms 1 and 2 have a problem. This is not an issue. Increase the number of zooms to 10 with **CAM 10 SET**.

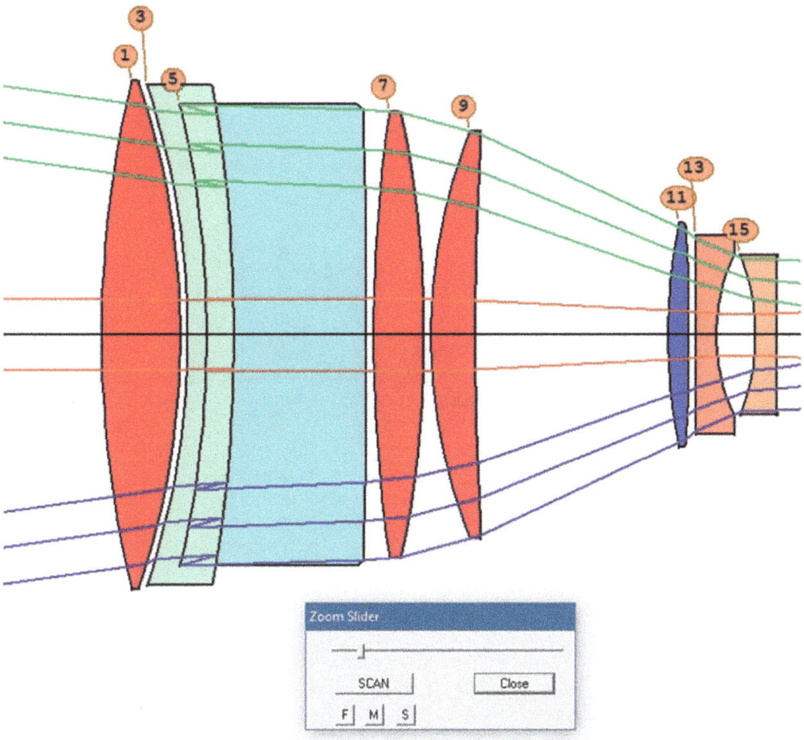

Figure 42.10. Interference between zoom groups when set in between defined zooms.

```
CAM CURVE
 ZOOM OFFSET, BY GROUP
 INDEX     1           2           3           4           5
 ___________________________________________________________
    1 6.217E-15 -3.553E-15
    2  -13.1980    12.6053
 *** OVERLAP DETECTED *** GROUP  1
    3   -9.0336    24.2955
    4   -0.9846    35.1351
    5    8.2221    45.1886
    6   17.6874    54.5231
    7   27.0623    63.1998
    8   36.2087    71.2743
    9   45.0758    78.8046
   10   53.6615    85.8488
```

The program reports that there is an overlap at the vertex in one of the ten new zoom settings. Now we can optimize all ten and eliminate the overlap.

To do so, we have to edit the optimization MACro. At the moment, it calls for correcting ray sets at each of the five zooms at both conjugates. Change the AANT file so it corrects *all* defined zooms instead of each one individually. Also, comment out the AEI line. Note that the ZROUP sections require their own END line, in addition to the END that ends the AANT file:

```
!AEI 9 1 123 0 0 0 20 2
PANT
VLIST RD ALL
VLIST TH ALL
VLIST GLM ALL
VLIST ZDATA ALL
END
AANT
AEC
ACC
AZA 3 10 5
ACA
ACM 3 1 1
ADT 7 .1 10
M     5.00000        0.100000E-01 A BACK
AAC 35 1 5              ! REQUEST A MAXIMUM SEMI APERTURE ON ALL ELEMENTS OF 30
ACA 50 1 1              ! MONITOR RAYS TO KEEP AW
ZGROUP ALL
M    0.500000E+01    0.100000E+02 A GIHT
GSR       0.500000       5.000000        4   M      0.000000
GNR       0.500000       4.000000        4   M      0.400000
GNR       0.500000       3.000000        4   M      0.600000
GNR       0.500000       3.000000        4   M      0.850000
GNR       0.500000       3.000000        4   M      1.000000
END
 ZFOCUS    0.400000E+04    4    0.150000E+02    5
ZGROUP ALL
GSR       0.500000       5.000000        4   M      0.000000
GNR       0.500000       4.000000        4   M      0.400000
GNR       0.500000       3.000000        4   M      0.600000
GNR       0.500000       3.000000        4   M      0.850000
GNR       0.500000       3.000000        4   M      1.000000
END
END
SNAP/DAMP 1
SYNOPSYS     50
```

Change the AZA monitor to require 3 mm clearance at the vertex at all defined zooms, with a higher weight:

```
AZA 3 10 5
```

Now optimize and anneal again. The lens is corrected at ten zooms, and there is no overlap anymore. This version is shown in figure 42.11 (**C42L1**).

We now have a pretty good lens, but elements 4 and 5 look somewhat strange and might be redundant. We might be able to remove one of them. The search routines do not care if something looks strange; they just crunch the numbers and call home. We have to try. Add a line before the PANT file

```
AED 5 QUIET 1 123
```

and reoptimize. The program reports that the best place to remove an element is at surface 23, but the lens is then not as good, so we reject this solution, shown in figure 42.12.

Execute your BUMP MACro, and you see that the near conjugate is also quite well corrected. It is time for some finishing touches.

Since we have come a long way, it is time to save this lens so you can easily get back to it if necessary. Click the button in the top toolbar, and the lens is saved with a name taken from the current log number. ZSEARCH has already added a LOG command to your MACro, so each run increases this number. That way,

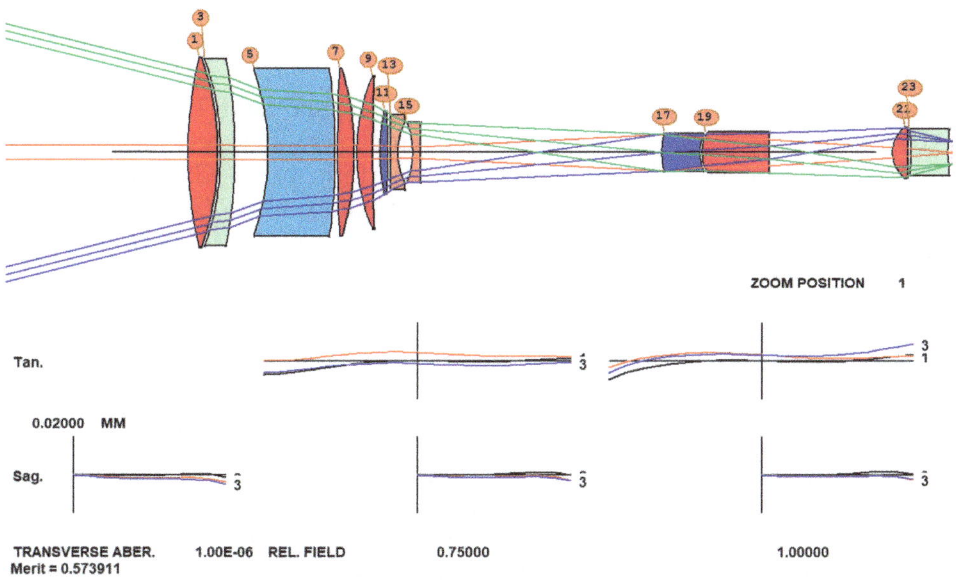

Figure 42.11. Lens corrected at ten zooms, with no interference.

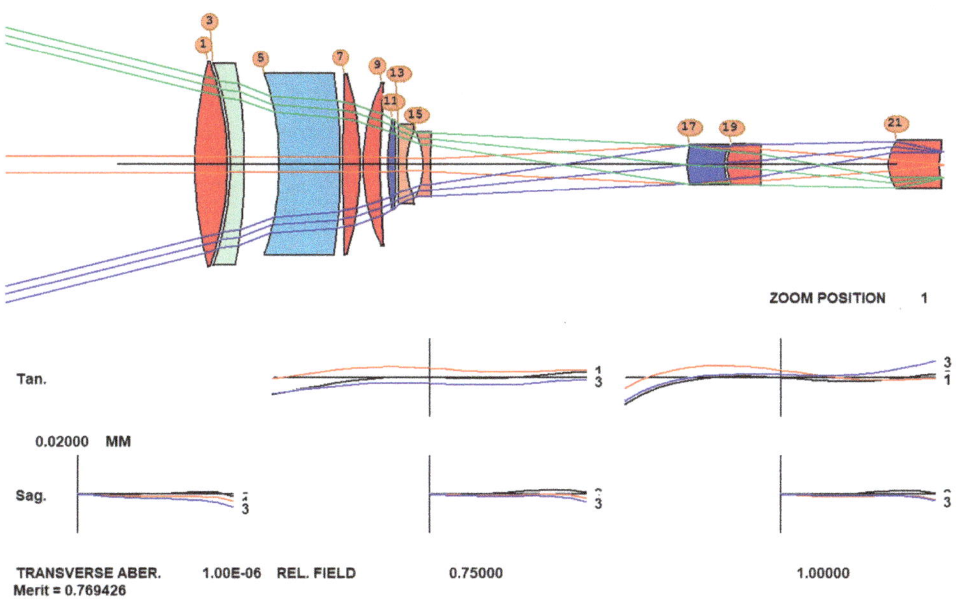

Figure 42.12. Zoom lens with element removed by AED, in zoom 1. The quality is degraded, so we revert to the previous design.

whenever you obtain a version that you might want to go back to, just click that button and it is saved.

We have to replace the model glasses with real glass, using ARGLASS (Automatic Real GLASS). It is a good idea to run the optimization once more before running ARGLASS, since it reuses the same merit function and variable list, and you want those to be current. (If you changed ACONs, you *must* run the optimization again in the current one, since ARGLASS uses the variables and merit function that apply to *that* ACON.)

Type **MRG** and select the Ohara catalog, 'QUIET', and 'SORT'. Since geometric aberrations are the main issue here, not secondary color, we expect that the simpler ARGLASS will suffice and we do not need the performance of GSEARCH for this lens. The MRG dialog will prepare the input for, and run, ARGLASS.

Click 'OK', and you obtain the lens shown in figure 42.13 (**C42L2**).

We know the last group needs four elements. Let us run ZSEARCH again, ask for groups of 2, 3, 3, 4, and allow an aperture of 35 mm radius instead of 30 mm, instead of working our way there in stages as we did above. Will the result be as good? Better?

Doing just that, then assigning a real stop to surface 17, adding monitors for lens thicknesses, optimizing and annealing once again, we obtained the lens in figure 42.14. In this case, there was no overlap of zoom groups, and the ray fans were somewhat better—except at full field in zoom 5, where the TFAN rises sharply at one end. Since the aberrations everywhere else are significantly better than before, we elect to accept this solution and reduce the clear aperture of surfaces 11 and 12 slightly to vignette the problem rays. Checking the near conjugate (with the BUMP MACro) we see that things are well corrected there too. This version is shown in figure 42.15.

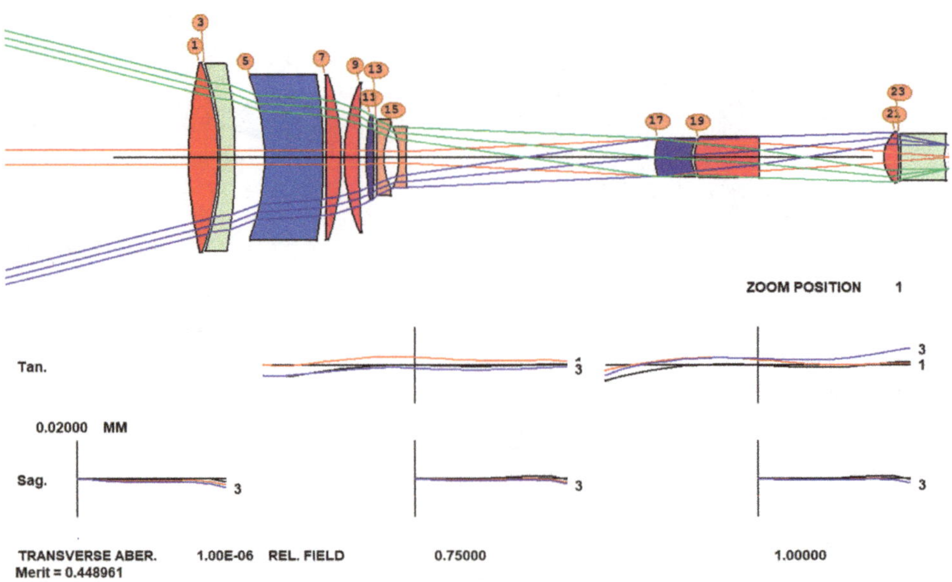

Figure 42.13. Zoom lens, with real glass.

How did we know which CAO to reduce? It is simple: run the ZoomSlider and watch the lowest of the rays shown in blue. That is the full field upper rim ray, defined according to the logic explained in chapter 22. The surface requiring the largest aperture for that ray in that section appears to be number 11.

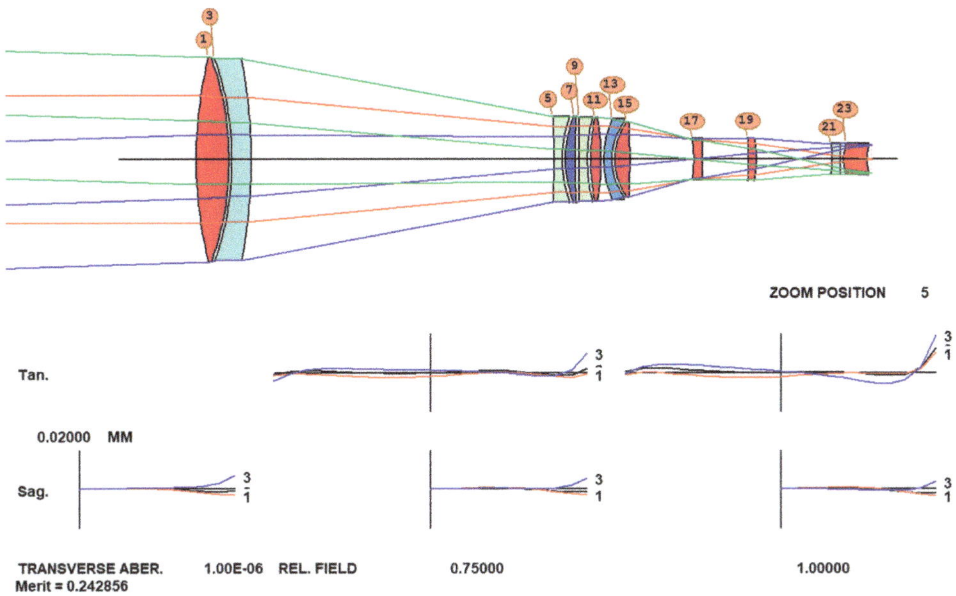

Figure 42.14. Zoom lens found when input parameters for ZSEARCH were altered per previous results.

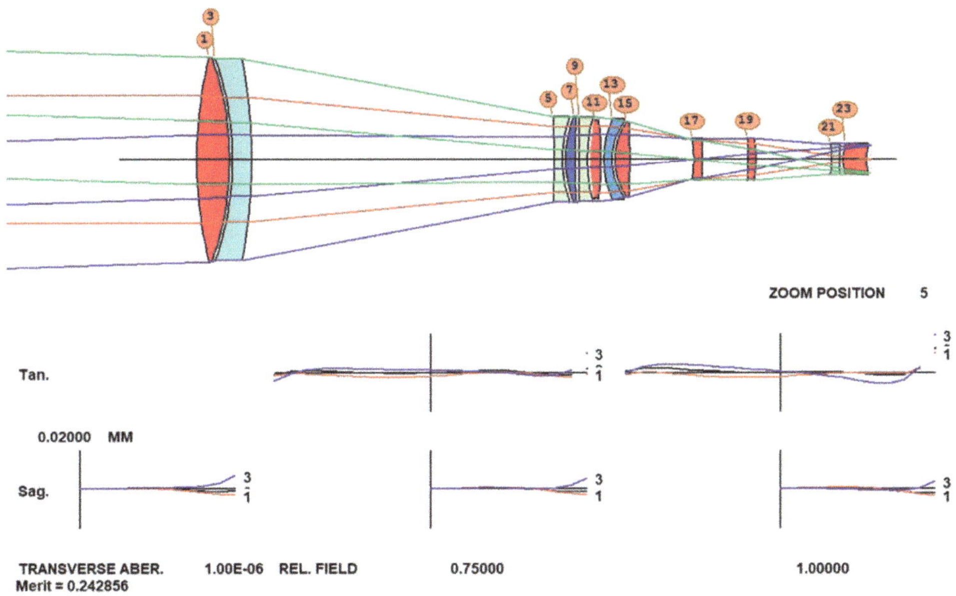

Figure 42.15. Lens modified by vignetting slightly at full field in zoom 5 by reducing the aperture on surface 11.

You will notice that, when ARGLASS matched real glass to the lens in figure 42.13, it selected glass type S-LAH58 for some elements, including the second, and this is a very expensive glass and that is a large element. So this time, specify a price limit no more than 12 times that of BK7 when you run MRG.

Do just that, and when it finishes, tweak up the aperture on 11 as you did above. You obtain the lens in figure 42.16 (**C42L3**).

Now check the spot sizes. First, type

```
OFF 27
SSS .003
```

to show spot symbols at the entered size. Then type `MSF` to open the dialog shown in figure 42.17.

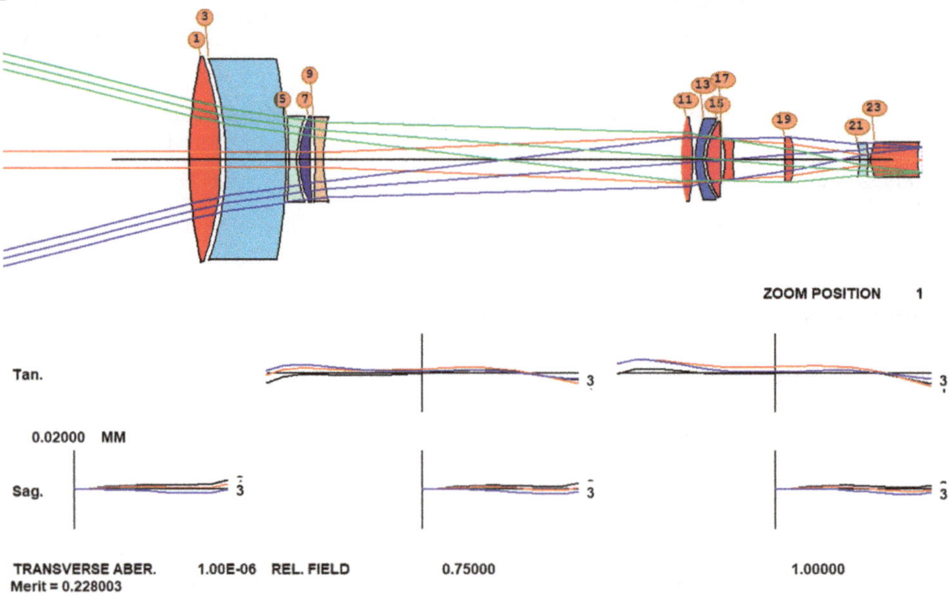

Figure 42.16. Final zoom lens, with real glass.

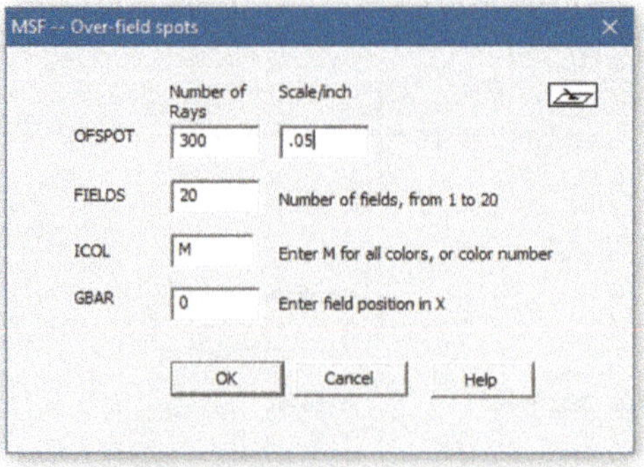

Figure 42.17. The **MSF** dialog for plotting spot diagrams over the field.

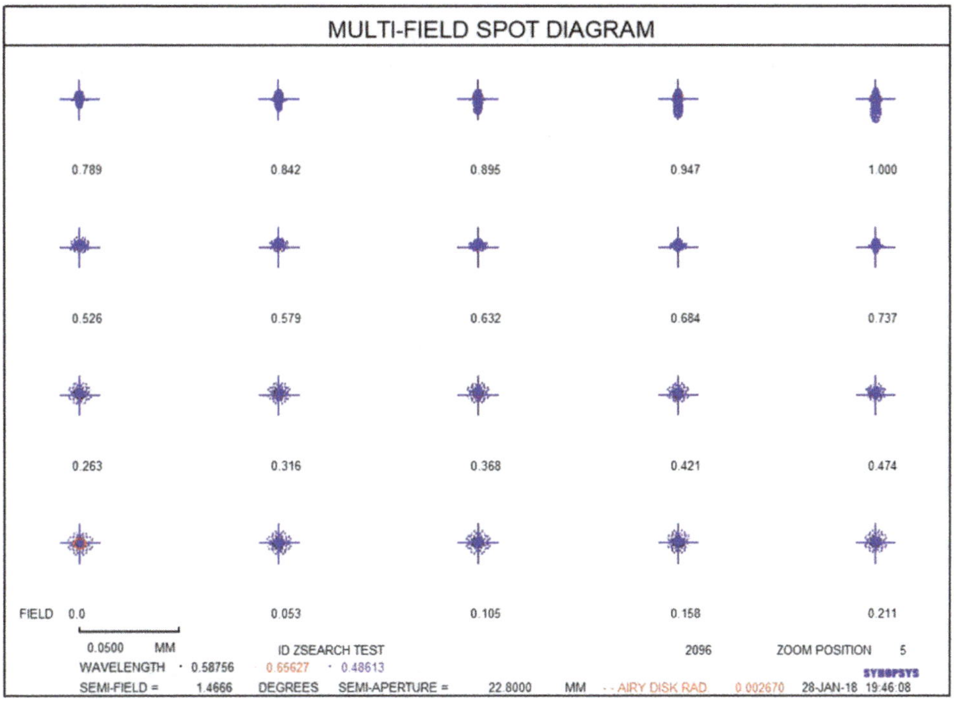

Figure 42.18. Spots plotted over field for zoom 5 at near conjugate.

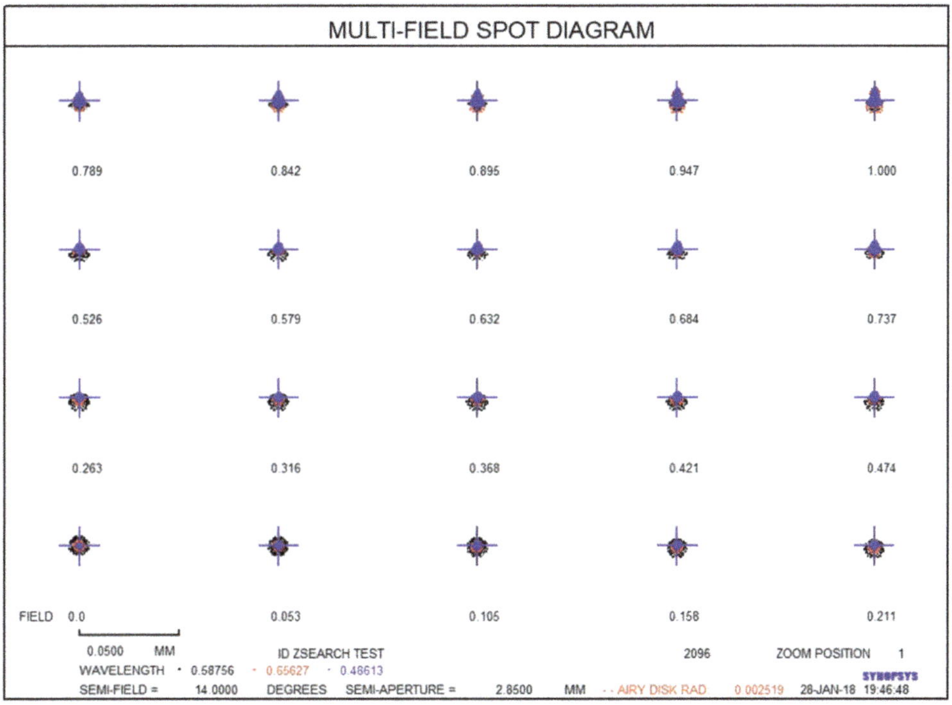

Figure 42.19. Spots in zoom 1 at infinity conjugate.

Change the scale to 50 μm as shown and click 'OK'. Do this for all zooms at both conjugates. The spot sizes are fairly constant over all cases. Figure 42.18 shows the spots at near conjugate in zoom 5, which gives the largest spots, and figure 42.19 shows the spots at infinity conjugate in zoom 1, which is more representative of the other cases.

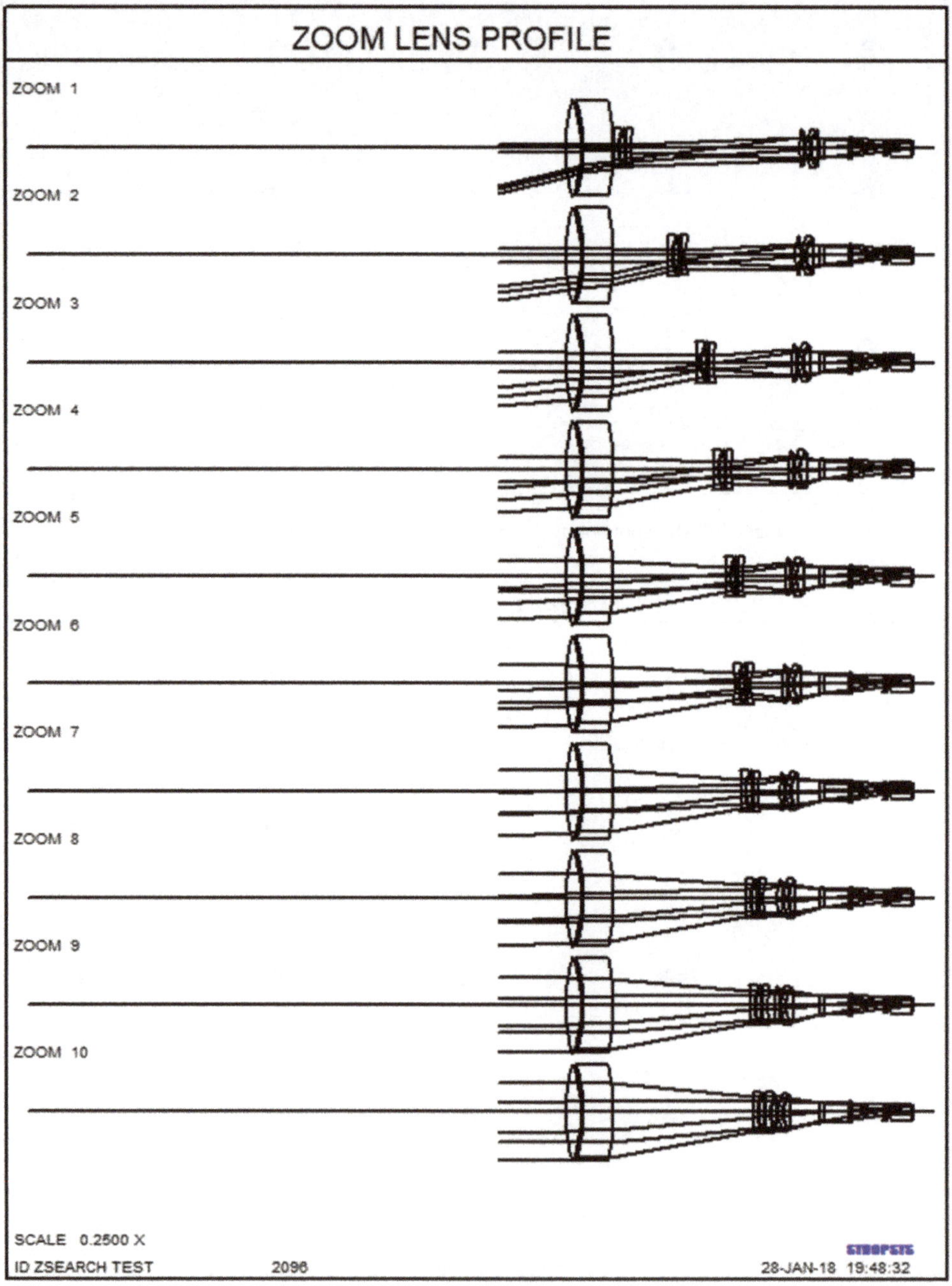

Figure 42.20. The zoom lens at ten zoom positions.

You quickly learn that the benefit of adding yet another element goes down as the number of elements becomes larger. That makes sense: adding an element to a doublet increases the variable count by 50%. Adding one to a ten-element lens increases the count by 10%, and so on.

Type

```
OFF 65
ZDWG .25
```

to see a picture of the lens at ten zooms, as shown in figure 42.20.

That is enough for this lesson—you get the idea. If the lens is still not good enough, another run with AEI should help. Your boss will be delighted to learn that you have a good design—and it is still only 3:00 on Friday afternoon. You can head out to the golf course.

Chapter 43

Designing a free-form mirror system

Correcting the image; avoiding beam interference

Optics fabrication is becoming more and more sophisticated as better methods of shaping and measuring optical surfaces become available. One result is the emergence of 'free-form' optics, which consists of lenses or mirrors whose shape is not axially symmetric about the center of the part. A simple example is the off-axis paraboloid, where the parent is polished to the required aspheric shape and then the desired part is cut from that parent. More elaborate shapes may involve high-order aspheric terms described with either power series, Zernike, or Forbes polynomials. As interest shifts to such systems, it becomes important to be able to design them.

SYNOPSYS provides features that can simplify that process, and you should read about **FFBUILD** in the help file before you go any further. This lesson assumes you have read that chapter and shows how to design a system with free-form mirrors.

The first step is to lay out the rough geometry you are after. Here is an example with three mirrors, shown in figure 43.1.

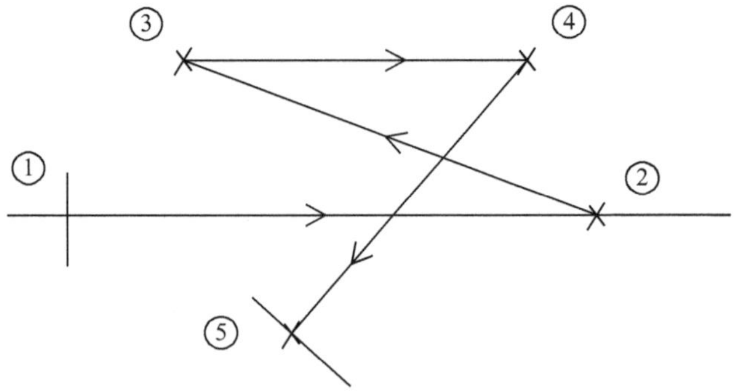

Figure 43.1. Proposed location of free-form mirrors.

doi:10.1088/978-0-7503-1611-8ch43

Light will come in from the left at surface 1, hit mirrors located at 2, 3, and 4, and then go to the image plane at 5. Here is the input for FFBUILD (**C43M1**):

```
FFBUILD
SYSTEM
ID EXAMPLE FFBUILD
OBB 0 2 12 0
WAVL CDF
UNI MM
CFOV
END

GEOM
2 MIRROR  0 0 140
3 MIRROR  0 40 30
4 MIRROR 0 40 120
5 IMAGE  0 -30 60 -7 7
 END

SHAPES
2 ZERN
3 ZERN
4 ZERN
END
```

In this example, the mirrors will be assigned Zernike polynomials, which accept up to 36 coefficients that are functions of polar coordinates on the surface. Since FFBUILD only supports designs with bilateral symmetry, terms asymmetric in x will not be used.

The above input contains part of the requirements: the semi-field angle is two degrees, the field is circular, and the semi-aperture is 25 mm; other requirements will be added as we go along. However, first run the above input file, which produces two results: a mirror system (with flat surfaces at the moment) and an optimization MACro that contains most of the input needed to refine this design. Here is that system, shown in figure 43.2.

The optimization MACro is quite long and contains variables for the angles and global positions of the mirrors and image plane in y and z, as well as for the Zernike coefficients on the mirrors. Most of the variables are commented out, however, since the process works better if you first rough out the design with only the radii and angles varying, and then add the other variables gradually as they are needed. Here is part of that MACro; the lines in green are commented out:

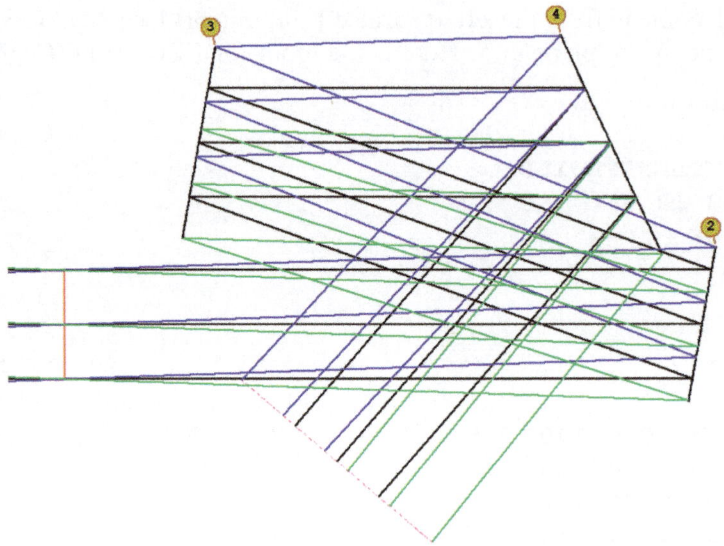

Figure 43.2. System returned by FFBUILD, before optimization.

```
PANT
 SKIP
VY    2 YG
VY    2 ZG
VY    3 YG
VY    3 ZG
VY    4 YG
VY    4 ZG
VY    5 YG
VY    5 ZG
 EOS
VY    2 RAD
! VY    2 CC  10  -10
! VY    2 G    3
! VY    2 G    4
! VY    2 G    7
! VY    2 G    8
! VY    2 G   10
! VY    2 G   11
! VY    2 G   14
! VY    2 G   15
! VY    2 G   16
! VY    2 G   19
! VY    2 G   20
! VY    2 G   23
! VY    2 G   24
! VY    2 G   26
! VY    2 G   27
! VY    2 G   30
! VY    2 G   31
! VY    2 G   34
! VY    2 G   35
! VY    2 G   36
...
```

Most of the file consists of operands that will control the clearance of the beam as it bounces between mirrors. Here is an example portion:

```
LLL 1.0000 1 1.0000
A P CCLEAR 1 0 1 0  1 3
S CAO  3
LLL 1.0000 1 1.0000
A P CCLEAR 1 0 -1 0  1 3
S CAO  3
LLL 1.0000 1 1.0000
A P CCLEAR -1 0 1 0  1 3
S CAO  3
LLL 1.0000 1 1.0000
A P CCLEAR -1 0 -1 0  1 3
S CAO  3
LLL 1.0000 1 1.0000
A P CCLEAR 0 0 1 0  1 3
S CAO  3
LLL 1.0000 1 1.0000
A P CCLEAR 0 0 -1 0  1 3
S CAO  3
LLL 1.0000 1 1.0000
A P CCLEAR 1 0 1 0  1 4
S CAO  4
LLL 1.0000 1 1.0000
A P CCLEAR 1 0 -1 0  1 4
S CAO  4
...
```

The first three lines of this excerpt tell the program to trace the full-field upper rim ray and then look at the segment of the ray path between surfaces 1 and 2. Calculate where that segment intersects surface 3, find the absolute distance from the vertex, and subtract the clear aperture radius of surface 3. If the difference is more than 1 mm, the intersection is outside the clear aperture and the aberration is zero. However, if it is less than 1 mm, the merit function receives a penalty. The program has also assigned the DCCR surface property to the mirrors, so the default clear aperture is centered between the extreme points on the surface required by the rays in the meridional plane, instead of at the vertex, which is the default.

The remaining **CCLEAR** entries control the clearance between upper and lower rim rays at the top and bottom of the field between each mirror pair and the other mirrors. There are many combinations and they all must be controlled.

The merit function contains GNR requests for seven points in the y–z plane and three in the skew field direction (because of the **CFOV** directive in the **SYSTEM** file, declaring a circular field), and controls distortion in both x and y with the **GDR** request (according to the desired image size in words 6 and 7 of the **IMAGE** line in the **GEOM** section):

```
GDR 0 10 4 P  0.700000E+01 -0.700000E+01
```

It is a good idea to run this MACro just as it is, the first time around. That will get the design roughed out well enough so you can gradually improve things. After running it, the system starts to look more reasonable, as shown in figure 43.3.

The image is formed in the right place and the beams clear all of the mirrors rather well.

We have specified the focal length we are after indirectly in this example. It would not be a good idea to control FOCL itself, since that is a paraxial property that does not have much meaning with a folded system such as this one. What we want is for the distance between the upper and lower field points at the image to be 14 mm. The program controls that with the **GDR** request, as noted above.

The system is roughed out, but the image quality is not at all good; now you have to free up some of the other variables. It is wise to sneak up on the system little by little, so go slowly. If you go too fast, the system will sometimes jump to a bizarre configuration that is nowhere near the nice region you are in now. Control low-order aberrations with low-order terms and save the higher orders for when you need them. Remove the comment character ('`!`') before the variables for `G 3` through `G 10` on each surface by deleting the '`!`' character on those lines:

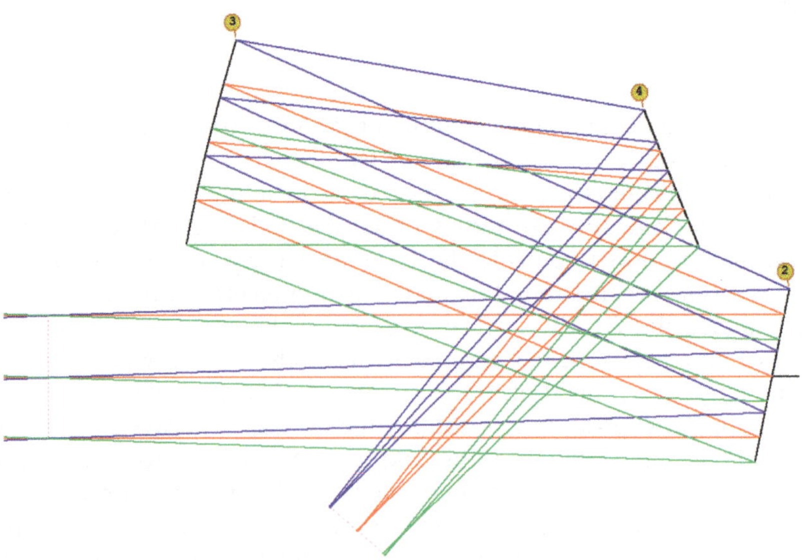

Figure 43.3. Free-form system after initial optimization.

```
VY   2 RAD
! VY   2 CC 10 -10
 VY   2 G   4
 VY   2 G   7
 VY   2 G   8
 VY   2 G  10
! VY   2 G  11
! VY   2 G  14
```

Do the same for surfaces 3 and 4.

Also, comment out the `SKIP` directive that bypassed the global *y*- and *z*-location variables, so they will become active,

```
PANT
! SKIP
VY   2 YG
VY   2 ZG
VY   3 YG
VY   3 ZG
VY   4 YG
VY   4 ZG
VY   5 YG
VY   5 ZG
```

and then run the MACro and anneal (**50, 2, 50**). The MF comes down. Now free up the remaining G variables and optimize again. The MF goes down to 0.000 69. Anneal, and the MF comes down to 0.000 52—pretty good.

You are probably wondering why we did not let the CC (conic constant) vary. You can experiment with that too, if you want—but we find that the existing Zernike terms can generate much the same shape as the conic constant, and it is not wise to use duplicate variables.

It is time to assess where we are. Go to the MAP dialog (**MMA**) and ask for a map of the wavefront variance over a grid of object points, object points 'CREC', ray pattern 'CREC 9', 'Show circles', EANALOG scale 0.01, and 'Execute'. The result is in figure 43.4.

Let us examine the results. The worst field point is now at GBAR 0.66, as indicated by the MAP analysis. Here is that image, created by the **MDI** dialog, shown in figure 43.5.

All other points are better. This is a superb design. Let us assume for this application we will use a CCD array sensor with pixels of 10 μm on a side, so this looks good.

You can obtain a better view with RSOLID, which shows only the portion of the surfaces within the decentered CAO. However, first you should go to the Edge Wizard (**MEW**), select 'Create All', and adjust the mirror thicknesses as you want, as you did in chapter 40. Now the mirrors are assigned realistic edges and thicknesses. Then create an RSOLID picture, shown in figure 43.6. (Type **MPE** and select that option or click the button.)

Your free-form system is designed (**C43L1**).

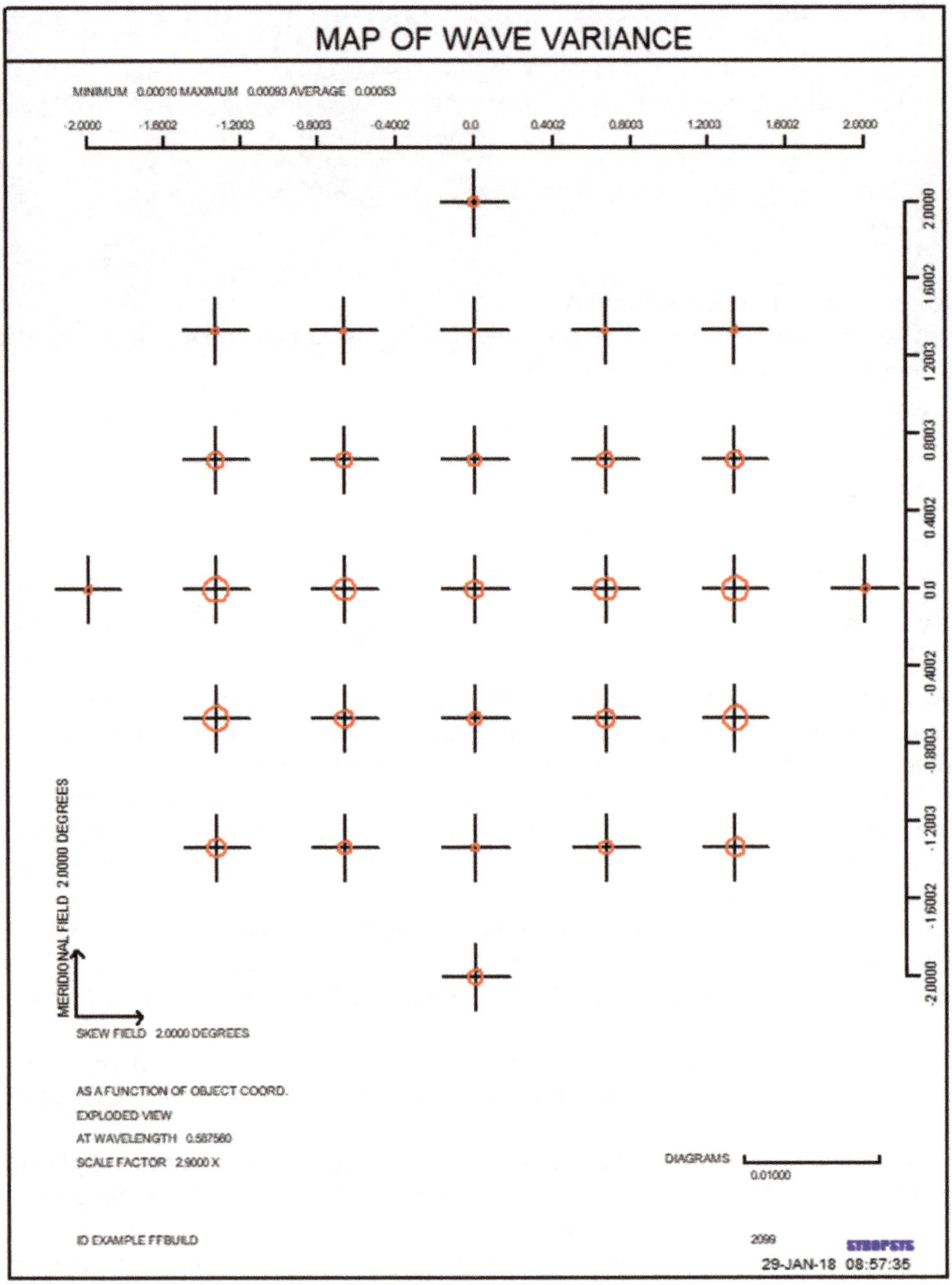

Figure 43.4. Wavefront variance over field.

Now you can look at the shapes that resulted, with the FreeForm Analysis tool (**FFA**). The command

```
FFA 2 0 RSAG SURF
```

produces the picture in figure 43.7, which shows the shape when all rotationally symmetric terms are excluded. This tells you how much the surface varies from a symmetric curve.

To see contours instead, type **FFA 2 0 RSAG CONTOUR** and you obtain the picture in figure 43.8.

The shape of the actual surface is given by **FFA 2 0 SAG CONTOUR**, shown in figure 43.9. It is rather close to a spherical surface. Proceeding in this manner, you can look at the shapes of all of the mirrors.

What about distortion? The GDR request has handled that nicely too. Here is the picture produced by the command **GDIS 31**, in figure 43.10—not bad at all.

One more question remains: how does one test these mirrors? The easiest way is to look at the fringes when tested in double-pass in an interferometer against a reference surface of a known radius. **FFA** can show that too. Here is the output from the command **FFA 2 0 RFRINGES**, in figure 43.11.

If you see that fringe pattern, the mirror is perfect.

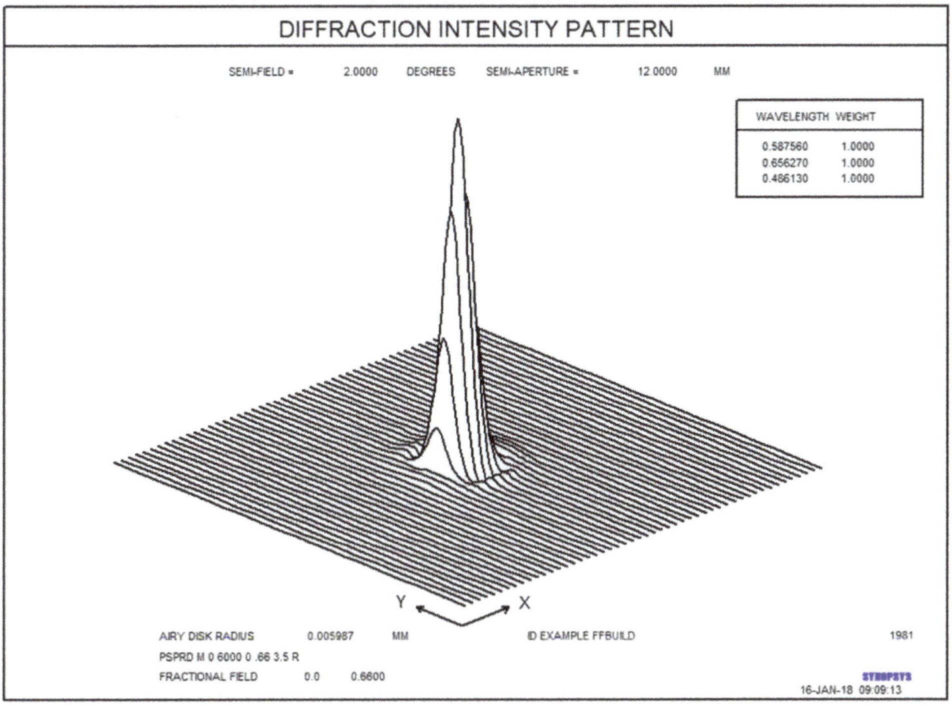

Figure 43.5. Diffraction pattern at worst field point of free-form mirror design.

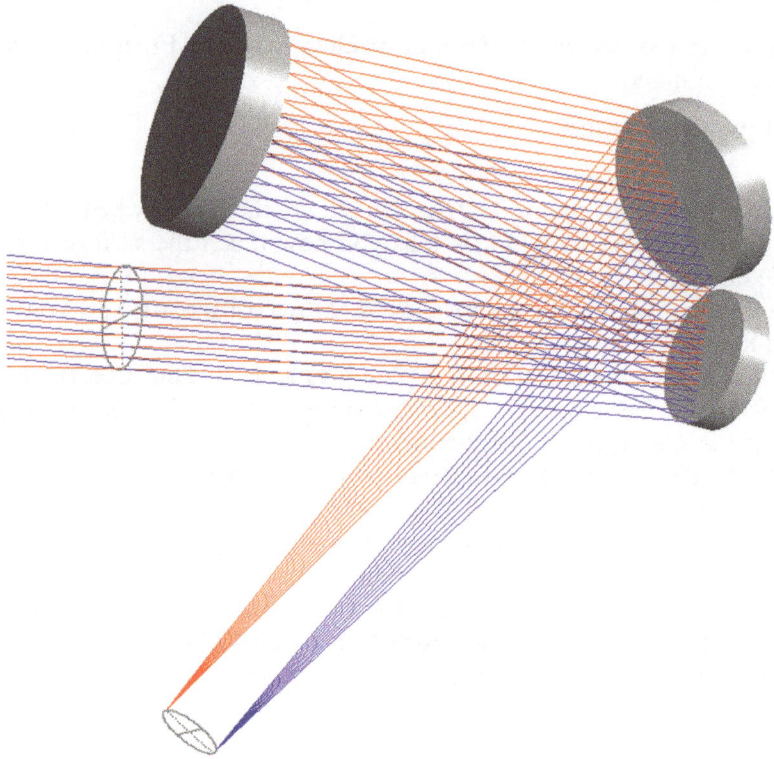

Figure 43.6. RSOLID view of the final design.

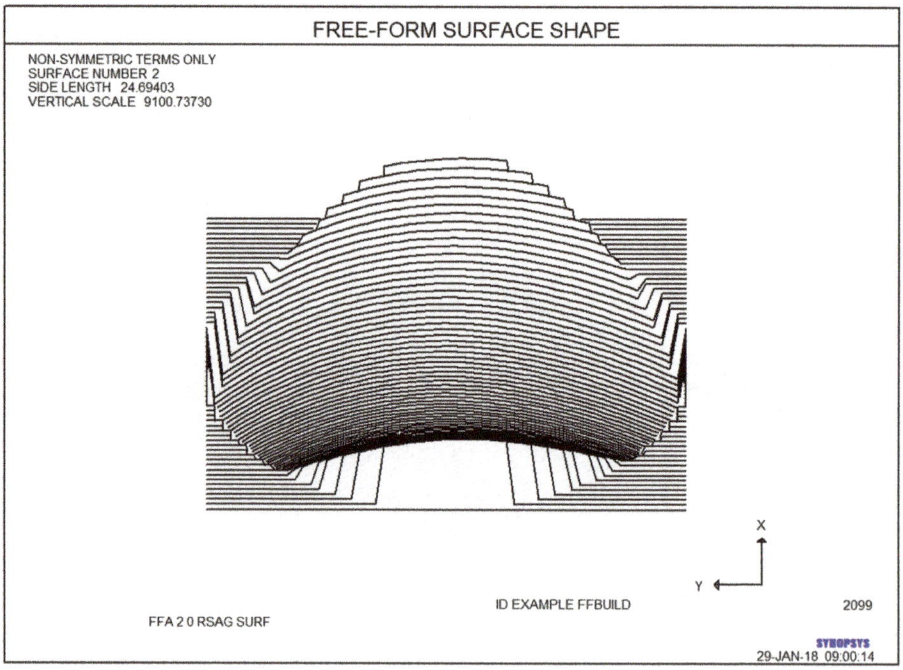

Figure 43.7. Shape of free-form mirror on surface 2, with symmetric terms subtracted.

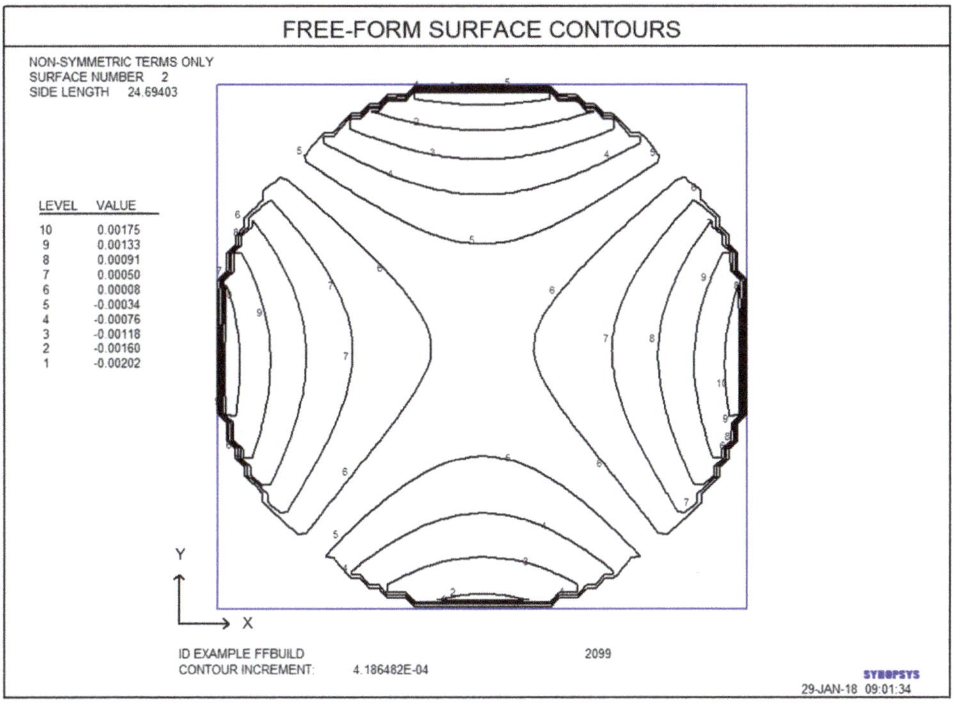

Figure 43.8. Contour plot of the non-symmetric terms on surface 2.

One more task: the mechanical engineers will want to model the system, and probably will need to know the location of many points on each mirror in global coordinates. Here is the input that will generate a table of coordinates over the surface of surface 4:

```
MAP GSAG OVER SURFACE ON SURFACE 4
FGRID POINT 0 0
RGRID CREC 7 7
SCALE AUTO
DIGITAL
ACTUAL
PRINT FULL
```

The output is:

```
MAPPING PROGRAM OUTPUT
   X-COORD.    Y-COORD.     DATA

 -0.234203E-06 0.187556E+02 0.144825E+03
 -0.157171E+02 0.260566E+02 0.141918E+03
 -0.785854E+01 0.260802E+02 0.141977E+03
```

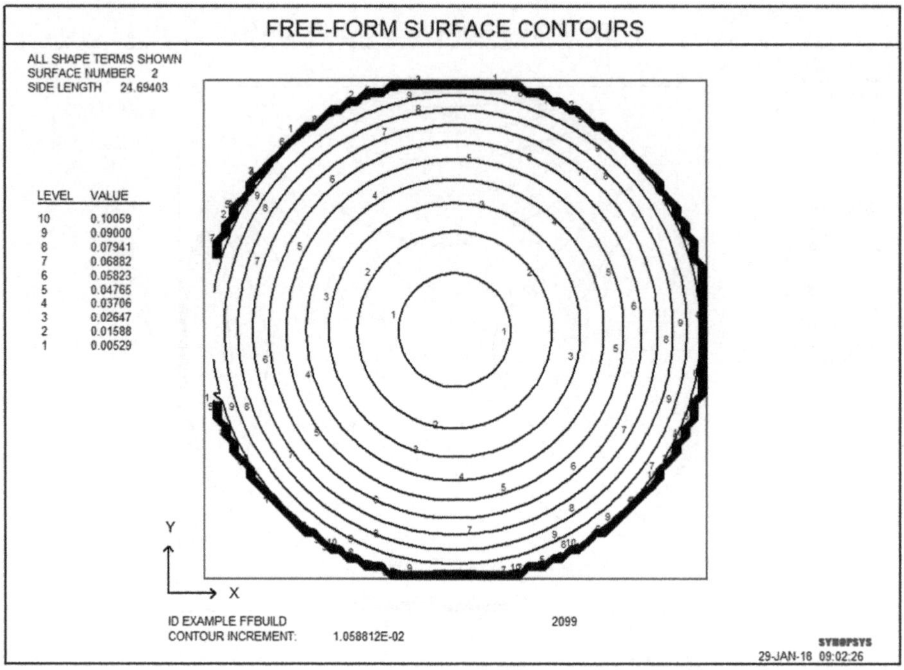

Figure 43.9. Contours of complete shape of surface 2.

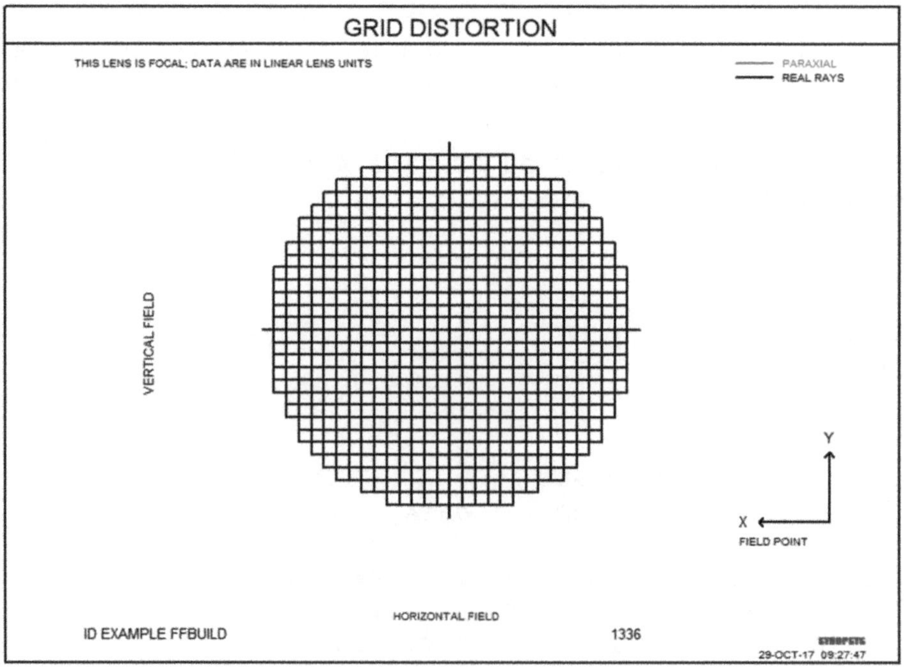

Figure 43.10. Grid distortion of the free-form mirror design.

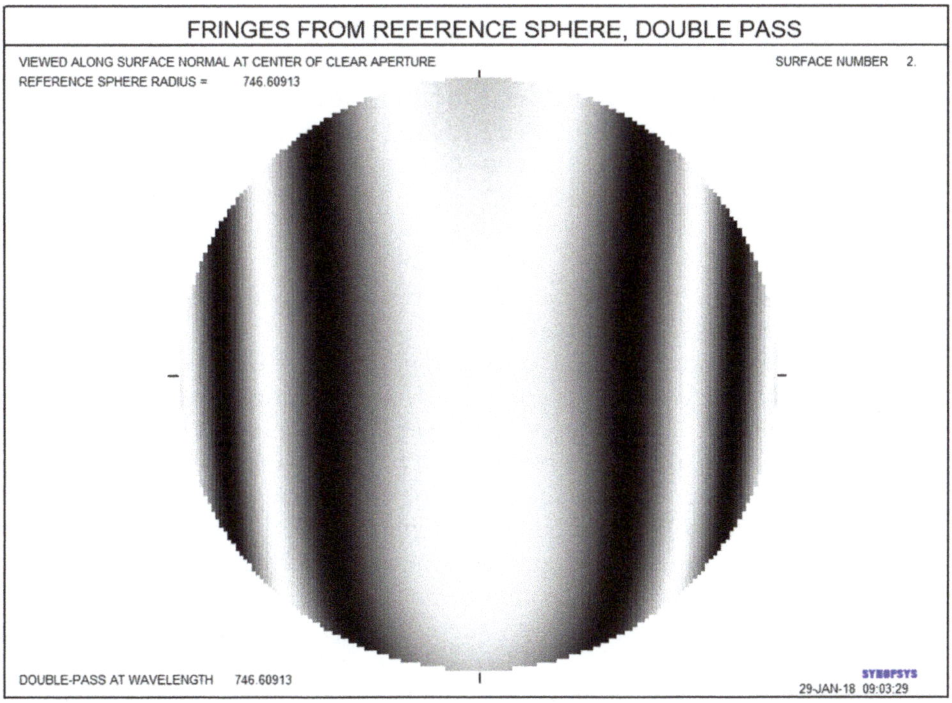

Figure 43.11. Fringe pattern of free-form mirror on surface 2 relative to a reference sphere.

-0.234203E-06	0.260882E+02	0.141997E+03
0.785854E+01	0.260802E+02	0.141977E+03
0.157171E+02	0.260566E+02	0.141918E+03
-0.157171E+02	0.333759E+02	0.139056E+03
-0.785854E+01	0.333994E+02	0.139115E+03
-0.234203E-06	0.334073E+02	0.139135E+03
0.785854E+01	0.333994E+02	0.139115E+03
0.157171E+02	0.333759E+02	0.139056E+03
-0.235756E+02	0.406448E+02	0.136069E+03
-0.157171E+02	0.406823E+02	0.136163E+03
-0.785854E+01	0.407055E+02	0.136221E+03
-0.234203E-06	0.407133E+02	0.136240E+03
0.785854E+01	0.407055E+02	0.136221E+03
0.157171E+02	0.406823E+02	0.136163E+03
0.235756E+02	0.406448E+02	0.136069E+03
-0.157171E+02	0.479763E+02	0.133238E+03
-0.785854E+01	0.479990E+02	0.133295E+03
-0.234203E-06	0.480067E+02	0.133314E+03
0.785854E+01	0.479990E+02	0.133295E+03
0.157171E+02	0.479763E+02	0.133238E+03

```
-0.157171E+02 0.552588E+02 0.130284E+03
-0.785854E+01 0.552808E+02 0.130339E+03
-0.234203E-06 0.552882E+02 0.130358E+03
 0.785854E+01 0.552808E+02 0.130339E+03
 0.157171E+02 0.552588E+02 0.130284E+03
-0.234203E-06 0.625590E+02 0.127375E+03
IMAGE>
```

So that is how you can design a free-form mirror system with these advanced tools. The computer does most of the work for you.

Now it is up to you and the shop to communicate well enough so they understand the results and can make the parts correctly. Here are some pointers:

1. In this example, the surfaces are defined by Zernike terms, as requested. Variable G 39 can vary the center point of the expansion, but we did not use that variable here. Although it is sometimes helpful, the center of the expansion would then not be at the vertex, which is a complication we like to avoid. The vertex of the surface is not at the center of the clear aperture either, which is something we cannot avoid. So be careful here. There are two center points to consider. Also, variable G 51 can vary the y-scale of the expansion, which distorts the Zernike zones and can sometimes be helpful. However, this is also to be avoided unless it really makes a difference.
2. When presenting these data to the shop, be sure they understand the coordinate systems and locations of the relevant parameters.
3. Check out the other features of the **FFA** program. You can create a table of sags across the surface parallel to the surface normal at the center of the CAO that will be of great interest to the technician running the precision milling equipment.

Chapter 44

An aspheric camera lens from scratch

Spy camera; aspheric plastic elements; restricting glass model

When developing a modern cell-phone camera lens or a pinhole spy camera, designers are resorting more and more to using multiple aspheric surfaces. These are typically embodied as small plastic elements, and even though the molds are expensive to make, the lenses can be produced in quantity at very low cost. It is even possible to mold mounting flanges directly onto the elements, simplifying assembly and enabling some dimensions to be held to very tight tolerances.

To help in designing such systems, DSEARCH can do a global search for systems with aspherics. Here is an example, a five-element lens with plastic elements and a glass cover plate before the image.

This will be the input to DSEARCH (**C44M1**):

```
TIME
CCW              ! clear command window
CORE 14 ! use 14 cores for speed

DSEARCH 1 QUIET         ! start DSEARCH; put best lens in library location 1
SYSTEM                  ! define the system specs
ID DSEARCH ASPHERIC CAMERA LENS        ! identification
OBB 0 41.3 .285         ! infinite object, semi field 41.3 degrees, semi ap. 0.285
UNI MM                  ! lens will be in millimeters
WAVL CDF                ! use visual wavelengths at C, d, and F lines
END                     ! end of system section

GOALS                   ! define the goals here
ELEMENTS 5              ! we want a four-element lens with a cover glass
BACK 0.4 SET     ! ask for 0.4 mm back focus distance
FNUM 2.7 10      ! ask for F/2.7, weight of 10
THSTART .5              ! global search use thicknesses .5 mm
RSTART 30               ! and starting radius of 30 mm
```

doi:10.1088/978-0-7503-1611-8ch44

```
ASPH R          ! use all terms in real mode; quick mode spherical
ASPH Q          ! vary CC in quick mode too
ASPH 3          ! allow three aspheric terms: CC, 4th, 6th power of aperture

ANNEAL 10 1 Q 30  ! anneal each case, temp 10 degrees, cool 1, including quick
SNAP 5                  ! redraw PAD screen every five passes
STOP FIRST              ! put the stop in front
STOP FIXED              ! and keep it there
RT 0.5
QUICK 50 50             ! run quick mode 50 passes, then real mode 50
NGRID 6         ! 6x6 grid of rays in pupil
NPASS 50        ! 50 passes in the MACro when finished
TOPD                    ! correct both transverse aberrations and OPDs
FOV 0 .2 .4 .6 .8 1     ! correct six field points
FWT 5 4 3 3 3 3         ! with these weights
COVER .3 1.51872 64     ! the cover glass will be 0.3 mm thick with this GLM
PLASTIC 1 3 5 7         ! the four elements will be plastic
END                     ! end of goals section

SPECIAL PANT    ! special PANT section starts here
RDR .001                ! these are tiny lenses; reduce derivative increments
TLIMIT 3 .1             ! limits on thicknesses and spaces
SLIMIT 5 .1
END                     ! end of PANT section

SPECIAL AANT    ! start of special AANT section; these go into the merit fn.
ACC 1.0         ! center thickness no more than 1.0 mm
ACM .2 .1 .2    ! and no thinner than 0.2 mm
ACA 55 1 1              ! avoid critical angle; 55 degrees from surface normal
AEC .1 .1 .1            ! keep edges over 0.1 mm
M 1.35 10 A P YA 1      ! target the chief ray at three field points
M .945 10 A P YA .7     ! to control distortion
M .54 10 A P YA .4
END                     ! end of AANT section
GO                      ! DSEARCH runs
TIME                    ! when it is finished, see how long the run took.
```

There are some subtle considerations here. First, these will be very tiny lenses, and the default edge-control target (1 mm) that DSEARCH puts in its optimization MACro is too large; that is changed with the AEC monitor. Also, the default minimum airspace and thickness monitor of 1 mm (also too thick) is overridden with the ACM of 0.2 mm. The added ACC monitor will not let thickness grow to more than 1.0 mm, overriding the default of 25.4 mm, and the ACA monitor will penalize solutions where rays encounter too steep an angle anywhere.

Those monitors are very weak as entered, and this is done on purpose: if you strongly control those items, DSEARCH will tend to favor designs that do not offend them—but we want the program to favor designs with small *image errors*, and do not really care much about the mechanical properties at first. When you obtain a good design, you can easily modify those monitors later on, increasing the weights to make the design practical.

DSEARCH lets you manage the image distance in two ways: if you just give a distance, such as **`BACK 0.4`**, the program adds a YMT solve at the end and includes a target in the AANT file to control the resulting value. If you add a weighting factor, such as **`BACK 0.4 100`**, that weight is applied to the target. The other way is to request an exact value, in this case with **BACK 0.4 SET**. Now the program will simply set the back-focus distance to the entered value, 0.4 in this case, and will *not* add a YMT solve. This is often a good choice for difficult designs, especially when the other options return systems with a virtual image.

Since we are enabling use of aspherics, we have to be careful to give a grid higher than the default **NGRID** 4, and to correct at six field points instead of the default three. Otherwise there will likely be intermediate pupil and field zones that fly away out of control.

The bounds on the glass variables also need some attention. Those will be replaced with plastics from the U (Unusual materials) catalog when the design looks good, and we want the model glass to fall in the area where plastics are to be found. This is the purpose of the **PLASTIC** declaration in the input file. Any surface so designated is restricted to the area on the glass map shown in figure 44.1.

The red dots are the plastics currently in the U catalog. The program will keep those glass models that are declared **PLASTIC** within the area shown. Those that reach the boundary (which is all of them since the area is so small) will slide up and down along those boundaries.

Run the DSEARCH MACro listed above and in a few minutes you see the best design the program found, shown in figure 44.2.

This is amazing. The lens is already diffraction limited, straight out of DSEARCH. The OPD errors are less than 1/4 wave almost everywhere—but the last lens is too close to the cover plate. The edge-control monitor AEC works well at the edge of the lens, but this problem seems to occur only partway out, where the monitor does not see it. (You could also just *move* the cover plate, since the aberrations of a flat window are independent of location, but let us assume the image distance is fixed.)

So we have to fix that. Add to the AANT section in the file DSEARCH_OPT, which is in a new editor window, i.e. the lines

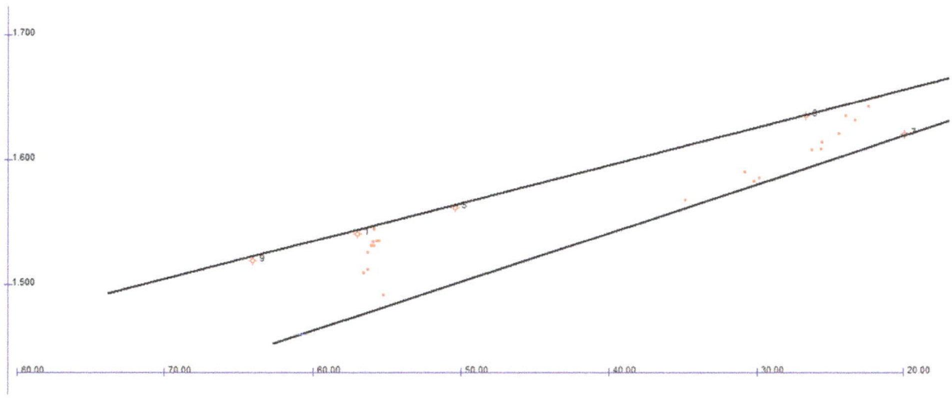

Figure 44.1. Area of glass map valid for plastic materials.

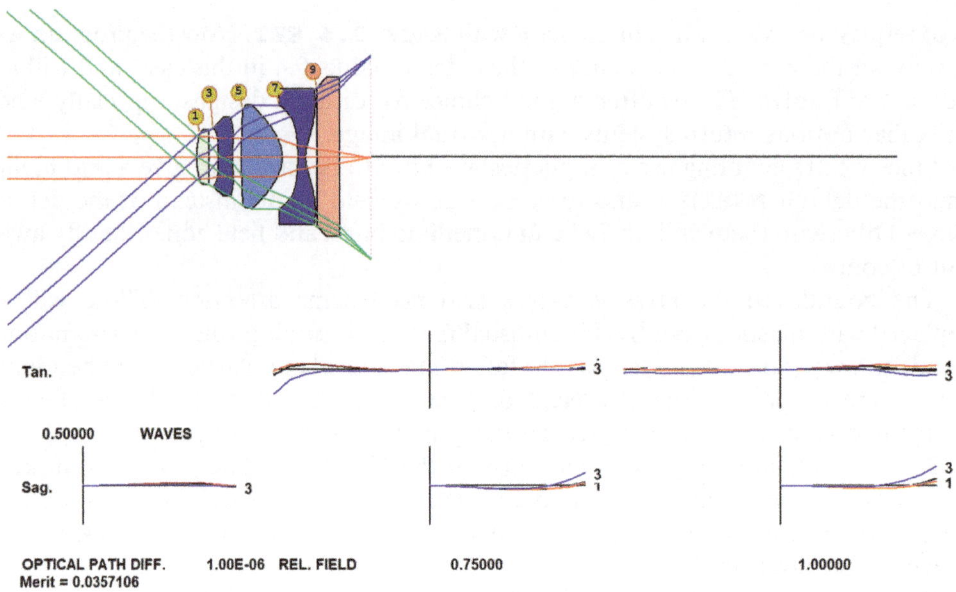

Figure 44.2. Aspheric pinhole lens found by DSEARCH, before optimizing.

```
LLL .1 5 .05
A P ZG .8 0 0 0 9
S P ZG .8 0 0 0 8
```

This entry sets a lower limit of 0.1 to the difference between the global *z*-coordinates of the chief ray on surfaces 8 and 9 at the 0.8 field.

Now run this file and anneal (**20, 2, 50**). The lens is improved, as shown in figure 44.3, and the clearance issue is corrected.

Let us see if we can improve it some more. At the moment, the aspheres use G-terms 3 and 6 only, which vary the fourth- and sixth-power terms in the polynomial expansion. Add to the PANT file the variables

```
VY 1 G 10
VY 2 G 10
VY 3 G 10
VY 4 G 10
VY 5 G 10
VY 6 G 10
VY 7 G 10
VY 8 G 10
```

to also vary the eight-power terms, and reoptimize and anneal. The MF comes down to 0.028, as shown in figure 44.4.

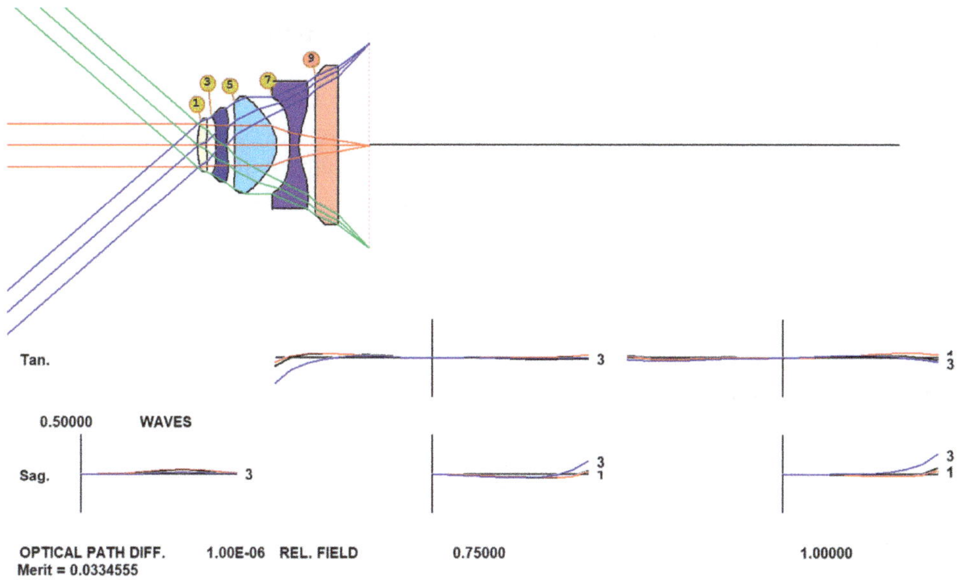

Figure 44.3. Lens reoptimized with altered MF. Clearance issues have been resolved.

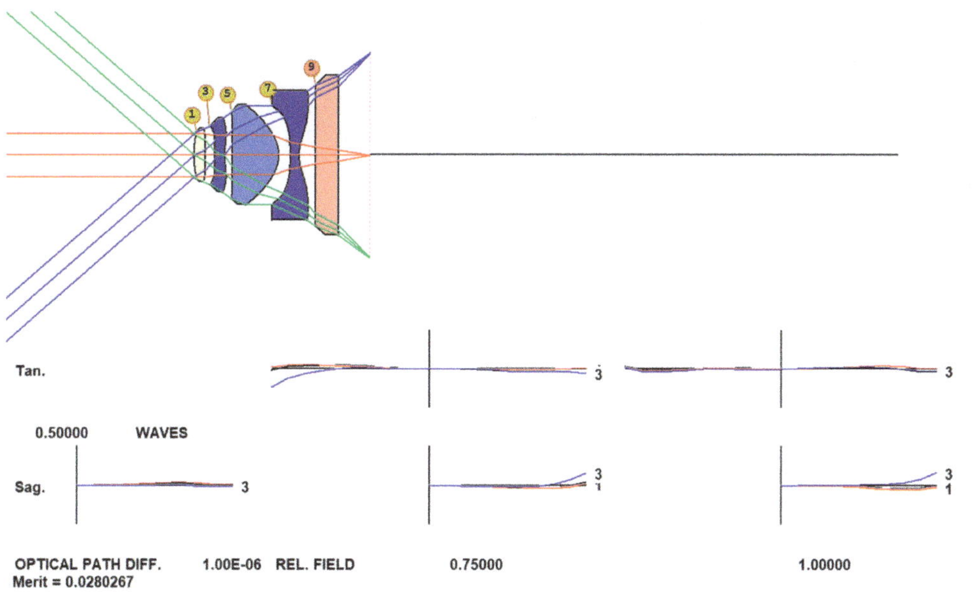

Figure 44.4. Lens optimized and annealed with added eighth-power terms.

Now it is time to switch to real plastics—but first change the material on surface 9, the cover plate, to the real glass the customer wants to use: Hoya type BSC7. To do so, open the WorkSheet and type in the edit pane

```
9 GTB H
BSC7
```

click ‘Update’ and save a checkpoint. The model glass on the cover plate is replaced. Optimize once again.

Now open the Real-Glass Menu (**MRG**) and select the U catalog. That catalog does not have ordinary optical glasses, but it does have the plastic materials. When you specify that catalog, the ARGLASS program (which is run from the **MRG** dialog) automatically selects only plastics and only replaces GLMs designated **PLASTIC** in the RLE file. It has three modes; it can replace the lenses in numerical order, reverse order, or it can sort them so it replaces the ones furthest from a real material first. The latter option is often better, so check the ‘SORT’ option. When this run finishes, delete the GLM variables from the optimization file, optimize, and anneal again.

Sometimes changing to real glass causes ray failures. The program adjusts the curvatures to maintain element powers, but if aspheric terms are present, some rays can still fail. If this happens, run ARGLASS again after the other materials are changed. This usually works; if not, try a different lens from DSEARCH.

Now there are real materials everywhere as shown in figure 44.5.

The MTF curves for this design (**C44L1**), which is close to perfect, are shown in figure 44.6. (To obtain these MTF curves, go to the **MMF** dialog, select the ‘Multicolor’ option, and click ‘Execute’.)

44.1 Encore

That is a start, and now you understand how to use the program, but what could we have done differently? This design is at the diffraction limit, but the MTF at full field is much lower than on-axis. Why is that?

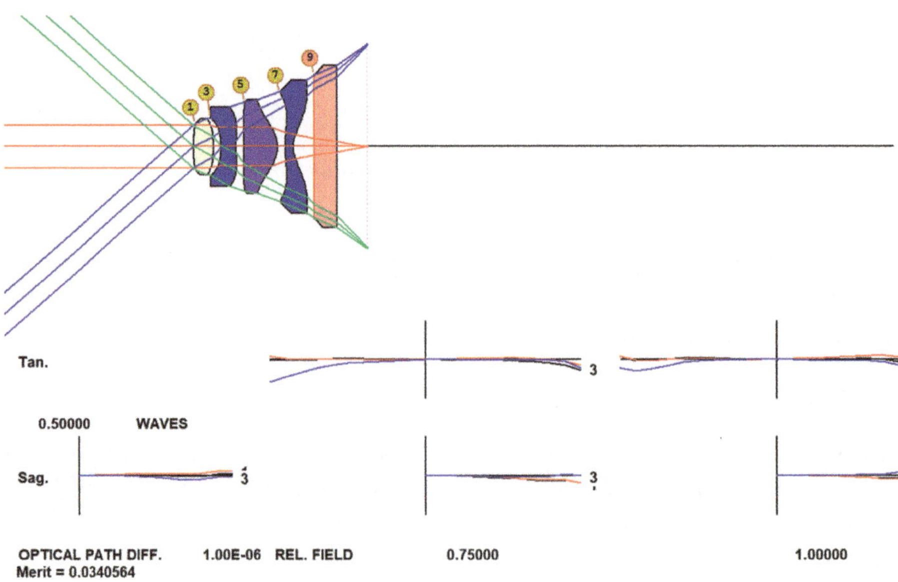

Figure 44.5. Lens optimized with real plastics.

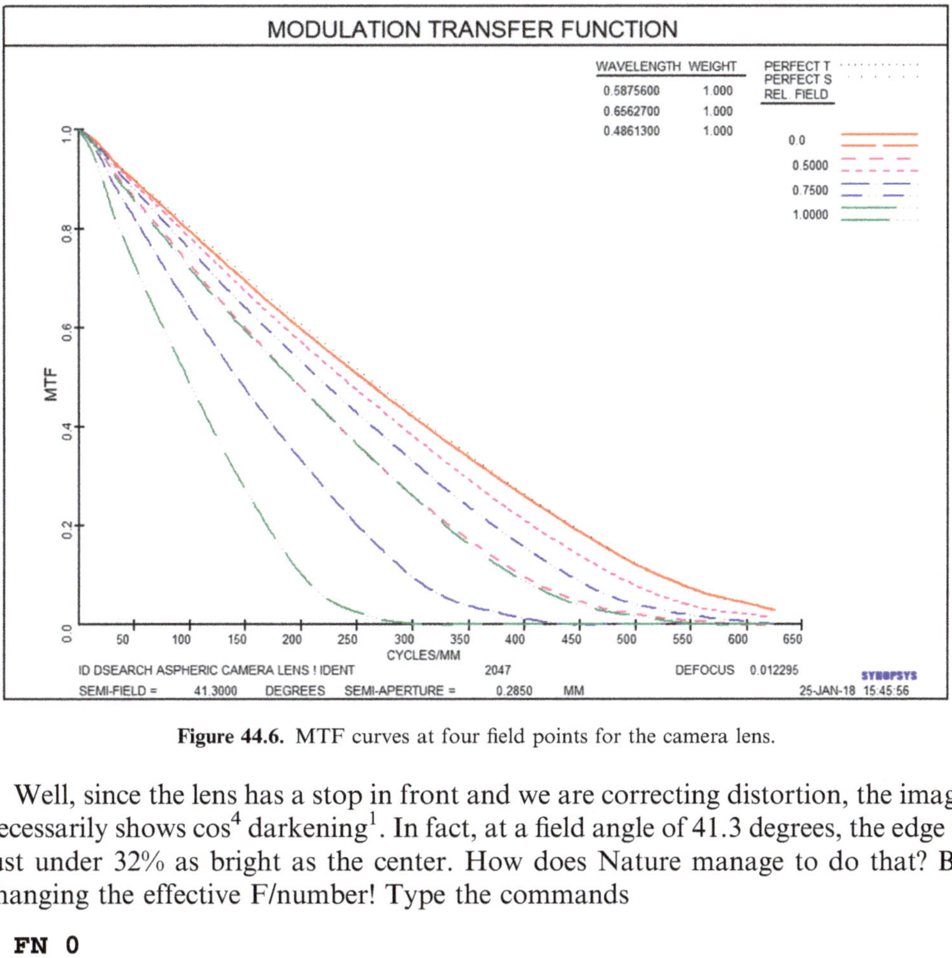

Figure 44.6. MTF curves at four field points for the camera lens.

Well, since the lens has a stop in front and we are correcting distortion, the image necessarily shows $\cos^4$ darkening[1]. In fact, at a field angle of 41.3 degrees, the edge is just under 32% as bright as the center. How does Nature manage to do that? By changing the effective F/number! Type the commands

```
FN 0
FN 1
```

and you see that while the on-axis F/number is indeed about 2.7, at the edge it is 5.3 in the tangential direction and 3.54 in the sagittal. The higher F/number increases the size of the Airy diffraction disk, which lowers the cutoff frequency in the y-direction. That is what the MTF curves tell us.

If that situation is satisfactory, you are done. However, let us assume you really want uniform illumination over the field. You cannot achieve that unless you let the distortion grow, which may not be a problem if you plan to compensate electronically afterwards. Here is what you do:

[1] The illumination at the image is reduced by three factors: the obliquity of the entering beam, the obliquity of the chief ray at the image, and the greater distance of the edge of the image than the center from the exit pupil. The first two each contribute a cosine, and the last a cosine squared. The result is four powers of the cosine.

1. Delete (or comment out) the lines in the SPECIAL AANT section of the DSEARCH input that give targets to the chief-ray YA at three field points. Those were there to control distortion:

```
SKIP
M 1.35 10 A P YA 1
M .945 10 A P YA .7
M .54 10 A P YA .4
EOS
```

2. Add some new requirements in the same section. These will control the relative illumination at five field points, and the distortion will be free to grow to satisfy them:

```
M 1 1 A P ILLUM .2
M 1 1 A P ILLUM .4
M 1 1 A P ILLUM .6
M 1 1 A P ILLUM .8
M 1 1 A P ILLUM 1
```

3. Since the F/number at the edge of the field will now be smaller than before—which is harder to correct—increase the weights on the outer two fields from 3.0 to 4.0 (on the **FWT** line):
4. Change the AEC monitor to **AEC .1 1 .1** to better avoid overlap. Note that DSEARCH always puts in a default AEC monitor, and we want to override that one with this one.
5. Comment out the `QUICK` directive. Some tasks work better that way, and it is a variation you will want to explore.

Run this version on DSEARCH, add the tenth-power variables, then optimize and anneal. The construction is different, and the MF comes down to 0.0325, as shown in figure 44.7. Save this version.

Then again assign the real glass to the cover plate and insert real plastics with the **MRG** dialog—the quality is much worse. What happened?

If you look at the glass table display showing plastics in the U catalog, you see that there is a large gap between a group of plastics to the left and another to the right, as shown in figure 44.8. If a model plastic happens to fall inside that gap, or if the optimization program moves it there when other models are matched, the program must make a significant change when it then converts it to a real material. Sometimes the direction it chooses works well and sometimes not.

However, there may be a solution. Go back to the version you saved, run the optimization MACro once again so the data are current, and then go to **MRG** once more. This time, select '**REVERSE ORDER**' and run the program again—much better.

When it finishes, delete the GLM variables and optimize again. The lens (**C44L2**) is excellent, as shown in figure 44.9.

The MTF is superb, as shown in figure 44.10.

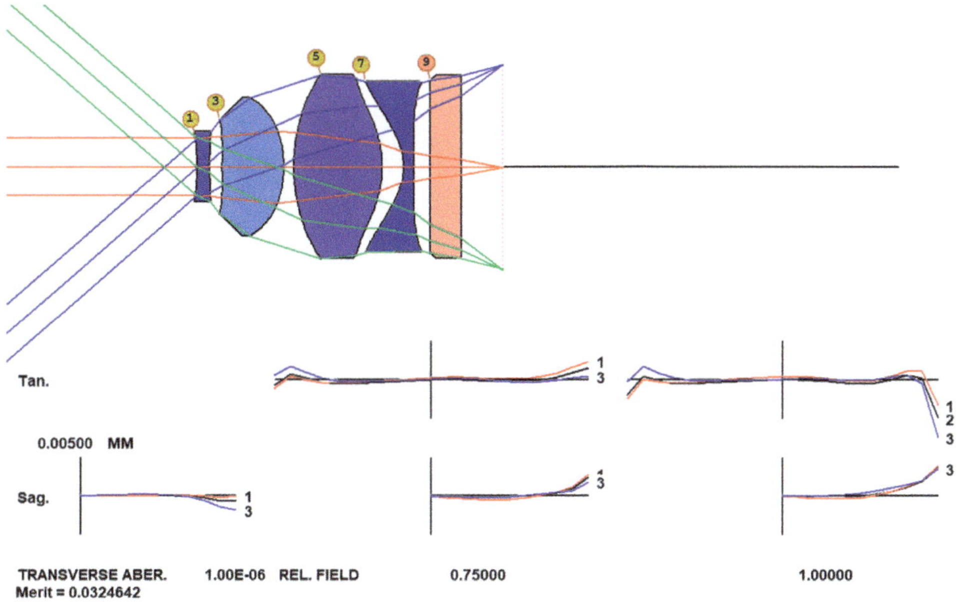

Figure 44.7. Lens returned by DSEARCH with uniform illumination as a goal, then optimized and annealed.

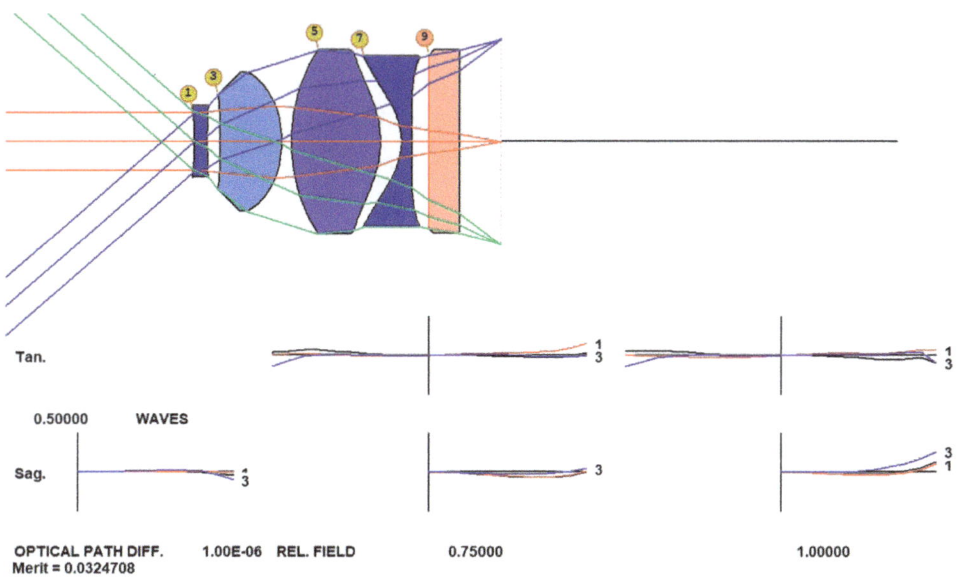

Figure 44.8. The plastic materials in the U catalog leave a gap in the area where the *V*-number is about 45. This can cause difficulties when models in that region are matched to real plastics. Changing the matching order can sometimes overcome them.

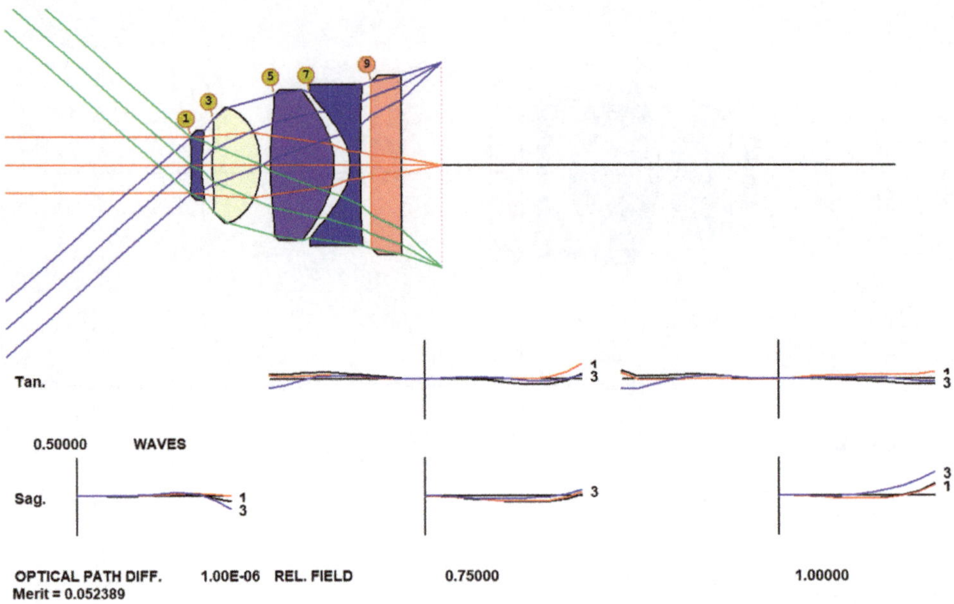

Figure 44.9. Uniform-illumination lens with real plastics.

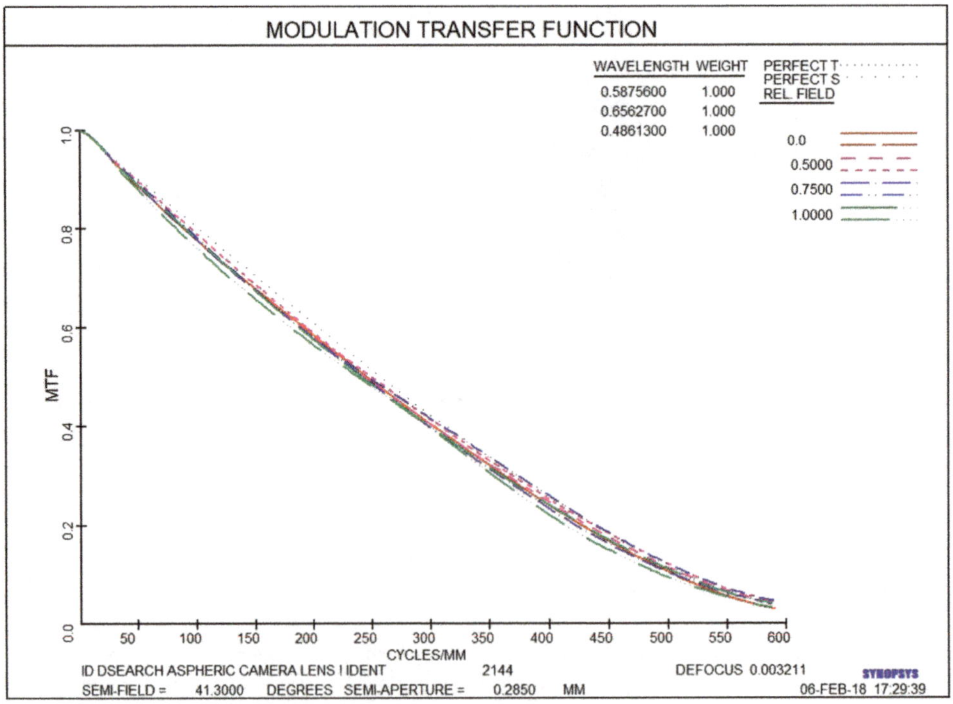

Figure 44.10. MTF curves for the uniform-illumination lens.

The illumination is quite uniform, shown in figure 44.11, plotted with the command

```
ILLUM 500 P
```

The program has indeed introduced significant distortion, however, as shown in figure 44.12, in a plot produced by the command

```
GDIS 21 G
```

44.2 Coda

We made it look easy, and it is if you follow the steps above. But of course, lens design has pitfalls all over the place, and things do not always work perfectly the first

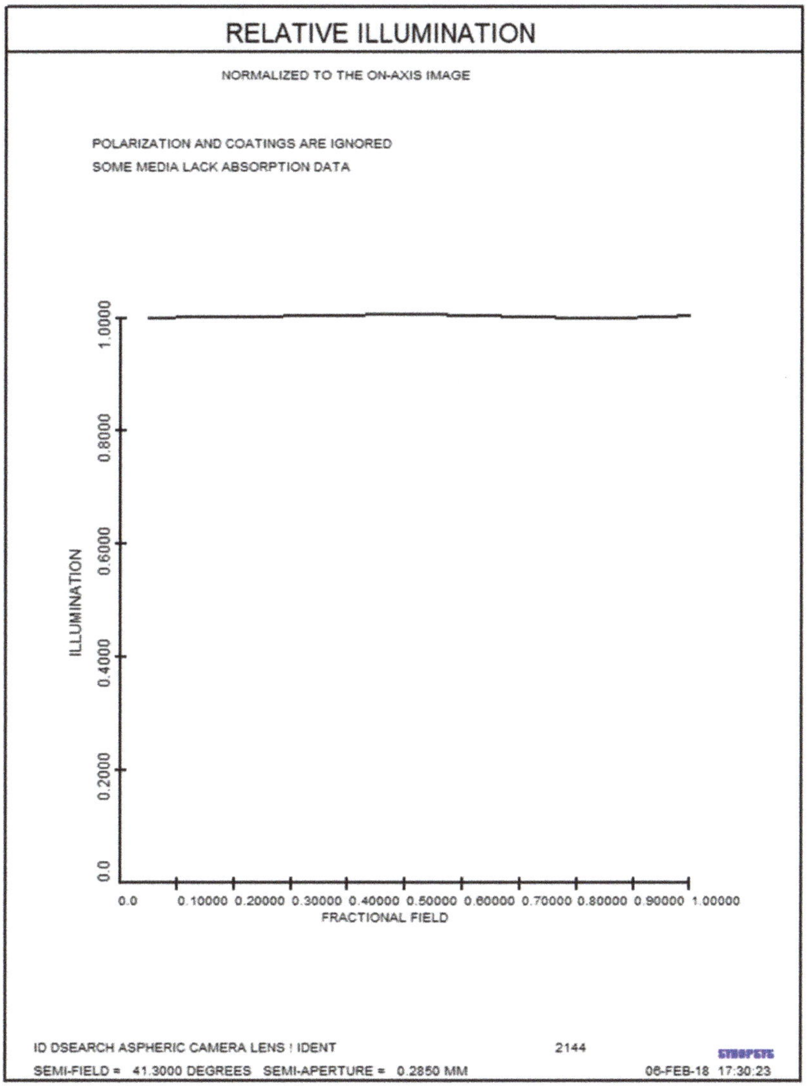

Figure 44.11. Illumination uniformity curve.

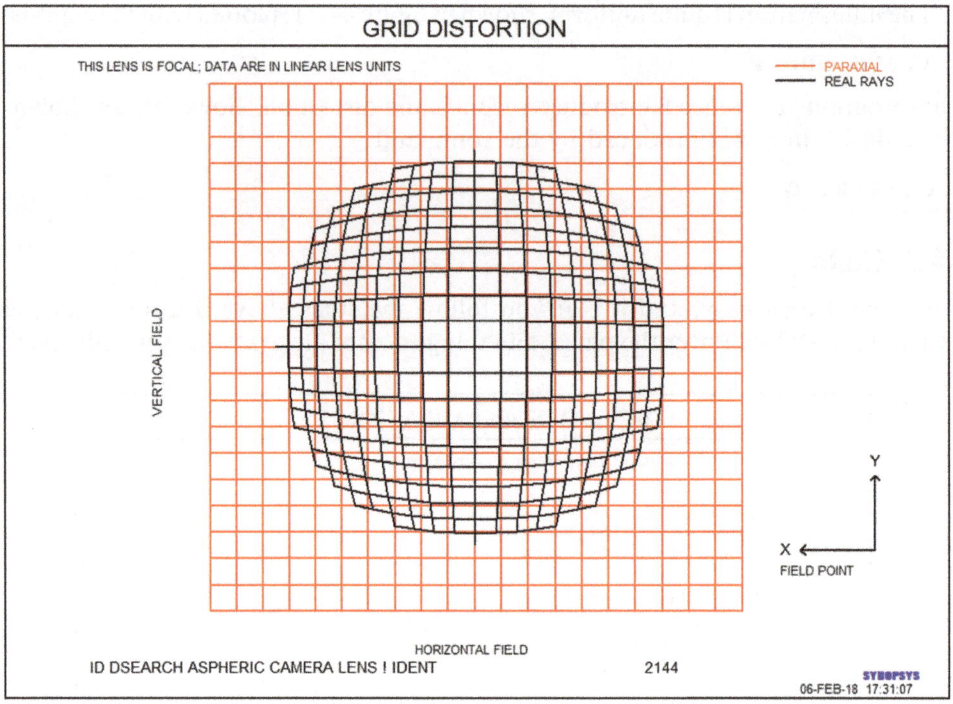

Figure 44.12. Distortion plot of the uniform-illumination lens.

time. Here are some of the problems that you may encounter and how to deal with them:

1. We specified an aspheric count of 3 in this example, which assigns terms up to R^6 to the surfaces. What happens if you use fewer or more terms than this? As a rule, it is better to start with a smaller number and then add more after you have optimized the result as well as possible, as we did above. Too many terms right at the start can send the design to a region where the terms are fighting each other and become too large. Also, ray tracing can prove a problem with many high-order terms, since the beam can exhibit caustics or steep ray angles where you do not want them. Sometimes you can obtain excellent results by starting with only two terms and then adding more later. The **ASPH** directive in the DSEARCH input file tells the program how to use the conic constant and higher-order aspheric terms: (**ASPH Q**) uses a conic constant even during the quick mode, (**ASPH R**) will use all the requested G-coefficients during the real mode (instead of just the conic constant), and (**ASPH Q R**) will do both. Changing any of these will send the program up different branches of the design tree.
2. Note that the **FNUM** request in the DSEARCH input file specified a weight of 10; this is more important than meets the eye. Chapter 35 explained how, if you left the weighting factor off, the program would control the F/number exactly with a paraxial solve—which can lead to ray failures if the radius that

results is too steep. So, for fast lenses such as this one, it is a good idea to add a weight so there will be no UMC solve. Then the program adds a requirement to the merit function to control the F/number, with a starting radius on that surface given by the RSTART value. In the second example, where we did not target the image height, the F/number would probably have grown larger than the target value if we had assigned a lower weight. The program will do absolutely anything to reduce the merit function, and giving up a little on that score would likely bring down the other aberrations significantly, resulting in a great image at a higher F/number. To prevent that, we specified a weight of 10 so that solution would not look so attractive.

3. We chose to give the back focus distance a fixed value in this example. If you instead input a weighting factor on the **BACK** line, the program would assign a YMT solve to the last surface, so the image would always be at the paraxial focus, and then add a target to the AANT file to drive it to the requested value. Both methods work, but when you are targeting the YA of selected rays in order to control the image height, as we are here, it is best to set the value yourself. Otherwise the program may fail to correct a virtual image since the image height has to change when it moves it from inside to outside the lens.
4. Remember that DSEARCH uses the annealing feature (if you request it, which is nearly always a good idea), and that feature makes small random changes to the lens, over and over. This exercise was prepared with switch 98 turned on—so the reader could obtain similar results by using the same random numbers. However, if you run it with that switch turned off, your results will differ and will sometimes be better. It is often a good idea to run DSEARCH more than once (with *random* random numbers) and look at some of the other configurations it returns each time.
5. These designs met our goals very nicely, but suppose you do not want the cost of a four-element lens. What can you do with three elements? Try it and find out. It will probably not be as good, but then, maybe you do not need that level of resolution with your sensor.
6. Remember that DSEARCH is searching a very bushy design tree and cannot examine every branch every time. If you change almost anything in the DSEARCH input, such as the **RT** parameter, field weights, monitor targets, number of iterations, etc—the program will search a different set of branches and return different results. The power of this method is that it can search a huge number of branches simultaneously, and most runs return at least one lens that is at or close to your requirements. By all means experiment with the input and keep the better results in the library so you can examine them at your leisure.

This lesson used plastics for all the lens elements except the cover glass. What if you want some elements to be made of glass and the others of plastic? Simple: just declare which elements are of plastic in the DSEARCH input file, and the program will restrict them to the smaller range where plastics are to be found. Glass elements, on the other hand, will still be free to move over the usual range of the glass map. When the design is satisfactory and you run **ARG**, the program will match only the plastic elements if the 'U' catalog is selected—and will match only glass elements with any other catalog—simple indeed. An example is given in chapter 45.

Let us also observe that the lenses shown above are *very tiny*, measuring just over 2 mm from first to last element. On paper they work fine, but one must ask just how small an element one can reasonably expect from lens makers. Figure 44.13 shows a design by Irina Livshits of ITMO University in St Petersburg, Russia. This lens uses glass elements, with no aspherics, and shows that the technology for making microscope objectives, which are also very small, can be applied here as well. In any case, one always wants to engage the shop and verify that what is on the print is within the capability of their equipment and experience.

Also, in case it is not obvious, the presence of the cover glass matters. Even though all surfaces are flat, a window such as that has aberrations of its own, acting in that regard like a negative lens, the effect depending on the thickness and the convergence angle of the beam. So if the application will involve a cover glass—or a beamsplitter prism, for example—it is wise to include a cover glass of that thickness in the DSEARCH input so the correction will take account of the effect.

44.3 Tolerancing the aspheric lenses

It will be instructive to ask for a tolerance budget for the lens we designed above (**C44L2**). To do so, we have to understand how aspheric surfaces are tested. These elements are assigned general aspheric terms G 3, 6, and 10, which modify the shape

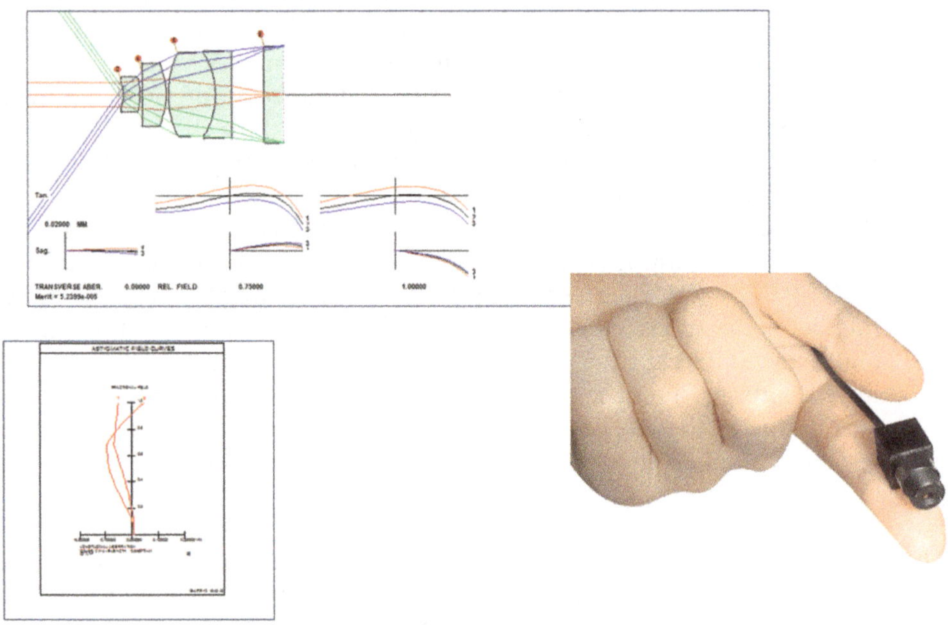

Figure 44.13. Example of a very small spy camera lens, successfully manufactured in Russia. Reproduced with permission from Irina Livshits.

given by the radius and conic constant—but it makes little sense to calculate a tolerance for the individual coefficients since that is not what the technician measures in the shop anyway.

There are two common techniques for testing aspherics: one can design a null system, or simply measure the surface sag at an array of points and calculate some statistics that tell how far from the nominal shape the actual surface is. The first, the null test, is difficult and expensive to set up. One designs a second system that produces exactly the opposite aberrations of the surface under test. Then one reflects a beam off that surface and examines the fringes with an interferometer. If done correctly, this test is extremely sensitive and can show figure errors of a small fraction of a wavelength. However, the null test must first be designed and then manufactured to its own tolerance budget. In this example, where the aspheric curves are steep, the only practical null test would be the other elements of the design itself—which are no easier to make than the surface you want to test.

So this is a good example of a lens for which one would prefer to use a profilometer. Then one can measure the actual surface sags at many points and compare the results with the ideal shape. Here is our BTOL MACro (**C44M2**). The **PFTEST** directive tells the program to assign this attribute to all aspheric shapes in the lens:

```
BTOL 2
EXACT INDEX 1 3 5 7 9
EXACT VNUM 1 3 5 7 9
ADJ 10 TH 100
PFTEST ALL
TOL WAVE 0.05
GO
```

This MACro produces a budget, part of which is shown below:

```
BUDGET TOLERANCE ANALYSIS                                                  -----B-----
 EL. SURF        RADIUS          RADIUS TOLERANCE          THICKNESS THICKNESS TOL
                                (RADIUS)       (FRINGES)

_____________________________________________________________________________________

   1    1        3.72587        0.00000         0.00000         0.10717         0.00427
   1    2        1.02836        0.00000         0.00000         0.10647         0.00351
   2    3        1.66021        0.00000         0.00000         0.49337         0.00213
   2    4       -0.82156        0.00000         0.00000         0.10000         0.00836
   3    5        3.72508        0.00000         0.00000         0.62444         0.00945
   3    6       -0.75690        0.00000         0.00000         0.16803         0.00218
   4    7       -0.49490        0.00000         0.00000         0.10000         0.00723
   4    8       14.84252        0.00000         0.00000         0.10000         0.12700
   5    9       INFINITE        0.00000         2.08787         0.30000         0.01490
   5   10       INFINITE        0.00000         2.26536         0.40000         0.00000
       11       INFINITE        0.00000         0.00000         0.00000         0.00000
```

In this case, the aspheric surfaces are not assigned a radius tolerance (or an irregularity, conic constant, or rolled-edge tolerance), because those are all figure errors that are included in a later section that applies to the aspherics, shown below:

```
. . .
SURFACE TESTED WITH PROFILOMETER: TOLERANCE IS STANDARD DEVIATION OF SAG ERROR
OVER SURFACE
```

ELE	SURF	ST. DEV. TOL	G 15 CHANGE
1	1	7.71846E-05	0.00427
1	2	0.00011	0.00401
2	3	0.00018	0.00399
2	4	0.00047	0.00645
3	5	0.00057	0.00538
3	6	0.00080	0.00670
4	7	0.00079	0.00710
4	8	0.00213	0.01532

Here you see the tolerances that BTOL assigned to the aspheric figure of each element. On surface 1, for example, the program found that a value of 0.004 27 for term G 15 would be part of a reasonable budget. That term creates an astigmatism error in the shape, which is an error that does not affect focus or magnification but increases the wavefront variance. The standard deviation (**SD**) of the surface error is given as 7.718E−5 mm. This is what the technician can deduce, once the profilometer measurements are in hand. He calculates the measured SD by subtracting each measured surface sag from that calculated for the ideal shape, then finding the variance of those data (the average of the squares minus the square of the average). The SD is just the square root of the variance.

However, what if the figure error is not simple astigmatism? As explained earlier, if the aberrations are small, the effect on the system MTF is a function of the wavefront variance and is not very sensitive to which aberrations may be present. So we use term G 15 as a proxy for any kind of small figure error.

IOP Publishing

Lens Design
Automatic and quasi-autonomous computational methods and techniques
Donald Dilworth

Chapter 45

Designing a very wide-angle lens

Creating a wide-angle front end so DSEARCH works over a very wide field

Designing a wide-angle lens with DSEARCH presents a new complication: if you enter a wide-angle object specification in the SYSTEM section of the DSEARCH file, it is likely that none of the candidate configurations will work because no light can get through at the extreme field angle. DSEARCH can correct for some ray failures, but usually cannot optimize such a system. So what do you do?

There is a rather simple trick that works in such cases: rough out a simple front end that converts the beam into one with a smaller angle, and then go from there, declaring that portion with **USE CURRENT**. Here is an example.

We want a lens with a semi-field angle of 92.4 degrees that works at F/2.0. The added elements will be made of plastic, which can be aspheric. First, we have to create a front end that will trace.

Enter a simple RLE file with two lenses and specify object type OBD, which is used for wide angles, declaring a paraxial stop on 5. Start with a moderate angle, say 50 degrees, and then, using the WorkSheet sliders, give the elements some negative power and bend them to the right. When that looks good, increase the OBD field angle, continuing in this manner until you reach the desired angle of 92.4 degrees. Here is a suitable front end, in figure 45.1:

```
RLE
ID WIDE-ANGLE DESEARCH
WAVL .6562700 .5875600 .4861300
 APS        5
 UNITS MM
 OBD 1.00000E+09  92.4 0.2887 -11.0345861 0 0 0.2887
```

doi:10.1088/978-0-7503-1611-8ch45

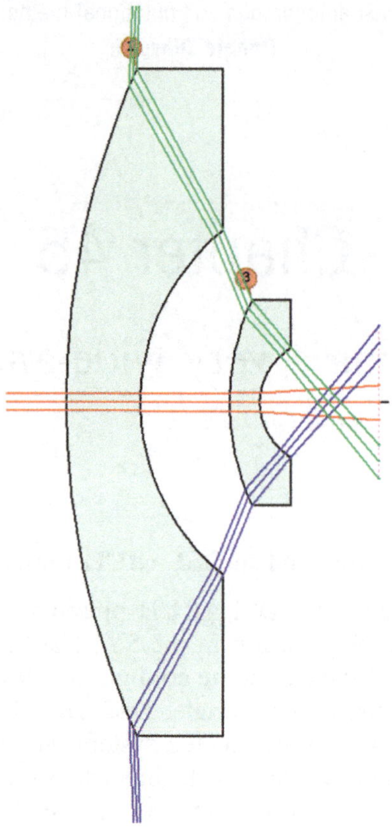

Figure 45.1. The front end roughed out so light can get through at a shallower angle.

```
 0 AIR
 0 CV 1.0000000000000E-09 AIR
 1 CV    0.0356159993000  TH    2.50000000
 1 GLM    1.50000000           55.00000000
 2 CV    0.1318873610000  TH    2.99808431 AIR
 3 CV    0.1145140002814  TH    1.00000000
 3 GLM    1.50000000           55.00000000
 4 CV    0.4600712360000  TH    4.00383115 AIR
 5 CV    0.0000000000000  TH    0.00000000 AIR
END
```

The 92.4 degree entering beam exits at a reasonable angle. Now create a DSEARCH input MACro (**C45M1**):

```
CORE 14
TIME
DSEARCH 2 QUIET
USE CURRENT 5 ALL

GOALS
ELEMENTS 5
FNUM 2 1
BACK 5 SET
STOP MIDDLE
STOP FREE
ASPHERIC 3 5 6 7 8 9 10 11 12 13 14
FOV 0 .2 .4 .6 .8 1
NGRID 6
SNAP 10
RT 0.5
PLASTIC 5 7 9 11 13
!QUICK 30 40
ANNEAL 50 10 Q 40
NPASS 50
END
SPECIAL AANT
ACC 10 1 1
ACA 65 1 10
LUL 90 .1 1 A TOTL
END

GO
TIME
```

This file says to use the current system (the two lenses you adjusted above) and start adding elements at surface 5. All surfaces are to be variables, including the current ones. It specifies a back-focus distance of 5 mm, fixed with the **SET** directive. We can free up that thickness later, if it does not have to be exactly that value.

This input requests a maximum element thickness of 10 mm and an upper limit on the total length of 90 mm, to keep things reasonable. Also, it restricts ray intercept angles to no more than 65 degrees. Otherwise, for steep field angles like this, one can obtain grazing-incidence rays, which are impractical because of coating concerns and can lead to ray failures while optimizing.

Note that we are *not* using the QUICK option in this case. That is a powerful tool for simpler jobs, but this one is not simple. Third-order aberrations have little meaning with so wide a field, and we need the power of the full optimization on each candidate system. That line is commented out above, to emphasize the point.

Your input is now prepared, so run this DSEARCH file. In about two minutes you see the results shown in figure 45.2.

At this stage, the lens has only conic constants assigned to the surfaces, since that is the default unless you enter **ASPH R** in the DSEARCH file, which we did not do here. This makes sense; it is a good idea not to use high-order terms when you are

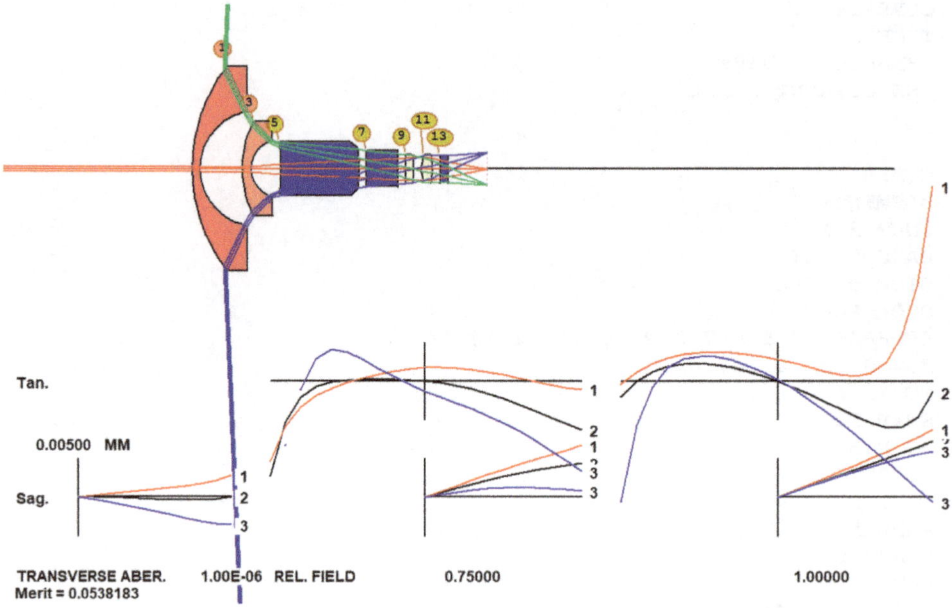

Figure 45.2. First results from DSEARCH for a wide-angle lens.

roughing out a lens. Save those for when the lens is as good as you can get it with only spherical or conic surfaces. Even so, this is an excellent start.

However, it needs improvement. Run the optimization MACro that DSEARCH created and the lens is better. The MF comes down to 0.026 and the surfaces now have higher-order aspheric terms.

Now change the thickness variable declaration to **VLIST TH ALL**. This will let the back focus vary now, since we are close to a solution.

Optimize and anneal (**20, 2, 50**); the result is shown in figure 45.3.

This is real progress. Now the design has taken shape and we see that the stop naturally wants to be near the last elements. Assign a real stop to surface 11 with WS:

```
APS -11
```

delete the variable for YP1 and reoptimize. The MF is now at 0.0007.

Now let us insert real materials. Make a checkpoint, open the **MRG** dialog, select the 'U' catalog (which will match only plastic elements), select 'QUIET', 'SORT', and click 'OK'. The lens now has real plastics, shown in figure 45.4.

The aberrations look great—but the TFAN is entirely vignetted at full field. What happened? The edge from 2 to 3 is feathered. Isn't the AEC monitor supposed to take care of that?

Here is a problem we have not seen yet: surface 2 has become a hyperhemisphere, meaning it extends beyond the hemisphere point, where AEI cannot see it, and it

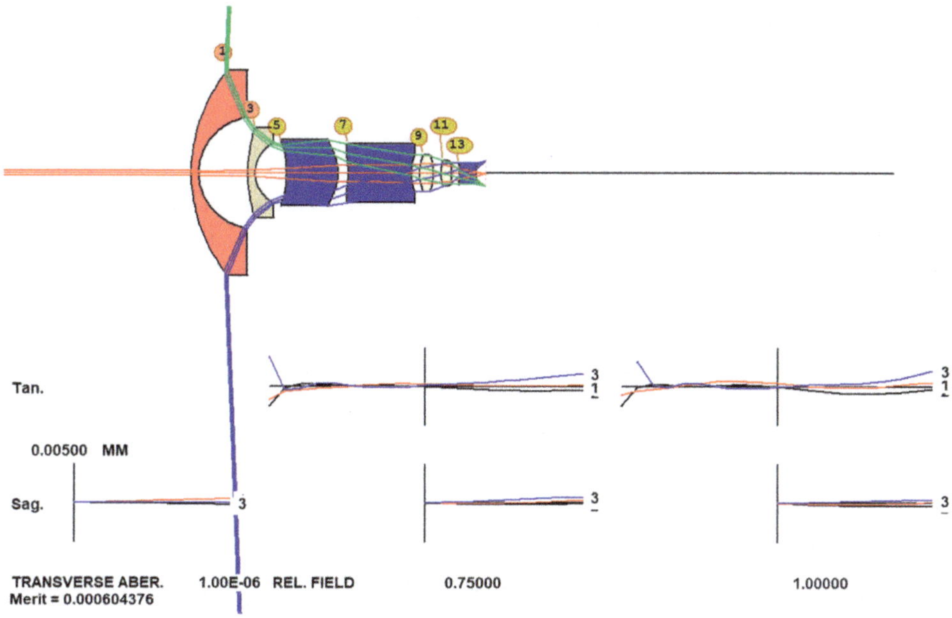

Figure 45.3. Wide-angle lens reoptimized.

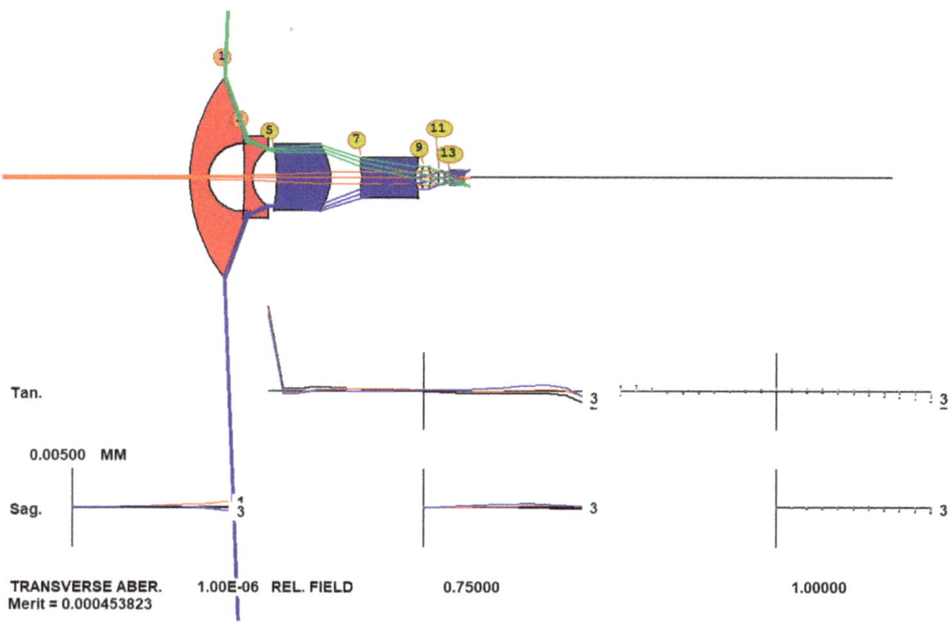

Figure 45.4. Lens with real plastics replacing the glass models on the last five elements.

now overlaps with surface 3. However, the overlap is easily fixed. Add two lines to the AANT file:

```
M 1 1 A P ZG 1 0 -1 0 3
S P ZG 1 0 -1 0 2
```

and reoptimize. This input says to find the global *z*-coordinate of the full-field lower rim ray on surfaces 2 and 3 and target the difference to 1 mm. The problem is fixed and now the MF is 0.000 58.

Now replace the first two elements with glass. Run **MRG** again, this time selecting the Ohara catalog. The program matches only the first two elements, which are glass, not plastic, and the design comes back just as good as before, shown in figure 45.5.

Now for the finishing touches. In WS, enter the lines

```
CSTOP
WAP 2
```

and reoptimize. Now the stop will be nicely filled at all field points. This version is shown in figure 45.6 (**C45L1**).

How well did we do? Let us look at the diffraction pattern over the field. Go to the **MPF** dialog, select 'Show' visual appearance, and click 'Execute'. The result, in figure 45.7, is perfect over the whole field.

Let me add some words of wisdom. Note that we did not use a curvature or thickness solve in this exercise, since a common problem with very wide-angle lenses

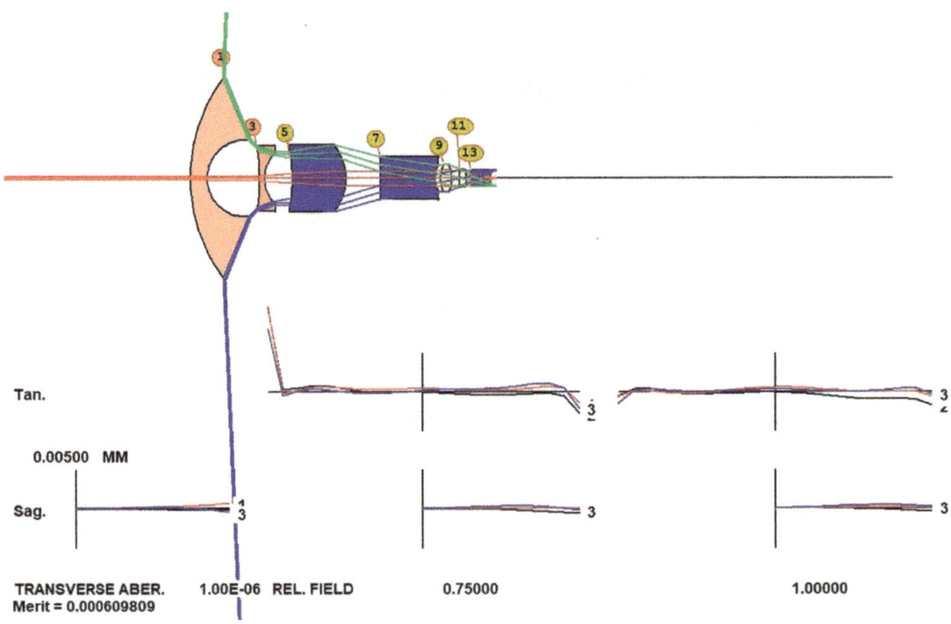

Figure 45.5. Wide-angle lens with all real materials.

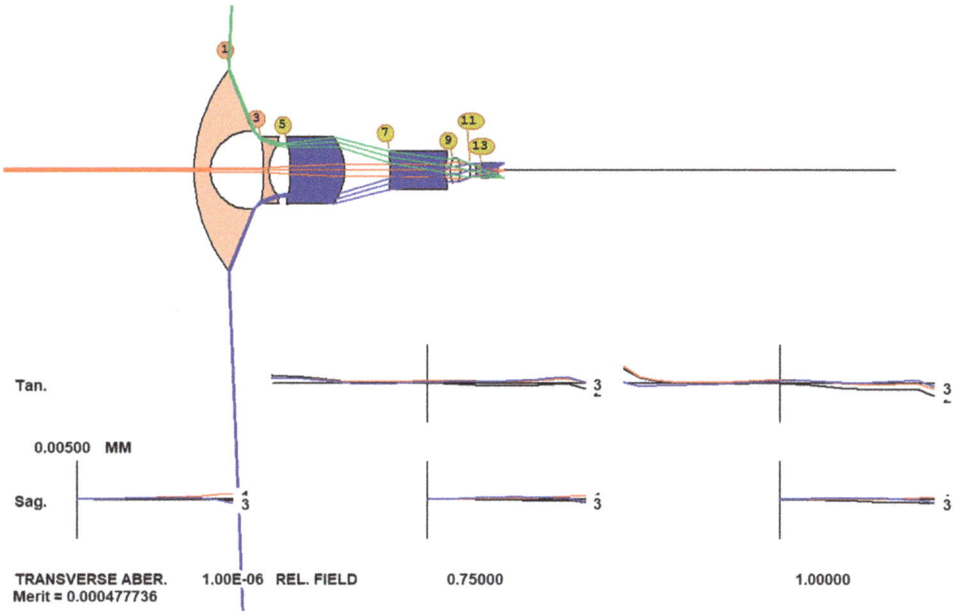

Figure 45.6. The final wide-angle lens.

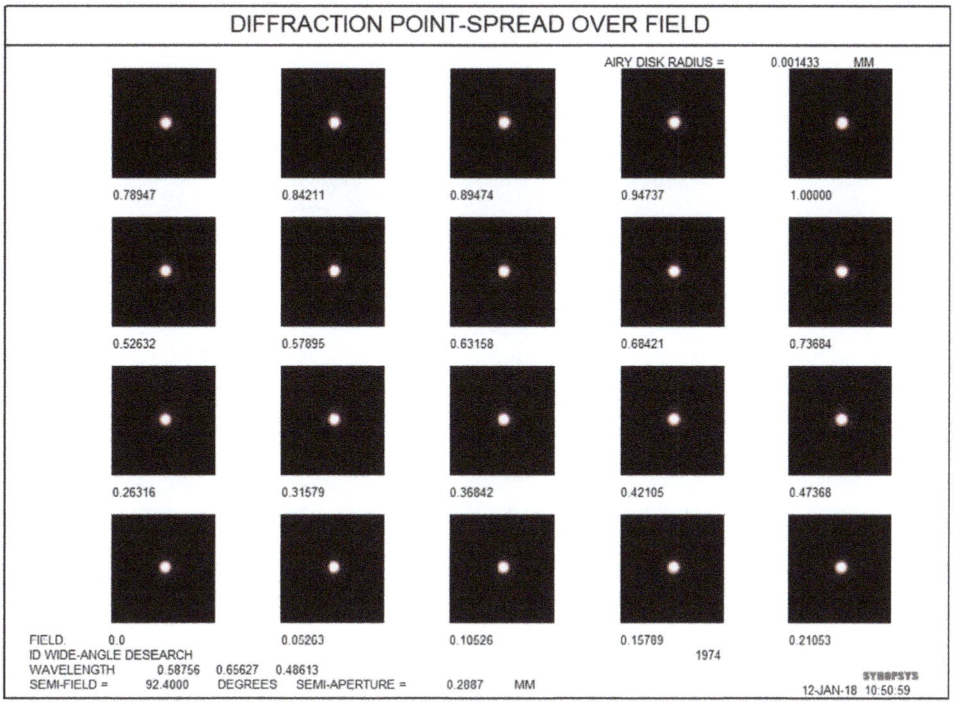

Figure 45.7. Diffraction pattern over field of the final wide-angle lens.

is trying to avoid ray failures. While using solves makes perfect sense mathematically, they can cause just this kind of problem with this kind of lens. Also, we did not change to a real pupil until after the lens had its final form. The real-pupil search is robust but not infallible, and with this kind of steep ray angle and power-series aspherics it is possible to obtain to a configuration where there is no solution to the search. Even worse, sometimes there are *two* solutions, and the program might choose the wrong one. All that can be avoided by using the implied pupil until the design is in good shape.

This lens is obviously very good, but do we really need seven elements? Also, can we obtain a shorter lens? How about one with a smaller first element? All these questions normally come up when you design a lens and all can be quickly answered. Just add the new requirements to the DSEARCH input file and find out.

Lastly, do not hesitate to investigate more than just the top lens returned by DSEARCH. We used the top one in this case, but that is not always the best one when you get to the final design. That is why DSEARCH returns more than just one solution.

IOP Publishing

Lens Design

Automatic and quasi-autonomous computational methods and techniques

Donald Dilworth

Chapter 46

A complex interferometer

Setting up an interferometer

Interferometers have two channels and the beams are combined at a beamsplitter. One often wants to see the difference in the shape of the two wavefronts, as when testing aspheric mirrors. To work properly, the shapes should be very similar.

In this example, fringes between the two channels give spectral information as the position of one of the mirrors is moved back and forth. In this configuration, the instrument is called a *Fourier-transform* spectrometer. Here one is not concerned with the shape of the wavefront, as long as the two channels match, but with its absolute phase.

We will set up one channel at first, entering those data that are easy to figure out, and then let the program calculate the rest for us. Here is the input for the first step (**C46M1**):

```
RLE
ID INTERFEROMETER EXAMPLE
WAVL 4.6 4.25 3.9
OBB 0 1 30 0 0 0 30
1 TH 100                ! DUMMY SURFACE FOR REFERENCE
2 AT -45 0 100          ! SCAN MIRROR
2 REFL
2 TH 0
3 AT -45 0 100          ! FOLD AXIS
3 TH -200               ! TO BEAMSPLITTER
4 TH -3 GTB U           ! THROUGH 3 MM THICK GERMANIUM
GE
5 REFL                  ! REFLECT AT BEAMSPLITTER
5 PTH -4 PIN 4          ! COMING BACK AGAIN
4 AT 30 0 100           ! TILT OF BEAMSPLITTER
7 AT 30 0 100           ! TILT AXIS
7 TH 70                 ! TO REFERENCE MIRROR
8 REFL                  ! HERE
8 PTH -7                ! BACK TO BEAMSPLITTER
```



```
9 AT -30 0 100          ! ENTER IT AGAIN
9 PTH 4 PIN 4           ! SAME SIZE AS BEFORE
10 TH -.1               ! SMALL AIRSPACE
11 PTH 4 PIN 4          ! COMPENSATOR DUPLICATES GEOMETRY
12
13 AT 30 0 100          ! DUMMY TO FOLLOW BEAM
13 TH -200              ! DISTANCE TO FOLD MIRROR 1
14 AT -30 0 100         ! RIGHT HERE
14 GID
FOLD 1                  ! IDENTIFY IT
14 REFL                 ! REFLECT THERE
15 AT -30 0 100         ! DUMMY TO FOLLOW BEAM
15 TH 250               ! DISTANCE TO PRIMARY MIRROR
16 RD -180 CC -1 TH -90   ! PARABOLOID HERE
16 REFL                 ! REFLECT THERE
16 GID
PRIMARY M               ! IDENTIFY IT
17 AT 45 0 100          ! SMALL FOLD MIRROR
17 REFL                 ! REFLECT HERE
17 GID
SECONDARY M             ! IDENTIFY IT
18 AT 45 0 100          ! DUMMY TO FOLLOW BEAM
18 TH 90                ! DISTANCE TO TERTIARY
19 RD -180 TH -350      ! DISTANCE TO FINAL IMAGE
19 REFL                 ! REFLECT AT TERTIARY
APS 19                  ! THE STOP IS HERE
19 GID
TERTIARY M
20
END
```

Run the MACro above and you obtain the PAD picture in figure 46.1.

To obtain this display, click on the PAD 'Top' button, select 'Custom rayset', HBAR 0.0 and 11 rays. Also select the 'Solo' top display option and turn on switch 38, which shows numbers for all surfaces, including dummies.

So far, you have the basic elements in place but do not yet know the details of the tertiary mirror at surface 19. You want a sharp image on surface 20, and you will insert additional fold mirrors to separate three wavelength regions onto different detectors later when you get to that step. Now you need to know the radius and conic constant on 19. Type the following into a new editor:

```
PANT
VY 19 ASPH
END
AANT
GSR 0 1 4 P
END
SYNO 10
```

After running this file, the system looks just the way it should, as shown in figure 46.2.

The command ASY now shows the shape of surface 19:

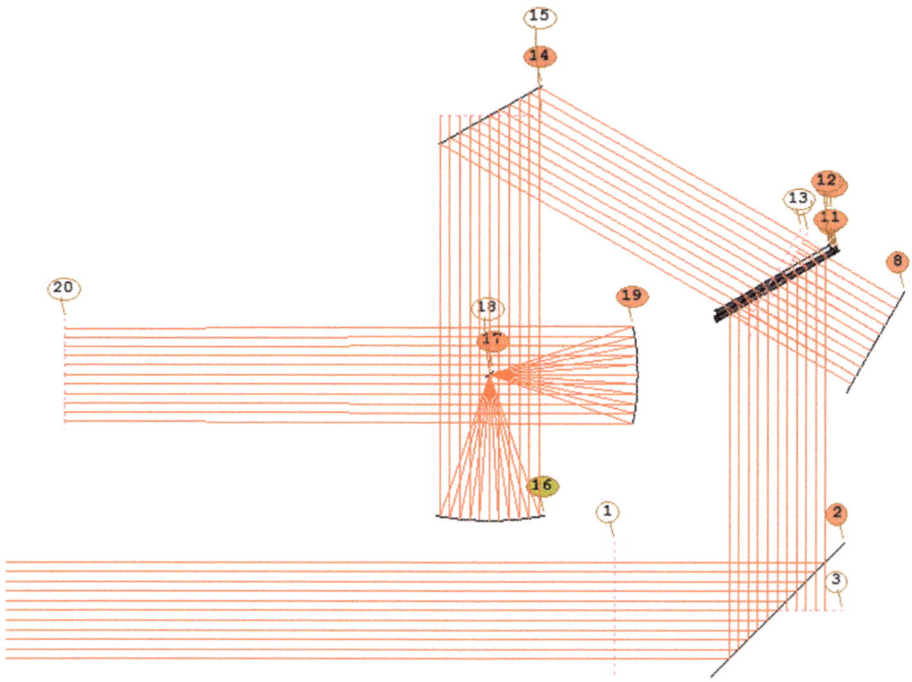

Figure 46.1. Interferometer, original setup.

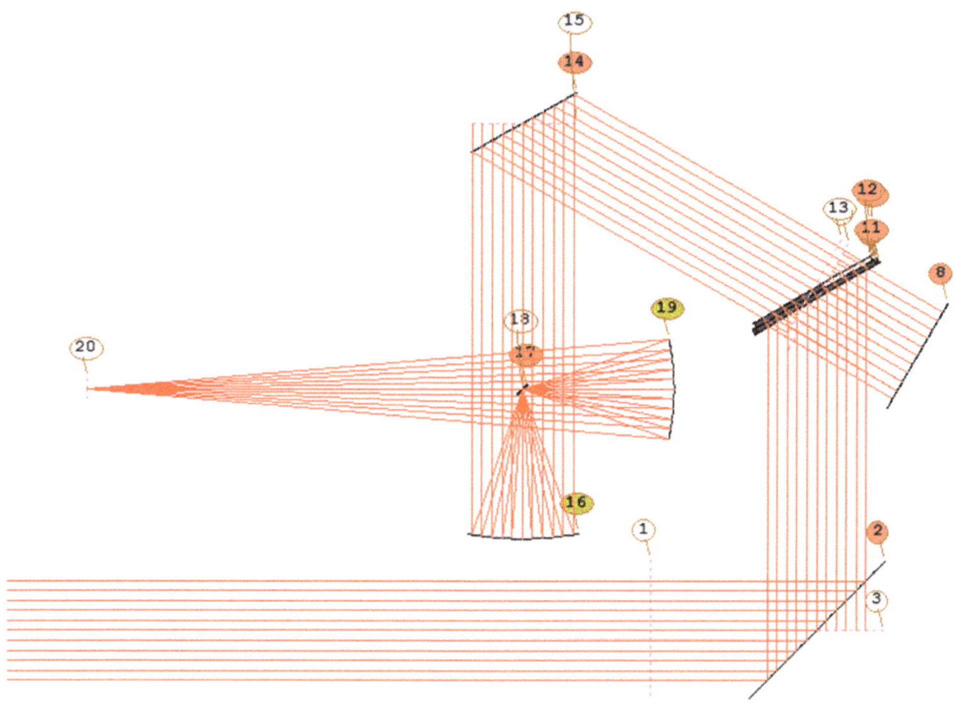

Figure 46.2. Interferometer after optimizing the shape of surface 19.

```
SYNOPSYS AI>ASY

SPECIAL SURFACE DATA
______________________________________________________________________________
 SURFACE NO.  16 -- CONIC SURFACE
 CONIC CONSTANT (CC)      -1.000000
 SEMI-MAJOR AXIS (b) INFINITE       SEMI-MINOR AXIS (a) INFINITE

 _______________
 SURFACE NO.  19 -- CONIC SURFACE
 CONIC CONSTANT (CC)      -0.349174
 SEMI-MAJOR AXIS (b)   -219.999999  SEMI-MINOR AXIS (a)    177.482393

TILT AND DECENTER DATA
LEFT-HANDED COORDINATES
______________________________________________________________________________
SURF TYPE          X            Y            Z        ALPHA      BETA     GAMMA
______________________________________________________________________________
  2 REL       0.00000      0.00000      0.00000   -45.0000    0.0000    0.0000
  3 REL       0.00000      0.00000      0.00000   -45.0000    0.0000    0.0000
  4 REL       0.00000      0.00000      0.00000    30.0000    0.0000    0.0000
  7 REL       0.00000      0.00000      0.00000    30.0000    0.0000    0.0000
  9 REL       0.00000      0.00000      0.00000   -30.0000    0.0000    0.0000
 13 REL       0.00000      0.00000      0.00000    30.0000    0.0000    0.0000
 14 REL       0.00000      0.00000      0.00000   -30.0000    0.0000    0.0000
 15 REL       0.00000      0.00000      0.00000   -30.0000    0.0000    0.0000
 17 REL       0.00000      0.00000      0.00000    45.0000    0.0000    0.0000
 18 REL       0.00000      0.00000      0.00000    45.0000    0.0000    0.0000

KEY TO SURFACE TYPES
____________________
GLB  GLOBAL COORDINATES                LOC  LOCAL COORDINATES
REL  RELATIVE COORDINATES              REM  REMOTE TILTS IN RELATIVE COORD.
SYNOPSYS AI>
```

One channel looks good; now let us set up the second. We can start with the above setup and just modify it as required. First, bump this setup into ACON 2, with the 'ACON' copy button, and then modify the geometry at the beamsplitter. Make a CHG file (**C46M2**):

```
CHG
13 SIN                 ! NEED TWO ADDITIONAL SURFACES
13 SIN
4 NAS                  ! REMOVE TILTS THERE NOW
7 NAS
9 NAS
13 NAS
4 TH -3 GTB U
GE
4 AT 30 0 100            ! TILT OF BEAMSPLITTER
5 TH -.1 TRANS
6 PTH 4 PIN 4
7 TH 0
8 AT -30 0 100
8 TRANS                 ! NOT REFLECTIVE ANYMORE
8 TH -70
```

```
9 REFL
9 PTH -8 AIR
10 AT 30 0 100
10 PTH -4 PIN 4
11 PTH -5
12 REFL
12 PTH 5
13 PTH 4 PIN 4
14 TH 0
15 AT 30 0 100
15 TH -200
END
```

This file first deletes most of the declarations assigned to the beamsplitter in channel 1, since reflections and tilts now occur on different surfaces. Then it replaces them with the data for the other channel. The new system is shown in figure 46.3.

(Here we have turned off switch 38 to make a cleaner picture.)

Both channels are defined, and they are both current, in ACONS 1 and 2. Now make a perspective drawing showing both at the same time. Create a MACro:

```
ACON 1
HPLOT 1
PER 0 0 .015 1 123
```

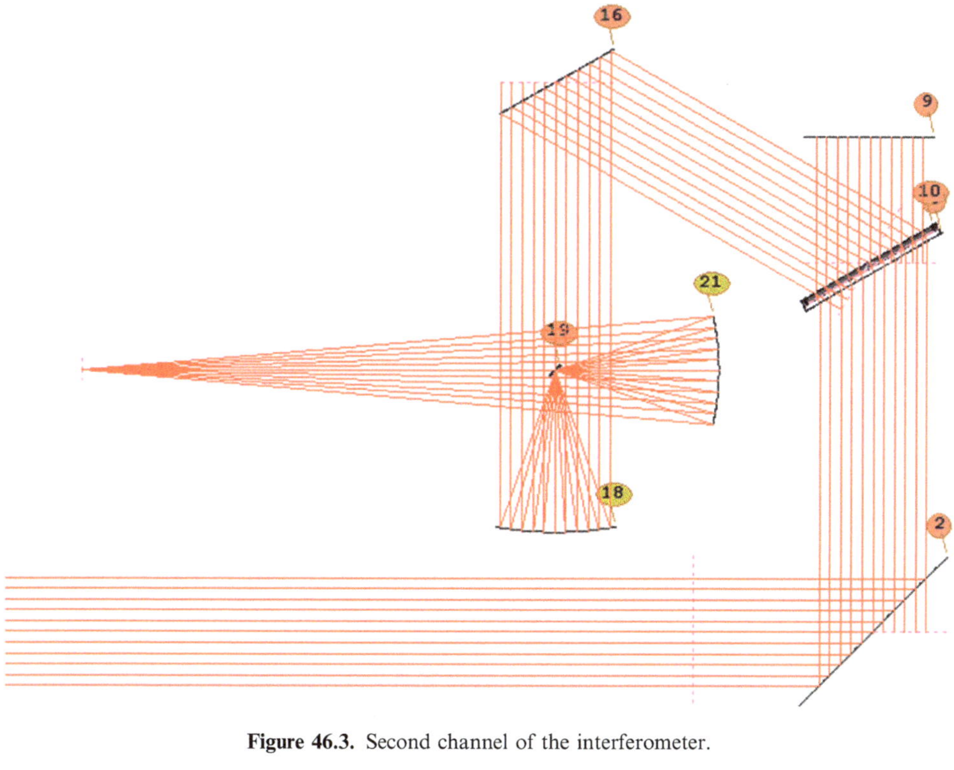

Figure 46.3. Second channel of the interferometer.

```
PUP 2 1 10
PLOT
RED
TRACE P 0 0 10
END

ACON 2
APLOT 1
PER 0 0 .015 1 123
PUP 2 1 10
PLOT
BLUE
TRACE P 0 0 10
END
```

This gives us the picture in figure 46.4.

Not bad—let us improve it some more. Open the Edge Wizard (**MEW**) and select 'Create All'. Do this for both ACONs. Now run the MACro above, after turning on switch 20, adding TRA requests for HBAR = 1 and −1, and changing the PER requests to 'RSOLID'. The picture is in figure 46.5. We are off to a very good start (**C46L1**).

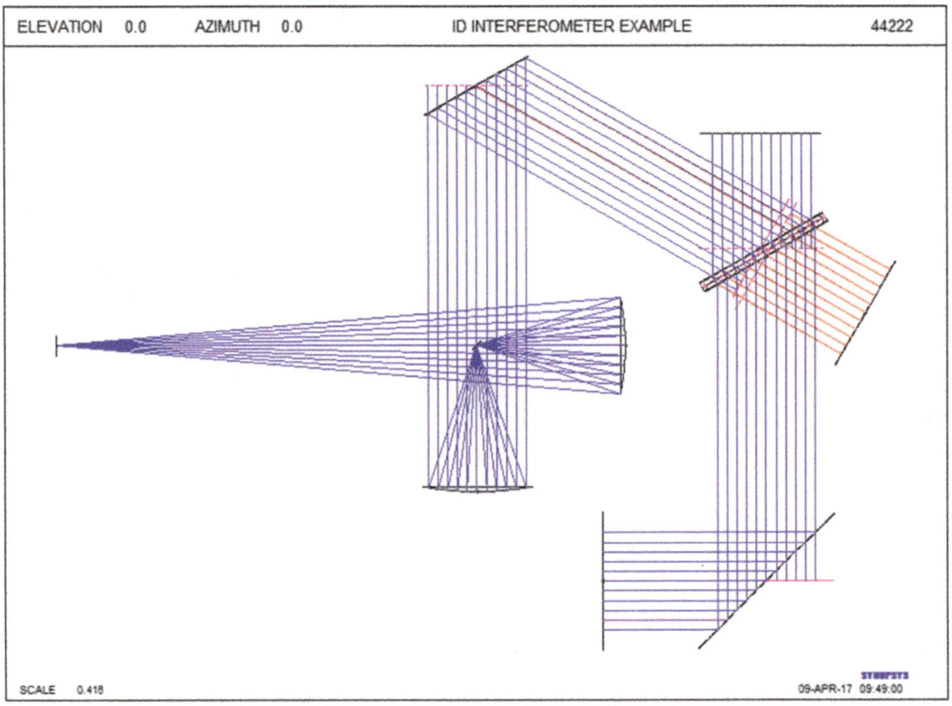

Figure 46.4. Both channels of the interferometer, shown by PERSPECTIVE.

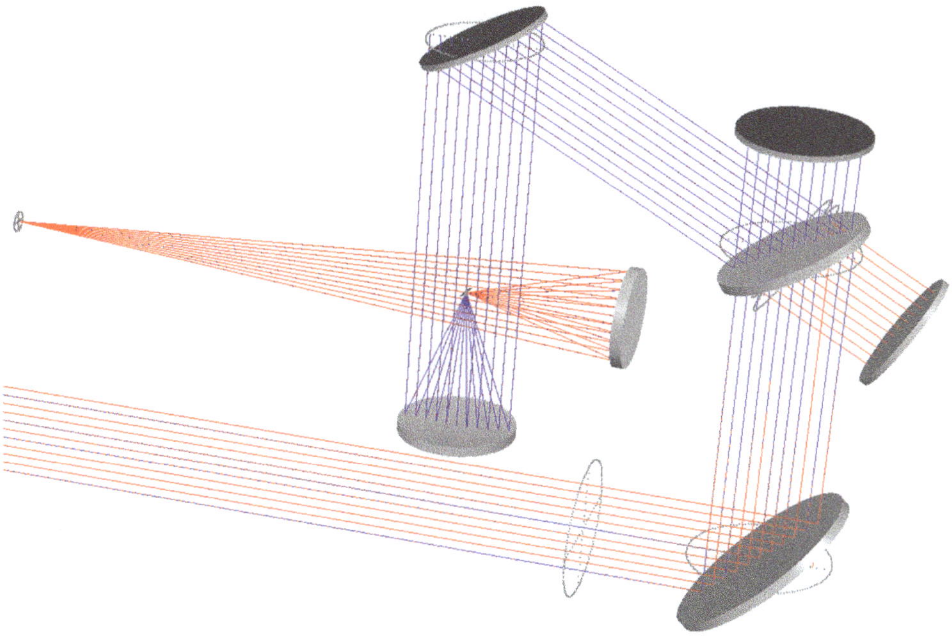

Figure 46.5. Both channels of the interferometer, with mirror thicknesses added, shown by RSOLID.

This is a brief lesson on how to set up systems like this. SYNOPSYS can show the system and the image quality very nicely, and it can even model the interference between two channels. Read about the commands IFR and IFP in the help file.

The next step will be to add additional fold mirrors and detector optics in the space before the final image. If beamsplitting cubes are to be used to separate different wavelengths, the systems should be designed with blocks of glass of the equivalent thickness before the image. However, now that you know how these things are done, we leave that as an exercise for the student.

If you are especially perceptive, you will have noticed that there is a small decenter in the beam as it goes through the beamsplitter—which we have ignored. That is not difficult to model either; just adjust a decenter on the primary mirror to compensate, if you really want to get that precise.

Lastly, we want to observe that, in SYNOPSYS, you can design up to six configurations at the same time, and they are completely separate systems unless you declare otherwise. This is in contrast to the practice in some other codes, where there is actually only a single configuration, and when you ask for a different configuration you obtain the same system unless you spell out the differences. The result can be much the same, but the philosophy is reversed.

Chapter 47

A four-element astronomical telescope

Global search for a telescope design with no secondary color

The goal in this lesson is to design a very good astronomical telescope for advanced amateur astronomers. The input file for DSEARCH is as follows (**C47M1**):

```
CORE 14
TIME
DSEARCH 1 QUIET
 SYSTEM
 ID DSEARCH TELESCOPE
 OBB 0 0.7 75
 WAVL 0.6563 0.5876 0.4861

 UNITS MM
 END
 GOALS
 ELEMENTS 4
 FNUM 8
 TOTL 0 0
 STOP FIRST
 STOP FIX
 TSTART 25
 ASTART 50
 RT 0
 OPD
 FOV 0.0 0.75 1.0
 FWT 5.0 3.0 1.0
 NPASS 40
```

doi:10.1088/978-0-7503-1611-8ch47

```
  ANNEAL 50 10 Q
  COLORS 3
  SNAPSHOT 10
  QUICK 40 40
  END
  SPECIAL PANT

  END
  SPECIAL AANT
  ADT 7 .1 10
  LUL 400 .1 1 A TOTL
  END
  GO
 TIME
```

The reader should need no coaching by this time, but some points may not be obvious. We want a telescope with extremely good image quality at F/8. This DSEARCH input will target only OPD errors in the real merit function (plus the usual monitors and first-order goals). Run this MACro, and DSEARCH returns the lens in figure 47.1 in about 11 seconds.

This is essentially perfect already, but it has model glasses, and we need to replace them with real glass. Run the optimization MACro prepared by DSEARCH, make a checkpoint, and then use **MRG** to replace them with glass from the Guangming catalog. The lens is just as good, but one of the elements received a very expensive glass. Can we achieve better quality and a lower price?

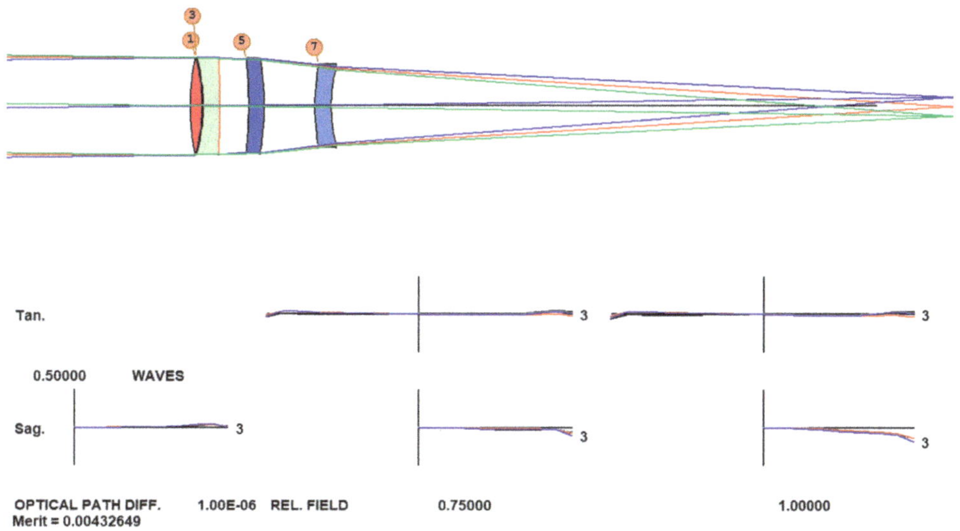

Figure 47.1. Lens returned by DSEARCH for a telescope example.

It is time to use another tool. Restore the checkpoint and save the optimization MACro with the name GSOPT.MAC.

Now make a new MACro to run GSEARCH, containing the following:

```
GSEARCH 3 QUIET
SURF
1 3 5 7
END
NEAREST 5 P 10
G
END
GO
```

Now the program will search 81 combinations of glass (3^4) and only investigate glass types not more than ten times the price of BK7. The result (**C47L1**) is superb, after more optimization and annealing, shown in figure 47.2.

If you have been reading carefully, you will notice that here we have violated some of the rules we mentioned in previous chapters. We optimized this lens with OPD errors but not TAP. Yes, that sometimes falls into a hole where the output is collimated instead of focusing, but not very often. One has to keep one's eyes open, and do not be afraid to experiment.

Also, we put an **ADT** monitor in the DSEARCH file, which is not always a good idea, but we obtained excellent results anyway. Running SYNOPSYS is a lot like driving a Maserati: if you know how to drive it, it works better. To prepare this lesson we tried a variety of input variations, many of which produced a lens with about one wave of secondary color. The trick here seems to be the positive flint element—but not all of the DSEARCH runs found that solution. We have a very

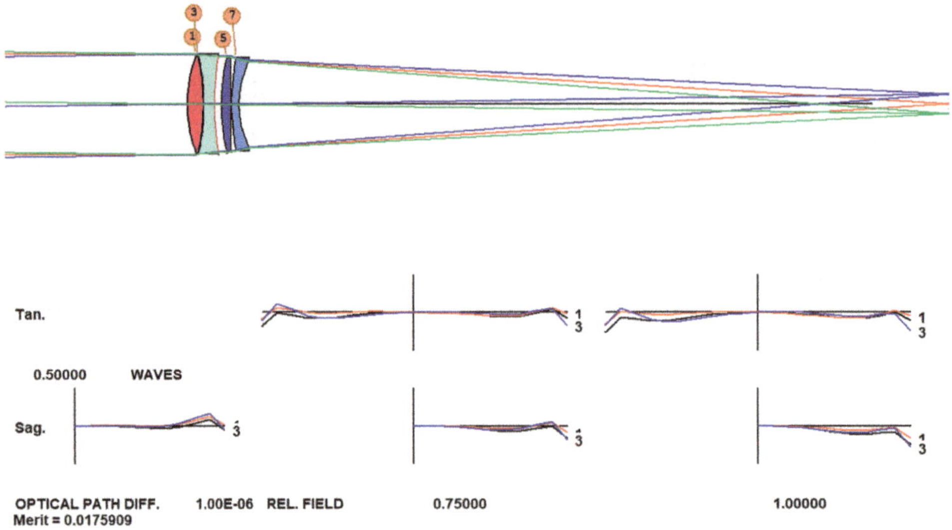

Figure 47.2. Telescope design with real glass.

bushy bush here, and it is a good idea to run the program several times, as we did, and select the best lens of the lot. (We also did this lesson again with switch 98 turned off, which yields different results each time, and got different lenses that also had no secondary color.) There are lots of solutions out there, and if you master these search tools you will be an expert explorer in no time.

We also ran this exercise with an F/number of 7 instead of F/8 and obtained very similar results. How low do you think you can go? Try 6 and see how close you can come. Then use AEI to try to improve that design. That is what lens design is all about: see what works, find a way to improve it, and be very pleased when you succeed.

IOP Publishing

Lens Design
Automatic and quasi-autonomous computational methods and techniques
Donald Dilworth

Chapter 48

A sophisticated merit function

The earlier chapters tell how one can (and should) incorporate all the goals that have to be satisfied by the lens into the merit function, mechanical as well as optical. This lesson gives an example of such a design, one that required extensive control of both characteristics. It will be instructive for the reader to study how this was achieved. The system (**C48L1**) is shown in figure 48.1.

This lens works in the thermal infrared, from 8 to 12 μm, and must be corrected to 1/4 wave or better. The merit function that optimized it was rather complicated, owing mostly to the need to keep the obscuration to a minimum. You saw in section 8.1 how introducing an obscuration into the entrance pupil affects the MTF and learned that one does not want it to be any larger than absolutely necessary. The lens in this lesson is a *catadioptric* design, meaning it has both lenses and mirrors, and the effect of the obscuration is just as important here.

Such systems can be arranged in one of two ways: axially symmetric (which is easier to design and build but is obscured) or with off-axis components, which are more difficult but can avoid an obscuration. A symmetric design is obscured since one has to collect the light from the first mirror and send it back through a hole in that mirror. In this example, we want to keep that obscuration to not more than 45%. Here is our MACro (**C48M1**):

```
LOG
STO 2
AWT: 0
BB: 0.45

PANT
VLIST RAD  2  4 5 6 7 8 9 10
VY 2 CC 100 -100
VY 4 CC 100 -100
VY 1 TH
```



```
VY 2 TH 399 -300
VLIST TH 6 8 10
END

AANT
OBS .1 BB
AEC 1 .2
!M 95 .1 A TOTL
M 10.16 1 A GIHT

NAME BULGE
M 0 .1 A P YA 1 0 1 0 6
S P YA 1 0 1 0 7

NAME BULGE2
M 0 .01 A P YA 1 0 1 0 6
S P YA 1 0 1 0 9

NAME OBSC
M BB 50 A P YA 1 0 1 0 4
DIV YMP1

NAME SETBACK
M 19 .1 A ZG 14
S ZG 2

NAME FOCUSAIR
M 2 1 A TH 10
S P ZA 1 0 1 0 10

NAME CLEARCONE
LL 0 1 1 .01
A P YA 0 0 BB 0 3
S P YA 1 0 1 0 5

M 1.5 1 A BACK

GSR AWT 10 6 P
GSR AWT 10 6 1
GSR AWT 10 6 3
GNR AWT 6 6 P .7
GNR AWT 6 6 P 1
GNR AWT 6 6 1 1
GNR AWT 6 6 3 1
M 0 50 A 1 YA 1
S 3 YA 1
END

SNAP
SYNO 20
```

First, we have to explain how the system is set up.

1. Surface 1 is a dummy used for controlling geometry.
2. Surface 2 is the primary mirror, a conic.
3. Surface 3 is another dummy, whose position is variable and will be used to locate surface 5.
4. Surface 4 is declared coincident with surface 1, whose thickness is variable. This is the secondary mirror, also a conic.

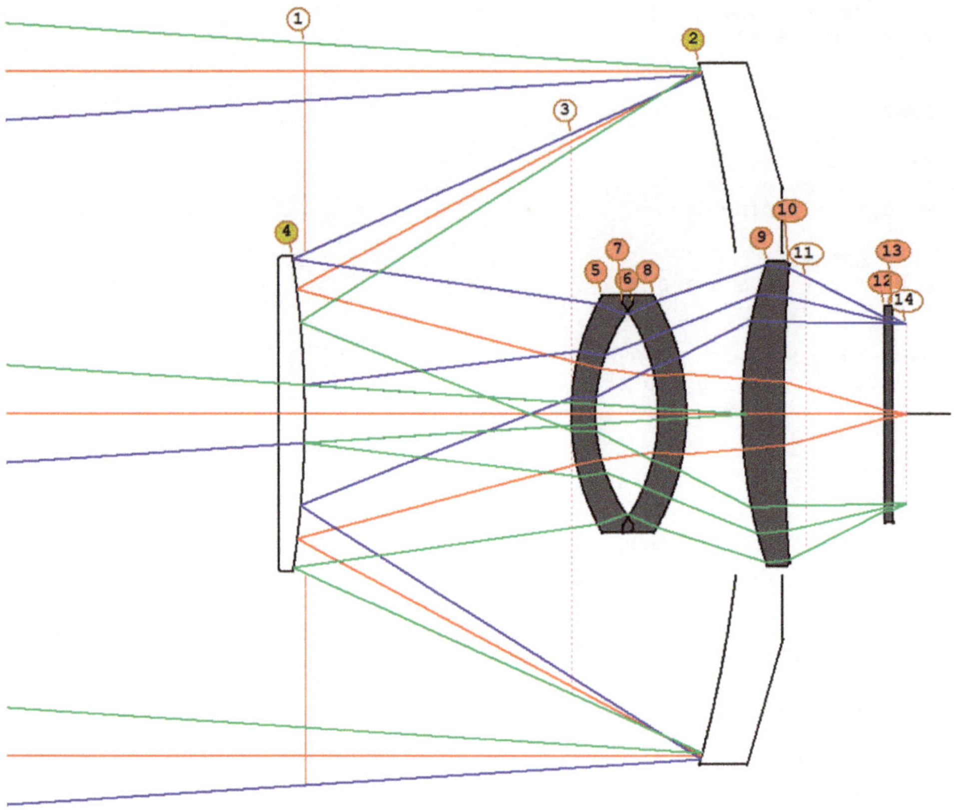

Figure 48.1. Catadioptric system requiring a sophisticated merit function.

5. Surface 5 is declared coincident with surface 3. As 3 moves, 5 will move with it.
6. Surfaces 5 through 10 are a set of Germanium corrector elements.
7. Surface 11 represents the location of a shutter.
8. Surfaces 12 and 13 are a Germanium window in front of the detector at 14.

Look at the items in the above MACro.

1. The symbol **AWT** is assigned a value of 0. This is the aperture weighting and can be altered if you want to do trade-off studies.
2. Symbol **BB** is assigned the value 0.45. This is the allowed obscuration, and can also be varied for trade-off studies.
3. The AANT file declares **OBS .1 BB**, which causes the ray grids generated by the **GSR** and **GNR** requests to delete any rays falling in a circle of radius **BB** (0.45 in this case) and decentered by 0.1 in the *y*-direction at full field. That portion will be obscured anyway, and it makes no sense to correct rays that will not get through.
4. The **AEC** monitor keeps edges to no less than 1 mm.

5. The target on **GIHT** controls the focal length.
6. The next aberration is assigned the name **BULGE**. This name will show up on aberration listings, making it easy to identify the values later.
7. The **BULGE** aberration is defined as the difference between the *y*-coordinate of the upper rim ray (URR) at full field on surface 6, and the same ray on surface 7. This keeps the first two lenses about the same size. We do not want either one to be much larger than the other.
8. The aberration named **BULGE2** is assigned a low weight (and is not met exactly) and is there to ensure that the lens at 9 is not very much larger than the lens at 6. The weight is adjusted to give the best balance. We do not want the hole in the primary to be larger than the secondary mirror and this is one of several ways we might control that.
9. Aberration **OBSC** takes the URR (upper rim ray) on 4 and divides by the paraxial **YMP1**, assigning a target value of **BB** (which is the number 0.45) and a high weight of 50. Now, if the extreme ray on 4 wants a larger aperture than **BB**, the MF gets a penalty. This will keep the obscuration due to the secondary mirror under control.
10. Aberration **SETBACK** gives a target of 19 mm to the difference between the global *z*-coordinate of surface 14 and that of surface 2. This controls the distance from the primary mirror to the image plane.
11. Aberration **FOCUSAIR** gives a target of 2 mm to the difference between thickness 10 and the *z*-coordinate of the URR on that surface. If that lens were strongly curved toward the image, it would encroach on the airspace between 10 and 11, and we have to leave room in that space for the shutter. This will control that.
12. Aberration **CLEARCONE** gives a lower limit to the difference between the *y*-coordinate of the on-axis ray at a fractional entrance coordinate of **BB** on surface 3 and the URR (upper rim ray) on surface 5. The first of these rays comes in at the edge of the obscuration, and we do not want it obscured after the first reflection by the lens at 5, whose aperture is given by the second ray.
13. We want to maintain a clearance of 1.5 mm between the cover plate and the sensor. Thickness 13 is governed by a YMT solve and can change.
14. The rest of the MF consists of rays to be traced and corrected. Lateral color is corrected by taking the difference between the chief rays in colors 1 and 3, with a target of 0.

With a merit function like this, one can optimize the system and see how it works with a given obscuration. Then you can reduce the obscuration, in stages, by redefining symbol **BB**, and optimize each case. When the results start to degrade, you know how far you can push that parameter.

You see how it is sometimes necessary to control individual rays on individual surfaces, along with global positions of selected surfaces. If your project has complex requirements such as these, do not be afraid to create a complex merit function to design it. That is what those options are there for.

Readers with sharp eyes will notice that surfaces 6 and 7 overlap as shown in figure 48.1. Those edges result from the default calculation and of course one has to adjust them before the lenses are fabricated. This is easily done with the procedures outlined in chapter 40. We leave this as an exercise for the student, but here is a hint: type

```
FEATHER 6
```

and the program reports the *y*-coordinate of the feathering point as well as the surface sags at 6 and 7. Assign that coordinate to the points E on 6 and A on 7. Then the sag tolerance on those surfaces determines the airspace tolerance too.

Chapter 49

When automatic methods do not apply

The previous chapters of this book presented numerous examples of how autonomous methods can find excellent lens configurations more quickly and easily than was possible using classic techniques. The design process has been shortened from weeks to minutes, and you might think there is no need for the kind of expertise that was important a generation ago—but you would be wrong. These new methods cover a wide range of design problems, but there are situations where they are not appropriate; then it is up to you. Here is an example, call it a 'final exam'.

49.1 The 'final exam' problem

Some years ago, I was asked to devise a problem to be given to an advanced lens design class, one that would test their knowledge of optics. I came up with a task that cannot be solved by a computer program. You have to think, you have to be clever, and you have to know your optics. I urge the reader to first try to solve it before reading the solution. If you can do it, congratulations. If not, go back to the first chapters and study them again. Here is a revised version of that problem, shown in figure 49.1.

The rules are as follows:

1. The images must fall on the slits of a spectrometer. They must be narrow in one direction but long in the other.
2. The orientation of the two images must be at 90 degrees, as shown in figure 49.2.

There are other requirements:

1. Images must be narrow. 90% of the energy must pass through a slit of 13 μm width.
2. The image length must be between 0.1 and 0.2 mm.
3. Chromatic aberration must be corrected. Analysis will be performed at the C, d, and F Fraunhofer lines.

doi:10.1088/978-0-7503-1611-8ch49

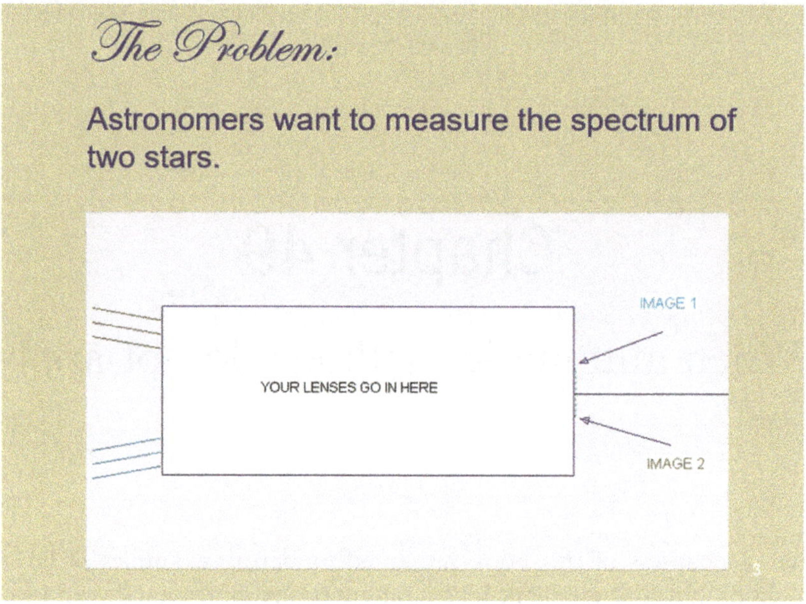

Figure 49.1. Outline of the 'final exam' problem.

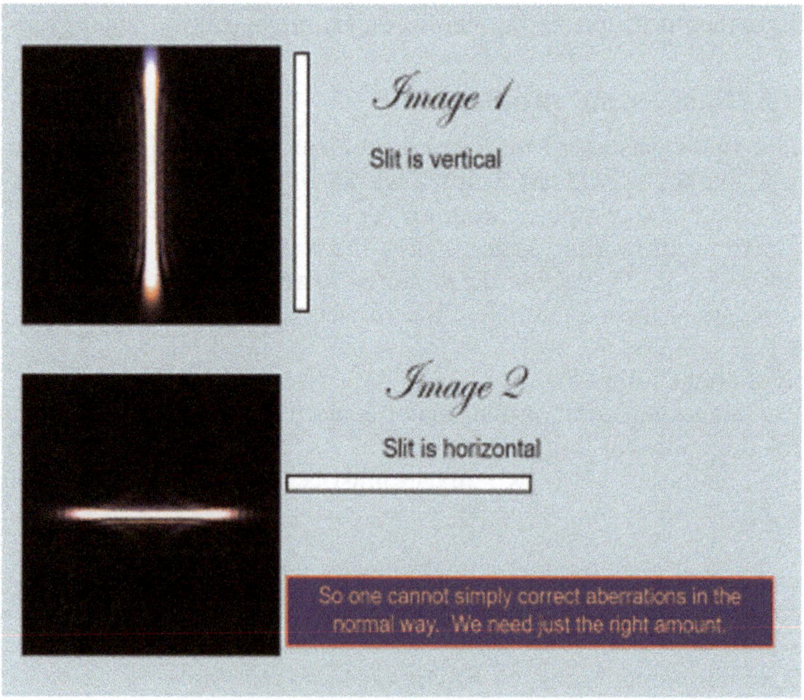

Figure 49.2. The orientation of the two slits.

4. The stars are at the top and bottom of a field of 20 degrees total angle. Pupil diameter is 20 mm.
5. All surfaces must be spherical (no cylinder lenses).
6. The simplest lens system that meets the requirements is the winner. If a tie, then the highest energy wins.
7. Two versions will be submitted, at F/10 and F/5.

So that is the problem. Can you solve it? How many elements do you need? Give it a try before you read the solution below.

49.2 The solution

Understanding the source of aberrations and how they are affected by design variables is the key. What we need is a controlled amount of astigmatism. What does that aberration look like? figure 49.3 shows an example.

What properties of astigmatism should one know about?

- It is always zero at the center of the field if the lenses are spherical and centered.
- It is symmetric with field angle.

Well, that is no good. We need the top and bottom of the field to be different. Therefore, the system cannot be centered.

In figure 49.4 you see an example where tangential rays are defocused, while sagittal rays are almost in focus. The difference in *angle* is the astigmatism. In figure 49.5 you see that if you take only a short, decentered section of the spherical aberration curve, what you wind up with looks a lot like astigmatism. That is a clue.

Figure 49.3. Illustration of the properties of astigmatism.

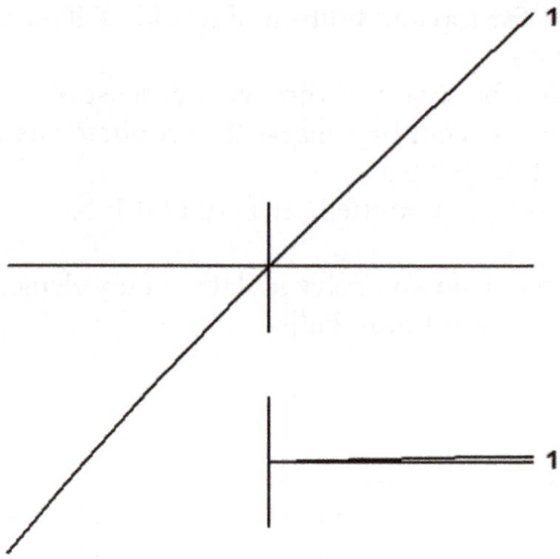

Figure 49.4. TFAN and SFAN plots showing severe astigmatism.

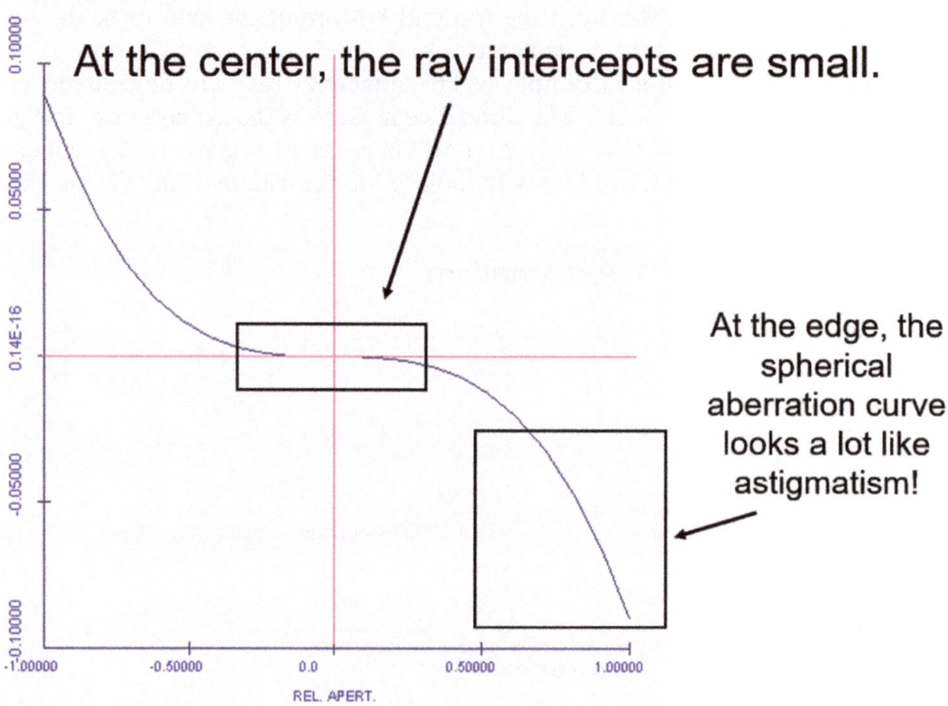

Figure 49.5. A decentered plot of spherical aberration looks a lot like astigmatism.

If you *decenter* the pupil, then spherical aberration looks like astigmatism. So a decentered element somewhere might be of use.

What about astigmatism of a *centered* lens? figure 49.6 shows that, away from the optical axis, the sagittal and tangential focus surfaces separate. However, then the top and bottom of the field get the *same* astigmatism. That is no good; they have to be different.

However, if you *tilt* the focal surface, this lets one see the tangential image at one side of the field and the sagittal image at the other—so tilting something might work (see figure 49.7).

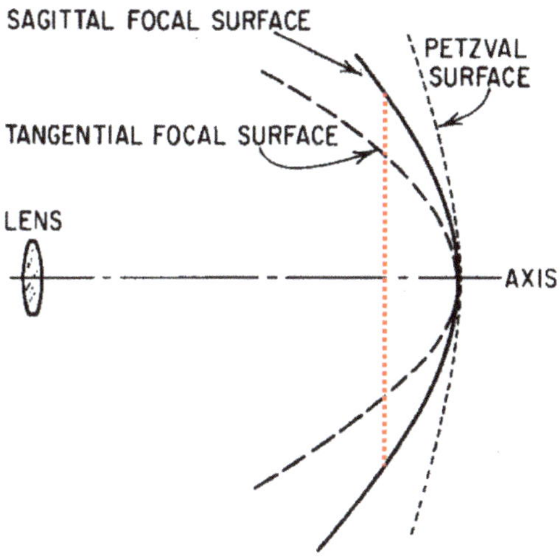

Figure 49.6. Astigmatism of a centered lens shows up off-axis.

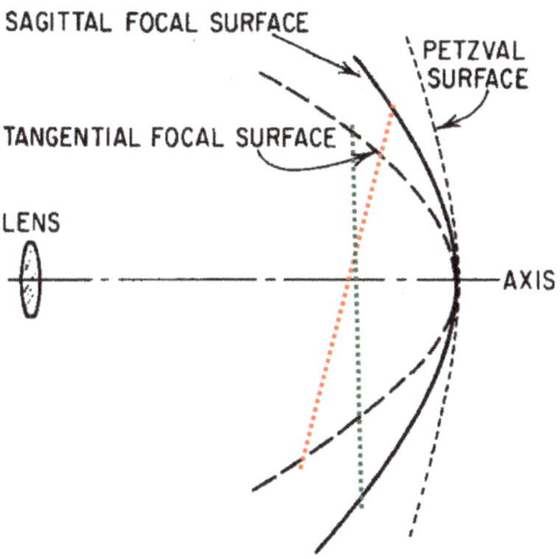

Figure 49.7. Tilting the focal surface.

Here is another possibility: shifting the stop *decenters* the off-axis beam in the pupil, as shown in figure 49.8. Look at the path of the full-field chief ray. It is decentered in the box to the left. If the stop were put there, you would get a different chief ray. That is another clue.

So we have identified three possible ways to make the sagittal and tangential astigmatism show up at the focal plane as we want:

- Decenter an element.
- Tilt something.
- Shift the stop surface.

So far, so good. Let us guess that we can do the job with a four-element lens. We run DSEARCH with the following input file (**C49M1**),

```
CORE 14
DSEARCH 3 QUIET
SYSTEM
ID FINAL EXAM PROBLEM
OBB 0 10 10
WAVL 0.6563 0.5876 0.4861

UNITS MM
END
GOALS
ELEMENTS 4
FNUM 10
BACK 0 0
TOTL 100 .1
STOP MIDDLE
STOP FREE
```

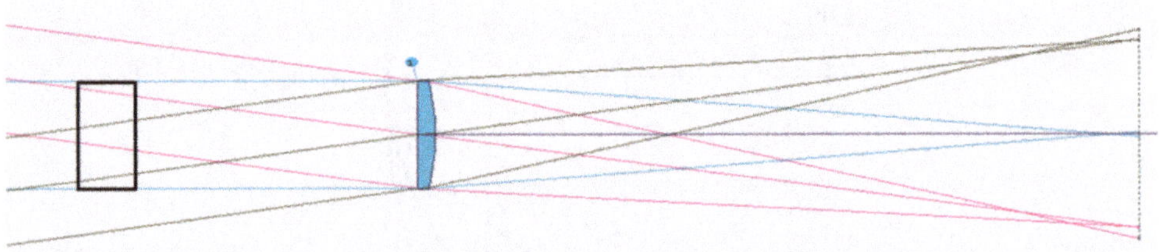

Figure 49.8. Shifting the stop causes the beam to be decentered in the pupil.

```
RT 0.5
FOV 0.0 0.75 1.0 0.0 0.0
FWT 5.0 3.0 1.0 1.0 1.0
NPASS 44
ANNEAL 200 20 Q
COLORS 3
SNAPSHOT 10
QUICK 44 44
END
SPECIAL PANT

END
SPECIAL AANT

END
GO
```

and it comes back with ten designs, well corrected but rotationally symmetric at the moment. The top one is shown in figure 49.9.

Now we have to modify this lens so it meets our specs. We will go in steps. Here are the variables:

```
PANT
VY 0 YP1
VLIST RAD ALL
VLIST TH ALL
SKIP
VY 3 AT 2
VY 5 YDC 2
VY 9 AT 1
EOS
VLIST GLM 1 3 5 7
END
```

This will vary the stop position, vary radii, thicknesses and airspaces, and the glass models, and we will activate the tilt and decenter variables later.

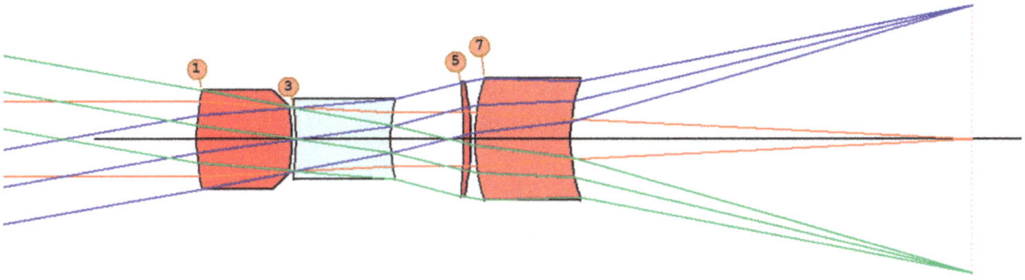

Figure 49.9. Top lens returned by DSEARCH.

The trick now is to create a merit function (MF) that leads to the solution we want. Here is an outline (**C49M2**):

```
AANT
AEC 3 1 1
ACM 3 1 1
LLL 1 1 1 A ETH 8
M 0 1 A 1 YA 1
S 3 YA 1
SKIP
LLL .11 1 .1
A P YA 1 0 1
S P YA 1 0 -1

LLL .11 1 .1
A P XA -1 -1
S P XA -1 1

LUL .19 1 .1
A P YA 1 0 1
S P YA 1 0 -1

LUL .19 1 .1
A P XA -1 -1
S P XA -1 1

(*** magic ***)
EOS
GXR 0 2 4 P 1
GXR 0 2 4 1 1
GXR 0 2 4 3 1
GYR 0 1 4 P -1
GYR 0 1 4 1 -1
GYR 0 1 4 3 -1
END
SNAP
SYNO 30
```

Ignoring for the moment the `(*** magic ***)` portion, we try this MF, optimize and anneal, and find it does not work. We need to get 90% of the energy through our 13 μm slits, but this design only gets 89% through. Why? What can we do?

It is easy to show that, at F/10, a perfect image at these wavelengths *would* get only 89% of the energy through. So, on the face of it, the problem is impossible!

Except, it is not. Here is where your knowledge of optics is essential. The software has corrected the aberrations to the degree we specified, the image is diffraction limited in the cross-slit direction, and we still do not meet the specs. What can we do?

We have to be clever. Here is the thought process:

- The specification says F/10.
- However, then the diffraction pattern is already too large to fit inside the slit.
- (Clever trick) the spec says nothing about distortion!
- What does *distortion* have to do with the size of the diffraction pattern?

Can you answer that last question? figure 49.10 offers a clue.

That should be enough of a hint. From chapter 2 you learned about the Lagrange invariant. You learned that, if you change the value of y_B, you also change the value of u_A. Got it yet?

As figure 49.11 shows, a steeper angle creates a smaller Airy diffraction disk—and a smaller disk can get more energy through a small slit.

So the trick is to get just the right amount of distortion:

- Then the cone angle of the rays will change at the edge of the field.
- Then the radius of the Airy diffraction disk will change.
- So we can make it smaller.
- So we can get more energy inside the slit.

That is the `(*** magic ***)` in the AANT file: ask for some barrel distortion. So we add to the AANT file the lines

```
M .90 1 A P YA 1
S P YA -1
DIV CONST 2
DIV GIHT
```

This says the actual image scale should be 90% as large as the paraxial scale. Free up the tilt and decenter variables and the slit-length aberrations and optimize again. Then anneal (**55, 2, 50**).

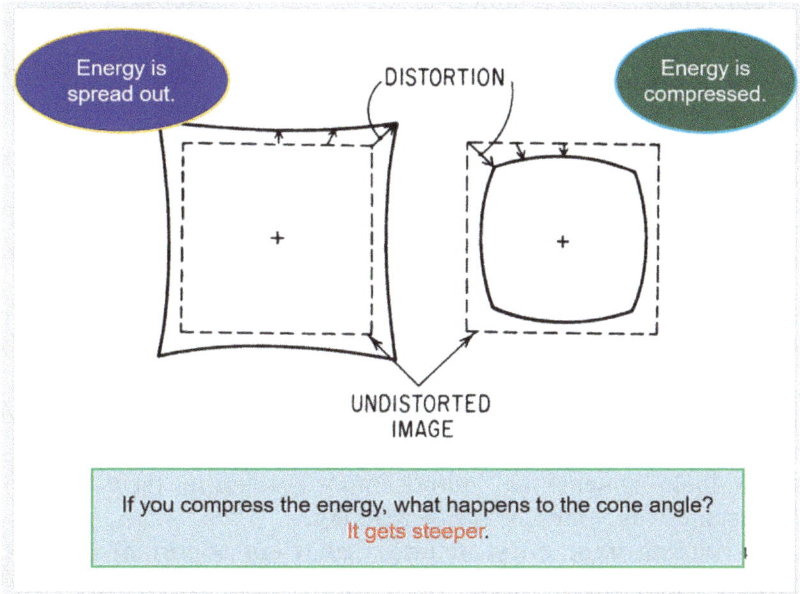

Figure 49.10. Distortion alters the energy density at the image.

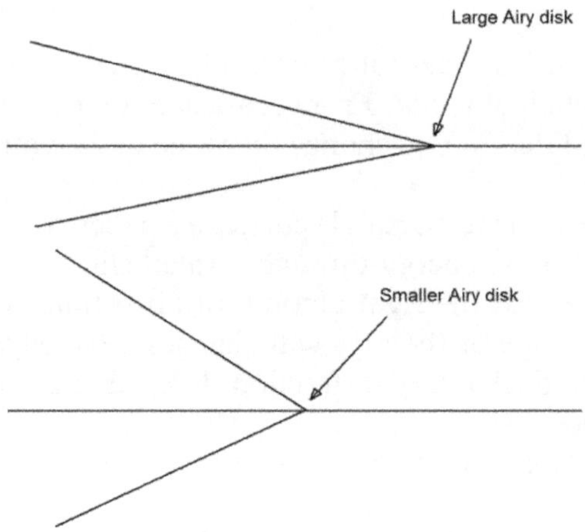

Figure 49.11. Relation between cone angle and Airy diffraction disk.

The solution to the F/10 case is shown in figure 49.12 (**C49L1**). This lens puts 90.7% through the *y* slit and 92.7% through the *x*.

The distortion of this lens is about –0.363 mm, as is apparent in figure 49.13.

With this trick, all the goals were met. Figure 49.14 shows the images at top and bottom of the field, as displayed by the Image Tools Menu, `MIT`.

The other half of the problem, meeting the same goals at F/5 is in fact easier, since the Airy disk is already much smaller. Figure 49.15 shows the wavefront fringes for both designs—a classic example of astigmatism.

Part of this problem is to calculate the fraction of the energy that gets through the slit. Here is a simple way to obtain it:

```
WMODEL M 1 9999
FOR SLIT
SIZE .3 .013
VARY X POS FROM -.05 TO .05
PLOT
```

This produces a table and a plot of the energy as the slit is moved past the image. A similar calculation, substituting *y* for *x* and calculating the model at a field of –1.0 gives the slit trace at the other side of the field.

So that is how one approaches a problem that does not lend itself to the powerful search tools we have used in earlier chapters. To summarize:

- Always look for items that are not important and set them free.
- Understand why your lens is not working.
- See if you can trade off something that is not important in order to gain something that is important.

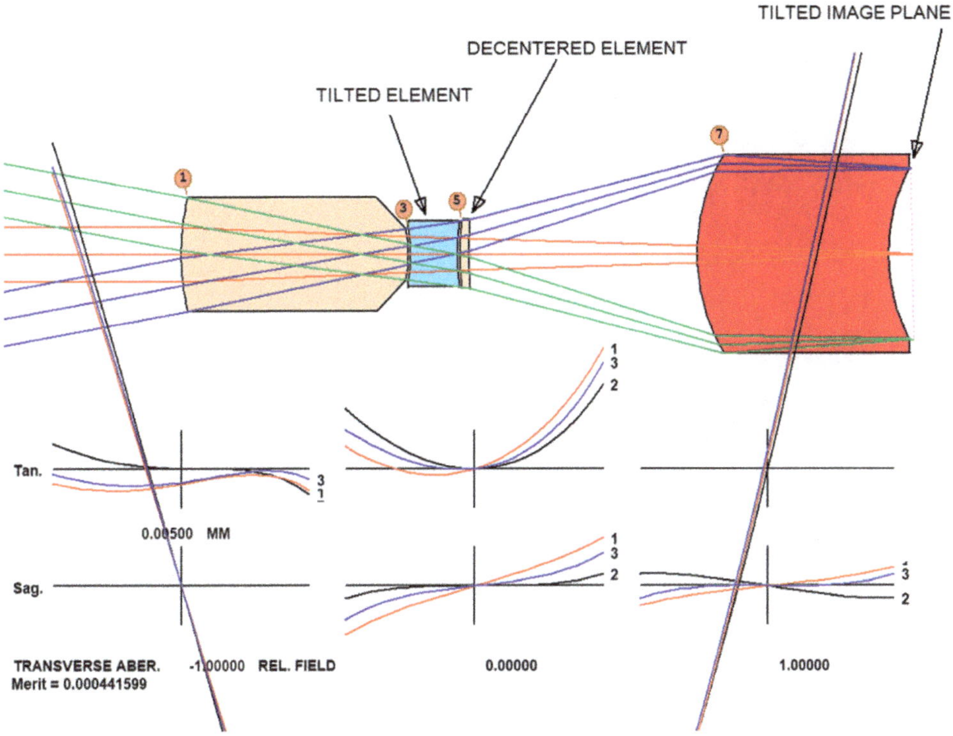

Figure 49.12. Final solution to the 'final exam' problem.

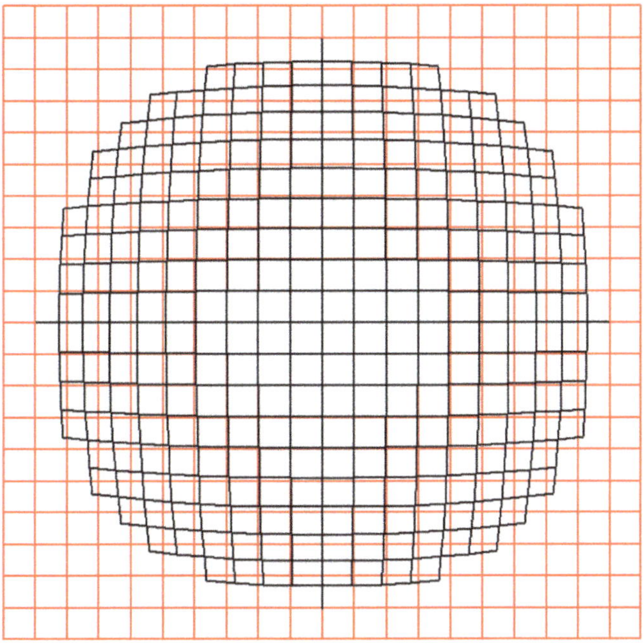

Figure 49.13. Distortion of the final lens.

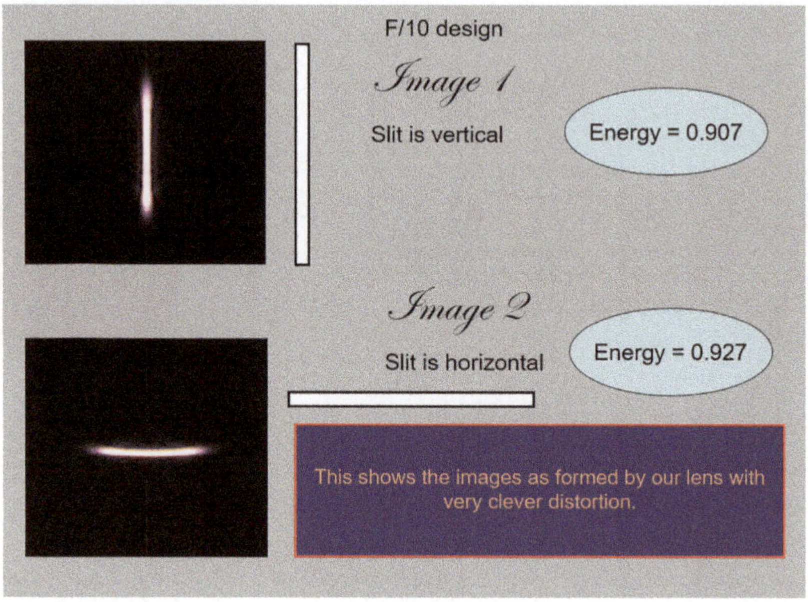

Figure 49.14. Image at top and bottom of the field of the final design.

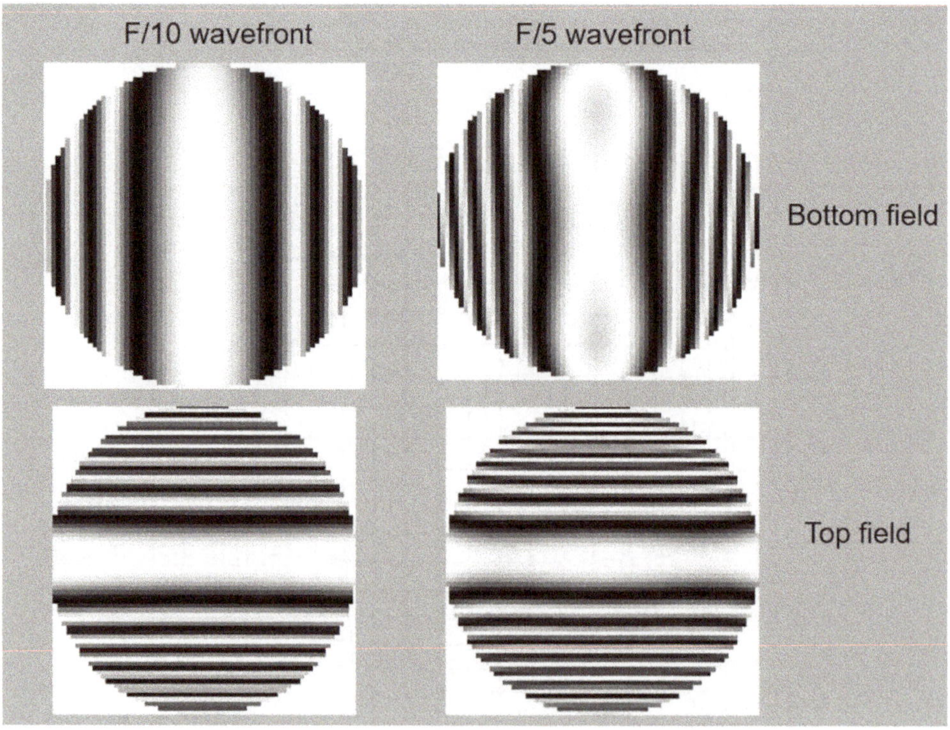

Figure 49.15. Wavefront fringes of both solutions.

- Check with the customer. He may have forgotten to tell you about some other requirements.

By the way, none of the students in the class found this solution. If you figured it out, great. I hope this book helped you to learn what you needed to know.

IOP Publishing

Lens Design

Automatic and quasi-autonomous computational methods and techniques

Donald Dilworth

Chapter 50

Other automatic methods

The previous chapters have illustrated the use of many autonomous features that make lens designing much faster and easier than in the past. We finish with examples of three additional tasks, that of matching testplates, designing thin-film coatings, and clocking the wedge error of lens elements, all of which can be done automatically by the software.

50.1 Testplate matching

In chapter 34, we designed a very good wide-band objective lens. The next step, before asking for a tolerance budget and making element drawings, is to match the design to the testplates of a selected vendor. Chapter 4 explained why this step is important. Proceed as follows.

Fetch the lens (**C34L2**) and run the optimization MACro once more (**C50M1**). The testplate-matching program will reuse the most recent parameter definition and merit function, so those have to be current. Now delete the curvature solve on surface 6. You want to match all surfaces, but cannot match one where the radius keeps changing. Then open the dialog MMT and enter the data shown in figure 50.1. For this lens we will use the testplates of JML, Inc. Click the 'OK' button.

The program runs **TPMATCH**, which matches every surface, lists the radii it found, and then shows the results. These testplates are measured in millimeters, and this lens is in units of inches, so the radii are first scaled by 0.039 37:

```
RESULTS OF TEST-PLATE FIT

 SURF. NO. FINAL RADIUS ACTION

    1  0.321082E+02 SUCCESSFUL MATCH FOUND
    2  0.806770E+01 SUCCESSFUL MATCH FOUND
```



```
3  0.867898E+01 SUCCESSFUL MATCH FOUND
4  0.935313E+01 SUCCESSFUL MATCH FOUND
5  0.893746E+01 SUCCESSFUL MATCH FOUND
6 -0.361589E+02 SUCCESSFUL MATCH FOUND
7                BYPASSED
```

Now, when you run BTOL, be sure to declare those surfaces matched, with the **TPR** designation, so BTOL will assign a tighter radius tolerance to them and everything else will be looser.

50.2 Automatic thin-film design

Thin films are widely used in optics, both for antireflection coatings on lenses and high-reflection coatings for mirrors. They also show up in dichroic mirrors, which separate spectral bands, and for other uses. They are deposited onto an optical surface in a vacuum chamber where the material is heated and evaporated, or sputtered onto, the substrate.

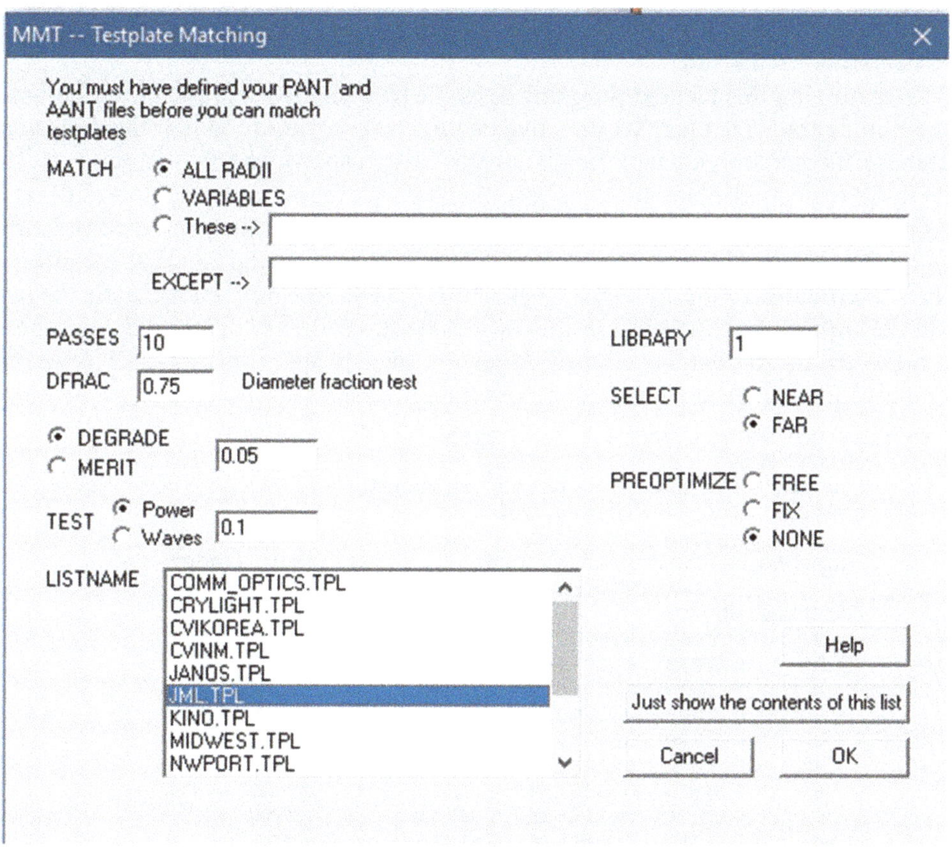

Figure 50.1. The MMT dialog for matching testplates.

Here we will create a custom thin-film design of 25 layers to reflect red light and transmit blue when the filter is used at an angle of 45 degrees. The input file is as follows (**C50M2**):

```
FILM
DESIGN
BUILD 30 1 1.62
ID TEST CUTOFF FILTER
AANT
GRW 0 45 25 .4 .6
GRW 1 45 25 .62 .8
END
FIX
SYNO 10
ANALY
LAM .4 .8 100 45
PLOT
RETURN
```

This file requests an average reflectance at 45 degrees of zero from 0.4 to 0.6 μm, and 1.0 from 0.62 to 0.8 μm.

After running this file, the program plots the characteristics of the film it designed, shown in figure 50.2. The **FIX** directive causes the program to match the design to a database of commonly used materials, which it also displays with the output. Note that the performance of the film stack varies somewhat with the polarization state of the light.

The design and manufacture of thin films is a mature discipline, and much expertise has been developed by many vendors. Of particular note is the fact that the effective index of refraction of the materials when deposited as a thin film is not exactly the same as in the bulk material, so these results, while adequate for computer modeling, should be adjusted by the vendor according to his own proprietary material properties database, and the design adjusted, before manufacture.

```
ID TEST CUTOFF FILTER             2153
     STACK DATA
     CONTROL WAVELENGTH =   0.5876 MICRONS
     CONTROL ANGLE =    0.000 DEG.
             OPTICAL  PHYSICAL
      SURF NO.   THICKNESS THICKNESS   INDEX   IMAG. INDEX
                 (WAVES)   (MICRONS)
     INCIDENT MEDIUM                1.0000
        2      0.4096  0.120339  1.9729         HFO2
        3      0.2863  0.121923  1.3655         MGF2
        4      0.3838  0.104692  2.1535         TAO5
        5      0.3005  0.127965  1.3655         MGF2
        6      0.3720  0.101486  2.1535         TAO5
        7      0.3028  0.128944  1.3655         MGF2
        8      0.3525  0.096174  2.1535         TAO5
        9      0.3220  0.128695  1.4585         SIO2
       10      0.3282  0.089524  2.1535         TAO5
```

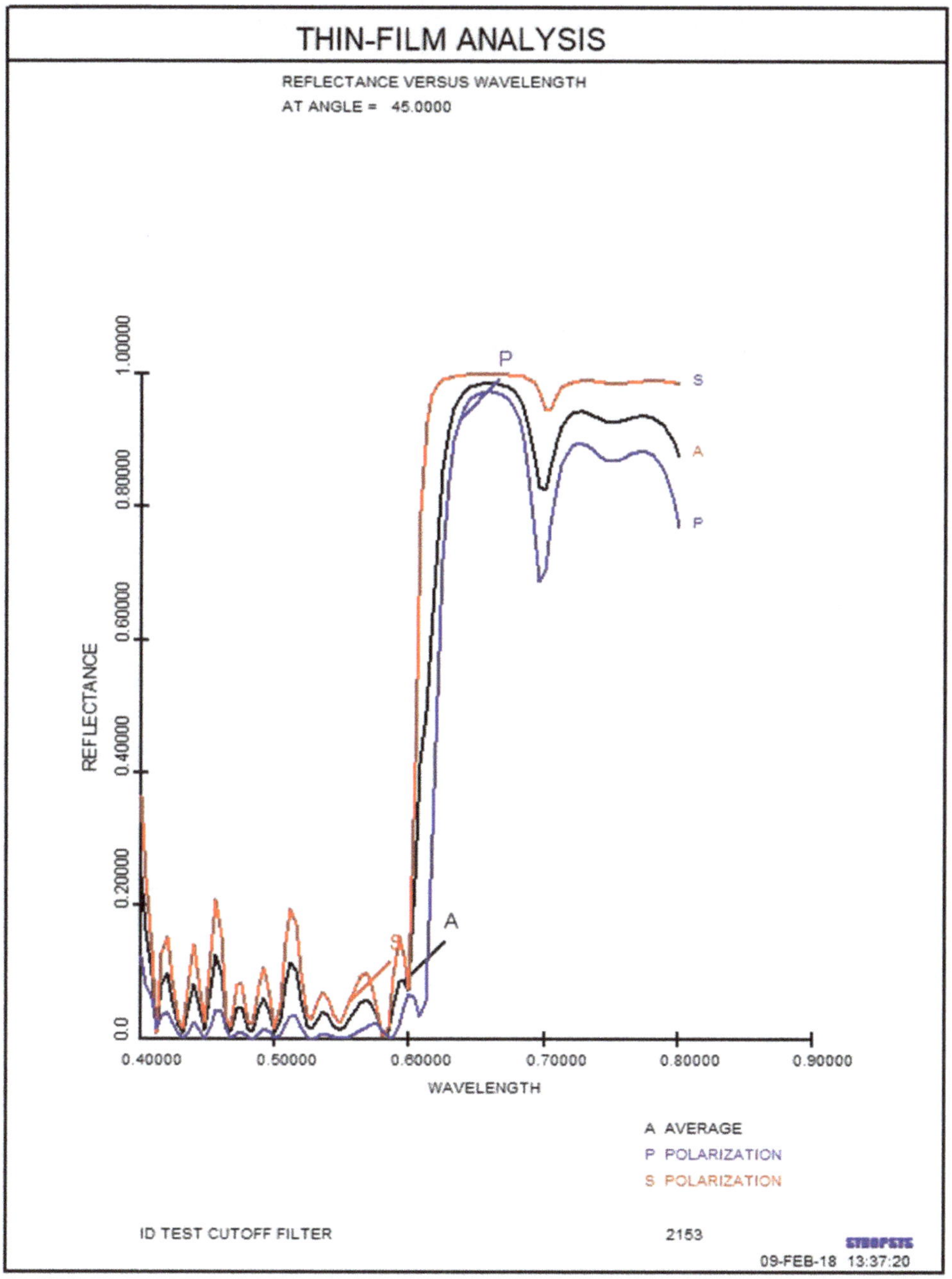

Figure 50.2. Example analysis of custom thin-film stack designed automatically.

11	0.3990	0.169883	1.3655		MGF2
12	0.8142	0.222101	2.1535		TAO5
13	0.2027	0.068438	1.7479		CEO2
14	0.4088	0.120109	1.9729		HFO2
15	0.1895	0.063979	1.7479		CEO2
16	0.3956	0.116242	1.9729		HFO2
17	0.1973	0.066629	1.7479		CEO2
18	0.3956	0.107916	2.1535		TAO5
19	0.1999	0.067490	1.7479		CEO2
20	0.3976	0.108471	2.1535		TAO5
21	0.1994	0.067346	1.7479		CEO2
22	0.4097	0.114626	2.0980		ZRO2
23	0.2042	0.073164	1.6210		CEF3
24	0.4209	0.123648	1.9729		HFO2
25	0.1743	0.062443	1.6210		CEF3
26	0.3726	0.125824	1.7479		CEO2
27	0.1865	0.066831	1.6210		CEF3
28	0.4156	0.122097	1.9729		HFO2
29	0.2285	0.077178	1.7479		CEO2
30	0.4248	0.118851	2.0980		ZRO2
31	0.0000	0.000000	1.6200		
SUBSTRATE			1.6200	0.0000	

50.3 Automatic clocking of wedge errors

Another automatic feature comes in handy when the elements of a lens are manufactured and measured, and each is found to have a small wedge error. Good shop practice can minimize such errors, but they never go to zero. Also, more accurate 'dewedging', as it is called, is more expensive. Thus you want to see if you can compensate at assembly for such errors. The Monte-Carlo evaluation program **MC** can model cases where lenses are mounted with the wedges alternating, up and down, between elements, and that usually helps a good deal. The **UCLOCK** program can do even better.

Here is an example. Get out the lens stored as 1.RLE, run UCLOCK, and specify a small wedge error on each of the four elements (**C50M3**):

```
FET 1
UCLOCK
WEIGHT 1 1 1
2 1
4 2
6 3
8 4
GO

UCLOCK LIST
UCLOCK PLOT
```

Here, element 1 has a wedge of 1 arcminute assigned to surface 2, element 2 has 2 minutes of wedge on surface 4, and so on. The shop must accurately measure these wedges and mark the edge where it is thickest so they know how the wedge is to be oriented when the lens is assembled.

Run this job, and the program finds the best clocking angles, shown in figure 50.3. Now the program has assigned an alpha tilt to the second side of each element, modeling the wedge, and a gamma tilt of two surfaces to the first side, modeling the clocking angle of that element in the cell. Element 1 is unclocked, giving a reference direction to the others. The program lists the results:

```
--- UCLOCK LIST

OPTIMUM CLOCKING OF WEDGED ELEMENTS IS AS FOLLOWS:
BORESIGHT WEIGHT =    1.0000
DISPERSION WEIGHT =    1.0000
AXIAL COMA WEIGHT =    1.0000

  No.   SURF   WEDGE, MIN     RADIANS       DEGREES
   1      2     1.00000       0.00029       0.01667
   2      4     2.00000       0.00058       0.03333
   3      6     3.00000       0.00087       0.05000
   4      8     4.00000       0.00116       0.06667

RESIDUAL BORESIGHT ERROR, IN LENS UNITS:    0.04879882
RESIDUAL DISPERSION ERROR, IN LENS UNITS:    0.00111729
RESIDUAL AXIAL COMA, IN WAVES:    0.02229963

UNCLOCKED BORESIGHT ERROR, IN LENS UNITS:    0.10480944
UNCLOCKED DISPERSION ERROR, IN LENS UNITS:    0.00174071
UNCLOCKED AXIAL COMA, IN WAVES:    1.13678441


 TILT AND DECENTER DATA
 LEFT-HANDED COORDINATES
```

SURF	TYPE	X	Y	Z	ALPHA	BETA	GAMMA
2	REL	0.00000	0.00000	0.00000	0.0167	0.0000	0.0000
3	REL	0.00000	0.00000	0.00000	0.0000	0.0000	-132.4438
4	REL	0.00000	0.00000	0.00000	0.0333	0.0000	0.0000
5	REL	0.00000	0.00000	0.00000	0.0000	0.0000	53.4173
6	REL	0.00000	0.00000	0.00000	0.0500	0.0000	0.0000
7	REL	0.00000	0.00000	0.00000	0.0000	0.0000	48.4099
8	REL	0.00000	0.00000	0.00000	0.0667	0.0000	0.0000

```
 KEY TO SURFACE TYPES
 ____________________
 GLB  GLOBAL COORDINATES                  LOC  LOCAL COORDINATES
 REL  RELATIVE COORDINATES                REM  REMOTE TILTS IN RELATIVE COORD.

 SURF MESSAGES
 __________________________________________________________________________
   3  UNDO TILTS/DECENTERS OF SURFACE NO.    2
   5  UNDO TILTS/DECENTERS OF SURFACE NO.    4
   5  UNDO TILTS/DECENTERS OF SURFACE NO.    3
   7  UNDO TILTS/DECENTERS OF SURFACE NO.    6
   7  UNDO TILTS/DECENTERS OF SURFACE NO.    5
   9  UNDO TILTS/DECENTERS OF SURFACE NO.    8
   9  UNDO TILTS/DECENTERS OF SURFACE NO.    7
 --- UCLOCK PLOT
```

Now you know that element 2 should be oriented so that the thickest edge is −132.4438 degrees rotated with respect to element 1, and so on.

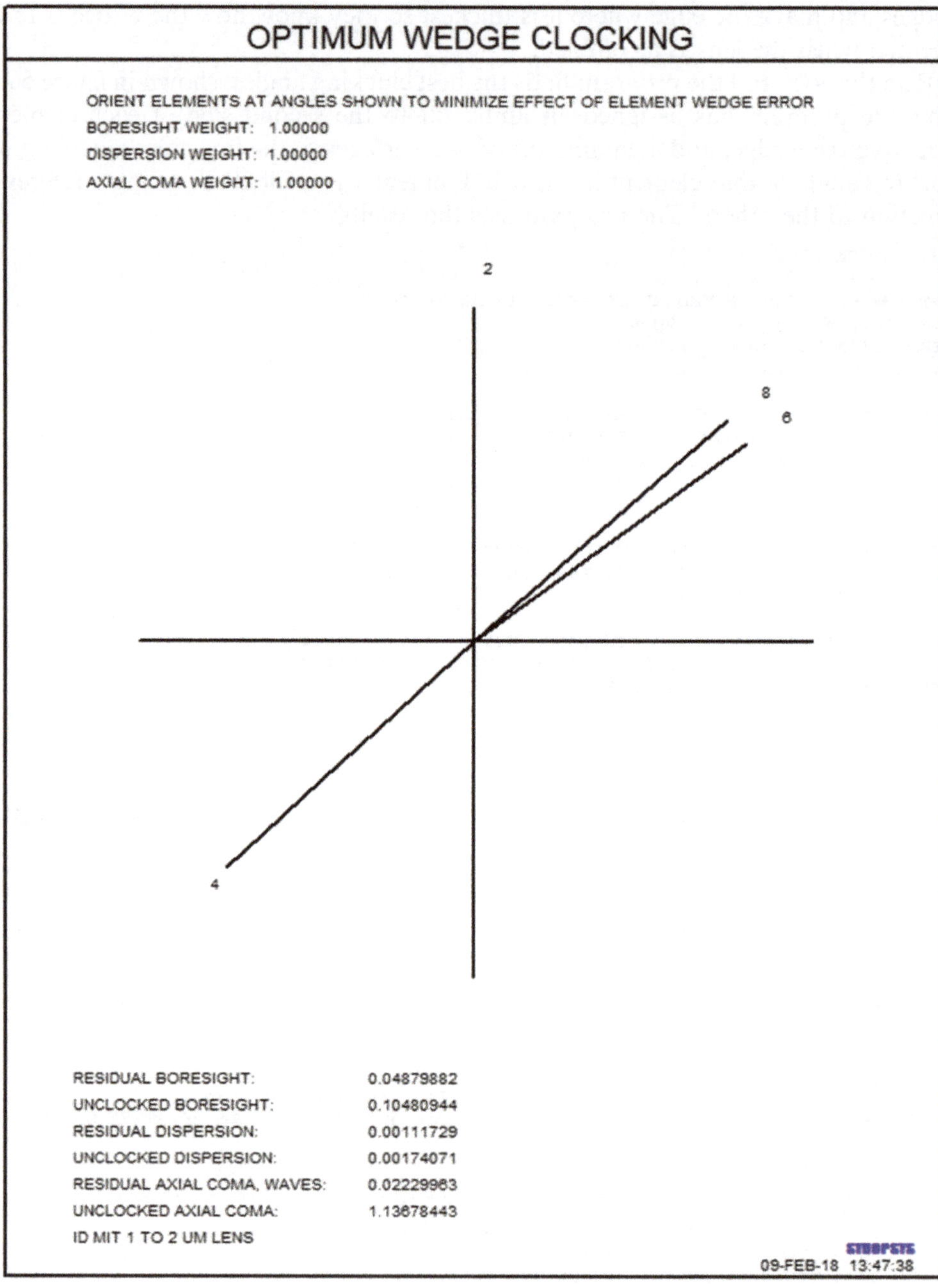

Figure 50.3. Output from UCLOCK, showing optimum clocking angle of each element of the lens.

The program finds that, if you do *not* clock the elements in this way, you will obtain an axial coma of 1.14 waves—which is pretty bad—but if you clock the elements as directed above, this goes down to 0.022 waves. The new geometry is shown in the perspective drawing in figure 50.4. Note how the surface curves, which are lined up in the local y-direction at each surface, are all rotated with respect to each other. This is an example of how the power of automatic lens design methods can improve yields.

One final note: dewedging is done by mounting the lens on a precision spindle, running a dial indicator on both sides, and adjusting the centration until both run true. Then the edge is ground down until the desired element diameter is reached. At this point the wedge error should be very small, but of course still nonzero.

However, this does not work for some meniscus elements, where the centers of curvature of the two sides are located close together on the optical axis. Such elements have to be ground and polished with careful attention to the wedge error right from the start, which is difficult and expensive. Keep this in mind and try to avoid such elements. The monitor **AMS** can control the separation.

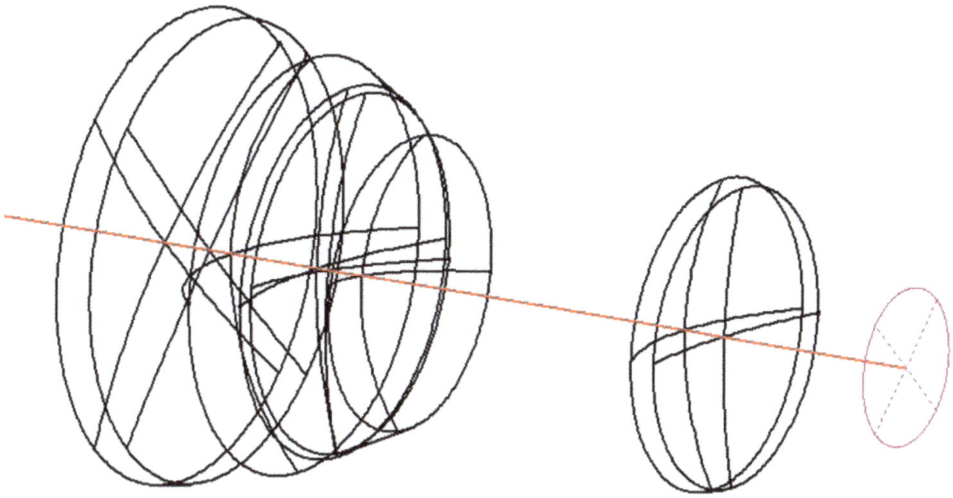

Figure 50.4. Rotation of elements shown in perspective after calculating with UCLOCK.

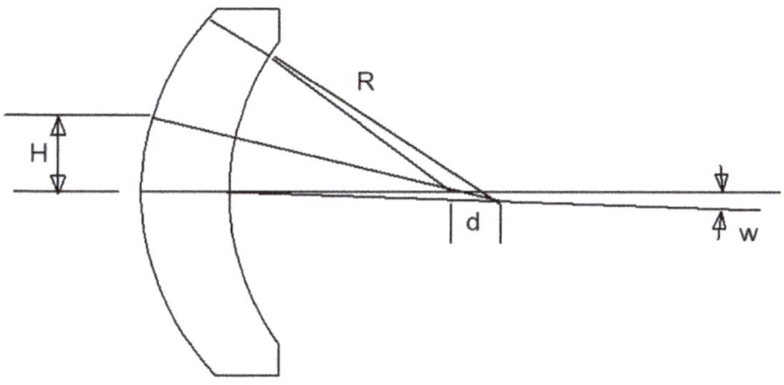

Figure 50.5. Removing the wedge of a meniscus lens becomes more difficult as the centers become closer together.

In figure 50.5, a lens has two radii with almost the same center location, the centers separated by a small distance d, and a wedge error of W. In order to remove the wedge, the lens has to be edged by an amount H, given by

$$H = W * R^2 / d.$$

If d goes to zero, H becomes infinite. Since one usually has only a small amount of extra glass that can be removed in edging, the quantity d has to be greater than some amount. That amount is the target for **AMS**.

Appendix A

A brief history of computer-aided lens design

Newcomers to the field of computer-aided lens design are probably familiar with but two or three commercial programs, which are the survivors of a long development effort by many researchers at many institutions. The author's career has spanned over 50 years and utilized a wide variety of computers, operating systems, and programming languages, including the following:

- 1961: At MIT, the IBM 650, a vacuum-tube processor, batch mode with punched cards.
- 1962: the Honeywell 800, a solid-state CPU, also batch mode. Programming was in the 'MAC' language, developed at MIT.
- 1963: the Honeywell 1800, a faster CPU.
- 1967: an IBM 1130, an early minicomputer, punch-card input, and 8 K words of 16 bits. First use of Fortran.
- 1971: the CDC 3300 and then 6600, batch mode via telephone connection. Programmed in Fortran.
- 1977: the Altos PC came out, with an 8080 CPU chip. Programmed in assembly language.
- 1983: the VAX 11/730, with 8 K of memory, programmed in Fortran.
- 1987: the PC became available and, with a version of Unix installed, could do interactive lens design.
- 1992: the SYNOPSYS program was ported to DOS, programmed in Fortran.
- 1999: the first native Windows version, fully interactive, programmed in both C++ and Fortran.

Of course, we were not the only ones developing lens design software. Other programs were developed by other authors as listed below; some were written for proprietary use in industry:

- Slams (C G Wynne)
- Ordeals (Tropel)

doi:10.1088/978-0-7503-1611-8ch51

- Flair (Radkowski)
- COP (Grey)
- Lead (Kodak)
- Father (B & L)
- Spade (Sperry)
- Optik V (Texas Institute)
- Alsie (Osaka, Suzuki)
- SIGMA (Kidger)
- Bathos (Blandford)
- ACCOS (Spencer)
- CERCO (French)
- Cool Genii (Genesee)
- CODE n (Harris)
- Oslo (Sinclair)
- ZEMAX (Moore)
- SYNOPSYS (Dilworth)

These programs employed a variety of optimization methods, of which the following are noteworthy:

- Correction (Itek)
- Orthonormalization (Grey, Unvala)
- Damped least-squares (Levinberg)
- Steepest descent
- Simplex (Bathos)
- Random search (Texas Instruments)
- Adaptive (Glatzel)
- Metric schemes
- Solution scaling
- Pseudo second derivatives (PSD; Dilworth)

Appendix B

Optimization methods

B.1 Mathematical methods of lens optimization

The program at Itek, mentioned in appendix A, seems to be unique; it used a correction algorithm instead of the minimization methods employed by almost everyone else. In that method, the number of targets in the merit function could not exceed the number of variables. So you selected a few carefully chosen rays, gave target values slightly smaller than the current ones, and submitted your batch run. If that converged, you reduced the targets and tried again. With much user intervention and many iterations, one could come up with a good design. That program was used on the recently declassified Corona project, which designed aerial reconnaissance cameras during the Cold War. The first illustration in chapter 38 shows a lens from that project.

The technique of orthonomalization was interesting to some researchers early on. This is a method that works via linear algebraic manipulations of the Jacobian matrix, with the goal of mapping the current set of variables into a different set, where the derivatives would all be the same magnitude and the effects of each variable independent of the others. While interesting from a mathematical standpoint, we observe that the process does not introduce any new information into the problem. At best, it might avoid numerical difficulties due to matrix conditioning—but more recent methods have avoided that problem in other ways.

The PSD methods, developed by the author, began in the 1980s with the goal of improving the then-standard damped-least-squared method, which converges very slowly for many problems. This difficulty arises because that method calculates only the first derivatives of the items in the merit function with respect to the design

doi:10.1088/978-0-7503-1611-8ch52

variables. The text below explains the mathematics of the least-squares method and why it performs poorly.

B.2 The DLS method and descendants

The merit function φ is the sum of the squares of the errors to be corrected, defined by the vector f. One creates the matrix L from the derivatives, and then the required changes in the variables are easily calculated after the matrix is inverted.

However, this solution is usually far off the mark, owing to the nonlinear nature of the design landscape. To improve its performance, the concept of 'damping' was introduced, the effect of which is to reduce the length of the solution vector. If the solution stays in the region of approximate linearity, it should be an improvement, and with many iterations one hopes to find a good result.

The improvement resulting from the use of a damping factor was only modest, and the damped-least-squares method still converged very slowly. In an effort to speed things up, many schemes were devised for utilizing the damping, D, in various ways. These include the following:

- Additive (many)
- Multiplicative (Meiron)
- Search for best (Dilworth)
- Different classes of variables
- Homogeneous second derivatives (Buchele, Feder)
- PSD (Dilworth)

The following mathematics summarizes this development (the least-squares optimization method):

$$\varphi = \sum f_i^2$$

$$G_j = \frac{1}{2}\frac{\partial \varphi}{\partial x_j}$$

$$L_{jk} = \frac{\partial G_j}{\partial x_k}$$

$$0 = G_j + L_{jk}\Delta x_k$$

$$\delta_j = -\, L_{jk}^{-1} G_k$$

The merit function φ is the sum of the squares of the image defects, f_i; the gradient G_j is one-half the derivative with respect to the design variable x_j, and the set of derivatives of G_j with respect to variable X_k gives us the Jacobian matrix, L_{jk}.

To find the optimum, set the gradient to zero and solve for the variable change δ_j.

Adding a damping term, D, as shown below, reduces the magnitude of δ_j, and one hopes the solution will then remain in the region of approximate linearity; This is the classical DLS method. Note that all variables obtain the same damping, which is applied to the diagonal of the matrix L_{jk}.

$$\varphi = \sum f_i^2 + \sum D^2 \delta_j^2$$

$$G_j = \sum f_i \frac{\partial f_i}{\partial x_j} + \sum D^2 \delta_j$$

$$L_{jk} = \sum \frac{\partial f_i}{\partial x_j} \frac{\partial f_i}{\partial x_k} + D^2 \bigg|_{j=k}$$

B.3 The PSD methods

Some of the methods mentioned above were markedly better than the raw DLS algorithm, but most failed to deal with the underlying source of the problem, which can be seen if we expand the Jacobian matrix to two derivatives, as follows (the PSD I method):

$$L_{jk} = \sum \frac{\partial f_i}{\partial x_j} \frac{\partial f_i}{\partial x_k} + \sum f_i \frac{\partial^2 f_i}{\partial x_j \partial x_k}$$

$$\frac{\partial^2 f_i}{\partial x_j^2} \approx \frac{\left.\frac{\partial f_i}{\partial x_j}\right|_{\Delta x_j} - \frac{\partial f_i}{\partial x_j}}{\Delta x_j + \varepsilon}$$

Here it is apparent that the second derivative, were its value known, should be added to the matrix L in precisely the place where the damping D appears on the diagonal in the older DLS method. In other words, the purpose of introducing D is to replace the unknown values of the second derivatives. That insight led to the first form of the PSD method, called PSD I[1].

The idea is quite simple; keep track of the changes in the first derivatives from one iteration to the next, divide by the variable change, and the result is just the second derivative (ignoring higher and mixed orders). Experience showed that the method was remarkably better than DLS, but only if a stabilizing factor ε was added where shown. That insight led to the PSD II method, which is based on the statistically expected influence of the higher-order derivatives[2], and worked better than did PSD I:

---

[1] Dilworth D C 1978 Pseudo-second-derivative matrix and its application to automatic lens design *Appl. Opt.* **17** 3372.

[2] Dilworth D C 1983 Improved convergence with the pseudo-second-derivative (PSD) optimization method *Proc. SPIE* **399** 159.

$$\left.\frac{\partial f_i}{\partial x_j}\right|_{\Delta x_j} - \frac{\partial f_i}{\partial x_j} \approx \frac{\partial^2 f_i}{\partial x_j^2}\Delta x_j + \frac{\partial^2 f_i}{\partial x_j^2}\left.\sqrt{\sum\nolimits_k \Delta x_k^2}\right|_{k\neq j},$$

$$\frac{\partial^2 f_i}{\partial x_j^2} \approx \frac{\left.\frac{\partial f_i}{\partial x_j}\right|_{\Delta x_j} - \frac{\partial f_i}{\partial x_j}}{|\Delta x_j| + \left.\sqrt{\sum\nolimits_k \Delta x_k^2}\right|_{k\neq j}},$$

$$L_{jk} = \sum \frac{\partial f_i}{\partial x_j}\frac{\partial f_i}{\partial x_k} + \sum f_i \left.\frac{\partial^2 f_i}{\partial x_j \partial x_k}\right|_{k=j}.$$

A further refinement leads to the PSD III method. On average, we assumed that the mixed second partial is roughly the same as the homogeneous second partial—but the latter involves both j and k. Which one should we use? Define $\sec_j$ as shown below, and then combine the influence of both j and k by the ratio of the second-order terms from the previous iteration. Now we have a better approximation to the second partials:

$$\frac{\partial^2 f_i}{\partial x_i \partial x_j} \approx \frac{\partial^2 f_i}{\partial x_j^2}$$

$$\sec_j = \sum \left.\left(f_j \frac{\partial^2 f_j}{\partial x_j \partial x_k}\right)\right|_{k=j}$$

$$\frac{\partial^2 f_i}{\partial x_j \partial x_k} \approx \sqrt{\frac{\partial^2 f_i}{\partial x_j^2}\frac{\partial^2 f_i}{\partial x_j^2}\frac{\sec_k}{\sec_j}}$$

$$\frac{\partial^2 f_i}{\partial x_j^2} \approx \frac{\left.\frac{\partial f_i}{\partial x_j}\right|_{\Delta x_j} - \frac{\partial f_i}{\partial x_j}}{|\Delta x_j| + \left.\sqrt{\sum\nolimits_k \Delta x_k^2 \frac{\sec_k}{\sec_j}}\right|_{k\neq j}}$$

This PSD method adds a term to the matrix where the previous D was added, but now the value of that term differs, in practice, from one variable and the next, by as much as 14 orders of magnitude—a far cry from the constant D used in the standard DLS method.

Since the derivation involves some crucial approximations and assumptions, one must test it before declaring success. To that end, we carried out a simple design job (a triplet) and plotted the logarithm of the value of φ as a function of iteration number, repeating the exercise with several different algorithms. The result is shown in figure B.1.

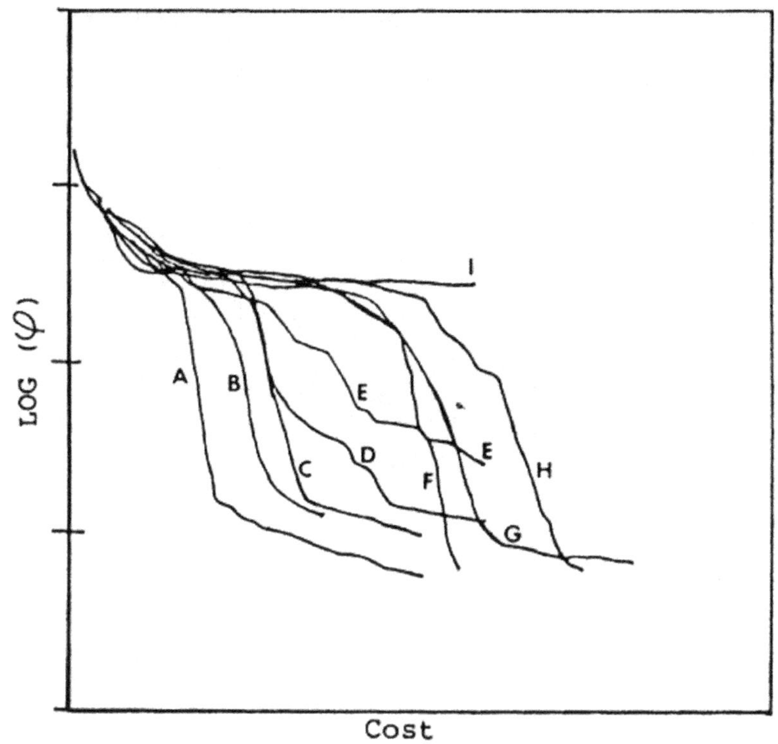

Figure B.1. Comparison of convergence rate for several optimization algorithms. Curve I is classic DLS, curve C is PSD I, and curve A is for PSD III.

The convergence rate of the PSD III method is orders of magnitude faster than the classic DLS method, and the assumptions therefore appear to be valid.

B.4 Global search algorithms

Many researchers have attempted to devise a method by which the computer could find the 'global optimum', a task both extremely complicated and fortunately unnecessary. As when searching for the end of a rainbow, you never reach your goal. One always suspects there may be a better solution yet to be found. The most popular search method involves defining a multi-dimensional grid of designs, where each radius, thickness, airspace, index, and Abbe number takes on a discrete set of values, resulting in a search space of perhaps 200 000 or more cases in all, which, when optimized with the ordinary DLS method, can take many days to evaluate. Although in principle this method can find the best of that large number of possibilities, it is too slow to be practical.

The algorithms in DSEARCH and ZSEARCH work differently; the default method uses a binary search tree: for a lens of N elements, one can generate a binary number of N bits and then create candidate lenses in which each element is assigned either a positive or negative power according to the value of the corresponding bit. Thus, for a lens of seven elements, there are only 2^7 cases, or 128 in all—a far cry from the huge search space mentioned above. However, one has to ask: does this simple method work, and what should be the initial powers of those elements?

Referring again to our original mountain-range metaphor, each case starts at the top of a tall mountain (with plane-parallel plates), selects a direction given by the particular value of the binary number representing that case, alters the curvatures according to that direction, and then jumps downhill and begins optimizing. But how far down should the algorithm jump? That jump controls the initial powers. An interesting result came out when I examined a range of initial radius values and ran DSEARCH for each one, as shown in figure B.2. I discovered that, if the initial radius is too long, the lens often fails to trace, due to the curvature solve on the last element which becomes so steep that rays experience total internal reflection errors. If the initial radius is too short, many other elements have the same problem. The program automatically corrects most of those ray failures, but in the process it might

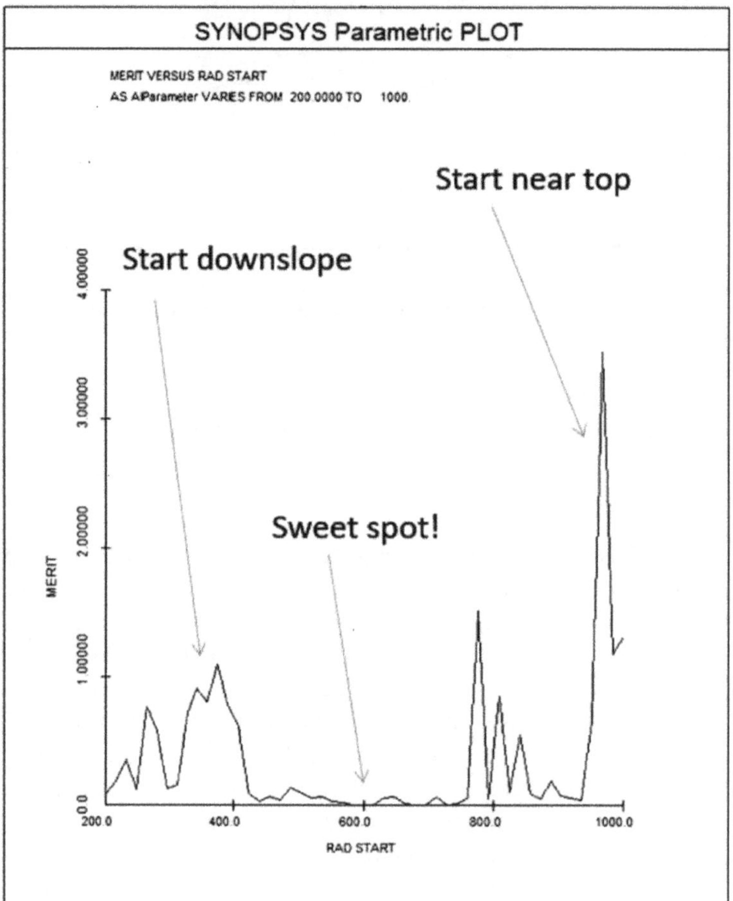

Figure B.2. Best merit function as a function of initial element radii.

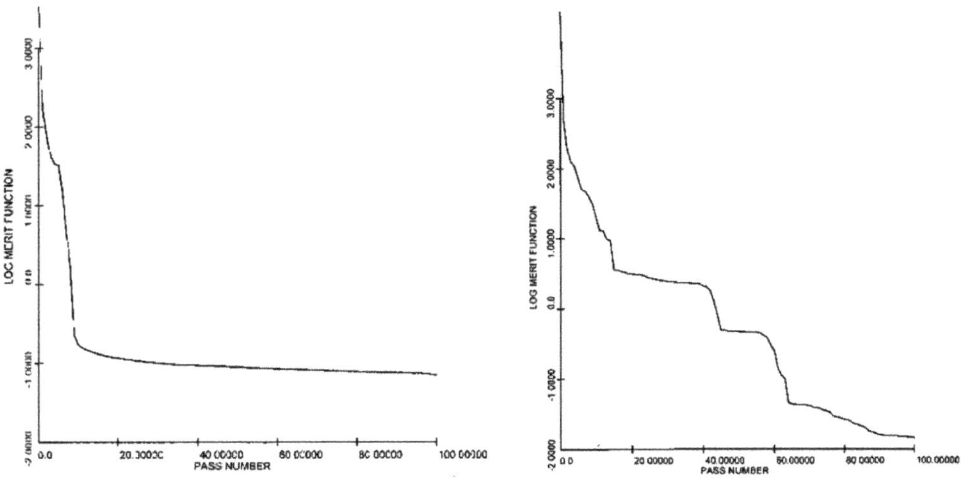

Figure B.3. Convergence rates for a well-behaved and an erratic lens; the latter is the better design.

move the design toward a better or a poorer solution, which accounts for the chaotic behavior at both ends of the curve. Fortunately, there seems to be a wide range where the initial value is not critical. So the initial powers can be assigned more or less at will, and perhaps more than one value can be tested.

Another question is, how many iterations should one request for each candidate solution? Fewer passes will run faster, but one does not want to miss a good solution that might be found with more. The goal is to devise a method that works reliably and very quickly. Figure B.3 shows the convergence history for two different lenses. On the left, the lens reached an excellent solution in only 30 or so iterations. The case on the right was more erratic; the MF came down in several steps, and if one stopped optimizing after 30 iterations that lens would not score as highly as the previous one and the search algorithm might reject it—even though after 80 iterations it is clearly superior. These results can influence the parameters one submits to the search routines.

A major goal of the search methods is to achieve maximum speed, and the optional QUICK mode enhances it significantly. That step optimizes each candidate lens with a special MF containing only the first-order goals and all of the third- and fifth-order aberrations (plus any SPECIAL AANT requirements the user may have submitted). This can be evaluated many times faster than can an MF with a grid of real rays, and it quickly weeds out the candidates that performed poorly on that score. Then there are fewer remaining cases to be optimized thoroughly with real rays.

B.5 Why are DSEARCH and ZSEARCH so powerful?

It is the speed of convergence of the PSD III method that makes the search programs DSEARCH and ZSEARCH practical. Each case can be optimized in a matter of seconds or less, and within a few minutes one can explore hundreds or thousands of different branches of the lens design tree. The PSD III algorithm seems to be key to the success of this new paradigm and, together with the binary search protocol, it is a very efficient way to explore the landscape.

It is remarkable that, if one actually *calculates* the second derivatives—which was not practical decades ago with more primitive computational tools—and optimizes the lens with those instead of the 'pseudo' second derivatives approximated by the PSD algorithms, the results are *not as good as those from PSD*! This tells us that the logic of the PSD calculation applies to *all* the higher-order derivatives, not just the second, and the matrix in fact more closely approaches what you would obtain with more than just two terms, as ideally it should. Nature is not always so kind to computer programmers, and it is wonderful when a program works better than one expects. Lens design is a very rewarding field.

B.6 Adding and deleting lens elements automatically

If a lens does not perform adequately, a time-honored strategy is to add an element somewhere. Then the powers of the other elements can be reduced as well as their aberration contributions. But where should one add the new element? This seems like a very complex problem, requiring deep insight into lens design theory, but it can in fact be solved by a rather simple algorithm that started with an idea of Florian Bociort[3], called the *saddle-point theory*.

The idea is that, if one adds a thin shell adjacent to an existing lens element, as shown in figure B.4, the ray paths do not change, so the MF is also unchanged—but now there are six new degrees of freedom, and it is likely that optimizing with these additional variables will lead to an improved design. No deep theory required.

That is the principle behind **AEI**, which you have used in several of the previous chapters. The program can test where adding an element does the most good, and the result is usually a better lens. Here is another instance where pure number crunching can yield results as good as or better than can be obtained by a human expert.

AEI is a special case of the more general saddle-point build (**SPB**), which can build up an entire lens one element at a time using the same method. That feature is similar in some respects to DSEARCH, but where it is applicable, the latter is often better since it can draw on a larger set of possibilities.

The reverse also works: try to reduce each lens element to a thin shell with zero power, and if the MF is not seriously degraded, that shell can simply be removed with little loss of quality. This is exactly what **AED** does, often yielding a lens that is almost as good as before but requiring one less element. These tools, together with the search routines, help you explore the very complex lens design landscape quickly and easily.

[3] Bociort F, Serebriakov A and van Turnhout M 2004 Saddle points in the merit function landscape of systems of thin lenses in contact *Proc. SPIE* **5523** 174–84.

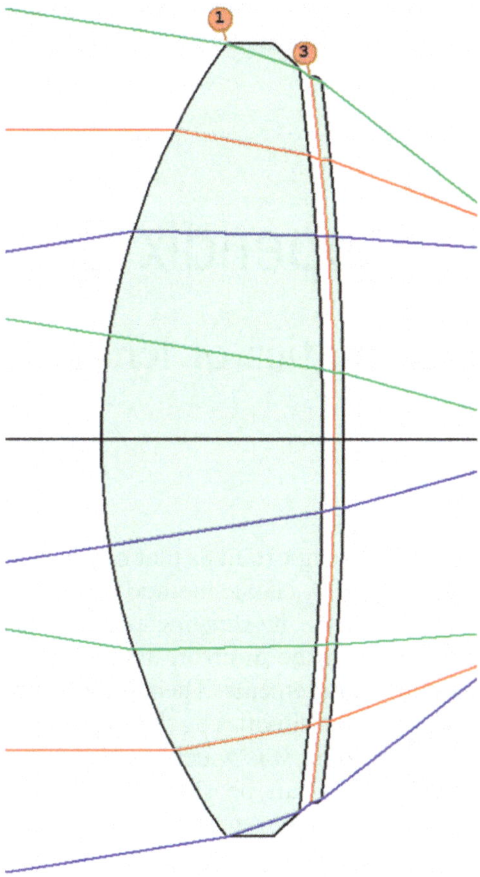

Figure B.4. A thin shell added adjacent to a lens element, part of the AEI algorithm.

B.7 What about traditional methods?

Open any of the classic texts on lens design and you will find pages of math, equations that can lead to a lens configuration with certain desirable properties, perhaps suitable for input to an optimization program, along with many examples of classic design forms that include specific data for those lenses. The goal has long been to provide starting points that have a good chance of yielding a good design when optimized. All this made sense in the days when having that starting point was the key to success, but today it is less useful since the new search methods can produce many excellent configurations in minutes, given just the design goals. This book has illustrated the importance and power of these new tools and justifies our assertion that the new paradigm has radically changed what lens designers do and how they do it. We believe that careful study of the examples presented here will prepare the student to fully utilize these new tools and to be more effective at their task than any of the old masters ever were.

Appendix C

The mathematics of lens tolerances

The subject of lens tolerances is as important as that of lens optimization, although the mathematics is somewhat simpler. A classic method of generating a tolerance budget was first to obtain a table of *inverse sensitivities*, which gives the amount by which every manufactured parameter can be in error, assuming everything else is perfect, while just meeting the imaging requirements. Then, if there are N such parameters, the practice was to divide each of the sensitivities by the square root of N, and that became the tolerance budget. This procedure was widely used and worked very well.

In fact, it worked *too well*. It can be shown that the budget so prepared is appropriate for the case when every parameter is always to be found exactly *at one end or the other* of its tolerance range. Most parameters are likely to be found at a random place *within* the budget, not exactly at the end, and the image degradation was therefore usually less than the budget allowed for, and that is why it worked so well. However, there was a price: the lenses were more expensive than they had to be.

The whole purpose of generating a budget is, of course, because everything made by human hands is imperfect. Every dimension of a finished lens is slightly different from the numbers on the drawing. How much different can they be? The budget provides an answer to that question.

The tolerance budget prepared by BTOL first calculates a set of *standard deviations* (**SD**) of each parameter so that, if the lens were made to that budget, it would meet the design goals. Then, when it prints the actual budget, it accounts for the difference between the standard deviation and the actual tolerance range of that parameter. Let me explain.

For a one-dimensional parameter such as element thickness, it can be shown that the SD equals the tolerance limit divided by the square root of 3. So the printed budget gives the desired SD multiplied by that factor. The budget is therefore somewhat looser than that given by the root-of-N rule, the image will be as desired within the stated confidence level, and the lens will be less expensive. For two-dimensional parameters, such as lens decenter, the adjustment factor is the square root of 2 instead of 3.

doi:10.1088/978-0-7503-1611-8ch53

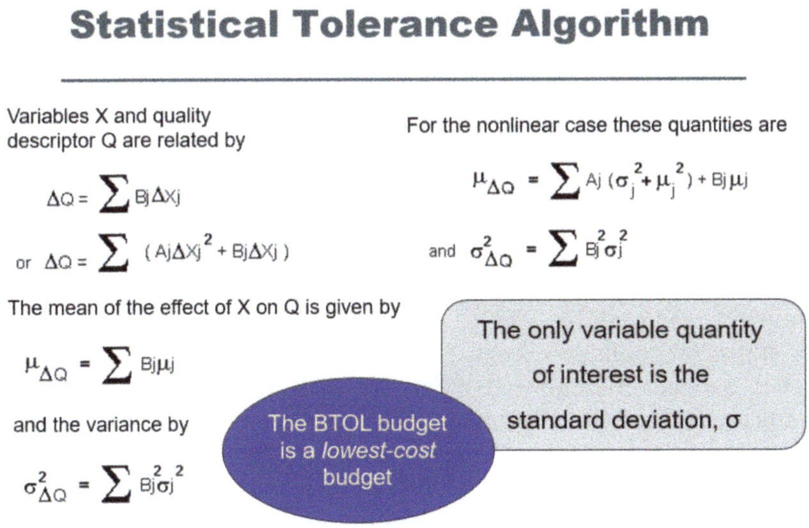

Figure C.1. Relation between the standard deviation of variables and degradation of image quality.

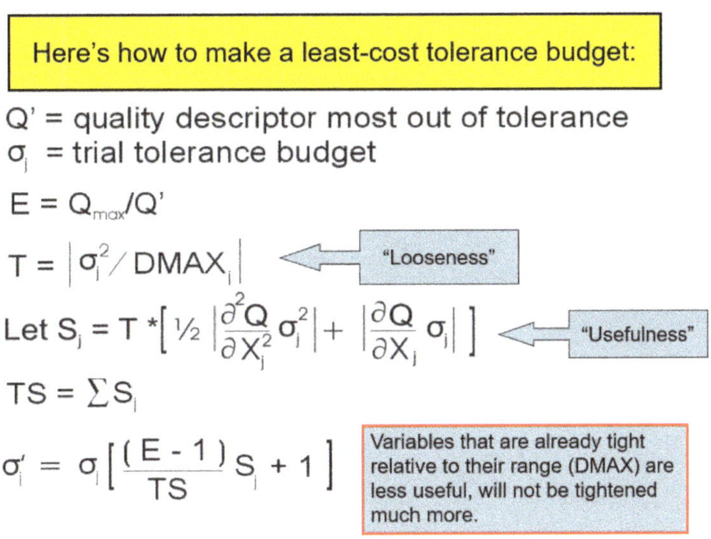

Figure C.2. Rules for deriving a least-cost tolerance budget.

Figure C.1 shows how the SD of the toleranced parameters influences the final image quality. It turns out that the only quantity of interest is the SD array. What is the rule for calculating those values?

There are an infinite number of budgets, all mathematically correct, that predict the same image quality, and the task is to find one that minimizes overall lens cost. Any parameter can be assigned a tighter tolerance, and some of the others loosened —but how do you know which ones? figure C.2 shows the logic of that calculation.

The program proceeds in steps, initially with all tolerances as loose as economically practical (given by the value of **DMAX** in the BTOL input, which defines the 'range' of each variable), evaluates the image quality, finds the image point most out of spec, identifies those parameters that most strongly affect that image point, and reduces the tolerance on those parameters. Then it iterates. If a given parameter starts to become excessively tight compared to its range, the program tries to leave it alone and tighten up others instead. The result is a least-cost budget. As the figure shows, one calculates the 'looseness' and 'usefulness' of each parameter and then modifies the SD according to a simple formula.

The budget produced by BTOL tells you that the lens will perform at the requested statistical confidence level, assuming every parameter is somewhere within its tolerance budget, preferably at a random location, and provided all modeled adjustments are performed at assembly. This calculation is not as mystical as it may seem at first; the program makes the allowed adjustments every time it calculates the derivatives of the quality with respect to the parameters. So the adjustments are built into the derivatives and the result comes about automatically.

Appendix D

Things every lens designer should understand

Here is a list of concepts and practices every lens designer should be aware of and follow. Those who want to explore the topics in more depth are encouraged to read any of the recent books on the topic, which develop the mathematical theory behind these concepts. A practicing designer does not need to dive that deep, in our opinion.

1. Lenses are designed by observing the paths of 'light rays', which do not really exist but are useful and quite accurate as long as one does not examine details with dimensions comparable to the wavelength of light.
2. The performance of a complex lens depends on the aberration contributions of individual elements. Positive and negative elements contribute aberrations with differing signs, so a combination of powers is usually required.
3. Aberrations arise because of three situations:
    a. A steep angle of incidence of a ray relative to a surface normal contributes high-order aberrations, because Snell's law then departs more strongly from the paraxial version. Such angles are usually to be avoided. Those aberrations are more difficult to correct and require a more complicated lens. Sometimes they cannot be avoided, and one must then strive to balance many orders of aberrations. This usually requires many lens elements.
    b. The bending of individual elements influences whether the Abbe sine condition is satisfied and is a useful variable for correcting many aberrations.
    c. The dispersion of the glass creates, and can be used to correct, chromatic aberration.
4. If your lens shows aberrations that vary over the aperture but are relatively constant over the field, correct them with elements near a pupil or stop.
5. If aberrations vary strongly with field angle but not with aperture position, correct them with elements near an image.

doi:10.1088/978-0-7503-1611-8ch54

6. Both of the above situations can be detected and handled by the Automatic Element Insertion feature (AEI).
7. If your lens has good performance, it is sometimes possible to remove an element with little loss of quality. The Automatic Element Deletion feature (AED) can test this possibility.
8. One can sometimes improve a lens by running AEI and AED repeatedly, thereby altering the lens construction in stages.
9. If an element is strongly curved, you can sometimes find a different solution region by flipping the bending. The bend-flip optimization (BFO) feature can do this automatically.
10. If the elements of your lens are each contributing large amounts of aberration, even though the final image looks great, you will likely see tight tolerances because even a small misalignment of such a design will throw the balance off. The techniques of tolerance desensitization are then called for. You want the elements as weak as possible, as long as the imaging goals are met. The command **THIRD CPLOT** will show the third-order contributions of every surface, and you can see where they are largest. Desensitization targets in the merit function can often relax tolerances, as discussed in chapters 10 and 13.
11. Distortion is corrected by elements not very close to the stop, as per item 5 above.
12. Try to use DSEARCH or ZSEARCH, if possible. Those features will discover lens configurations quickly and well. (Unless you already have a good configuration and the problem is simple.)
13. If the geometry of the problem makes it possible, aim for a lens with some degree of symmetry. That will make many field aberrations easier to correct.
14. If secondary color is an issue, try to favor glass types as described in chapters 12 and 34. The search routines will sometimes find those combinations automatically, and you can steer the process as instructed in those chapters.
15. Forget much of what the early texts advise. One of them says to select the glass types first when starting a design. Today, one lets the program find the best region of the glass map with the GLM variables or the search routines, and then fits the design to real glasses in the *last* step. Another text says that when designing a four-element lens one should design two doublets, correct them individually, and then combine them. This is nonsense. DSEARCH can design the lens by itself, and it will be much better.
16. If performance at or near the diffraction limit is required, be sure your merit function contains OPD targets. Sometimes one obtains the best results with a combination of OPD and TAP targets. Sometimes GO2 targets will yield a better image than GNO. That option targets the square of the OPD, which tends to ignore small errors and concentrate on larger ones. One cannot predict which will work the best, and you simply have to try them.

17. To peak the performance at a given MTF frequency, you should first get the design as close as you can with OPD aberrations, and then use the GSHEAR ray-grid option. When the MTF is very close to your target, you can switch to MTF aberrations to see if things improve. GSHEAR targets the difference in the OPD at separated points in the pupil, and that difference is what governs the MTF at that shear frequency.
18. Do not try to correct third-order aberrations to zero; you need them to balance the higher orders. You can sometimes reduce spacing and alignment sensitivity by reducing the contributions of a given element or group, but proceed with caution because the aberration balance may be thrown off.
19. If aspheric surfaces are allowed in your design, creep up on them gradually. Use the Automatic Aspheric Assignment (**AAA**) to determine which surface should be aspheric. That feature will add a conic constant where it will do the most good, and then you can run the Automatic G-variable Test (**AGT**) to determine which aspheric coefficients will be most useful on that surface. It makes no sense to use a high-order aspheric term to correct defocus, for example, so only add aspherics when the lens is already as good as possible without them. The exception is when designing molded plastic elements, which are usually aspheric early on—but even then, start off with just a few terms and add more as you need them.
20. Your merit function should contain a complete description of the problem, and this includes mechanical requirements as well as optical. Some designers only want image quality in their MF—but it makes no sense to obtain a great image if the lens will not fit in the required box. If the program knows about all your goals, it will favor designs that tend to meet them. That is exactly what you want.
21. One cannot always predict the results of a given optimization run, and often a new problem develops that was not anticipated beforehand. The process of lens design consists mostly of modifying the MF as required when one discovers a shortcoming. So change the MF and keep going. If you run into a blind alley and nothing seems to work, it is time to try a different search result.
22. Make frequent use of checkpoints, and save intermediate versions of your lens whenever you make a substantial improvement. You may well want to go back to that version later if unanticipated problems show up in later versions.
23. Your job is not complete until you make a table of tolerances. We know an expert designer who sends his design to a customer—expecting the *customer* to calculate the tolerance budget. This is unprofessional. If tolerances can be relaxed by a suitable change to the design, it is your job to figure that out. The customer will not know how. BTOL is the tool to use in most cases.
24. It is a good idea to become familiar with shop practice. Watching an optician make a precision surface has a way of humbling the designer—who

may then be more sensitive to the challenges inherent in the designs he sends to the shop. Learn what they find easy and what hard, and try to make their life less stressful. Very thin edges, for example present a challenge to the lens maker, even though the optimization program has no problem with them. Watch for meniscus lenses where the centers of curvature of the two sides are very close to each other. Such lenses are very difficult to manufacture since the methods used to eliminate wedge errors do not work very well, as explained in chapter 50. The **AMS** monitor can help in this situation. If a surface is almost flat, make it exactly flat. If a lens has two radii that are almost equal, make them exactly equal. Then there is no chance of the lens being inserted into the cell backwards.

25. If your DSEARCH or ZSEARCH run does not return a design you like, it is time to vary some of the input parameters. Even a small change can have a big effect. Items to consider include the following:
    - STOP FIX or Free
    - RSTART value, maybe more than one value
    - TSTART value
    - ASTART value
    - RT value
    - Include OPD, TOPD, OPSHEAR, or TOSHEAR
    - Number of FOVs
    - Weights on the fields
    - Number of rays in the grid
    - Number of passes
    - Enable or disable QUICK mode
    - Number of annealing passes
    - Try changing switches 95 and 67. They generally take a different path.

    It might seem daunting to think of exploring all the potential combinations of these parameters—but you should not have to. In our experience, most combinations return excellent starting points, and the purpose of trying still others is to give you more options. We generally obtain excellent results in from one to four attempts.

IOP Publishing

Lens Design
Automatic and quasi-autonomous computational methods and techniques
Donald Dilworth

Appendix E

Useful formulas

One degree = 0.017 453 29 radians.
One arcminute = 0.000 290 888 radians = 0.016 666 67 degrees.
The human eye can resolve about one arcminute.
One arcsecond = 4.848 14E−6 radians = 2.777 777E−4 degrees.
One milliradian = 0.052 9578 degrees = 3.437 75 arcminutes = 206.2648 arcseconds.
One micrometer = 0.000 039 37 inches.
For a lens in the visible spectrum, the diameter of the Airy diffraction disk = the F/number, approximately, in micrometers.
The radius of the first dark ring in the Airy disk = 1.22λF/number, in air, with an unobscured aperture.
The numerical aperture (NA) of a lens = $n\sin(\theta)$, where θ is the convergence angle of the marginal ray and n is the index in image space.
F/number = 0.5*NA; sinθ = −0.5*F/number.
Longitudinal aberration = 2 * transverse aberration * F/number.
Focus shift from plane parallel plate of index n = thickness * $(1 - 1/n)$.
Cutoff frequency F_{co} = 1743/(F/number) in lines/mm at $\lambda = 0.574$ µm.
Density of a filter D = log(1/transmission).
Diameter of geometric spot giving MTF cutoff of F_{co} = 0.039 37 inches/ F_{co}(c mm^{-1}).
Sag of sphere of radius R, at height s: $z = R - \sqrt{R^2 - s^2}$.
Radius of sphere with sag z, at height s: $R = \frac{s^2}{2z} - \frac{z}{2}$.
For lens of magnification m and focal length f, focal point $s_1 = (m + 1)f/m$; $S_2 = (m + 1)f$.
Reflection loss, uncoated surface of index n, Refl = $(\frac{n-1}{n+1})^2$.
Magnifying power, lens of focal length f, $M = (10 + f)/f$.

To defocus a lens of focal length f by one diopter, the image shift in inches $\Delta S = f^2/39.37$.
Strehl ratio = $\exp(-4\pi^2 \text{ variance})$, approximately.
Thin lens; s_1 and s_2 measured from lens $1/s_1 + 1/s_2 = 1/f$.
Thin lens; s_1 and s_2 measured from focal points $s_2\, s_2 = f^2$.
Angular resolution of a telescope in visual light = 4.66/diameter of objective in inches, in arcseconds.

Bibliography

These are all classic texts on lens design, and most contain extensive mathematical developments that support some insight. That insight was important in the days before routines like DSEARCH and ZSEARCH became available but is less so today. Some of the information is still useful, however, particularly in Warren Smith's books. Others are listed for those who want a deeper picture of the complex landscape that is optics and lens design. As a fringe benefit, you will gain an appreciation for the amount of labor that was once required of the lens designer—and be gratified at the power of the new automatic features, which relieve you of most of that labor.

Kingslake R and Johnson R B 2010 *Lens Design Fundamentals* (Bellingham, WA: SPIE)

Geary J M 2011 *Introduction to Lens Design* (Richmond, VA: Willmann-Bell)

Dilworth D C *SYNOPSYS Supplement to Joseph M Gary's Introduction to Lens Design* (Richmond, VA: Willmann-Bell)

Smith G H 2007 *Practical Computer-Aided Lens Design* (Richmond, VA: Willmann-Bell)

Laiken M 1991 *Lens Design* (New York: Marcel Dekker)

Smith W J 1966 *Modern Optical Engineering* (New York: McGraw-Hill)

Born M and Wolf E 1980 *Principles of Optics* 6th edn (Oxford: Pergamon)

Benford J R, Cook G H, Hass G, Hopkins R E, Kingslake R, Lueck I B, Rosin S, Scott R M and Shannon R R 1965 *Applied Optics and Optical Engineering* vol 3 ed R Kingslake (New York: Academic)

Rutten H G J and van Venrooij M A M 2002 *Telescope Optics* ed R Berry (Richmond, VA: Willmann-Bell)

Flügge J 1955 *Die Wissenshaftliche und Angewandte Photographie* ed K Michel (Berlin: Springer)

O'Shea D C 1985 *Elements of Modern Optical Design* (New York: Wiley)

Cox A 1964 *A System of Optical Design* (Waltham, MA: Focal)

Kingslake R 1983 *Optical System Design* (New York: Academic)

Kingslake R 1978 *Lens Design Fundamentals* (New York: Academic)

Kingslake R 1989 *A History of the Photographic Lens* (New York: Academic)

Yoder P R 2002 *Mounting Optics in Optical Instruments* (Bellingham, WA: SPIE)

Wolf W and Zissis G (US Office of Naval Research) 1995 *The Infrared Handbook* (Ann Arbor, MI: Environmental Institute of Michigan)

Lightning Source UK Ltd.
Milton Keynes UK
UKHW050622060821
388415UK00003B/119

9 780750 316095